Symmetry and Fluid Mechanics

Symmetry and Fluid Mechanics

Special Issue Editor
Rahmat Ellahi

MDPI • Basel • Beijing • Wuhan • Barcelona • Belgrade • Manchester • Tokyo • Cluj • Tianjin

Special Issue Editor
Rahmat Ellahi
Department of Mechanical Engineering, University of California
USA

Editorial Office
MDPI
St. Alban-Anlage 66
4052 Basel, Switzerland

This is a reprint of articles from the Special Issue published online in the open access journal *Symmetry* (ISSN 2073-8994) (available at: https://www.mdpi.com/journal/symmetry/special_issues/Symmetry_Fluid_Mechanics).

For citation purposes, cite each article independently as indicated on the article page online and as indicated below:

LastName, A.A.; LastName, B.B.; LastName, C.C. Article Title. *Journal Name* **Year**, *Article Number*, Page Range.

ISBN 978-3-03928-426-9 (Pbk)
ISBN 978-3-03928-427-6 (PDF)

© 2020 by the authors. Articles in this book are Open Access and distributed under the Creative Commons Attribution (CC BY) license, which allows users to download, copy and build upon published articles, as long as the author and publisher are properly credited, which ensures maximum dissemination and a wider impact of our publications.

The book as a whole is distributed by MDPI under the terms and conditions of the Creative Commons license CC BY-NC-ND.

Contents

About the Special Issue Editor ix

Preface to "Symmetry and Fluid Mechanics" xi

Rahmat Ellahi
Special Issue on Symmetry and Fluid Mechanics
Reprinted from: *Symmetry* 2020, *12*, 281, doi:10.3390/sym12020281 1

Muhammad Adil Sadiq
MHD Stagnation Point Flow of Nanofluid on a Plate with Anisotropic Slip
Reprinted from: *Symmetry* 2019, *11*, 132, doi:10.3390/sym11020132 11

Muhammad Jawad, Zahir Shah, Saeed Islam, Jihen Majdoubi, I. Tlili, Waris Khan and Ilyas Khan
Impact of Nonlinear Thermal Radiation and the Viscous Dissipation Effect on the Unsteady Three-Dimensional Rotating Flow of Single-Wall Carbon Nanotubes with Aqueous Suspensions
Reprinted from: *Symmetry* 2019, *11*, 207, doi:10.3390/sym11020207 27

Rahmat Ellahi, Ahmed Zeeshan, Farooq Hussain and A. Asadollahi
Peristaltic Blood Flow of Couple Stress Fluid Suspended with Nanoparticles under the Influence of Chemical Reaction and Activation Energy
Reprinted from: *Symmetry* 2019, *11*, 276, doi:10.3390/sym11020276 45

Muhammad Suleman, Muhammad Ramzan, Shafiq Ahmad, Dianchen Lu, Taseer Muhammad and Jae Dong Chung
A Numerical Simulation of Silver–Water Nanofluid Flow with Impacts of Newtonian Heating and Homogeneous–Heterogeneous Reactions Past a Nonlinear Stretched Cylinder
Reprinted from: *Symmetry* 2019, *11*, 295, doi:10.3390/sym11020295 63

Ibrahim M. Alarifi, Ahmed G. Abokhalil, M. Osman, Liaquat Ali Lund, Mossaad Ben Ayed, Hafedh Belmabrouk and Iskander Tlili
MHD Flow and Heat Transfer over Vertical Stretching Sheet with Heat Sink or Source Effect
Reprinted from: *Symmetry* 2019, *11*, 297, doi:10.3390/sym11030297 77

Taza Gul, Muhammad Altaf Khan, Waqas Noman, Ilyas Khan, Tawfeeq Abdullah Alkanhal and Iskander Tlili
Fractional Order Forced Convection Carbon Nanotube Nanofluid Flow Passing Over a Thin Needle
Reprinted from: *Symmetry* 2019, *11*, 312, doi:10.3390/sym11030312 91

Zahir Shah, Asifa Tassaddiq, Saeed Islam, A.M. Alklaibi and Ilyas Khan
Cattaneo–Christov Heat Flux Model for Three-Dimensional Rotating Flow of SWCNT and MWCNT Nanofluid with Darcy–Forchheimer Porous Medium Induced by a Linearly Stretchable Surface
Reprinted from: *Symmetry* 2019, *11*, 331, doi:10.3390/sym11030331 105

Muhammad Mubashir Bhatti and Dong Qiang Lu
Analytical Study of the Head-On Collision Process between Hydroelastic Solitary Waves in the Presence of a Uniform Current
Reprinted from: *Symmetry* 2019, *11*, 333, doi:10.3390/sym11030333 117

Ilyas Khan and Aisha M. Alqahtani
MHD Nanofluids in a Permeable Channel with Porosity
Reprinted from: *Symmetry* **2019**, *11*, 378, doi:10.3390/sym11030378 147

Liaquat Ali Lund, Zurni Omar, Ilyas Khan, Jawad Raza, Mohsen Bakouri and I. Tlili
Stability Analysis of Darcy-Forchheimer Flow of Casson Type Nanofluid Over an Exponential Sheet: Investigation of Critical Points
Reprinted from: *Symmetry* **2019**, *11*, 412, doi:10.3390/sym11030412 165

Iskander Tlili
Effects MHD and Heat Generation on Mixed Convection Flow of Jeffrey Fluid in Microgravity Environment over an Inclined Stretching Sheet
Reprinted from: *Symmetry* **2019**, *11*, 438, doi:10.3390/sym11030438 183

Anwar Saeed, Saeed Islam, Abdullah Dawar, Zahir Shah, Poom Kumam and Waris Khan
Influence of Cattaneo–Christov Heat Flux on MHD Jeffrey, Maxwell, and Oldroyd-B Nanofluids with Homogeneous-Heterogeneous Reaction
Reprinted from: *Symmetry* **2019**, *11*, 439, doi:10.3390/sym11030439 195

Muhammad Asif, Sami Ul Haq, Saeed Islam, Tawfeeq Abdullah Alkanhal, Zar Ali Khan, Ilyas Khan and Kottakkaran Sooppy Nisar
Unsteady Flow of Fractional Fluid between Two Parallel Walls with Arbitrary Wall Shear Stress Using Caputo–Fabrizio Derivative
Reprinted from: *Symmetry* **2019**, *11*, 449, doi:10.3390/sym11040449 213

Nawaf N. Hamadneh, Waqar A. Khan, Ilyas Khan and Ali S. Alsagri
Modeling and Optimization of Gaseous Thermal Slip Flow in Rectangular Microducts Using a Particle Swarm Optimization Algorithm
Reprinted from: *Symmetry* **2019**, *11*, 488, doi:10.3390/sym11040488 225

Imran Ullah, Tawfeeq Abdullah Alkanhal, Sharidan Shafie, Kottakkaran Sooppy Nisar, Ilyas Khan and Oluwole Daniel Makinde
MHD Slip Flow of Casson Fluid along a Nonlinear Permeable Stretching Cylinder Saturated in a Porous Medium with Chemical Reaction, Viscous Dissipation, and Heat Generation/Absorption
Reprinted from: *Symmetry* **2019**, *11*, 531, doi:10.3390/sym11040531 239

Dianchen Lu, Mutaz Mohammad, Muhammad Ramzan, Muhammad Bilal, Fares Howari and Muhammad Suleman
MHD Boundary Layer Flow of Carreau Fluid over a Convectively Heated Bidirectional Sheet with Non-Fourier Heat Flux and Variable Thermal Conductivity
Reprinted from: *Symmetry* **2019**, *11*, 618, doi:10.3390/sym11050618 267

Taza Gul, Haris Anwar, Muhammad Altaf Khan, Ilyas Khan and Poom Kumam
Integer and Non-Integer Order Study of the GO-W/GO-EG Nanofluids Flow by Means of Marangoni Convection
Reprinted from: *Symmetry* **2019**, *11*, 640, doi:10.3390/sym11050640 283

Rahmat Ellahi, Ahmed Zeeshan, Farooq Hussain and Tehseen Abbas
Two-Phase Couette Flow of Couple Stress Fluid with Temperature Dependent Viscosity Thermally Affected by Magnetized Moving Surface
Reprinted from: *Symmetry* **2019**, *11*, 647, doi:10.3390/sym11050647 297

Muhammad Salman Kausar, Abid Hussanan, Mustafa Mamat and Babar Ahmad
Boundary Layer Flow through Darcy–Brinkman Porous Medium in the Presence of Slip Effects and Porous Dissipation
Reprinted from: *Symmetry* **2019**, *11*, 659, doi:10.3390/sym11050659 . . . . **311**

Ali Saleh Alshomrani and Malik Zaka Ullah
Significance of Velocity Slip in Convective Flow of Carbon Nanotubes
Reprinted from: *Symmetry* **2019**, *11*, 679, doi:10.3390/sym11050679 . . . . **323**

Khalil Ur Rehman, M.Y. Malik, Waqar A Khan, Ilyas Khan and S.O. Alharbi
Numerical Solution of Non-Newtonian Fluid Flow Due to Rotatory Rigid Disk
Reprinted from: *Symmetry* **2019**, *11*, 699, doi:10.3390/sym11050699 . . . . **337**

Esmaeil Jalali, Omid Ali Akbari, M. M. Sarafraz, Tehseen Abbas and Mohammad Reza Safaei
Heat Transfer of Oil/MWCNT Nanofluid Jet Injection Inside a Rectangular Microchannel
Reprinted from: *Symmetry* **2019**, *11*, 757, doi:10.3390/sym11060757 . . . . **349**

Umar Khan, Adnan Abbasi, Naveed Ahmed, Saima Noor, Ilyas Khan, Syed Tauseef Mohyud-Din and Waqar A. Khan
Modified MHD Radiative Mixed Convective Nanofluid Flow Model with Consideration of the Impact of Freezing Temperature and Molecular Diameter
Reprinted from: *Symmetry* **2019**, *11*, 833, doi:10.3390/sym11060833 . . . . **369**

J. Prakash, Dharmendra Tripathi, A. K. Triwari, Sadiq M. Sait and Rahmat Ellahi
Peristaltic Pumping of Nanofluids through a Tapered Channel in a Porous Environment: Applications in Blood Flow
Reprinted from: *Symmetry* **2019**, *11*, 868, doi:10.3390/sym11070868 . . . . **385**

Yongou Zhang and Aokui Xiong
A Particle Method Based on a Generalized Finite Difference Scheme to Solve Weakly Compressible Viscous Flow Problems
Reprinted from: *Symmetry* **2019**, *11*, 1086, doi:10.3390/sym11091086 . . . . **411**

About the Special Issue Editor

Rahmat Ellahi, Ph.D., accomplished researcher, teacher, and a prolific scholar, plays a key role in promotion of science at national and international levels. He has successfully achieved distinction in academics, with the University of Punjab, Quaid-i-Azam University (Islamabad, Pakistan) and University of California (Riverside, CA, USA) as his alma mater. He has published more than 240 papers in journals in the USA, Germany, the U.K., Canada, etc. His work has been cited more than 13,000 times on Google Scholar having an h-index of 64. He is an editor or editorial board member for 20 international journals and referee for more than 300 international journals. He has edited six Special Issues for ISI impact factor journals. He has successfully supervised 28 research students (8 Ph.D. and 20 M.S.). His leadership in academics is further reflected through research collaboration with more than 70 international leading scientists all over the world. His research has help to upgrade the science capacity of several universities around the world including in the USA, Canada, Australia, U.K., France, China, Romania, Turkey, Iran, South Africa, Egypt, Saudi Arabia, Kuwait, India, Vietnam, and Pakistan. He is an author of six books published at national and international levels. He has organized 8 international conferences, delivered 20 seminars, and attended 25 conferences as a 2 key speaker and participant. As a referee, he has investigated more than 100 research projects submitted at HEC and USEFP under NRUP and Fulbright Grant (for Pakistan and Poland Scholars). He has been honored as a Highly Cited Researcher for the years 2018 and 2019 by Clarivate Analytics. He was among the top five Productive Scientists of Pakistan in Category A. In addition, he has received six (three international and there national) awards besides several honors. He is actively involved in different professional and academia bodies and institutes at national and international levels. In summary, Dr. Rahmat Ellahi is a superbly qualified individual in the field, encompassing all facets of a high-level educationist. He has provided remarkable contributions to society, teaching, research, development, and promotion of scientific cooperation, international collaboration, human resource development, and updating education systems with the latest trends. Consistent with his research work, the United States Education Foundation honor edhim with Fulbright Fellow for the years 2011 and 2012 and King Fahd University of Petroleum and Minerals, Dhahran, Saudi Arabia, honored him with the Chair Professor at the Research Institute KFUPM in 2018.

Preface to "Symmetry and Fluid Mechanics"

This book contains 25 chapters with the editorial in chapter 26. In this Special Issue, a total of 42 papers were submitted for possible publication. After a comprehensive peer review, only 25 papers were accepted for final publication. Extensive uses of realistic applications are commonly provided in each chapter. For the best possible reader understanding, a relevant list of references is also provided at the end of each chapter for further study. I wish to thank all reviewers for their excellent suggestions and critical reviews of the submitted manuscripts. I had many prominent scholars who contributed their original research work. I applaud them all on successful completion of this Special Issue. Errors and omissions, if any, are requested to be pointed out and will be acknowledged in the next possible Edition. Particularly, suggestions for improvement, scope, and format of the book are appreciated. I express my gratitude to MPDI for publishing this book and to Ms. Celina Si, Section Managing Editor, and my family and friends for their helpful cooperation. This Special Issue has already been cited more than 230 times in one year.

Rahmat Ellahi
Special Issue Editor

Editorial

Special Issue on Symmetry and Fluid Mechanics

Rahmat Ellahi [1,2,3]

[1] Department of Mathematics & Statistics, Faculty of Basic and Applied Sciences,
International Islamic University, Islamabad 44000, Pakistan; rahmatellahi@yahoo.com
[2] Department of Mechanical Engineering, University of California Riverside, Riverside, CA 92521, USA
[3] Center for Modeling & Computer Simulation, Research Institute, King Fahd University of Petroleum &
Minerals, Dhahran-31261, Saudi Arabia

Received: 11 February 2020; Accepted: 11 February 2020; Published: 13 February 2020

Abstract: This Special Issue invited researchers to contribute their original research work and review articles on "Symmetry and Fluid Mechanics" that either advances the state-of-the-art mathematical methods through theoretical or experimental studies or extends the bounds of existing methodologies with new contributions related to the symmetry, asymmetry, and lie symmetries of differential equations proposed as mathematical models in fluid mechanics, thereby addressing current challenges. In response to the call for papers, a total of 42 papers were submitted for possible publication. After comprehensive peer review, only 25 papers qualified for acceptance for final publication. The rest of the papers could not be accommodated. The submissions may have been technically correct but were not considered appropriate for the scope of this Special Issue. The authors are from geographically distributed countries such as the USA, Australia, China, Saudi Arabia, Iran, Pakistan, Malaysia, Abu Dhabi, UAE, South Africa, and Vietnam. This reflects the great impact of the proposed topic and the effective organization of the guest editorial team of this Special Issue.

Keywords: Newtonian and non-Newtonian fluids; nanofluids and particle shape effects; convective heat and mass transfer; steady and unsteady flow problems; multiphase flow simulations; fractional order differential equations; thermodynamics; physiological fluid phenomena in biological systems

1. Introduction

In fluid mechanics, the equations that govern the flows are usually very complex, involve a number of parameters, are inherited nonlinear in nature, and cannot be solved by traditional methods. Due to the complexity of these equations, finding analytical or semi-analytical or even numerical solutions is very important for many reasons, for instance, they provide a standard for checking the imperial results. Therefore, several new techniques have been developed. We hope that this issue will not only serve the said purpose but will also provide an overall picture and up-to-date findings to readers of the scientific community that ultimately benefits the industrial sector regarding its specific market niches and end users.

2. Methodologies and Usages

The Lagrangian meshfree particle-based method has advantages in solving fluid dynamics problems with complex or time-evolving boundaries for a single phase or multiple phases. A pure Lagrangian meshfree particle method based on a generalized finite difference (GFD) scheme is proposed by Zhang and Xiongin [1] to simulate time-dependent weakly compressible viscous flow. The flow is described with Lagrangian particles, and the partial differential terms in the Navier–Stokes equations are represented as the solution to a symmetric system of linear equations through a GFD scheme. In solving the particle-based symmetric equations, the numerical method only needs the kernel function itself instead of using its gradient, i.e., the approach is a kernel gradient free (KGF) method, which avoids the use of artificial parameters to solve the viscous term and reduces the limitations of using the kernel function. Moreover, the order of Taylor series expansion can easily be improved in the meshless algorithm. In this paper, the particle method is validated with several test cases, and the convergence, accuracy, and different kernel functions are evaluated.

In [2], an analytical study on blood flow analysis with a tapered porous channel is presented. The blood flow was driven by peristaltic pumping. Thermal radiation effects were also taken into account. The convective and slip boundary conditions were also applied in this formulation. These conditions are very helpful for carrying out the behavior of particle movement which may be utilized for cardiac surgery. The tapered porous channel had an unvarying wave speed with dissimilar amplitudes and phase. The non-dimensional analysis was utilized for some approximations, for example, the proposed mathematical modelling equations were modified by using a lubrication approach and the analytical solutions for stream function, nanoparticle temperature, and volumetric concentration profiles were obtained. The impacts of various emerging parameters on the thermal characteristics and nanoparticle concentration were analyzed with the help of computational results. The trapping phenomenon was also examined for relevant parameters. It was also observed that the geometric parameters, like amplitudes, non-uniform parameters, and phase difference, play important roles in controlling the nanofluid transport phenomena. The outcomes of the present model may be applicable for the smart nanofluid peristaltic pump which may be utilized in hemodialysis.

As magnetohydrodynamics (MHD) deals with the analysis of electrically conducting fluids and the study of nanofluids, considering the influence of MHD phenomena is a topic of great interest from industrial and technological points of view. Thus, the modified MHD mixed convective, nonlinear, radiative, and dissipative problem was modelled over an arc-shaped geometry for $Al_2O_3 + H_2O$ nanofluid at 310 K with a freezing temperature of 273.15 K by Khan et al. [3]. Firstly, the model was reduced into a coupled set of ordinary differential equations using similarity transformations. The impact of the freezing temperature and the molecular diameter were incorporated in the energy equation. Then, the Runge–Kutta scheme, along with the shooting technique, was adopted for the mathematical computations, and code was written in Mathematica 10.0. Further, a comprehensive discussion of the flow characteristics is provided. The results for the dynamic viscosity, heat capacity, and effective density of the nanoparticles were examined for various nanoparticle diameters and volume fractions

Jalali et al. [4] reported on laminar heat transfer and direct fluid jet injection of oil/MWCNT nanofluid using a numerical investigation with a finite volume method. Both slip and no-slip boundary conditions on solid walls were used. The objective of this study was to increase the cooling performance of heated walls inside a rectangular microchannel. The Reynolds numbers ranged from 10 to 50; slip coefficients were 0.0, 0.04, and 0.08; and nanoparticle volume fractions were 0%–4%. The results showed that using techniques for improving heat transfer, such as fluid jet injection with low temperature and adding nanoparticles to the base fluid, allowed for good results to be obtained. By increasing jet injection, areas with eliminated boundary layers along the fluid direction spread in the domain. Dispersing solid nanoparticles in the base fluid with higher volume fractions resulted in a better temperature distribution and Nusselt number. By increasing the nanoparticle volume fraction, the temperature of the heated surface penetrated the flow centerline, and the fluid temperature increased. Jet injection with higher velocity, due to its higher fluid momentum, resulted in a higher Nusselt number and affected lateral areas. The fluid velocity was higher in jet areas, which diminished the effect of the boundary layer.

The non-Newtonian fluid model named Casson fluid was considered by Rehman et al. [5]. The semi-infinite domain of the disk was fitted out with magnetized Casson liquid. The roles of both thermophoresis and Brownian motion were inspected by considering nanosized particles in a Casson liquid spaced above the rotating disk. The magnetized flow field was framed with Navier's slip assumption. The Von Karman scheme was adopted to transform flow narrating equations in terms of a reduced system. For better depiction, a self-coded computational algorithm was executed rather than the move-on with build-in array. Numerical observations via magnetic elements, Lewis numbers, Casson, slip, Brownian motion, and thermophoresis parameters subject to radial and tangential velocities, temperature, and nanoparticle concentration are reported. The validation of the numerical method being used is given through a comparison with existing work. Comparative values of the local Nusselt number and local Sherwood number are provided for the involved flow controlling parameters.

Alshomrani and Ullah [6] inspected velocity slip impacts in the three-dimensional flow of water-based carbon nanotubes because of a stretchable rotating disk. Nanoparticles like single and multi-walled carbon nanotubes (CNTs) were utilized. Graphical outcomes were acquired for both single-walled carbon nanotubes (SWCNTs) and multi-walled carbon nanotubes (MWCNTs). The heat transport system was examined in the presence of thermal convective condition. Proper variables led to a strong nonlinear standard differential framework. The associated nonlinear framework was tackled by an optimal homotopic strategy. Diagrams were plotted so as to examine how the temperature and velocities are influenced by different physical variables. The coefficients of skin friction and the Nusselt number are exhibited graphically. Our results indicate that the skin friction coefficient and Nusselt number are enhanced for larger values of nanoparticle volume fraction.

The Darcy–Brinkman flow over a stretching sheet in the presence of frictional heating and porous dissipation was examined in [7]. The governing equations were modeled and simplified under boundary layer approximations, which were then transformed into a system of self-similar equations using appropriate transformations. The resulting system of nonlinear equations was solved numerically under velocity and thermal slip conditions by the fourth-order Runge–Kutta method and built-in routine bvp4c in Matlab. Under special conditions, the obtained results were compared with the results available in the literature. An excellent agreement was observed. The variation of parameters was studied for different flow quantities of interest, and the results are presented in the form of tables and graphs.

The Couette–Poiseuille flow of couple stress fluid with a magnetic field between two parallel plates was investigated by Ellahi et al. [8]. The flow was driven due to the axial pressure gradient and uniform motion of the upper plate. The influence of heating at the wall in the presence of spherical and homogeneous Hafnium particles was taken into account. The temperature dependent viscosity model, namely, Reynolds' model was utilized. The Runge–Kutta scheme with shooting was used to tackle a non-linear system of equations. It was observed that the velocity decreased by increasing the values of the Hartman number, as heating of the wall reduced the effects of viscous forces; therefore, the resistance of magnetic force reduced the fluid velocity. However, due to shear thinning effects, the velocity was increased by increasing the values of the viscosity parameter, and as a result, the temperature profile also declined. The suspension of inertial particles in an incompressible turbulent flow with Newtonian and non-Newtonian base fluids can be used to analyze the biphase flows through diverse geometries that could possibly be future perspectives of the proposed model.

Characteristically, most fluids are not linear in their natural forms and therefore, fractional order models are very appropriate for handling these kinds of phenomena. In [9], the authors studied the base solvents of water and ethylene glycol for the stable dispersion of graphene oxide to prepare graphene oxide–water (GO-W) and graphene oxide–ethylene glycol (GO-EG) nanofluids. The stable dispersion of the graphene oxide in the water and ethylene glycol was taken from the experimental results. The combined efforts of the classical and fractional order models were imposed and compared under the effect of the Marangoni convection. The numerical method for the non-integer derivative that was used in this research is known as a predictor corrector technique of the Adams–Bashforth–Moulton method (Fractional Differential Equation-12 or shortly, FDE-12). The impacts of the modeled parameters were analyzed and compared for both GO-W and GO-EG nanofluids. The diverse effects of the parameters were observed through a fractional model rather than using the traditional approach. Furthermore, it was observed that GO-EG nanofluids are more efficient due to their high thermal properties compared with GO-W nanofluids.

In [10], instead of the more customary parabolic Fourier law, the hyperbolic Cattaneo–Christov (C–C) heat flux model was used to jump over the major hurdle of "parabolic energy equation". The more realistic three-dimensional Carreau fluid flow analysis was conducted with the attendance of temperature-dependent thermal conductivity. The other salient impacts affecting the considered model are the homogeneous-heterogeneous (h-h) reactions and magnetohydrodynamics (MHD). The boundary conditions supporting the problem are convective heat and h-h reactions. The considered boundary layer problem was addressed via similarity transformations to obtain the system of coupled differential equations. The numerical solutions were attained by undertaking the MATLAB built-in function bvp4c. To comprehend the consequences of assorted parameters on involved distributions, different graphs were plotted and are accompanied by requisite discussions in light of their physical significance. To substantiate the presented results, a comparison with the already conducted problem is also given. It is envisaged that there is a close correlation between the two results. This shows that dependable results are being submitted. It is noticed that h-h reactions depict an opposite behavior compared with the concentration profile. Moreover, the temperature of the fluid augments the values of thermal conductivity parameters.

The aim of [11] was to present an analysis of local similarity solutions of Casson fluid over a non-isothermal cylinder subject to suction/blowing. The cylinder was placed inside a porous medium and stretched in a nonlinear way. Further, the impacts of chemical reactions, viscous dissipation, and heat generation/absorption on flow fields were also investigated. Similarity transformations were employed to convert the nonlinear governing equations to nonlinear ordinary differential equations and then solved via the Keller box method. The findings demonstrate that the magnitude of the friction factor and mass transfer rate are suppressed with an increment in the Casson parameter, whereas the heat transfer rate was found to be intensified. An increase in the curvature parameter enhanced the flow field distributions. The magnitude of wall shear stress was found to be higher with an increase in the porosity and suction/blowing parameters.

The pressure-driven flow in the slip regime was investigated in rectangular microducts by Ullah et al. [12]. In this regime, the Knudsen number lay between 0.001 and 0.1. The duct aspect ratio was taken as $0 \leq \varepsilon \leq 1$. Rarefaction effects were introduced through the boundary conditions. The dimensionless governing equations were solved numerically using MAPLE, and MATLAB was used for artificial neural network modeling. Using a MAPLE numerical solution, the shear stress and heat transfer rate were obtained. The numerical solution could be validated for special cases when there was no slip (continuum flow), $\varepsilon = 0$ (parallel plates), and $\varepsilon = 1$ (square microducts). An artificial neural network was used to develop separate models for the shear stress and heat transfer rate. Both physical quantities were optimized using a particle swarm optimization algorithm. Using these results, the optimum values of both physical quantities were obtained in the slip regime. It is shown that the optimal values ensue for the square microducts at the beginning of the slip regime.

Unidirectional flows of fractional viscous fluids in a rectangular channel were studied in [13]. The flow was generated by the shear stress given on the bottom plate of the channel. The authors developed a generalized model on the basis of constitutive equations described by the time-fractional Caputo–Fabrizio derivative. Many authors have published different results by applying the time-fractional derivative to the local part of acceleration in the momentum equation. This fractional model approach does not have a sufficient physical background. By using fractional generalized constitutive equations, they developed a proper model to investigate exact analytical solutions corresponding to the channel flow of a generalized viscous fluid. The exact solutions for the velocity field and shear stress were obtained by using Laplace transform and Fourier integral transformation for three different cases, namely, (i) constant shear, (ii) ramped type shear, and (iii) oscillating shear. The results are plotted and discussed.

The research article in [14] deals with the determination of magnetohydrodynamic steady flow of three combined nanofluids (Jefferey, Maxwell and Oldroyd-B) over a stretched surface. The surface was considered to be linear. The Cattaneo–Christov heat flux model was considered necessary to study the relaxation properties of the fluid flow. The influence of homogeneous-heterogeneous reactions (active for auto catalysts and reactants) were taken into account. The modeled problem was solved analytically. The impressions of the magnetic field, Prandtl number, thermal relaxation time, Schmidt number, and homogeneous-heterogeneous reaction strength were considered through graphs. The velocity field diminished with an increasing magnetic field. The temperature field diminished with an increasing Prandtl number and thermal relaxation time. The concentration field upsurged with an increasing Schmidt number, which decreased with an increasing homogeneous-heterogeneous reaction strength. Furthermore, the impacts of these parameters on skin fraction, Nusselt number, and Sherwood number are also accessible through tables. A comparison between analytical and numerical methods is presented both graphically and numerically.

In the study presented in [15], Jeffrey fluid was studied in a microgravity environment. Unsteady two-dimensional incompressible and laminar g-Jitter mixed convective boundary layer flow over an inclined stretching sheet was examined. Heat generation and magnetohydrodynamic (MHD) effects were also considered. The governing boundary layer equations together with boundary conditions were converted into a non-similar arrangement using appropriate similarity conversions. The transformed system of equations was resolved mathematically by employing an implicit finite difference pattern through a quasi-linearization method. Numerical results of temperature, velocity, local heat transfer and local skin friction coefficient were computed and plotted graphically. It was found that local skin friction and local heat transfer coefficients increased for an increasing Deborah number when the magnitude of the gravity modulation was unity. Assessment with previously published results showed an excellent agreement.

The two-dimensional laminar incompressible magnetohydrodynamic steady flow over an exponentially shrinking sheet with the effects of slip conditions and viscous dissipation was examined by Lund et al. [16]. An extended Darcy Forchheimer model was considered to observe the porous medium embedded in a non-Newtonian Casson-type nanofluid. The governing equations were converted into nonlinear ordinary differential equations using an exponential similarity transformation. The resultant equations for the boundary values problem (BVPs) were reduced to initial values problems (IVPs) and then the shooting and fourth-order Runge–Kutta methods (RK-4th method) were applied to obtain numerical solutions. The results reveal that multiple solutions occur only for the high suction case. The results of the stability analysis showed that the first (second) solution is physically reliable (unreliable) and stable (unstable).

A mathematical model of a convection flow of magnetohydrodynamic (MHD) nanofluid in a channel embedded in a porous medium is introduced by Khan and Alqahtni [17]. The flow along the walls, characterized by a non-uniform temperature, is under the effect of the uniform magnetic field acting transversely to the flow direction. The walls of the channel are permeable. The flow is due to convection combined with uniform suction/injection at the boundary. The model was formulated in terms of unsteady, one-dimensional partial differential equations (PDEs) with imposed physical conditions. The cluster effect of nanoparticles was demonstrated in the $C_2H_6O_2$, and H_2O base fluids. The perturbation technique was used to obtain a closed-form solution for the velocity and temperature distributions. Based on numerical experiments, it was concluded that both the velocity and temperature profiles are significantly affected by ϕ. Moreover, the magnetic parameter retards the nanofluid motion whereas porosity accelerates it. Each H_2O-based and $C_2H_6O_2$-based nanofluid in the suction case had a higher magnitude of velocity as compared to the injections case.

An analytical simulation of the head-on collision between a pair of hydroelastic solitary waves propagating in opposite directions in the presence of a uniform current was proposed by Bhatti and Qiang [18] for an infinite, thin, elastic plate floating on the surface of water. The mathematical modeling of the thin elastic plate was based on the Euler–Bernoulli beam model. The resulting kinematic and dynamic boundary conditions were highly nonlinear and were solved analytically with the help of a singular perturbation method. The Poincaré–Lighthill–Kuo method was applied to obtain the solution of the nonlinear partial differential equations. The resulting solutions are presented separately for the left- and right-going waves. The behaviors of all emerging parameters are presented mathematically and discussed graphically for the phase shift, maximum run-up amplitude, distortion profile, wave speed, and solitary wave profile. It was found that the presence of a current strongly affects the wavelength and wave speed of both solitary waves. A graphical comparison with pure-gravity waves is also presented as a particular case study.

The 3D magnetohydrodynamic (MHD) rotational nanofluid flow through a stretching surface was investigated in the study presented in [19]. Carbon nanotubes (SWCNTs and MWCNTs) were used as nano-sized constituents, and water was used as a base fluid. The Cattaneo–Christov heat flux model was used for the heat transport phenomenon. This arrangement had remarkable visual and electronic properties, such as strong elasticity, high updraft stability, and natural durability. The heat interchanging phenomenon was affected by updraft emission. The effects of nanoparticles such as Brownian motion and thermophoresis were also included in the study. By considering the conservation of mass, motion quantity, heat transfer, and nanoparticle concentration, the whole phenomenon was modeled. The modeled equations were highly non-linear and were solved using the homotopy analysis method (HAM). The effects of different parameters are described in the tables, and their impacts on different state variables are displayed in graphs. Physical quantities like the Sherwood number, Nusselt number, and skin friction are presented through tables with the variations of different physical parameters.

The fractional order model is the most suitable model for representing such phenomena compared with other traditional approaches. The forced convection fractional order boundary layer flow comprising single-wall carbon nanotubes (SWCNTs) and multiple-wall carbon nanotubes (MWCNTs) with variable wall temperatures passing over a needle was examined in [20]. The numerical solutions for the similarity equations were obtained for the integer and fractional values by applying the Adams-type predictor corrector method. A comparison of the SWCNTs and MWCNTs for the classical and fractional schemes was investigated. The classical and fractional order impacts of the physical parameters such as skin fraction and Nusselt number are presented physically and numerically. It was observed that the impacts of the physical parameters over the momentum and thermal boundary layers in the classical model were limited; however, while utilizing the fractional model, the impacts of the parameters varied at different intervals.

A steady laminar flow over a vertical stretching sheet with the existence of viscous dissipation, heat source/sink, and magnetic fields was numerically inspected in [21] through a shooting scheme based Runge–Kutta–Fehlberg-integration algorithm. The governing equation and boundary layer balance were expressed and then converted into a nonlinear normal system of differential equations using suitable transformations. The impacts of the physical parameters on the dimensionless velocity, temperature, the local Nusselt, and skin friction coefficient are described. The results show good agreement with recent research. The findings reveal that the Nusselt number at the sheet surface augments, since the Hartmann number, stretching velocity ratio A, and Hartmann number Ha increase. Nevertheless, it reduces with respect to the heat generation/absorption coefficient δ.

The impacts of Newtonian heating and homogeneous-heterogeneous (h-h) reactions on the flow of Ag–H_2O nanofluid over a cylinder which is stretched in a nonlinear way are discussed in [22]. The additional effects of magnetohydrodynamics (MHD) and nonlinear thermal radiation are also added features of the problem under consideration. The shooting technique is used to obtain the numerical solution to the problem which comprises highly nonlinear system ordinary differential equations. The sketches of different parameters versus the involved distributions are given with requisite deliberations. The obtained numerical results are matched with those of an earlier published work, and an excellent agreement exists between them. From our obtained results, it is concluded that the temperature profile is enriched with augmented values of radiation and curvature parameters. Additionally, the concentration field is a declining function of the strength of h-h reactions.

The research conducted in [23] gives a remedy for the maligned tissues, cells, or clogged arteries of the heart by means of permeating a slim tube (i.e., catheter) in the body. The tiny size gold particles drift in the free space of catheters with flexible walls with coupled stress fluid. To improve the efficiency of curing and to speed up the process, activation energy was added to the process. The modified Arrhenius function and Buongiorno model, respectively, moderated the inclusion of activation energy and nanoparticles of gold. The effects of the chemical reaction and activation energy on peristaltic transport of nanofluids were also taken into account. It was found that the golden particles encapsulate large molecules to transport essential drugs efficiently to the affected part of the organ.

The aim of [24] was to study time-dependent, rotating, single-wall, electrically conducting carbon nanotubes with aqueous suspensions under the influence of nonlinear thermal radiation in a permeable medium. The impact of viscous dissipation was taken into account. The basic governing equations, in the form of partial differential equations (PDEs), were transformed into a set of ordinary differential equations (ODEs) suitable for transformations. The homotopy analysis method (HAM) was applied to obtain the solution. The effects of numerous parameters on the temperature and velocity fields are explained by graphs. Furthermore, the action of significant parameters on the mass transportation and the rates of fiction factor were determined and discussed by plots in detail. The boundary layer thickness was reduced by a greater rotation rate parameter in our established simulations. Moreover, velocity and temperature profiles decreased with increases in the unsteadiness parameter. The action of the radiation phenomena acts as a source of energy to the fluid system. For a greater rotation parameter value, the thickness of the thermal boundary layer decreases. The unsteadiness parameter rises with velocity and the temperature profile decreases. A higher value of ϕ augments the frictional force strength within a liquid motion. For greater R and θw, the heat transfer rate rises. The temperature profile reduces with rising values of Pr.

An axisymmetric three-dimensional stagnation point flow of a nanofluid on a moving plate with different slip constants in two orthogonal directions in the presence of a uniform magnetic field was considered by Sadiq [25]. The magnetic field was considered along the axis of the stagnation point flow. The governing Navier–Stokes equation, along with the equations of nanofluid for three-dimensional flow, was modified using similarity transform, and reduced nonlinear coupled ordinary differential equations were solved numerically. It was observed that the magnetic field M and slip parameter $\lambda 1$ increased the velocity and decreased the boundary layer thickness near the stagnation point. Also, a thermal boundary layer was achieved earlier than the momentum boundary layer with increases in the thermophoresis parameter Nt and Brownian motion parameter Nb. Important physical quantities, such as skin friction and the Nusselt and Sherwood numbers, were also computed and are discussed through graphs and tables.

3. Future Trends in Fluid Mechanics

Even with the completion of this Special Issue, the material that advances the state-of-the-art experimental, numerical, and theoretical methodologies or extends the bounds of existing methodologies through new contributions in fluid mechanics is still insufficient. Nanofluid technology can also help with the development of better industrial applications in fluid mechanics. As in the fields of fluid dynamics and mechanical engineering, most nanofluids are generally not linear in character; therefore, techniques like symmetry methods can help us to find meaningful solutions to nonlinear resulting equations.

Funding: This research received no external funding.

Acknowledgments: The guest editorial team of *Symmetry* would like to thank all authors for contributing their original work to this Special Issue, no matter what the final decision on their submitted manuscript was. The editorial team would also like to thank all anonymous professional reviewers for their valuable time, comments, and suggestions during the review process. We also acknowledge the entire staff of the journal's editorial board for providing their cooperation regarding this Special Issue. We hope that the scientists who are working in the same regime will not only enjoy this Special Issue but also appreciate the efforts made by the entire team.

Conflicts of Interest: The author declares no conflict of interest.

References

1. Zhang, Y.; Xiong, A. A Particle Method Based on a Generalized Finite Difference Scheme to Solve Weakly Compressible Viscous Flow Problems. *Symmetry* **2019**, *11*, 1086. [CrossRef]
2. Prakash, J.; Tripathi, D.; Tiwari, A.K.; Sait, S.M.; Ellahi, R. Peristaltic Pumping of Nanofluids through a Tapered Channel in a Porous Environment: Applications in Blood Flow. *Symmetry* **2019**, *11*, 868. [CrossRef]
3. Khan, U.; Abbasi, A.; Ahmed, N.; Alharbi, S.O.; Noor, S.; Khan, I.; Mohyud-Din, S.T.; Khan, W.A. Modified MHD Radiative Mixed Convective Nanofluid Flow Model with Consideration of the Impact of Freezing Temperature and Molecular Diameter. *Symmetry* **2019**, *11*, 833. [CrossRef]
4. Jalali, E.; Ali Akbari, O.; Sarafraz, M.M.; Abbas, T.; Safaei, M.R. Heat Transfer of Oil/MWCNT Nanofluid Jet Injection Inside a Rectangular Microchannel. *Symmetry* **2019**, *11*, 757. [CrossRef]
5. Rehman, K.U.; Malik, M.Y.; Khan, W.A.; Khan, I.; Alharbi, S.O. Numerical Solution of Non-Newtonian Fluid Flow Due to Rotatory Rigid Disk. *Symmetry* **2019**, *11*, 699. [CrossRef]
6. Alshomrani, A.S.; Ullah, M.Z. Significance of Velocity Slip in Convective Flow of Carbon Nanotubes. *Symmetry* **2019**, *11*, 679. [CrossRef]
7. Kausar, M.S.; Hussanan, A.; Mamat, M.; Ahmad, B. Boundary Layer Flow through Darcy–Brinkman Porous Medium in the Presence of Slip Effects and Porous Dissipation. *Symmetry* **2019**, *11*, 659. [CrossRef]
8. Ellahi, R.; Zeeshan, A.; Hussain, F.; Abbas, T. Two-Phase Couette Flow of Couple Stress Fluid with Temperature Dependent Viscosity Thermally Affected by Magnetized Moving Surface. *Symmetry* **2019**, *11*, 647. [CrossRef]
9. Gul, T.; Anwar, H.; Khan, M.A.; Khan, I.; Kumam, P. Integer and Non-Integer Order Study of the GO-W/GO-EG Nanofluids Flow by Means of Marangoni Convection. *Symmetry* **2019**, *11*, 640. [CrossRef]
10. Lu, D.; Mohammad, M.; Ramzan, M.; Bilal, M.; Howari, F.; Suleman, M. MHD Boundary Layer Flow of Carreau Fluid over a Convectively Heated Bidirectional Sheet with Non-Fourier Heat Flux and Variable Thermal Conductivity. *Symmetry* **2019**, *11*, 618. [CrossRef]
11. Ullah, I.; Alkanhal, T.A.; Shafie, S.; Nisar, K.S.; Khan, I.; Makinde, O.D. MHD Slip Flow of Casson Fluid along a Nonlinear Permeable Stretching Cylinder Saturated in a Porous Medium with Chemical Reaction, Viscous Dissipation, and Heat Generation/Absorption. *Symmetry* **2019**, *11*, 531. [CrossRef]
12. Hamadneh, N.N.; Khan, W.A.; Khan, I.; Alsagri, A.S. Modeling and Optimization of Gaseous Thermal Slip Flow in Rectangular Microducts Using a Particle Swarm Optimization Algorithm. *Symmetry* **2019**, *11*, 488. [CrossRef]
13. Asif, M.; Ul Haq, S.; Islam, S.; Abdullah Alkanhal, T.; Khan, Z.A.; Khan, I.; Nisar, K.S. Unsteady Flow of Fractional Fluid between Two Parallel Walls with Arbitrary Wall Shear Stress Using Caputo–Fabrizio Derivative. *Symmetry* **2019**, *11*, 449. [CrossRef]
14. Saeed, A.; Islam, S.; Dawar, A.; Shah, Z.; Kumam, P.; Khan, W. Influence of Cattaneo–Christov Heat Flux on MHD Jeffrey, Maxwell, and Oldroyd-B Nanofluids with Homogeneous-Heterogeneous Reaction. *Symmetry* **2019**, *11*, 439. [CrossRef]
15. Tlili, I. Effects MHD and Heat Generation on Mixed Convection Flow of Jeffrey Fluid in Microgravity Environment over an Inclined Stretching Sheet. *Symmetry* **2019**, *11*, 438. [CrossRef]
16. Ali Lund, L.; Omar, Z.; Khan, I.; Raza, J.; Bakouri, M.; Tlili, I. Stability Analysis of Darcy-Forchheimer Flow of Casson Type Nanofluid Over an Exponential Sheet: Investigation of Critical Points. *Symmetry* **2019**, *11*, 412. [CrossRef]
17. Khan, I.; Alqahtani, A.M. MHD Nanofluids in a Permeable Channel with Porosity. *Symmetry* **2019**, *11*, 378. [CrossRef]
18. Bhatti, M.M.; Lu, D.Q. Analytical Study of the Head-On Collision Process between Hydroelastic Solitary Waves in the Presence of a Uniform Current. *Symmetry* **2019**, *11*, 333. [CrossRef]
19. Shah, Z.; Tassaddiq, A.; Islam, S.; Alklaibi, A.M.; Khan, I. Cattaneo–Christov Heat Flux Model for Three-Dimensional Rotating Flow of SWCNT and MWCNT Nanofluid with Darcy–Forchheimer Porous Medium Induced by a Linearly Stretchable Surface. *Symmetry* **2019**, *11*, 331. [CrossRef]
20. Gul, T.; Khan, M.A.; Noman, W.; Khan, I.; Abdullah Alkanhal, T.; Tlili, I. Fractional Order Forced Convection Carbon Nanotube Nanofluid Flow Passing Over a Thin Needle. *Symmetry* **2019**, *11*, 312. [CrossRef]

21. Alarifi, I.M.; Abokhalil, A.G.; Osman, M.; Lund, L.A.; Ayed, M.B.; Belmabrouk, H.; Tlili, I. MHD Flow and Heat Transfer over Vertical Stretching Sheet with Heat Sink or Source Effect. *Symmetry* **2019**, *11*, 297. [CrossRef]
22. Suleman, M.; Ramzan, M.; Ahmad, S.; Lu, D.; Muhammad, T.; Chung, J.D. A Numerical Simulation of Silver–Water Nanofluid Flow with Impacts of Newtonian Heating and Homogeneous–Heterogeneous Reactions Past a Nonlinear Stretched Cylinder. *Symmetry* **2019**, *11*, 295. [CrossRef]
23. Ellahi, R.; Zeeshan, A.; Hussain, F.; Asadollahi, A. Peristaltic Blood Flow of Couple Stress Fluid Suspended with Nanoparticles under the Influence of Chemical Reaction and Activation Energy. *Symmetry* **2019**, *11*, 276. [CrossRef]
24. Jawad, M.; Shah, Z.; Islam, S.; Majdoubi, J.; Tlili, I.; Khan, W.; Khan, I. Impact of Nonlinear Thermal Radiation and the Viscous Dissipation Effect on the Unsteady Three-Dimensional Rotating Flow of Single-Wall Carbon Nanotubes with Aqueous Suspensions. *Symmetry* **2019**, *11*, 207. [CrossRef]
25. Sadiq, M.A. MHD Stagnation Point Flow of Nanofluid on a Plate with Anisotropic Slip. *Symmetry* **2019**, *11*, 132. [CrossRef]

 © 2020 by the author. Licensee MDPI, Basel, Switzerland. This article is an open access article distributed under the terms and conditions of the Creative Commons Attribution (CC BY) license (http://creativecommons.org/licenses/by/4.0/).

Article

MHD Stagnation Point Flow of Nanofluid on a Plate with Anisotropic Slip

Muhammad Adil Sadiq

Department of Mathematics, DCC-KFUPM, KFUPM Box 5084, Dhahran 31261, Saudi Arabia; adilsadiq@kfupm.edu.sa; Tel.: +966-598-658-229

Received: 1 January 2019; Accepted: 22 January 2019; Published: 24 January 2019

Abstract: In this article, an axisymmetric three-dimensional stagnation point flow of a nanofluid on a moving plate with different slip constants in two orthogonal directions in the presence of uniform magnetic field has been considered. The magnetic field is considered along the axis of the stagnation point flow. The governing Naiver–Stokes equation, along with the equations of nanofluid for three-dimensional flow, are modified using similarity transform, and reduced nonlinear coupled ordinary differential equations are solved numerically. It is observed that magnetic field M and slip parameter λ_1 increase the velocity and decrease the boundary layer thickness near the stagnation point. Also, a thermal boundary layer is achieved earlier than the momentum boundary layer, with the increase in thermophoresis parameter N_t and Brownian motion parameter N_b. Important physical quantities, such as skin friction, and Nusselt and Sherwood numbers, are also computed and discussed through graphs and tables.

Keywords: stagnation point flow; numerical solution; magnetic field; nanofuid

1. Introduction

The phenomenon of stagnation point flow has various uses in and aerodynamic industries. Such flows mainly compact with the movement of fluid close to the stagnated region of a rigid surface flowing in the fluid material, or retained with dynamics of fluid. Stagnation point has been studied by many researchers in the past because of its wide range of applications in engineering. Initially, stagnation point flow was analyzed by Hiemenz in 1911. He studied the two-dimensional stagnation point flow on a stationary plate. Stagnation point flow applications include cooling of electronic devices by fans, cooling of nuclear reactors, polymer extrusion, wire drawing, drawing of plastic sheets, and many hydrodynamic processes in engineering applications. Stagnation point flow possesses much physical significance, as it is used to calculate the velocity gradients and the rate of heat and mass transfer abutting to stagnation area of frames in high-speed flows, cooling of transpiration, rustproof designs of bearings, etc.

Recently, Borrelli et al. [1] deliberated over the impact of buoyancy on three-dimensional (3D) stagnation point flow. They stated that the buoyancy forces tend to favor an opposite flow. Later, Lok et al. [2] expanded on the work of Weidman [3] with buoyancy forces. They observed the discrete results for free convection and forced convection due to a singularity rising in the convection term. Steady oblique stagnation point flow of a viscous fluid was studied by Grosan et al. [4]. They solved the nonlinear coupled differential equation numerically using the Runge–Kutta method. It is observed that the location of the stagnation point depends strongly on the value of the shear parameter and magnetic parameter. Wang [5–7] discussed the three-dimensional stagnation flow in the absence of MHD and nanofluids on a flat plate, shrining disk, and rotating disk. Two-dimensional (2D) stagnation flow was discussed by Nadeem et al. [8] using HAM on a stretchable surface.

A fluid, heated by electric current in the presence of strong magnetic field, for example crystal growth in melting, has relevance in manufacturing industries. During the fluid motion, the association of electric current and magnetic field produces a divergence of Lorentz forces. This phenomenon prevents the convective motion of fluid and heat transfer characteristic changes accordingly. Ariel [9] investigated the flow near the stagnation point numerically for small magnetic fields; for large magnetic numbers, the perturbation technique was used. Raju and Sundeep [10] proved that with an increase in the magnetic number, there is an increase in the heat and mass transfer rates. They studied numerically the MHD flow of non-Newtonian fluid over a rotating cone or plate.

Generally, the size of nanoparticles is (1–100 nm). Currently, nanofluids are used for drug delivery in infected areas of the human body. Self-propagating objects containing drugs are used to remove blood clots in sensitive areas such as the brain, eye, heart, etc. Kleinstreuer [11] discussed the drug delivery system in humans at normal body temperature under the influence of some physical parameters such as nanoparticle length, artery diameter, and velocity of fluid. Recently, a mathematical model of nanofluid was developed by Choi [12]. Later, a contribution to heat transfer analysis in nanofluid was made by Buongiorno [13]. His mathematical model dealt with the non-homogeneous model for transport phenomena and heat transfer in nanofluids with applications to turbulence. Saleem et al. [14] discussed the effects of Brownian diffusion and thermophoresis on non-Newtonian fluid models, using HAM in the domain of a vertical rotating cone. Bachok et al. [15] studied the three-dimensional stagnation flow of a viscous fluid numerically, analyzed the velocity and heat transfer for different physical parameters, and compared three nanoparticles, namely C_u, Al_2O_3, T_iO_3. In [16] Ellahi et al. explored the heat and mass transfer of non-Newtonian fluid in an annulus in a porous medium using HAM. Recently, Sheikholeslami et al. [17] studied the effects of thermal radiation on steady viscous nanofluid in the presence of MHD numerically. Khan [18] explored Brownian diffusion and thermophoresis on stagnation point flow. He considered dual solutions for shrinking/stretching parameters and heat transfer in the presence of buoyancy forces on a stretchable surface. Mustafa et al. [19] investigated 3D nanofluid flow and heat transfer in two opposite directions on a plane horizontal stretchable surface. Thermal and momentum boundary layers were discussed using physical parameters such as Brownian motion and thermophoretic forces. Some more useful studies related to nanofluids can be found in [20–29].

In this article, an axisymmetric 3D stagnation point flow of a nanofluid on a moving plate with different slip constants in two orthogonal directions in the presence of uniform magnetic field has been considered and solved numerically.

2. Mathematical Formulation

Consider a stagnation point flow of a nanofluid over a plate with anisotropic slip in a Cartesian coordinate system, so that the x-axis is taken along the corrugations of plates, the y-axis is normal to the corrugations, and the z-axis is considered with the axis of stagnation flow. The velocities of the moving plate are (u, v) in (x, y) directions, respectively. A constant magnetic field is applied perpendicular to the corrugation along the axis of the stagnation flow in such a way that the magnetic Reynolds number is small. According to Wang [5], the potential flow far from the plate is defined as:

$$uu_x + vu_y + wu_z = -\frac{p_x}{\rho} + \nu\left(u_{xx} + u_{yy} + u_{zz}\right) - \frac{B_0^2}{\rho}u, \tag{1}$$

$$uv_x + vv_y + wv_z = -\frac{p_y}{\rho} + \nu\left(v_{xx} + v_{yy} + v_{zz}\right) - \frac{B_0^2}{\rho}v, \tag{2}$$

$$uw_x + vw_y + ww_z = -\frac{p_z}{\rho} + \nu\left(w_{xx} + w_{yy} + w_{zz}\right), \tag{3}$$

$$u\frac{\partial T}{\partial x}+v\frac{\partial T}{\partial y}+w\frac{\partial T}{\partial z} = \alpha_m\left(\frac{\partial^2 T}{\partial x^2}+\frac{\partial^2 T}{\partial y^2}+\frac{\partial^2 T}{\partial z^2}\right)$$
$$+\frac{(\rho C)_p}{(\rho C)_f}\left[D_B\left(\frac{\partial C}{\partial x}\frac{\partial T}{\partial x}+\frac{\partial C}{\partial y}\frac{\partial T}{\partial y}+\frac{\partial C}{\partial z}\frac{\partial T}{\partial z}\right)\right]$$
$$+\frac{D_T}{T_\infty}\left[\left(\frac{\partial T}{\partial x}\right)^2+\left(\frac{\partial T}{\partial y}\right)^2+\left(\frac{\partial T}{\partial z}\right)^2\right], \tag{4}$$

$$u\frac{\partial C}{\partial x}+v\frac{\partial C}{\partial y}+w\frac{\partial C}{\partial z} = D_B\left[\frac{\partial^2 C}{\partial x^2}+\frac{\partial^2 C}{\partial y^2}+\frac{\partial^2 C}{\partial z^2}\right]$$
$$+\frac{D_T}{T_\infty}\left[\frac{\partial^2 T}{\partial x^2}+\frac{\partial^2 T}{\partial y^2}+\frac{\partial^2 T}{\partial z^2}\right]. \tag{5}$$

and the boundary conditions are:

$$u - U = N_1\mu\frac{\partial u}{\partial z}, v - V = N_2\mu\frac{\partial v}{\partial z}, \quad T = T_w, \quad C = C_\infty \quad \text{at} \quad z = 0,$$
$$u \to ax, v \to ay, T \to T_\infty, C \to C_\infty \quad \text{at} \quad z \to \infty. \tag{6}$$

where (u, v) are the velocity components in the (x, y) directions, ν is the kinematic viscosity, T is the temperature, α_m is the thermal diffusivity, C is the volume of nanoparticles, $(\rho C)_p$ is the effective heat capacity of nanoparticles, $(\rho C)_f$ is the heat capacity of fluid, D_B is the Brownian diffusion coefficient and D_T is the thermophoretic diffusion coefficient. For the non-dimensionalization, we use the following similarity variables:

$$\begin{aligned} u &= axf'(\eta) + Uh(\eta),\\ v &= ayg'(\eta) + Vk(\eta),\\ w &= -\sqrt{a\nu}\left[f(\eta) + g(\eta)\right]. \end{aligned} \tag{7}$$

where $\eta = \sqrt{a/\nu}\, z$. Using Equation (7) in Equations (5) and (6) finally we get:

$$f''' + f''(f+g) - (f')^2 - M^2 f' = -(1+M^2) \tag{8}$$
$$g''' + g''(f+g) - (g')^2 - M^2 g' = -(1+M^2) \tag{9}$$
$$h'' + h'(f+g) - hf' - M^2 h = 0 \tag{10}$$
$$k'' + k'(f+g) - kg' - M^2 k = 0 \tag{11}$$
$$\theta'' + P_r(f+g)\theta' + P_r\left[N_t\theta'\phi' + N_b(\theta')^2\right] = 0 \tag{12}$$
$$\phi'' + S_c(f+g)\phi' + \frac{N_t}{N_b}\theta'' = 0. \tag{13}$$

and boundary conditions are:

$$f'(0) = \lambda_1 f''(0),\ g'(0) = \lambda_2 g''(0),\ h(0) = 1 + \lambda_1 h'(0),\ k(0) = 1 + \lambda_2 k'(0),\ f(0) = 0, g(0) = 0,$$
$$f'(\infty) \to 1,\ g'(\infty) \to 1,\ h(\infty) \to 1,\ k(\infty) \to 1,\ \theta(0) = 1,\ \theta(\infty) \to 0,\ \phi(0) = 1,\ \phi(\infty) \to 0. \tag{14}$$

here λ_1 and λ_2 are the slip parameters, P_r the prantle number, S_c the Schmidt number, N_t and N_b are thermophoresis parameter, Brownian motion parameters, respectively.

The expression for the skin friction coefficient, the local Nusselt number, and Sherwood number for second-grade fluid are defined as:

$$Re_x^{1/2} C_f = f''(0),\quad Nu_x\, Re_x^{-1/2} = -\theta'(0),\quad Sh_x\, Re_x^{-1/2} = -\varphi'(0), \tag{15}$$

where $Re_x = \frac{U_w x}{\nu}$ is the local Reynolds number. The solution of above coupled nonlinear differential equations are found numerically and discussed in the following section.

3. Result and Discussion

A system of nonlinear ordinary differential Equations (8)–(13) subject to the boundary conditions of Equation (14) are solved numerically using the Richardson extrapolation enhancement method. Richardson extrapolation is generally faster, and capable of handling BVP systems with unknown parameters. The values of these parameters can be determined under the presence of a sufficient number of boundary conditions. The solutions are discussed through graphs from Figures 1–10, and values of physical quantities, such as skin friction and Nusselt and Sherwood numbers, are presented in Tables 1–3.

Figures 1 and 2 show the variation of velocity profile f' and g' against η for different values of magnetic field M and slip parameter λ_1. It was observed that increasing in the values of M and λ_1 causes increase in the velocity profile, while boundary layer thickness reduces. Thus, these parameters cause a reduction in the momentum boundary layer. Analysis shows that increasing the values of these parameters to a sufficiently large level shows the monotonic behavior of velocity throughout the whole domain. Figures 3 and 4 shows the opposite behavior of h and k with the increment of M and λ_1, such that with the increase in value of these parameters, h and k decreases.

The temperature profile for the nanofluid against different values of thermophoresis parameter N_t and Brownian motion parameter N_b are plotted in Figures 5 and 6. As the temperature increase within the boundary layer, the values of these parameters increase. The thermal boundary layer is achieved earlier than the momentum boundary layer. The variation of nanoconcentration for different values of Schmidt number S_c and N_t is presented in Figures 7 and 8, respectively. It is observed that nanoconcentration ϕ decreases as the increase in S_c and boundary layer thickness decreases. Also, with the increase in N_t, the nanoconcentration decreases. Figures 9 and 10 show the velocity profile for different values of magnetic parameter $M = 0$ and for $M = 2$. It is observed that in the absence of magnetic parameter M, the boundary layer thickness is larger than while M is present. $M = 0$ in Figures 11 and 12 represents the results of Wang [5]. The slip parameter ratio can be defined as $\gamma = \frac{\lambda_2}{\lambda_1}$. Figures 13 and 14 describe the $f'(\eta), g'(\eta)$ for $\gamma = 0.5$. The range of γ varies from 0.2 to 10. $\gamma = 1$ represents the isotropic case where $f'(\eta) = g'(\eta)$ and $h(\eta) = k(\eta)$.

Table 1 shows local Nusselt number Nu_x and local Sherwood number Sh_x for the variation of P_r and thermophoresis parameter N_b. Here we see that with the increase of P_r, the local Nusselt number decreases, while local Sherwood number gives opposite results, meaning Sh_x increases. Moreover, with the increase of N_b, the results are again the opposite for Nu_x and Sh_x. Table 2 shows local Nusselt number and local Sherwood number for variations of slip parameter λ_1 and Brownian motion Nb. Here it is observed that with the increase of λ_1 both Nusselt number and local Sherwood number increase. Table 3 shows the skin friction coefficient C_f for different values of λ_1 and magnetic parameter M. Note that with the increment in λ_1, the value of skin friction decreases. A high value of M gives larger values of skin friction.

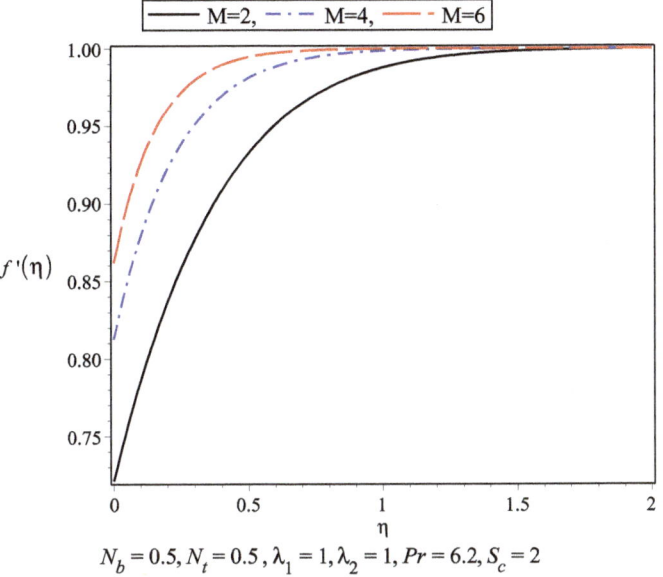

Figure 1. Variation of $f'(\eta)$ for different M.

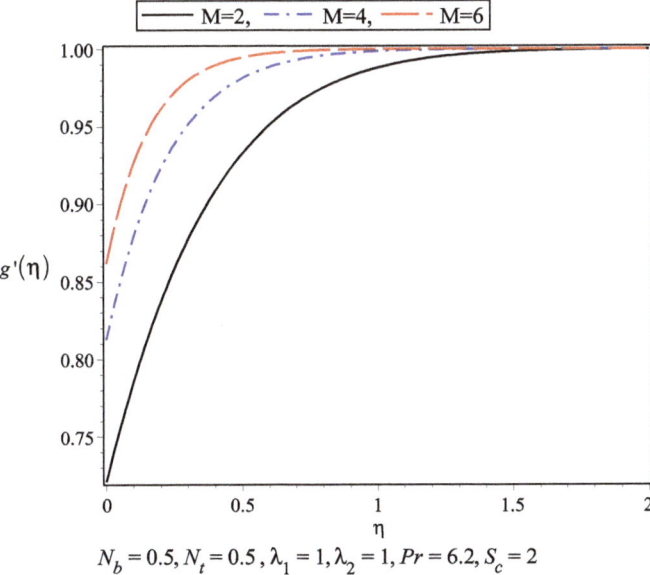

Figure 2. Variation of $g'(\eta)$ for different M.

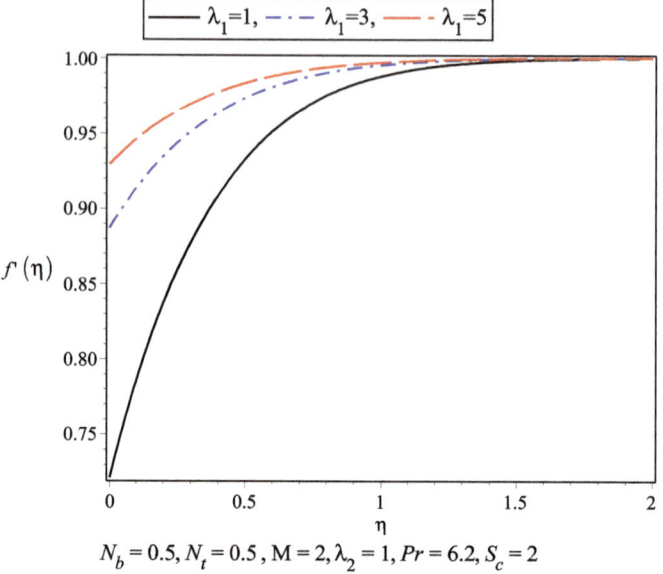

Figure 3. Variation of $f'(\eta)$ for different λ_1.

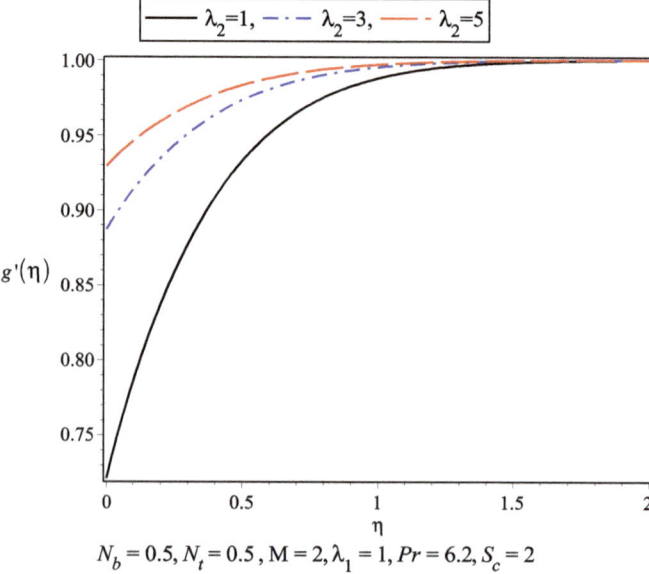

Figure 4. Variation of $g'(\eta)$ for different λ_2.

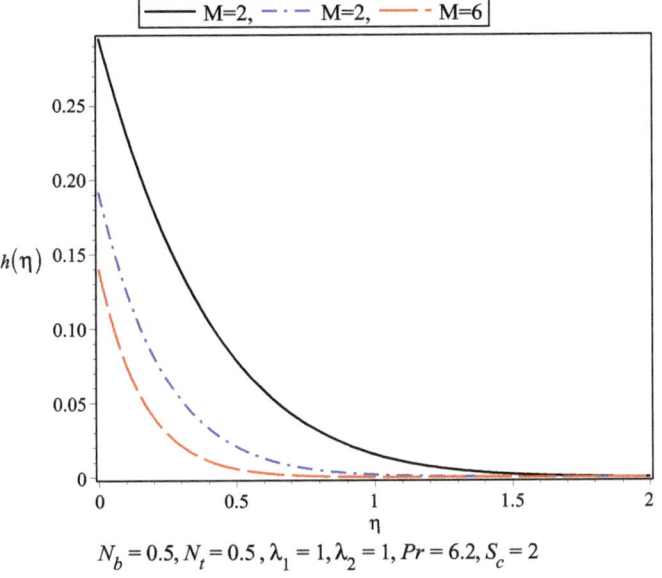

Figure 5. Variation of $h(\eta)$ for different M.

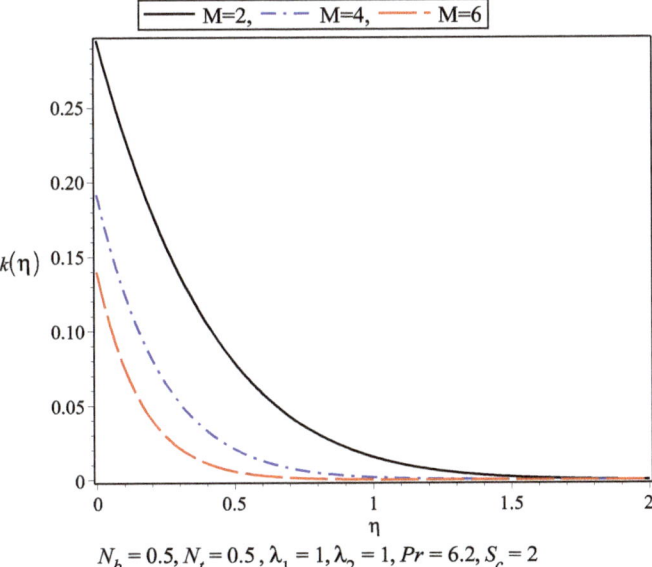

Figure 6. Variation of $k(\eta)$ for different M.

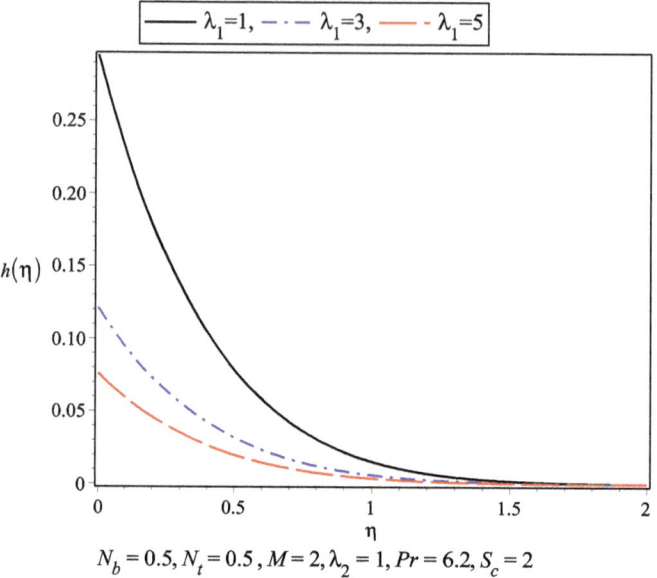

Figure 7. Variation of $h(\eta)$ for different λ_1.

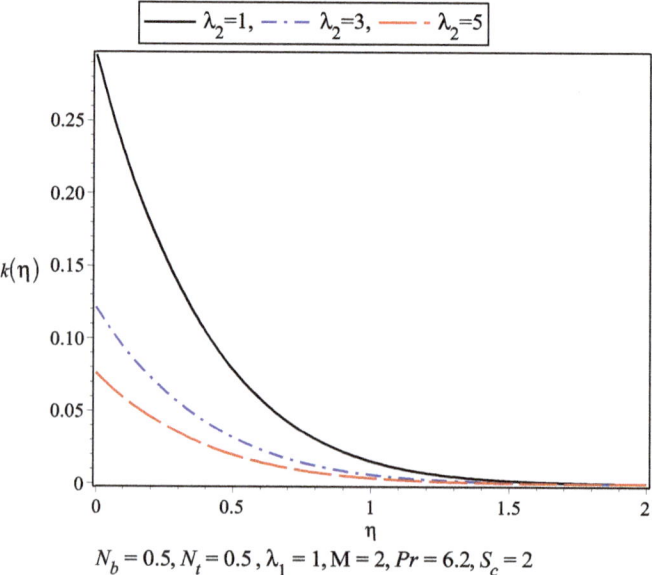

Figure 8. Variation of $k(\eta)$ for different λ_2.

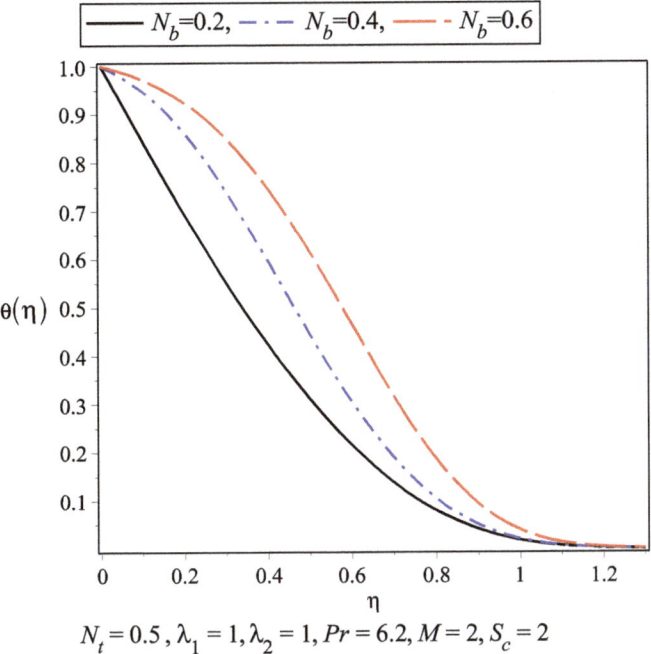

$N_t = 0.5, \lambda_1 = 1, \lambda_2 = 1, Pr = 6.2, M = 2, S_c = 2$

Figure 9. Variation of $\theta(\eta)$ for different N_b.

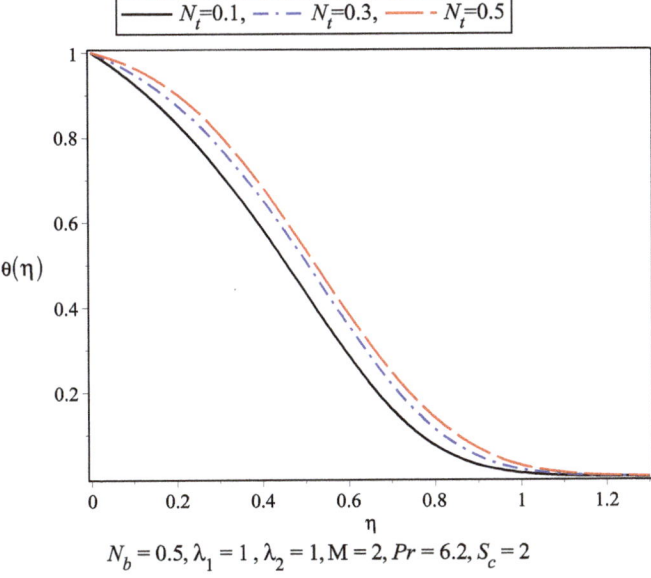

$N_b = 0.5, \lambda_1 = 1, \lambda_2 = 1, M = 2, Pr = 6.2, S_c = 2$

Figure 10. Variation of $\theta(\eta)$ for different N_t.

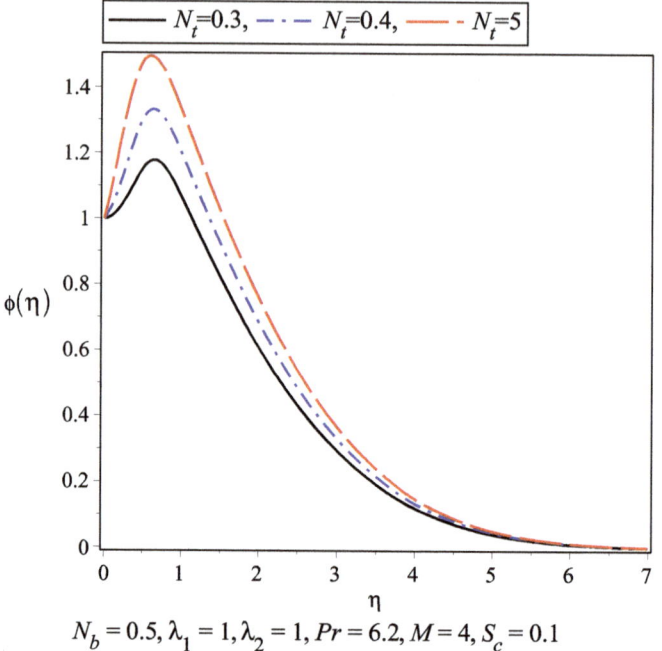

$N_b = 0.5, \lambda_1 = 1, \lambda_2 = 1, Pr = 6.2, M = 4, S_c = 0.1$

Figure 11. Variation of $\phi(\eta)$ for different N_t.

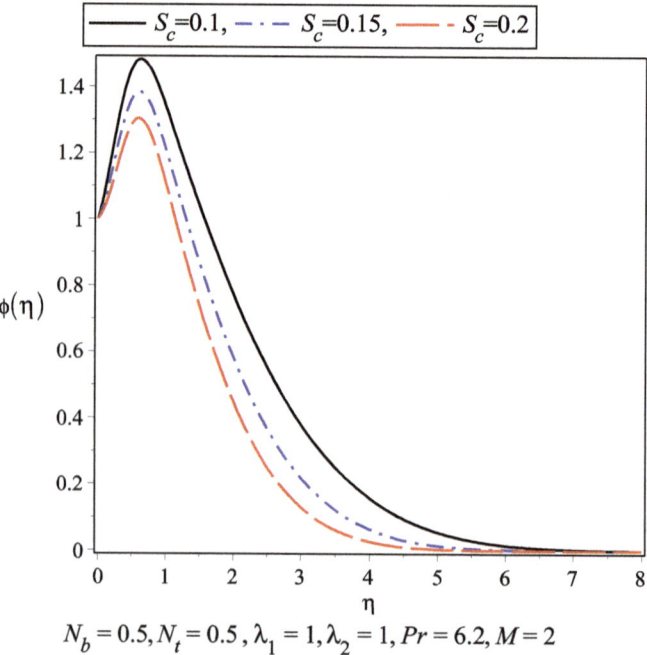

$N_b = 0.5, N_t = 0.5, \lambda_1 = 1, \lambda_2 = 1, Pr = 6.2, M = 2$

Figure 12. Variation of $\phi(\eta)$ for different S_c.

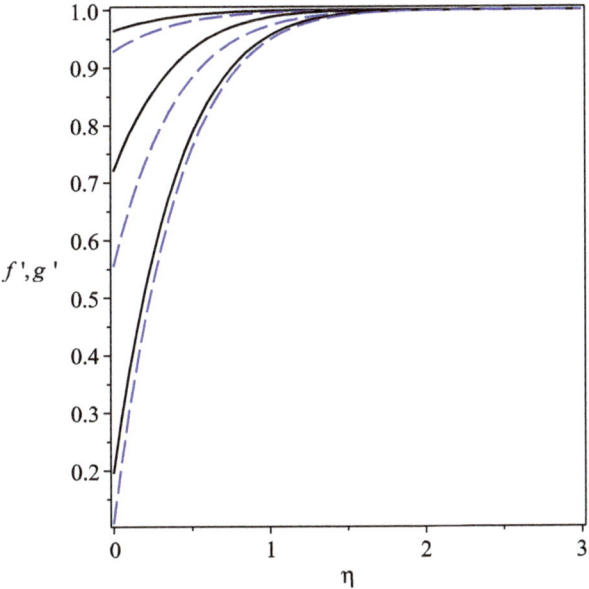

Figure 13. $f'(\eta)$ solid curves and $g'(\eta)$ dashed curves for $\gamma = \frac{\lambda_2}{\lambda_1} = 0.5$. From top: $\lambda_1 = 10, 1, 0.1$.

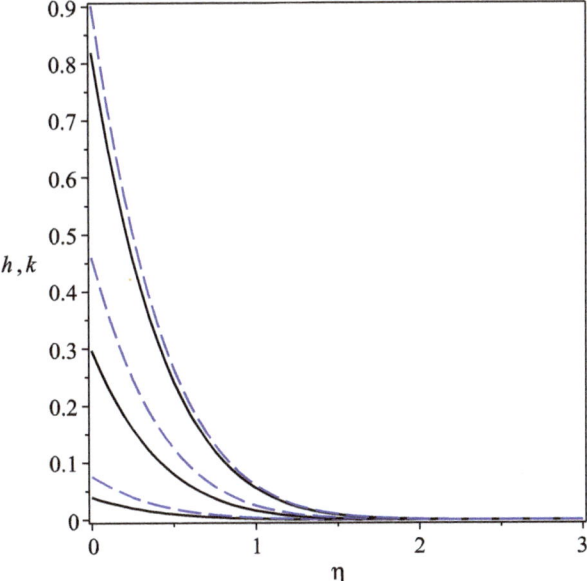

Figure 14. $h(\eta)$ solid curves and $k(\eta)$ dashed curves for $\gamma = \frac{\lambda_2}{\lambda_1} = 0.5$. From top: $\lambda_1 = 10, 1, 0.1$.

Table 1. Variation of Local Nusselt number Nu_x and Sherwood number Sh_x for different N_b and P_r.

	$\lambda_1 = 1, \lambda_2 = 1, S_c = 2, M = 2, N_t = 0.5$					
	$N_b = 0.1$		$N_b = 0.3$		$N_b = 0.5$	
P_r	Nu_x	Sh_x	Nu_x	Sh_x	Nu_x	Sh_x
5.5	0.67084	1.47332	0.45477	1.58078	0.30641	1.77542
5.6	0.66527	1.47440	0.44833	1.58438	0.29987	1.78168
5.7	0.65976	1.47546	0.44197	1.58792	0.29346	1.78781
5.8	0.65431	1.47651	0.43570	1.59139	0.28716	1.79381
5.9	0.64890	1.47754	0.42950	1.59481	0.28098	1.79968
6.0	0.64355	1.47856	0.42343	1.59816	0.27492	1.80543
6.1	0.63826	1.47956	0.41742	1.60145	0.26897	1.81105
6.2	0.63302	1.48055	0.41150	1.60468	0.26314	1.81655
6.3	0.62784	1.48152	0.40566	1.60785	0.25742	1.82192
6.4	0.62272	1.48248	0.39992	1.61096	0.25181	1.82718
6.5	0.61766	1.48342	0.39425	1.61401	0.24631	1.83232

Table 2. Variation of Local Nusselt number Nu_x and Sherwood number Sh_x for different M and λ_1.

	$\lambda_2 = 1, S_c = 2, N_b = 0.5, N_t = 0.5, P_r = 6.2$					
	$M = 2$		$M = 4$		$M = 6$	
λ_1	Nu_x	Sh_x	Nu_x	Sh_x	Nu_x	Sh_x
0.5	0.25129	1.78860	0.27363	1.85616	0.28527	1.88559
0.6	0.25465	1.79657	0.27603	1.86128	0.28698	1.88892
0.7	0.25737	1.80301	0.27791	1.86529	0.28829	1.89145
0.8	0.25962	1.80831	0.27943	1.86850	0.28933	1.89350
0.9	0.26152	1.81276	0.28068	1.87114	0.29018	1.89513
1.0	0.26314	1.81655	0.28172	1.87335	0.29087	1.89648
1.1	0.26453	1.81980	0.28261	1.87522	0.29146	1.89762
1.2	0.26575	1.82263	0.28337	1.87683	0.29196	1.89859
1.3	0.26682	1.82511	0.28403	1.87822	0.29239	1.89942
1.4	0.26776	1.82731	0.28461	1.87944	0.29277	1.90015
1.5	0.26861	1.82926	0.28513	1.88052	0.29310	1.90079

Table 3. Variation of Skin friction coefficient for different M and λ_1.

	$\lambda_2 = 1, S_c = 2, N_b = 0.5, N_t = 0.5, P_r = 6.2$		
	$M = 2$	$M = 4$	$M = 6$
λ_1	C_f	C_f	C_f
0.5	1.12177	1.36687	1.51354
0.6	1.00998	1.20285	1.31469
0.7	0.91823	1.07391	1..16200
0.8	0.84163	0.96991	1.04108
0.9	0.77675	0.88425	0.94294
1.0	0.72109	0.81248	0.86171
1.1	0.67283	0.75148	0.79336
1.2	0.63060	0.69899	0.73505
1.3	0.59334	0.65335	0.68473
1.4	0.56022	0.61330	0.64086
1.5	0.53059	0.57788	0.60227

4. Conclusions

The current paper investigated the effects of uniform magnetic field of axisymmetric three-dimensional stagnation point flow of a nanofluid on a moving plate with different slip constants.

The governing equations were made dimensionless and then solved using the Richardson extrapolation enhancement method. The following are the findings of the above work:

- An increase in the magnetic field M and slip parameter λ_1 causes an increase in the velocity profile and decrease in the boundary layer thickness near the stagnation point.
- It is observed that in the absence of magnetic parameter M the boundary layer thickness is larger than while M is present.
- The thermal boundary layer increases with an increase in the thermophoresis parameter N_t and Brownian motion parameter N_b. It is observed that the thermal boundary layer is achieved earlier compared to the momentum boundary layer.
- It is observed that with the increase in S_c and N_t the nanoconcentration ϕ decreases and vice versa.

Funding: The author wishes to express his thanks for financial support received from King Fahd University of Petroleum and Minerals.

Acknowledgments: The author wishes to express his thanks to King Fahd University of Petroleum and Minerals and reviewers to improve the manuscript.

Conflicts of Interest: The author declares that there is no conflict of interest.

Abbreviations

The following abbreviations are used in this manuscript:

(u,v)	velocity Components
ν	kinematic viscosity
N_1, N_2	slip coefficient
T	temperature
α_m	thermal diffusivity
C	volume of nano particles
$(\rho C)_f$	heat capacity of fluid
D_B	Brownian diffusion coefficient
D_T	thermophoretic diffusion coefficient
λ_1, λ_2	slip parameters
N_t	thermophoresis parameter
N_b	browning motion parameter
C_f	skin friction coefficient
Nu_x	local Nusselt number
Sh_x	Sherwood number
Re_x	local Reynolds number
S_c	Schmidt number
Pr	prantle number
γ	ratio of slip parameters
ϕ	nano concentration
M	magnetic parameter

References

1. Borrelli, A.; Giantesio, G.; Patria, M.C. Numerical simulations of three-dimensional MHD stagnation-point flow of a micropolar fluid. *Comput. Math. Appl.* **2013**, *66*, 472–489. [CrossRef]
2. Lok, Y.Y.; Amin, N.; Pop, I. Non-orthogonal stagnation point flow towards a stretching shee. *Int. J. Non Linear Mech.* **2006**, *41*, 622–627. [CrossRef]
3. Tilley, B.S.; Weidman, P.D. Oblique two-fluid stagnation-point flow. *Eur. J. Mech. B Fluids* **1998**, *17*, 205–217. [CrossRef]
4. Grosan, T.; Pop, I.; Revnic, C.; Ingham, D.B. Magnetohydrodynamic oblique stagnation-point flow. *Meccanica* **2009**, *44*, 565. [CrossRef]

5. Wang, C.Y. Stagnation flow on a plate with anisotropic slip. *Eur. J. Mech. B Fluids* **2013**, *38*, 73–77. [CrossRef]
6. Wang, C.Y. Off-centered stagnation flow towards a rotating disc. *Int. J. Eng. Sci.* **2008**, *46*, 391–396. [CrossRef]
7. Wang, C.Y. Stagnation flow towards a shrinking sheet. *Int. J. Non Linear Mech.* **2008**, *43*, 377–382. [CrossRef]
8. Nadeem, S.; Hussain, A.; Khan, M. HAM solutions for boundary layer flow in the region of the stagnation point towards a stretching sheet. *Commun. Nonlinear Sci. Numer. Simul.* **2010**, *15*, 475–481. [CrossRef]
9. Ariel, P.D. Hiemenz flow in hydromagnetics. *Acta Mech.* **1994**, *103*, 31–43. [CrossRef]
10. Raju, C.S.; Sandeep, N. Heat and mass transfer in MHD non-Newtonian bio-convection flow over a rotating cone/plate with cross diffusion. *J. Mol. Liq.* **2016**, *215*, 115–126. [CrossRef]
11. Kleinstreuer, C.; Li, J.; Koo, J. Microfluidics of nano-drug delivery. *Int. J. Heat Mass Trans.* **2008**, *51*, 5590–5597. [CrossRef]
12. Choi, S.U.S. Enhancing thermal conductivity of fluids with nanoparticles. *ASME Publ. Fed* **1995**, *231*, 99–106.
13. Buongiorno, J. Convective transport in nanofluids. *J. Heat Transf.* **2006**, *128*, 240–250. [CrossRef]
14. Nadeem, S.; Saleem, S. Analytical study of third grade fluid over a rotating vertical cone in the presence of nanoparticles. *Int. J. Heat Mass Transf.* **2015**, *85*, 1041–1048. [CrossRef]
15. Bachok, N.; Ishak, A.; Nazar, R.; Pop, I. Flow and heat transfer at a general three-dimensional stagnation point in a nanofluid. *Phys. B Condens. Matter* **2010**, *405*, 4914–4918. [CrossRef]
16. Ellahi, R.; Aziz, S.; Zeeshan, A. Non Newtonian nanofluids flow through a porous medium between two coaxial cylinders with heat transfer and variable viscosity. *J. Porous Media* **2013**, *16*, 205–216. [CrossRef]
17. Sheikholeslami, M.; Ganji, D.; Javed, M.Y.; Ellahi, R. Effect of thermal radiation on nanofluid flow and heat transfer using two phase model. *J. Magn. Magn. Mater.* **2015**, *374*, 36–43. [CrossRef]
18. Makinde, O.D.; Khan, W.A.; Khan, Z.H. Buoyancy effects on MHD stagnation point flow and heat transfer of a nanofluid past a convectively heated stretching/shrinking sheet. *Int. J. Heat Mass Transf.* **2013**, *62*, 526–533. [CrossRef]
19. Junaid Ahmad Khan, M.; Mustafa, T.; Hayat, A.; Alsaedi, A. Three-dimensional flow of nanofluid over a non-linearly stretching sheet: An application to solar energy. *Int. J. Heat Mass Transf.* **2015**, *86*, 158–164. [CrossRef]
20. Upadhya, M.; Mahesha, S.; Raju, C.S.K. Unsteady Flow of Carreau Fluid in a Suspension of Dust and Graphene Nanoparticles With Cattaneo–Christov Heat Flux. *J. Heat Transf.* **2018**, *140*, 092401. [CrossRef]
21. Li, Z.; Sheikholeslami, M.; Ahmad Shafee, S.; Ali J Chamkha, S. Effect of dispersing nanoparticles on solidification process in existence of Lorenz forces in a permeable media. *J. Mol. Liq.* **2018**, *266*, 181–193. [CrossRef]
22. Raju, C.S.K.; Saleem, S.; Mamatha, S.U. Iqtadar Hussain, Heat and mass transport phenomena of radiated slender body of three revolutions with saturated porous: Buongiorno's model. *Int. J. Therm. Sci.* **2018**, *132*, 309–315. [CrossRef]
23. Ram, P.; Kumar, A. Analysis of Heat Transfer and Lifting Force in a Ferro-Nanofluid Based Porous Inclined Slider Bearing with Slip Conditions. *Nonlinear Eng.* **2018**. [CrossRef]
24. Soomro, F.A.; Hammouch, Z. Heat transfer analysis of CuO-water enclosed in a partially heated rhombus with heated square obstacle. *Int. J. Heat Mass Transf.* **2018**, *118*, 773–784.
25. Hayat, T.; Qayyum, S.; Alsaedi, A.; Ahmad, B. Results in Physics, Significant consequences of heat generation/absorption and homogeneous-heterogeneous reactions in second grade fluid due to rotating disk. *Results Phys.* **2018**, *8*, 223–230. [CrossRef]
26. Hussain, S.; Aziz, A.; Aziz, T.; Khalique, C.M. Slip Flow and Heat Transfer of Nanofluids over a Porous Plate Embedded in a Porous Medium with Temperature Dependent Viscosity and Thermal Conductivity. *Appl. Sci.* **2016**, *6*, 376. [CrossRef]
27. Anuar, N.; Bachok, N.; Pop, I. A Stability Analysis of Solutions in Boundary Layer Flow and Heat Transfer of Carbon Nanotubes over a Moving Plate with Slip Effect. *Energies* **2018**, *11*, 3243. [CrossRef]
28. Fetecau, C.; Vieru, D.; Azhar, W.A. Natural Convection Flow of Fractional Nanofluids Over an Isothermal Vertical Plate with Thermal Radiation. *Appl. Sci.* **2017**, *7*, 247. [CrossRef]

29. Khan, N.S.; Gul, T.; Islam, S.; Khan, I.; Alqahtani, A.M.; Alshomrani, A.S. Alqahtani and Ali Saleh Alshomrani, Magnetohydrodynamic Nanoliquid Thin Film Sprayed on a Stretching Cylinder with Heat Transfer. *Appl. Sci.* **2017**, *7*, 271. [CrossRef]

© 2019 by the authors. Licensee MDPI, Basel, Switzerland. This article is an open access article distributed under the terms and conditions of the Creative Commons Attribution (CC BY) license (http://creativecommons.org/licenses/by/4.0/).

Article

Impact of Nonlinear Thermal Radiation and the Viscous Dissipation Effect on the Unsteady Three-Dimensional Rotating Flow of Single-Wall Carbon Nanotubes with Aqueous Suspensions

Muhammad Jawad [1], Zahir Shah [1], Saeed Islam [1], Jihen Majdoubi [2], I. Tlili [3], Waris Khan [4] and Ilyas Khan [5],*

1. Department of Mathematics, Abdul Wali Khan University, Mardan, Khyber Pakhtunkhwa 23200, Pakistan; muhammadjawad175@yahoo.com (M.J.); zahir1987@yahoo.com (Z.S.); saeedislam@awkum.edu.pk (S.I)
2. Department of Computer Science, College of Science and Humanities at Alghat Majmaah University, Al-Majmaah 11952, Saudi Arabia; j.majdoubi@mu.edu.sa
3. Department of Mechanical and Industrial Engineering, College of Engineering, Majmaah University, Al-Majmaah 11952, Saudi Arabia; l.tlili@mu.edu.sa
4. Department of Mathematics, Kohat University of Science and technology, Kohat, KP 26000, Pakistan; Wariskhan758@yahoo.com
5. Faculty of Mathematics and Statistics, Ton Duc Thang University, Ho Chi Minh City 72915, Vietnam
* Correspondence: ilyaskhan@tdt.edu.vn

Received: 30 December 2018; Accepted: 31 January 2019; Published: 12 February 2019

Abstract: The aim of this article is to study time dependent rotating single-wall electrically conducting carbon nanotubes with aqueous suspensions under the influence of nonlinear thermal radiation in a permeable medium. The impact of viscous dissipation is taken into account. The basic governing equations, which are in the form of partial differential equations (PDEs), are transformed to a set of ordinary differential equations (ODEs) suitable for transformations. The homotopy analysis method (HAM) is applied for the solution. The effect of numerous parameters on the temperature and velocity fields is explanation by graphs. Furthermore, the action of significant parameters on the mass transportation and the rates of fiction factor are determined and discussed by plots in detail. The boundary layer thickness was reduced by a greater rotation rate parameter in our established simulations. Moreover, velocity and temperature profiles decreased with increases of the unsteadiness parameter. The action of radiation phenomena acts as a source of energy to the fluid system. For a greater rotation parameter value, the thickness of the thermal boundary layer decreases. The unsteadiness parameter rises with velocity and the temperature profile decreases. Higher value of ϕ augments the strength of frictional force within a liquid motion. For greater R and θw; the heat transfer rate rises. Temperature profile reduces by rising values of Pr.

Keywords: unsteady rotating flow; porous medium; aqueous suspensions of CNT's; nonlinear thermal radiation; viscous dissipation effect; HAM

1. Introduction

The recent period of technology and science has been totally affected by nanofluids, and they possess an important role in various engineering and machinery uses, like biomedical applications, detergency, transferences, industrialized cooling, microchip technology and nuclear reactions. By adding nanoparticles, mathematicians and physicists developed to a way to increase the capacity of heat transfer in the base fluid. The investigators applied advanced techniques, like in the base fluids, by adding ultra-fine solid particles. Heat transfer and single-phase higher thermal conductivity coefficients are greater in nanofluids as compared to bottom liquids. Khan [1] has examined Buongiorno's model with heat transfer and mass for nanofluid flow. Mahdy et al. [2] have applied Buongiorno's model for nanofluid flow with heat transfer through an unsteady contracting cylinder. Malvandi et al. [3] have described flow through a vertical annular pipe. In [4–6], researchers have examined the flow of nanofluids through a stretching sheet. Ellahi [7] has discussed nanofluid flow through a pipe with the MHD effect. Nanofluid flow through a cone has been deliberated by Nadeem et al. [8]. Abolbashari et al. [9] have debated entropy generation for the analytical modelling of Casson nanofluid flow. A mixture of suspended metallic nanoscale particles with a base fluid is known as a mixture of nanoparticles, the word nanofluid was invented by Choi [10]. Nanofluids were formulated for heat transport, momentum and mass transfer by Buongiorno, in order to obtain four equations of two components for non-homogeneous equilibrium models [11]. Most of the research available on the nanofluid problem is cited in works [12–14]. Presently, in these studies, few significant attempts have been presented for nonlinear thermal radiation [15]. Kumar et al. [15] have reported the rotating nanofluid flow problem with deliberation of dissimilar types of nanoparticles, including entropy generation to use the second law of thermodynamics. Nadeem et al. [16] have studied the inclusion of nanoparticles of titanium and copper oxide, which related to the rotating fluid flow problem. Mabood et al. [17] have studied the flow of rotating nanofluid and the impact of magnetic and heat transfer. Shah et al. [18,19] have deliberated a rotating system nanofluid flow with hall current and thermal radiation. Gireesha et al. [20] have examined a single-wall nanotube in an unsteady rotating flow with heat transfer and nonlinear thermal radiation. Currently, Ishaq et al. [21] have discussed the thermal radiation effect with respect to the entropy generation of unsteady nanofluid thin film flow on a porous stretching sheet. In the field of nanoscience, the study of the flow of fluid with nanoparticles (nanofluids) holds a great amount of attention. Nanofluids scattered with 10^9 nm sized materials are arranged in fluids such as nanofibers, droplets, nanoparticles, nanotubes etc. Solid phase and liquid phase are the two period systems. To augment the thermal conductivity of fluids, nanofluids can be applied, and they thrive in stable fluids, exhibiting good writing and dispersion properties on hard materials [22,23]. Sheikholeslami et al. [24–26] have described the significance of nanofluids in in nanotechnology. Yadav et al. [27–29] have deliberated nanofluids with the MHD effect by using dissimilar phenomena, further study of heat transfer enhancement, numerical simulation, stability and instability with linear and non-linear flows of nanofluid. The Darcy–Forchheimer flow of radiative carbon nanotubes (CNTs) in nanofluid in a rotating frame has been investigated by Shah et al. [30–33]. Dawar et al. [33] have studied CNTs, Casson MHD and nanofluid with radiative heat transfer in rotating channels. Khan et al. [34] have studied three-dimensional Williamson nanofluid flow over a linear stretching surface. Shah et al. [35] have described the analysis of a micropolar nanofluid flow with radiative heat and mass transfer. Khan et al. [36,37] have discussed the Darcy–Forchheimer flow of micropolar nanofluid between two plates. Shah et al. [38] have described MHD thin film flow of Williamson fluid over an unsteady permeable stretching surface. Jawad et al. [39] have investigated the Darcy–Forchheimer flow of MHD nanofluid thin film flow with Joule dissipation and Navier's partial slip. Khan et al. [40] have investigated the slip flow of Eyring–Powell nanoliquid. Hammed et al. [41] have described a combined magnetohydrodynamic and electric field effect on an unsteady Maxwell nanofluid flow. Dawar et al. [42] have described unsteady squeezing flow of MHD CNT nanofluid in rotating channels. Khan et al. [43] have investigated the Darcy–Forchheimer flow of MHD CNT nanofluid with radiative thermal aspects. Sheikholeslami et al. [44] have investigated the uniform magnetic force impact on water based nanofluid with thermal aspects in a porous enclosure. Feroz et al. [45] have examined the entropy generation of carbon nanotube flow in a rotating channel. Alharbi et al. [46] have described entropy generation in MHD Eyring–Powell fluid flow over an unsteady oscillatory porous stretching surface with thermal radiation. Liao [47] learned in 1992 that this method was a fast way to find the approximate solution and that for the solution of nonlinear problems it was a better fit.

The main objective of this article is to investigate the augmentation of heat transfer in time dependent rotating single-wall carbon nanotubes with aqueous suspensions under the influence of nonlinear thermal radiation in a permeable medium. The impact of viscous dissipation is taken into account. For the solution, the homotopy analysis method (HAM) [48–50] is applied. The effect of numerous parameters on the temperature and velocity fields are explained by graphs.

2. Problem Formulation

We considered a three-dimensional time dependent electrically conducting incompressible laminar rotating flow of nanofluid. Aqueous suspensions of single-wall CNTs based on water were assumed as a nanoscale materials with a porous medium. The x, y and z variables are the Cartesian coordinates where $\Omega(t)$ is angular velocity of the rotating fluid, which is measured about the z-axis. The surface velocities in the x and y axes are taken as $u_w(x,t) = \frac{bx}{(1-\delta t)}$ and $v_w(x,t)$ respectively, and $w_w(x,t)$ is the velocity in the z-axis, known as the mass flux wall velocity. The governing equations under these assumptions are taken in the form as [4,14,18]

$$\frac{\partial u}{\partial x} + \frac{\partial v}{\partial y} + \frac{\partial w}{\partial z} = 0 \tag{1}$$

$$\frac{\partial u}{\partial t} + u\frac{\partial u}{\partial x} + v\frac{\partial u}{\partial y} + w\frac{\partial u}{\partial z} + \frac{2\Omega v}{1-\delta t} = -\frac{1}{\rho}\frac{\partial P}{\partial x} + \frac{\mu_{nf}}{\rho_{nf}}\frac{\partial^2 u}{\partial z^2} - \frac{\sigma \beta_0^2}{1-\delta t}u - \frac{\mu_{nf}}{k^*}\frac{u}{1-\delta t} \tag{2}$$

$$\frac{\partial v}{\partial t} + u\frac{\partial v}{\partial x} + v\frac{\partial v}{\partial y} + w\frac{\partial v}{\partial z} + \frac{2\Omega u}{1-\delta t} = -\frac{1}{\rho}\frac{\partial P}{\partial y} + \frac{\mu_{nf}}{\rho_{nf}}\frac{\partial^2 v}{\partial z^2} - \frac{\sigma \beta_0^2}{1-\delta t}v - \frac{\mu_{nf}}{k^*}\frac{v}{1-\delta t} \tag{3}$$

$$\frac{\partial w}{\partial t} + u\frac{\partial w}{\partial x} + v\frac{\partial w}{\partial y} + w\frac{\partial w}{\partial z} = -\frac{1}{\rho}\frac{\partial P}{\partial z} + \frac{\mu_{nf}}{\rho_{nf}}\frac{\partial^2 w}{\partial z^2} \tag{4}$$

$$\frac{\partial T}{\partial t} + u\frac{\partial T}{\partial x} + v\frac{\partial T}{\partial y} + w\frac{\partial T}{\partial z} = \alpha_{nf}\frac{\partial^2 T}{\partial z^2} + \frac{1}{(\rho c_p)_{nf}}\frac{\partial q_r}{\partial z} + \frac{\mu_{nf}}{(\rho c_p)_{nf}}\left[\left(\frac{\partial u}{\partial z}\right)^2 + \left(\frac{\partial v}{\partial z}\right)^2\right] \tag{5}$$

The corresponding boundary conditions are given as

$$\begin{aligned} u = u_w(x,t), v = 0, w = 0, T = T_w \quad &\text{at} \quad z = 0 \\ u \to 0, v \to 0, w \to 0, T \to T_\infty \quad &\text{at} \quad z \to \infty. \end{aligned} \tag{6}$$

where x, y and z, are the directions of the velocity components. The angular velocity of the nanofluid is denoted as Ω, nanofluid dynamic viscosity is denoted as μ_{nf}, the nanofluid density is denoted as ρ_{nf}, the nanofluid thermal diffusivity is denoted as α_{nf}, temperature of the nanofluid is denoted as T, T_w and T_∞ denotes the wall and outside surface temperature, respectively.

The expression of radiative heat flux in Equation (5) is written as:

$$q_r = -\frac{4\sigma^*}{3(\rho c_p)_{nf} k^*}\frac{\partial T^4}{\partial z} = -\frac{16\sigma^*}{3k^*}T^3\frac{\partial T}{\partial z} \tag{7}$$

where the mean absorption coefficient is denoted by k^* and the Stefan–Boltzman constant is denoted by σ^*. By using Equation (7) in Equation (5), this produces the following equation:

$$\begin{aligned} &\frac{\partial T}{\partial t} + u\frac{\partial T}{\partial x} + v\frac{\partial T}{\partial y} + w\frac{\partial T}{\partial z} \\ &= \alpha_{nf}\frac{\partial^2 T}{\partial z^2} + \frac{16\sigma^*}{(\rho c_p)_{nf} k^*}\left[T^3\frac{\partial^2 T}{\partial z^2} + 3T^2\left(\frac{\partial T}{\partial z}\right)^2\right] + \frac{\mu_{nf}}{(\rho c_p)_{nf}}\left[\left(\frac{\partial u}{\partial z}\right)^2 + \left(\frac{\partial v}{\partial z}\right)^2\right] \end{aligned} \tag{8}$$

α_{nf}, μ_{nf} and ρ_{nf} are interrelated with ϕ, which is denoted as:

$$\begin{aligned}
\rho_{nf} &= \left(1-\phi+\phi\left(\frac{(\rho_s)_{CNT}}{\rho_f}\right)\right), \mu_{nf} = \frac{\mu_f}{(1-\phi)^{2.5}}, \alpha_{nf} \\
&= (1-\phi)(\rho_f)_f + \phi(\rho_s)_{CNT}, \\
(\rho C_p)_{nf} &= (1-\phi)(\rho C_p)_f + \phi(\rho C_p)_{CNT}, \frac{k_{nf}}{k_f} \\
&= \frac{1-\phi+2\phi\left(\frac{k_{CNT}}{k_{CNT}-k_f}\right)\ln\left(\frac{k_{CNT}+k_f}{2k_f}\right)}{1-\phi+2\phi\left(\frac{k_f}{k_{CNT}-k_f}\right)\ln\left(\frac{k_{CNT}+k_f}{2k_f}\right)}
\end{aligned} \quad (9)$$

The volumetric heat capacity of the CNT and base fluid is denoted by $(\rho C_p)_{CNT}$ and $(\rho C_p)_f$ respectively. Nanofluid thermal conductivity is denoted as k_{nf}, base fluid thermal conductivity is denoted by k_f, CNT thermal conductivity is denoted by k_{CNT}, nanoparticle volume fraction is denoted by ϕ, the base fluid density viscosity is denoted by ρ_f and CNT density is denoted by ρ_{CNT}.

Similarity transformations are now introduce as:

$$u = \frac{bx}{(1-\delta t)}f'(\eta), v = \frac{bx}{(1-\delta t)}g(\eta), w = -\sqrt{\frac{bv}{(1-\alpha t)}}f(\eta), \eta = \sqrt{\frac{b}{v(1-\alpha t)}}z \\ T = T_\infty(1+(1-\delta t)\theta(\eta)) \quad (10)$$

Inserting Equation (10) in Equations (2)–(6), we have:

$$\frac{1}{(1-\phi)^{2.5}\left((1-\phi)+\phi\frac{(\rho_s)_{CNT}}{\rho_f}\right)} f^{iv} - \frac{\beta_0^2}{\lambda}f'' - \frac{\mu_{nf}}{(1-\phi)^{2.5}k^*b}f'' \\ - \left[\lambda\left(f'+\frac{\eta}{2}f''\right) + f'f'' - ff''' - 2\frac{\Omega}{b}g'\right] = 0 \quad (11)$$

$$\frac{1}{(1-\phi)^{2.5}\left((1-\phi)+\phi\frac{(\rho_s)_{CNT}}{\rho_f}\right)} g''' - \frac{\beta_0^2}{\lambda}g' + \frac{\mu_{nf}}{(1-\phi)^{2.5}k^*b}g' \\ - \left[\lambda\left(g+\frac{\eta}{2}g'\right) + gf'' - fg'' - 2\frac{\Omega}{b}f''\right] = 0 \quad (12)$$

$$\theta'' + R\left[(1+(\theta_w-1)\theta)^3\theta'' + 3(1+(\theta_w-1)\theta)^2(\theta_w-1)\theta'^2\right] + \frac{Ec}{Pr}\left(f''^2+g'^2\right) \\ -\frac{1}{Pr}\left[\lambda\frac{\eta}{2}\theta' - f\theta'\right] = 0 \quad (13)$$

The boundary conditions are written as:

$$\begin{aligned} f(0) &= 0, f'(0) = 1, g(0) = 0, \theta(0) = 1, \text{ at } \eta = 0 \\ f'(\eta) &\to 0, g(\eta) \to 0, f(\eta) \to 0, \theta(\eta) \to 0, \text{ at } \eta \to \infty. \end{aligned} \quad (14)$$

Here, Ω denotes the rotation parameter and is defined as $\Omega = \frac{\omega}{b}$, and λ represents the unsteadiness parameter, defined as $\lambda = \frac{\delta}{b}$. R is radiation parameter, defined as $R = \frac{16\sigma^* T_\infty^3}{3k_{nf}k^*}$. Pr is Prandtl number, defined as $Pr = \frac{\alpha_{nf}}{v_{nf}}$. Ec represents the Eckert number, defined as $Ec = \frac{u_w^2}{c_p(T-T_\infty)}$. The temperature ratio parameter is denoted by θ_w and $\theta_w = \frac{T_w}{T_a}$.

Physical Quantities of Interest

In the given problem, the physical quantities of interest are C_{fx}, C_{fy} and Nu_x, which is denoted as:

$$C_{fx}\frac{\tau_{wx}}{\rho_f u_w^2(x,t)}, C_{fy}\frac{\tau_{wy}}{\rho_f u_w^2(x,t)}, Nu_x = \frac{xq_w}{(T_w-T_\infty)} \quad (15)$$

The surface heat flux is denoted by q_w, and τ_{wx} and τ_{wy} are surface shear stress, which are written as:

$$\tau_{wx} = \mu_{nf}\left(\frac{\partial u}{\partial z}\right)_{z=0}, \tau_{wy} = \mu_{nf}\left(\frac{\partial v}{\partial z}\right)_{z=0} \text{ and } q_w = -k_{nf}\left(\frac{\partial T}{\partial z}\right) + (q_r)_{z=0}. \tag{16}$$

By the use of Equations (15) and (16), then we have:

$$\sqrt{Re_x}C_{fx} = \frac{1}{(1-\phi)^{2.5}}f''(0), \sqrt{Re_x}C_{fy} = \frac{1}{(1-\phi)^{2.5}}g'(0), \frac{Nu_x}{\sqrt{Re_x}} = \frac{k_{nf}}{k_f}\left(-\left[1+Rd\theta_w^3\right]\theta'(0)\right) \tag{17}$$

$Re_x = u_w x/v$ is the Reynolds number.

3. HAM Solution

Liao [46,47] proposed the homotopy analysis technique in 1992. In order to attain this method, he used the ideas of a topology called homotopy. For the derivation he used two homotopic functions, where one of them can be continuously distorted into another. He used Ψ_1, Ψ_2 for binary incessant purposes and X and Y are dual topological plane, additionally, if Ψ_1 and Ψ_2 map from X to Y, then Ψ_1 is supposed to be homotopic to Ψ_2. If a continuous function of ψ is produced:

$$\Psi : X \times [0, 1] \to Y \tag{18}$$

So that $x \in X$

$$\Psi[x, 0] = \Psi_1(x) \text{ and } \Psi[x, 1] = \Psi_2(x) \tag{19}$$

where Ψ is homotopic. Here, we use the HAM to solve Equations (11)–(13), consistent with the boundary restrain (14). The preliminary guesses are selected as follows:

The linear operators are denoted as $L_{\hat{f}}$ and $L_{\hat{\theta}}$, and are represented as

$$L_{\hat{f}}(\hat{f}) = \hat{f}''', L_{\hat{g}}(\hat{g}) = \hat{g}'', L_{\hat{\theta}}(\hat{\theta}) = \hat{\theta}'' \tag{20}$$

Which has the following applicability:

$$L_{\hat{f}}(e_1 + e_2\eta + e_3\eta^2 + e_4\eta^3) = 0, L_{\hat{g}}(e_5 + e_6\eta + e_7\eta^2) = 0$$
$$L_{\hat{\theta}}(e_8 + e_9\eta) = 0 \tag{21}$$

where the representation of coefficients is included in the general solution by $e_i (i = 1-> 7)$.

The corresponding non-linear operators are sensibly selected as $N_{\hat{f}}, N_{\hat{g}}$ and $N_{\hat{\theta}}$, and identify in the form:

$$N_{\hat{f}}\left[\hat{f}(\eta;\zeta), \hat{g}(\eta;\zeta)\right] = \frac{1}{(1-\phi)^{2.5}\left((1-\phi)+\phi\frac{(\rho_s)_{CNT}}{\rho_f}\right)}\hat{f}_{\eta\eta\eta\eta} - \frac{\beta_0^2}{\lambda}\hat{f}_{\eta\eta} - \frac{\mu_{nf}}{(1-\phi)^{2.5}k^*b}\hat{f}_{\eta\eta}$$
$$-\left[\lambda\left(\hat{f}_{\eta\eta} + \frac{\eta}{2}\hat{f}_{\eta\eta\eta}\right) + \hat{f}_{\eta}\hat{f}_{\eta\eta} - \hat{f}\hat{f}_{\eta\eta} - 2\frac{\Omega}{b}\hat{g}_{\eta}\right] \tag{22}$$

$$N_{\hat{g}}\left[\hat{f}(\eta;\zeta), \hat{g}(\eta;\zeta)\right] = \frac{1}{(1-\phi)^{2.5}\left((1-\phi)+\phi\frac{(\rho_s)_{CNT}}{\rho_f}\right)}\hat{g}_{\eta\eta\eta} - \frac{\beta_0^2}{\lambda}\hat{g}_{\eta} + \frac{\mu_{nf}}{(1-\phi)^{2.5}k^*b}\hat{g}_{\eta}$$
$$-\left[\lambda\left(\hat{g}_{\eta} + \frac{\eta}{2}\hat{g}_{\eta\eta}\right) + \hat{g}\hat{f}_{\eta\eta} - \hat{f}\hat{g}_{\eta\eta} - 2\frac{\Omega}{b}\hat{f}_{\eta\eta}\right] \tag{23}$$

$$N_{\hat{\theta}}\left[\hat{f}(\eta;\zeta), \hat{g}(\eta,\zeta), \hat{\theta}(\eta;\zeta)\right] = \hat{\theta}_{\eta\eta} + R\left[(1+(\theta_w-1)\hat{\theta})^3\hat{\theta}_{\eta\eta} + 3(1+(\theta_w-1)\hat{\theta})^2(\theta_w-1)\hat{\theta}_{\eta}^2\right]$$
$$+\frac{Ec}{Pr}\left(\hat{f}_{\eta\eta}^2 + \hat{g}_{\eta}^2\right) - \frac{1}{Pr}\left[\lambda\frac{\eta}{2}\hat{\theta}_{\eta} - \hat{f}\hat{\theta}_{\eta}\right] \tag{24}$$

For Equations (8–10), the 0th-order scheme takes the form

$$(1-\zeta)L_{\hat{f}}\left[\hat{f}(\eta;\zeta) - \hat{f}_0(\eta)\right] = p\hbar_{\hat{f}}N_{\hat{f}}\left[\hat{f}(\eta;\zeta), \hat{g}(\eta;\zeta)\right] \tag{25}$$

$$(1-\zeta)L_{\hat{g}}[\hat{g}(\eta;\zeta) - \hat{g}_0(\eta)] = p\hbar_{\hat{g}}N_{\hat{g}}\left[\hat{f}(\eta;\zeta), \hat{g}(\eta;\zeta)\right] \tag{26}$$

$$(1-\zeta)L_{\hat{\theta}}[\hat{\theta}(\eta;\zeta) - \hat{\theta}_0(\eta)] = p\hbar_{\hat{\theta}}N_{\hat{\theta}}\left[\hat{f}(\eta;\zeta), \hat{g}(\eta;\zeta), \hat{\theta}(\eta;\zeta)\right] \tag{27}$$

where the boundary constrains are:

$$\begin{array}{l} \hat{f}(\eta;\zeta)\big|_{\eta=0} = 0, \ \frac{\partial \hat{f}(\eta;\zeta)}{\partial \eta}\big|_{\eta=0} = 1, \ \hat{g}(\eta;\zeta)\big|_{\eta=0} = 0 \\ \hat{\theta}(\eta;\zeta)\big|_{\eta=0} = 1, \ \frac{\partial \hat{f}(\eta;\zeta)}{\partial \eta}\big|_{\eta\to\infty} \to 0, \ \hat{g}(\eta;\zeta)\big|_{\eta\to\infty} \to 0 \\ \hat{f}(\eta;\zeta)\big|_{\eta\to\infty} \to 0, \ \hat{\theta}(\eta;\zeta)\big|_{\eta\to\infty} \to 0. \end{array} \tag{28}$$

where the embedding restriction is $\zeta \in [0,1]$, to normalize for the solution convergence $\hbar_{\hat{f}}, \hbar_{\hat{g}}$ and $\hbar_{\hat{\theta}}$ are used. When $\zeta = 0$ and $\zeta = 1$, we get:

$$\hat{f}(\eta;1) = \hat{f}(\eta), \hat{g}(\eta;1) = \hat{g}(\eta), \hat{\theta}(\eta;1) = \hat{\theta}(\eta) \tag{29}$$

Expanding $\hat{f}(\eta;\zeta), \hat{g}(\eta;\zeta)$ and $\hat{\theta}(\eta;\zeta)$ through Taylor's series for $\zeta = 0$,

$$\begin{array}{l} \hat{f}(\eta;\zeta) = \hat{f}_0(\eta) + \sum_{n=1}^{\infty} \hat{f}_n(\eta)\zeta^n \\ \hat{g}(\eta;\zeta) = \hat{g}_0(\eta) + \sum_{n=1}^{\infty} \hat{g}_n(\eta)\zeta^n \\ \hat{\theta}(\eta;\zeta) = \hat{\theta}_0(\eta) + \sum_{n=1}^{\infty} \hat{\theta}_n(\eta)\zeta^n. \end{array} \tag{30}$$

$$\hat{f}_n(\eta) = \frac{1}{n!} \frac{\partial \hat{f}(\eta;\zeta)}{\partial \eta}\bigg|_{p=0}, \hat{g}_n(\eta) = \frac{1}{n!} \frac{\partial \hat{g}(\eta;\zeta)}{\partial \eta}\bigg|_{p=0}, \hat{\theta}_n(\eta) = \frac{1}{n!} \frac{\partial \hat{\theta}(\eta;\zeta)}{\partial \eta}\bigg|_{p=0}. \tag{31}$$

The boundary constrains are:

$$\hat{f}_w(0) = \hat{f}_w'(0) = \hat{g}(0) = 0, \hat{\theta}_w(0) = 0, \text{at} \eta = 0 \tag{32}$$

$$\hat{f}_w'(\infty) = \hat{g}_w(\infty) = \hat{\theta}_w(\gamma) = \hat{f}_w(\infty) \to 0 \text{at} \eta \to \infty. \tag{33}$$

Resulting in:

$$\begin{array}{l} \Re_n^{\hat{f}}(\eta) = \frac{1}{(1-\phi)^{2.5}\left((1-\phi) + \phi\frac{(\rho_s)_{CNT}}{\rho_f}\right)} \hat{f}_{n-1}^{iv} - \frac{\beta_0^2}{\lambda} \hat{f}_{n-1}^{\prime\prime} - \frac{\mu_{nf}}{(1-\phi)^{2.5}k^*b} \hat{f}_{n-1}^{\prime\prime} \\ - \left[\lambda\left(\hat{f}_{n-1}^{\prime\prime} + \frac{\eta}{2}\hat{f}_{n-1}^{\prime\prime\prime}\right) + \sum_{j=0}^{w-1} \hat{f}_{w-1-j}^{\prime}\hat{f}_j^{\prime\prime} - \sum_{j=0}^{w-1} \hat{f}_{w-1-j}\hat{f}_j^{\prime\prime} - 2\frac{\Omega}{b}\hat{g}_{n-1}^{\prime} \right] \end{array} \tag{34}$$

$$\begin{array}{l} \Re_n^{\hat{g}}(\eta) = \frac{1}{(1-\phi)^{2.5}\left((1-\phi) + \phi\frac{(\rho_s)_{CNT}}{\rho_f}\right)} \hat{g}_{n-1}^{\prime\prime\prime} - \frac{\beta_0^2}{\lambda}\hat{g}_{n-1}^{\prime} + \frac{\mu_{nf}}{(1-\phi)^{2.5}k^*b}\hat{g}_{n-1}^{\prime} \\ - \left[\lambda(\hat{g}_{n-1}^{\prime} + \frac{\eta}{2}\hat{g}_{n-1}^{\prime\prime}) + \sum_{j=0}^{w-1} \hat{g}_{w-1-j}\hat{f}_j^{\prime\prime\prime} - \sum_{j=0}^{w-1} \hat{f}_{w-1-j}\hat{g}_j^{\prime\prime} - 2\frac{\Omega}{b}\hat{f}_{n-1}^{\prime\prime} \right] \end{array} \tag{35}$$

$$R_n^{\hat{\theta}}(\eta) = (\hat{\theta}''_{n-1}) + R\left[(1 + (\theta_w - 1)\hat{\theta}_{n-1})^3 \hat{\theta}''_{n-1} + 3(\theta_w - 1)(1 + (\theta_w - 1)\hat{\theta}_{n-1})^2 \hat{\theta}'_{n-1}\right] + \frac{Ec}{Pr}\left(\hat{f}''^2 + \hat{g}'^2\right) - \frac{1}{Pr}\left[\lambda\frac{\eta}{2}\hat{\theta}'_{n-1} - \hat{f}\hat{\theta}'_{n-1}\right] \tag{36}$$

where

$$\chi_n = \begin{cases} 0, & \text{if } \zeta \leq 1 \\ 1, & \text{if } \zeta > 1. \end{cases} \tag{37}$$

4. Results and Discussion

In this section, we described the physical resources of the modeled problems and their impact on $f'(\eta)$, $g(\eta)$, and $\theta(\eta)$, which are identified in Figures 1–13. In Figure 1, a graphical representation of the problem is shown.

4.1. Velocity Profile $f'(\eta)$ and $g(\eta)$

The influence of Ω, β, ϕ and λ on the velocity profile is presented in Figures 1–8. The impact of Ω on $f'(\eta)$ and $g(\eta)$ is presented in Figures 1 and 2. For greater values of Ω, $f'(\eta)$ is increased. Actually, increasing the rotation parameter increases kinetic energy, which in result augmented the velocity profile. Indeed, $g(\eta)$ is decreased due to a greater rotation parameter rate, as compared to the stretching rate, which had a greater rotation rate for a larger value of Ω. Hence, the velocity field increases for larger rotation effects. Figures 3 and 4 represent the influence of ϕ on the $f'(\eta)$ and $g(\eta)$ profile. With an increase in ϕ, the velocity profile decreases. It is examined that $f'(\eta)$ and $g(\eta)$ decrease uniformly in nanofluids with a rise in ϕ. This is due to the detail that an increase in the ϕ increases the density of the nanofluid and this results in equally slowing the fluid $f'(\eta)$ and $g(\eta)$ profiles. Figures 5 and 6 describe the effect of λ on $f'(\eta)$ and $g(\eta)$. It is defined in the figures that with increases in λ, the velocity profiles decrease consistently. It is also indicated from the figures that the velocity profiles intensify for rising λ, while we can examine an opposing influence of λ on $f'(\eta)$ and $g(\eta)$ inside the nanofluid and in the thickness of the layer. Figures 7 and 8 represent the influence of β on $f'(\eta)$ and $g(\eta)$. With increases in β, the velocity profiles of fluid film decrease consistently. It was also detected that a rise in β resulted in a decrease in the fluid profiles $f'(\eta)$ and $g(\eta)$ of the nanofluid, as well as for the layer thickness. The purpose behind such an influence of β is for the stimulation of a delaying body force, stated as Lorentz force, due to the presence of β in an electrically conducting nanofluid layer. Since β suggests the ratio of hydromagnetic body force and viscous force, a larger value of β specifies a stronger hydromagnetic body force, which has a trend to slow the fluid flow.

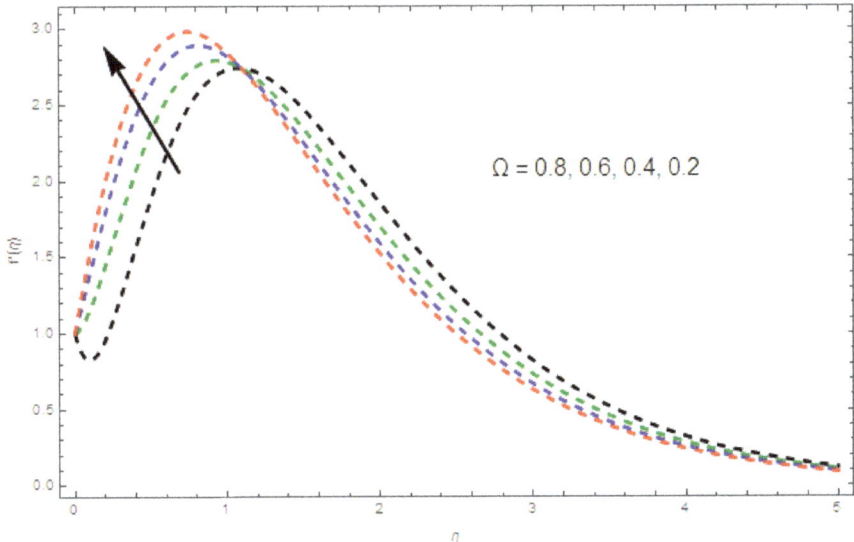

Figure 1. The influence of the rotation parameter Ω on $f'(\eta)$ when $\beta = 0.2, \phi = 0.1, \lambda = 0.3$.

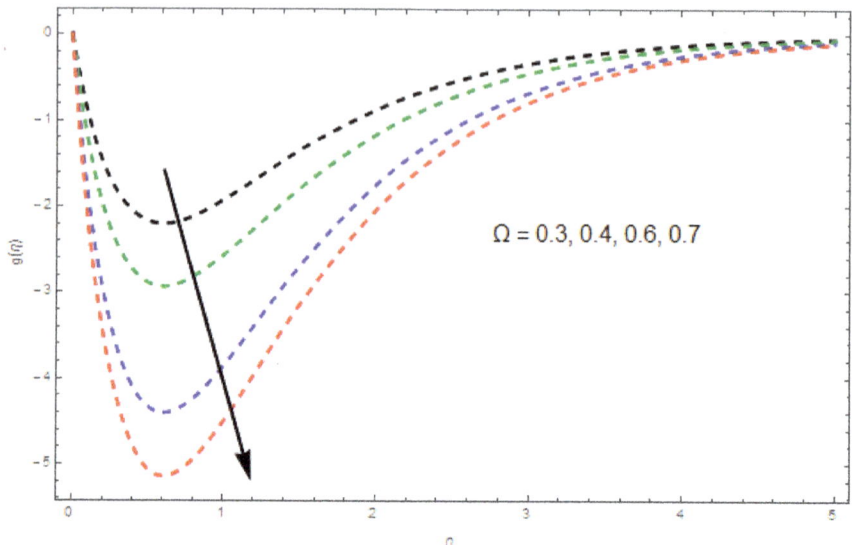

Figure 2. The effect of the rotation rate parameter Ω on $g(\eta)$ when $\beta = 0.2, \phi = 0.1, \lambda = 0.7$.

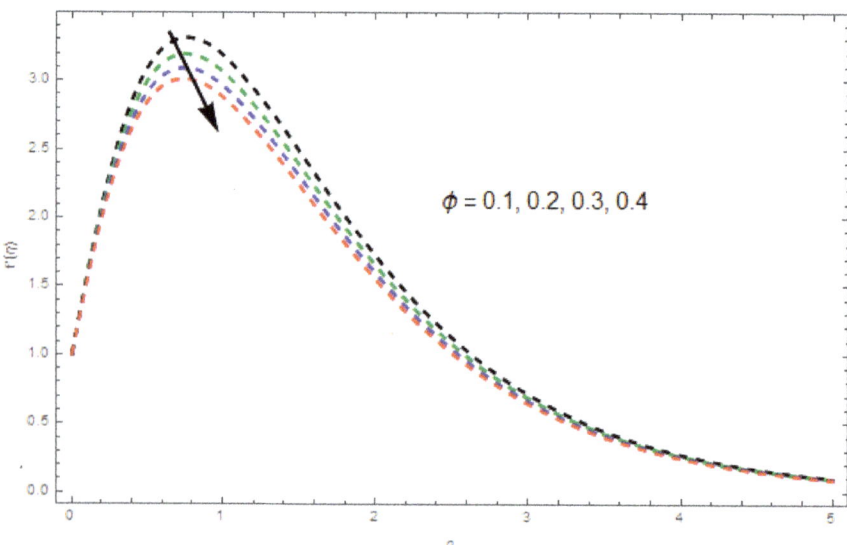

Figure 3. The effect of nanoparticle volume friction ϕ on $f'(\eta)$ when $\beta = 0.9, \lambda = 0.5, \Omega = 0.1$.

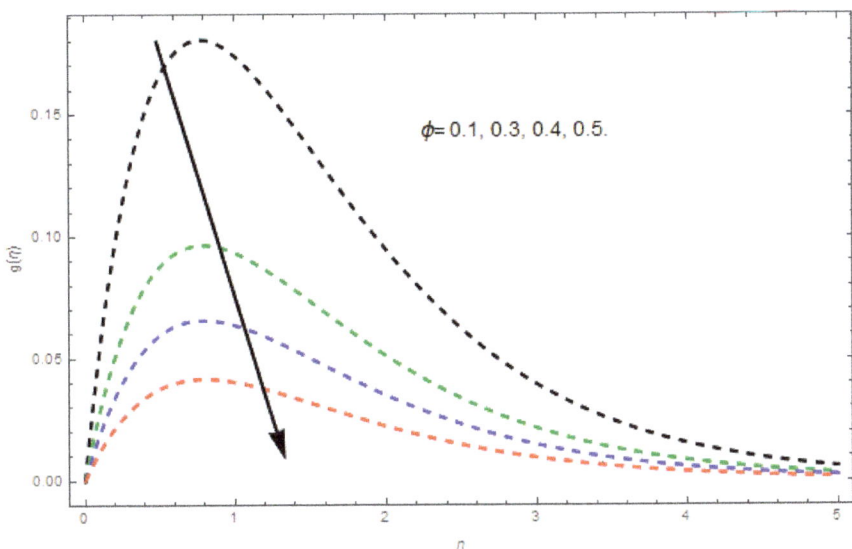

Figure 4. The impact of nanoparticle volume friction ϕ on $g(\eta)$ when $\beta = 0.1, \lambda = 0.2, \Omega = 0.7$.

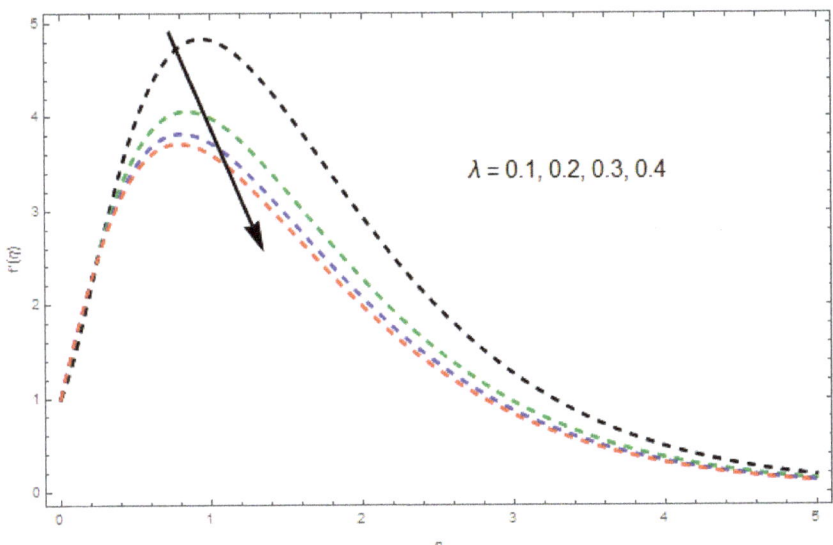

Figure 5. The influence of the unsteadiness parameter λ on $f'(\eta)$ when $\beta = 0.9, \phi = 0.1, \Omega = 0.1$.

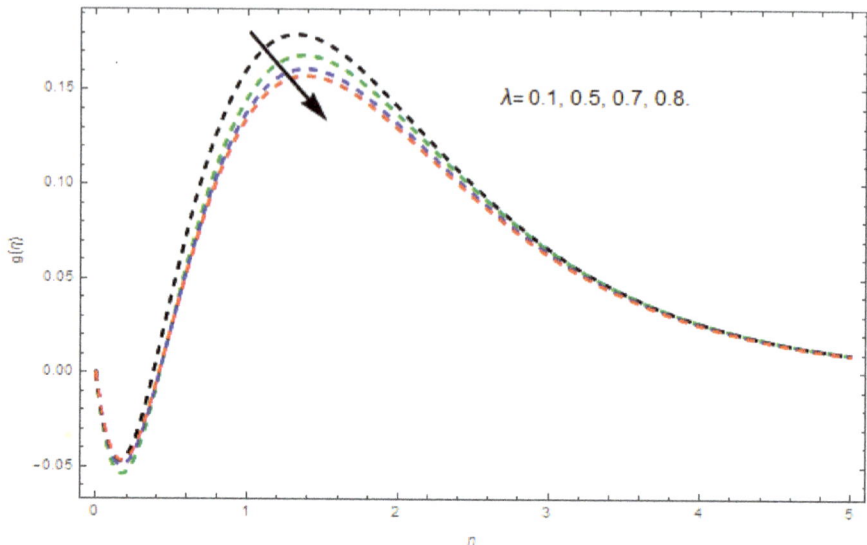

Figure 6. The effect of the unsteadiness parameter λ on $g(\eta)$ when $\beta = 0.2, \phi = 0.1, \Omega = 0.2$.

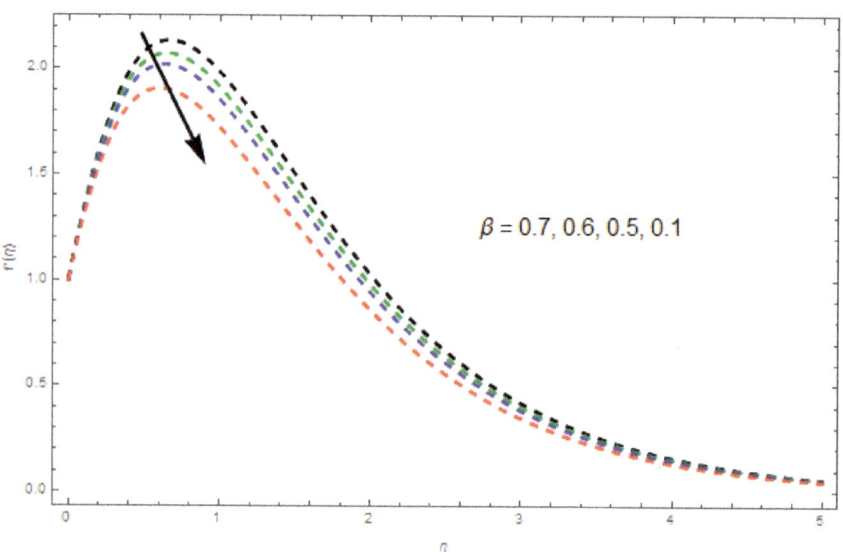

Figure 7. The effect of magnetic field β on $f'(\eta)$ when $\lambda = 0.2, \phi = 0.1, \Omega = 0.1$.

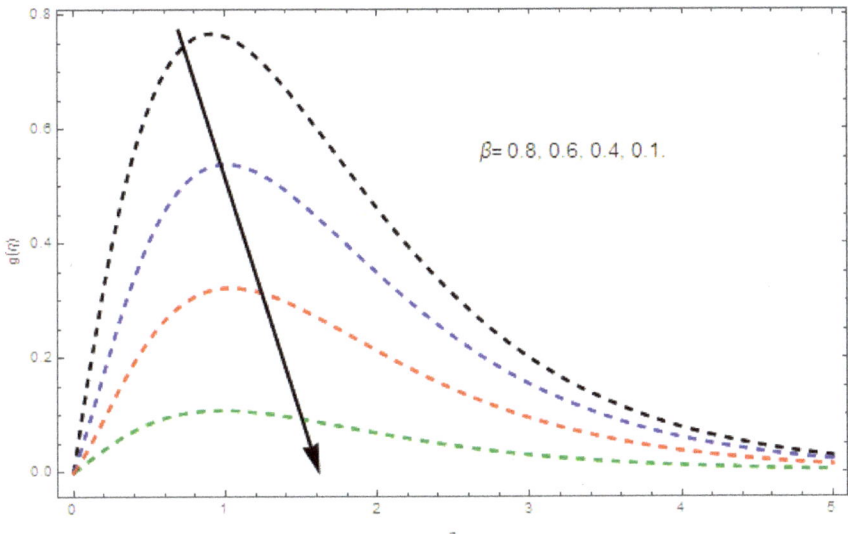

Figure 8. The effect of magnetic field β on $g(\eta)$ when $\lambda = 0.7, \phi = 0.1, \Omega = 0.1$.

4.2. Temperature Profile $\theta(\eta)$

The impact of $\theta(\eta)$ on physical parameters λ, Ec, θ_w, Rd and Pr is defined in Figures 9–13. Figure 9 represent the impact of λ on the $\theta(\eta)$ profile. Figure 9 represents that the increase in the λ momentum boundary layer thickness decreases. Figures 10 and 11 disclosed the responses of $\theta(\eta)$ for θ_w and Rd. The influence of Rd on $\theta(\eta)$ is presented in Figure 11. By increasing Rd, the temperature of the nanofluid boundary layer area is increased. It is observed in Figure 11 that $\theta(\eta)$ is augmented with a rise in the Rd. An enhanced Rd parameter leads to a release of heat energy in the flow direction, therefore the fluid $\theta(\eta)$ is increased. Graphical representation identifies that $\theta(\eta)$ is increased when we increase the ratio strength and thermal radiation temperature. The impact of $\theta(\eta)$ on Pr is given in Figure 12, where $\theta(\eta)$ decreases with large value of Pr and for smaller value increases. The variation of $\theta(\eta)$ for the variation of Pr is illustrated such that Pr specifies the ratio of momentum diffusivity to thermal diffusivity. It is realized that $\theta(\eta)$ is reduced with a rising Pr. Moreover, by suddenly rising Pr, the thermal boundary layer thickness decreases. Figure 13 identifies that for increasing Ec, the $\theta(\eta)$ attainment is enlarged, which supports the physics. By increasing Ec, the heat stored in the liquid is dissipated, due to the enhanced temperature, whereas $\theta(\eta)$ increases with increasing values of the Eckert number and the thermal boundary layer thickness of the nanofluid becomes larger.

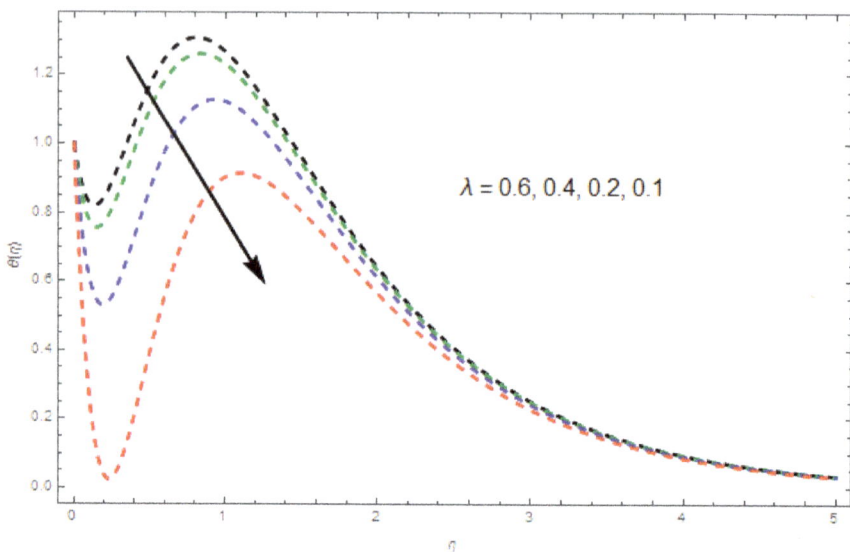

Figure 9. The effect of the unsteadiness parameter (λ) on $\theta(\eta)$ when $Rd = 0.1, Ec = 0.2, Pr = 7, \theta_w = 1.2$.

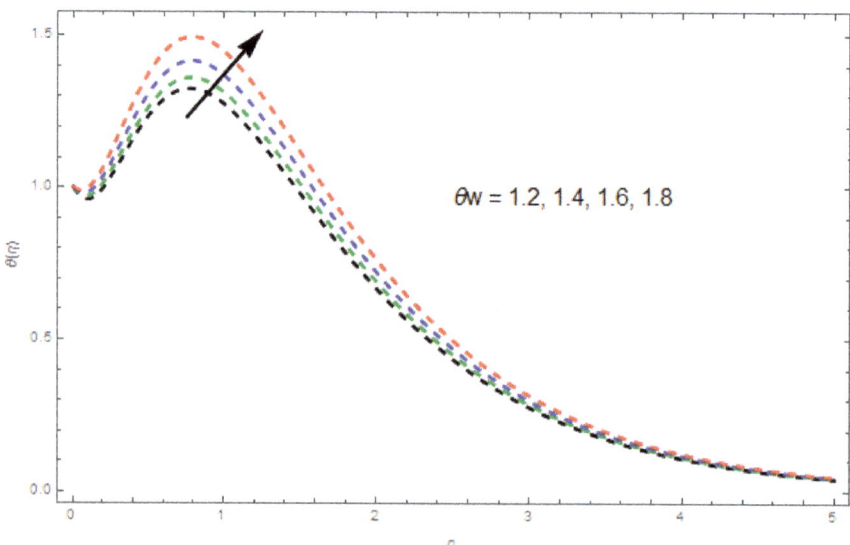

Figure 10. The influence of the temperature ratio parameter on $\theta(\eta)$ when $Rd = 0.5, \lambda = 0.6, Pr = 6.6, Ec = 0.2$.

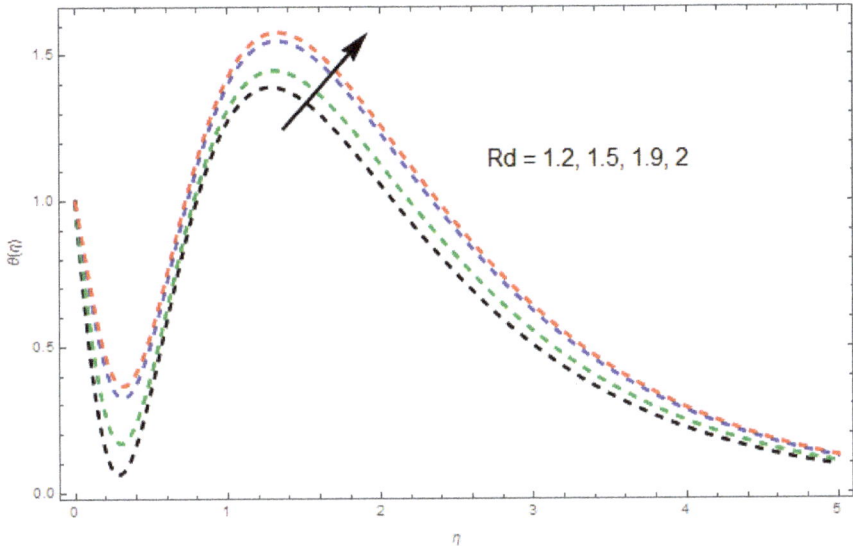

Figure 11. The influence of the radiation parameter (Rd) on $\theta(\eta)$ when $\theta_w = 0.5, \lambda = 0.6, \Pr = 6.6, Ec = 0.6$.

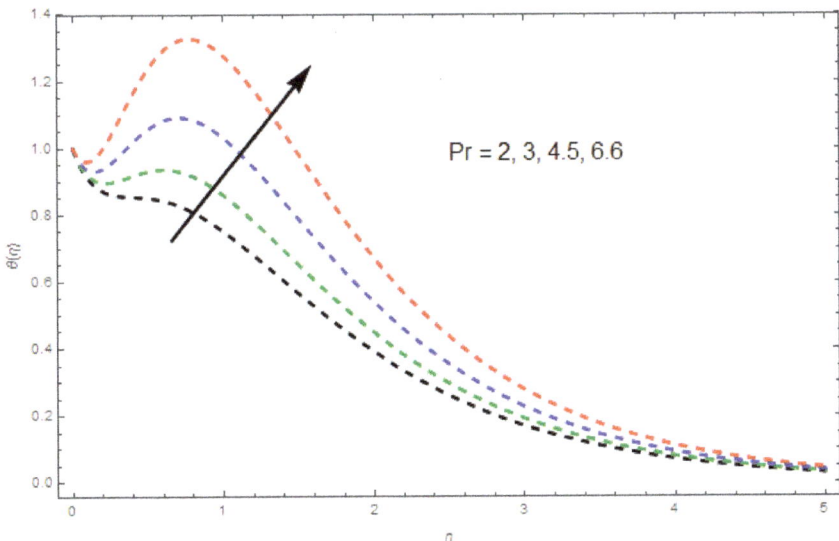

Figure 12. The influence of Prandtl number $(\Pr)$ on $\theta(\eta)$ when $Rd = 0.5, \lambda = 0.6, \theta_w = 6.6, Ec = 0.2$.

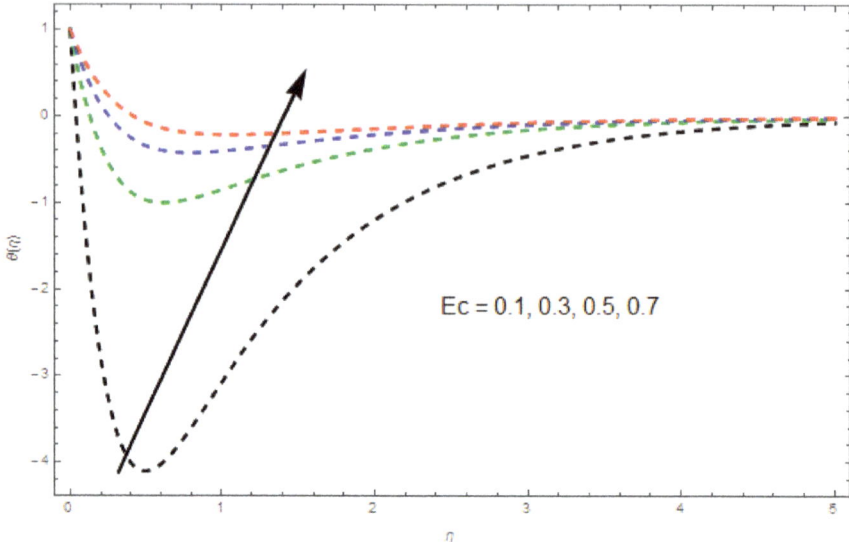

Figure 13. The impact of the Eckert number (Ec) on $\theta(\eta)$ when $Rd = 0.4, \lambda = 0.6, \theta_w = 0.6, \Pr = 1.8$.

4.3. Table Discussion

The effect of skin friction in the x and y directions is shown in Table 1, due to change of parameters $\phi, \beta, \lambda, \rho_s$ and ρ_f. We can see in Table 1 that the skin friction coefficient rises with rising values of λ, β, ρ_s and ϕ. The skin friction coefficient decreases with increasing values of ρ_f, as shown in Table 1. The impact of $Rd, \theta_w, \Pr$ and Ec on $-\frac{k_{nf}}{k_f}\left[1 + Rd\theta_w^3\right]\theta'(0)$ are calculated numerically. It was detected that a higher value of Rd and θ_w increases the heat flux, while a higher value of $\Pr$ and Ec decreases it, as shown in Table 2. Some physical properties of CNTs are shown in Tables 3 and 4.

Table 1. Skin friction $f''(0)$ versus various values of embedded parameters.

ϕ	β	λ	ρ_s	ρ_f	C_{fx}	C_{fy}
0.1	0.3	0.5	0.4	0.4	1.80946	1.80946
0.3					2.36646	2.36646
0.5					3.23183	3.23183
0.1	0.1				1.80946	1.80946
	0.3				1.83280	1.83280
	0.5				1.86280	1.86280
	0.1	0.1			1.80946	1.80946
		0.3			1.81835	1.81835
		0.5			1.82867	1.82867
			0.1	0.1	1.80946	1.80946
			0.3		1.82759	1.82759
			0.5		1.84571	1.84571
				0.1	1.80946	1.80946
				0.3	1.79496	1.79496
				0.5	1.78530	1.78530

Table 2. Nusslet number versus various values of embedded parameters.

R	θ_w	Pr	Ec	$-\frac{k_{nf}}{k_f}[1+Rd\theta_w^3]\theta'(0)$
0.1	0.5	0.7	0.6	0.408170
0.3				0.452415
0.5				0.497617
0.1	0.1			0.408170
	0.3			0.418514
	0.5			0.429786
	0.1	0.1		0.408170
		0.3		0.401541
		0.5		0.393572
		0.1	0.1	0.408170
			0.3	0.420824
			0.5	0.430319

Table 3. Thermophysical properties of different base fluids and carbon nanotubes (CNTs).

Physical Properties	Base Fluid	Nanoparticles	
	Water/Ethylene Glycol)	SWCNT	MWCNT
$\rho(kg/m^3)$	=	2,600	1,600
$c_p(J/kgK)$	=	425	796
$k(W/mK)$	=	6,600	3,000

Table 4. Variation of thermal conductivities of CNT nanofluids with the diameter of CNTs.

Diameter of CNTs	Thermal Conductivity of Nanofluids
50 mm	1.077
200 μm	1.078
20 μm	1.083
5 μm	1.085
50 nm	1.085
500 pm	1.087

5. Conclusions

Exploration on nanoparticle presentation has established more deliberation in mechanical and industrial engineering, due to the probable uses of nanoparticles in cooling devices to produce an increase in continuous phase fluid thermal performance. The radiation phenomena acts as a source of energy to the fluid system. In this research, the time dependent rotating single-wall electrically conducting carbon nanotubes with aqueous suspensions under the influence of nonlinear thermal radiation in a permeable medium was investigated. The basic governing equations, in the form of partial differential equations (PDEs), were transformed into to a set of ordinary differential equations (ODEs) suitable for transformations. The optimal approach was used for the solution. The main concluding points are given below:

- The thermal boundary layer thickness is reduced by a greater rotation rate parameter.
- Velocity and temperature profile decrease due to increases in the unsteadiness parameter.
- A greater ϕ increases the asset of frictional force within a fluid motion.
- The heat transfer rate rises for greater Rd and θ_w values.
- The skin friction coefficient increases with increasing values of ϕ and β.
- A greater value of Rd and θ_w increases the heat flux, while a greater value of Pr and Ec decreases it.
- By enhancing Pr, $\theta(\eta)$ is reduced.

Author Contributions: M.J. and Z.S. modeled the problem and wrote the manuscript. S.I., I.T. and I.K. thoroughly checked the mathematical modeling and English corrections. W.K. and Z.S. solved the problem using Mathematica software. M.J., J.M. and I.T. contributed to the results and discussions. All authors finalized the manuscript after its internal evaluation.

Acknowledgments: The fourth author would like to thank Deanship of Scientific Research, Majmaah University for supporting this work under the Project Number No 1440-43.

Conflicts of Interest: The authors declare no conflict of interest.

References

1. Khan, W.A. Buongiorno model for nanofluid Blasius flow with surface heat and mass fluxes. *J. Thermophys. Heat Transf.* **2013**, *27*, 134–141. [CrossRef]
2. Mahdy, A.; Chamkha, A. Heat transfer and fluid flow of a non-Newtonian nano fluid over an unsteady contracting cylinder employing Buongiorno'smodel. *Int. J. Numer. Method Heat Fluid Flow* **2015**, *25*, 703–723. [CrossRef]
3. Malvandi, A.; Moshizi, S.A.; Soltani, E.G.; Ganji, D.D. Modified Buongiorno's model for fully developed mixed convection flow of nanofluids in a vertical annular pipe. *Comput. Fluids* **2014**, *89*, 124–132. [CrossRef]
4. Hayat, T.; Ashraf, M.B.; Shehzad, S.A.; Abouelmaged, E.I. Three dimensional flow of Erying powell nanofluid over an exponentially stretching sheet. *Int. J. Numer. Method Heat Fluid Flow* **2015**, *25*, 333–357. [CrossRef]
5. Nadeem, S.; Haq, R.U.; Akbar, N.S.; Lee, C.; Khan, Z.H. Numerical study of boundary layer flow and heat transfer of Oldroyed-B nanofluid towards a stretching sheet. *PLoS ONE* **2013**, *8*, e69811. [CrossRef] [PubMed]
6. Rosmila, A.B.; Kandasamy, R.; Muhaimin, I. Lie symmetry groups transformation for MHD natural convection flow of nanofluid over linearly porous stretching sheet in presence of thermal stratification. *Appl. Math. Mech. Engl. Ed.* **2012**, *33*, 593–604. [CrossRef]
7. Ellahi, R. The effects of MHD an temperature dependent viscosity on the flow of non-Newtonian nanofluid in a pipe analytical solutions. *Appl. Math. Model.* **2013**, *37*, 1451–1467. [CrossRef]
8. Nadeem, S.; Saleem, S. Series solution of unsteady Erying Powell nanofluid flow on a rotating cone. *Indian J. Pure Appl. Phys.* **2014**, *52*, 725–737.
9. Abolbashari, M.H.; Freidoonimehr, N.; Rashidi, M.M. Analytical modeling of entropy generation for Casson nano-fluid flow induced by a stretching surface. *Adv. Powder Technol.* **2015**, *6*, 542–552. [CrossRef]
10. Choi, S.U.S.; Siginer, D.A.; Wang, H.P. Enhancing thermal conductivity of fluids with nanoparticle developments and applications of non-Newtonian flows. *ASME N. Y.* **1995**, *66*, 99–105.
11. Buongiorno, J. Convective transport in nanofluids. *ASME J Heat Transf.* **2005**, *128*, 240–250. [CrossRef]
12. Kumar, K.G.; Rudraswamy, N.G.; Gireesha, B.J.; Krishnamurthy, M.R. Influence of nonlinear thermal radiation and viscous dissipation on three-dimensional flow of Jeffrey nanofluid over a stretching sheet in the presence of Joule heating. *Nonlinear Eng.* **2017**, *6*, 207–219.
13. Rudraswamy, N.G.; Shehzad, S.A.; Kumar, K.G.; Gireesha, B.J. Numerical analysis of MHD three-dimensional Carreau nanoliquid flow over bidirectionally moving surface. *J. Braz. Soc. Mech. Sci. Eng.* **2017**, *23*, 5037–5047. [CrossRef]
14. Gireesha, B.J.; Kumar, K.G.; Ramesh, G.K.; Prasannakumara, B.C. Nonlinear convective heat and mass transfer of Oldroyd-B nanofluid over a stretching sheet in the presence of uniform heat source/sink. *Results Phys.* **2018**, *9*, 1555–1563. [CrossRef]
15. Kumar, K.G.; Gireesha, B.J.; Manjunatha, S.; Rudraswamy, N.G. Effect of nonlinear thermal radiation on double-diffusive mixed convection boundary layer flow of viscoelastic nanofluid over a stretching sheet. *IJMME* **2017**, *12*, 18.
16. Nadeem, S.; Rehman, A.U.; Mehmood, R. Boundary layer flow of rotating two phase nanofluid over a stretching surface. *Heat Transf. Asian Res.* **2016**, *45*, 285–298. [CrossRef]
17. Mabood, F.; Ibrahim, S.M.; Khan, W.A. Framing the features of Brownian motion and thermophoresis on radiative nanofluid flow past a rotating stretching sheet with magnetohydrodynamics. *Results Phys.* **2016**, *6*, 1015–1023. [CrossRef]
18. Shah, Z.; Islam, S.; Gul, T.; Bonyah, E.; Khan, M.A. The electrical MHD and hall current impact on micropolar nanofluid flow between rotating parallel plates. *Results Phys.* **2018**, *9*, 1201–1214. [CrossRef]

19. Shah, Z.; Gul, T.; Khan, A.M.; Ali, I.; Islam, S. Effects of hall current on steady three dimensional non-newtonian nanofluid in a rotating frame with brownian motion and thermophoresis effects. *J. Eng. Technol.* **2017**, *6*, 280–296.
20. Gireesha, B.J.; Ganesh, K.; Krishanamurthy, M.R.; Rudraswamy, N.G. Enhancement of heat transfer in an unsteady rotating flow for the aqueous suspensions of single wall nanotubes under nonlinear thermal radiation. *Numer. Study* **2018**. [CrossRef]
21. Ishaq, M.; Ali, G.; Shah, Z.; Islam, S.; Muhammad, S. Entropy Generation on Nanofluid Thin Film Flow of Eyring–Powell Fluid with Thermal Radiation and MHD Effect on an Unsteady Porous Stretching Sheet. *Entropy* **2018**, *20*, 412. [CrossRef]
22. Sarit, K.D.; Stephen, U.S.; Choi Wenhua, Y.U.; Pradeep, T. *Nanofluids Science and Technology*; Wiley-Interscience: New Your, NY, USA, 2007; Volume 397.
23. Wong, K.F.V.; Leon, O.D. Applications of nanofluids: Current and future. *Adv. Mech. Eng.* **2010**, *2*, 519659. [CrossRef]
24. Sheikholeslami, M.; Haq, R.L.; Shafee, A.; Zhixiong, L. Heat transfer behavior of Nanoparticle enhanced PCM solidification through an enclosure with V shaped fins. *Int. J. Heat Mass Transf.* **2019**, *130*, 1322–1342. [CrossRef]
25. Sheikholeslami, M.; Gerdroodbary, M.B.; Moradi, R.; Shafee, A.; Zhixiong, L. Application of Neural Network for estimation of heat transfer treatment of Al2O3-H2O nanofluid through a channel. *Comput. Methods Appl. Mech. Eng.* **2019**, *344*, 1–12. [CrossRef]
26. Sheikholeslami, M.; Mehryan, S.A.M.; Shafee, A.; Sheremet, M.A. Variable magnetic forces impact on Magnetizable hybrid nanofluid heat transfer through a circular cavity. *J. Mol. Liquids* **2019**, *277*, 388–396. [CrossRef]
27. Yadav, D.; Lee, D.; Cho, H.H.; Lee, J. The onset of double-diffusive nanofluid convection in a rotating porous medium layer with thermal conductivity and viscosity variation: A revised model. *J. Porous Media* **2016**, *19*, 31–46. [CrossRef]
28. Yadav, D.; Nam, D.; Lee, J. The onset of transient Soret-driven MHD convection confined within a Hele-Shaw cell with nanoparticles suspension. *J. Taiwan Inst. Chem. Eng.* **2016**, *58*, 235–244. [CrossRef]
29. Yadav, D.; Lee, J. The onset of MHD nanofluid convection with Hall current effect. *Eur. Phys. J. Plus* **2015**, *130*, 162–184. [CrossRef]
30. Shah, Z.; Dawar, A.; Islam, S.; Khan, I.; Ching, D.L.C. Darcy-Forchheimer Flow of Radiative Carbon Nanotubes with Microstructure and Inertial Characteristics in the Rotating Frame. *Case Stud. Therm. Eng.* **2018**, *12*, 823–832. [CrossRef]
31. Shah, Z.; Bonyah, E.; Islam, S.; Gul, T. Impact of thermal radiation on electrical mhd rotating flow of carbon nanotubes over a stretching sheet. *AIP Adv.* **2019**, *9*, 015115. [CrossRef]
32. Shah, Z.; Dawar, A.; Islam, S.; Khan, I.; Ching, D.L.C.; Khan, Z.A. Cattaneo-Christov model for Electrical MagnetiteMicropoler Casson Ferrofluid over a stretching/shrinking sheet using effective thermal conductivity model. *Case Stud. Therm. Eng.* **2018**. [CrossRef]
33. Dawar, A.; Shah, Z.; Islam, S.; Idress, M.; Khan, W. Magnetohydrodynamic CNTs Casson Nanofl uid and Radiative heat transfer in a Rotating Channels. *J. Phys. Res. Appl.* **2018**, *1*, 017–032.
34. Khan, A.S.; Nie, Y.; Shah, Z.; Dawar, A.; Khan, W.; Islam, S. Three-Dimensional Nanofluid Flow with Heat and Mass Transfer Analysis over a Linear Stretching Surface with Convective Boundary Conditions. *Appl. Sci.* **2018**, *8*, 2244. [CrossRef]
35. Shah, Z.; Islam, S.; Ayaz, H.; Khan, S. Radiative Heat and Mass Transfer Analysis of Micropolar Nanofluid Flow of Casson Fluid between Two Rotating Parallel Plates with Effects of Hall Current. *ASME J. Heat Transf.* **2019**, *141*, 022401. [CrossRef]
36. Khan, A.; Shah, Z.; Islam, S.; Khan, S.; Khan, W.; Khan, Z.A. Darcy–Forchheimer flow of micropolar nanofluid between two plates in the rotating frame with non-uniform heat generation/absorption. *Adv. Mech. Eng.* **2018**, *10*, 1687814018808850. [CrossRef]
37. Shah, Z.; Bonyah, E.; Islam, S.; Khan, W.; Ishaq, M. Radiative MHD thin film flow of Williamson fluid over an unsteady permeable stretching. *Heliyon* **2018**, *4*, e00825. [CrossRef] [PubMed]
38. Jawad, M.; Shah, Z.; Islam, S.; Islam, S.; Bonyah, E.; Khan, Z.A. Darcy-Forchheimer flow of MHD nanofluid thin film flow with Joule dissipation and Navier's partial slip. *J. Phys. Commun.* **2018**. [CrossRef]

39. Khan, N.; Zuhra, S.; Shah, Z.; Bonyah, E.; Khan, W.; Islam, S. Slip flow of Eyring-Powell nanoliquid film containing graphene nanoparticles. *AIP Adv.* **2018**, *8*, 115302. [CrossRef]
40. Hammed, K.; Haneef, M.; Shah, Z.; Islam, I.; Khan, W.; Asif, S.M. The Combined Magneto hydrodynamic and electric field effect on an unsteady Maxwell nanofluid Flow over a Stretching Surface under the Influence of Variable Heat and Thermal Radiation. *Appl. Sci.* **2018**, *8*, 160. [CrossRef]
41. Dawar, A.; Shah, Z.; Khan, W.; Idrees, M.; Islam, S. Unsteady squeezing flow of MHD CNTS nanofluid in rotating channels with Entropy generation and viscous Dissipation. *Adv. Mech. Eng.* **2019**, *10*, 1–18. [CrossRef]
42. Khan, A.; Shah, Z.; Islam, S.; Dawar, A.; Bonyah, E.; Ullah, H.; Khan, Z.A. Darcy-Forchheimer flow of MHD CNTs nanofluid radiative thermal behaviour andconvective non uniform heat source/sink in the rotating frame with microstructureand inertial characteristics. *AIP Adv.* **2018**, *8*, 125024. [CrossRef]
43. Sheikholeslami, M.; Shah, Z.; Shafi, A.; Khan, I.; Itili, I. Uniform magnetic force impact on water based nanofluid thermal behavior in a porous enclosure with ellipse shaped obstacle. *Sci. Rep.* **2019**. [CrossRef]
44. Feroz, N.; Shah, Z.; Islam, S.; Alzahrani, E.O.; Khan, W. Entropy Generation of Carbon Nanotubes Flow in a Rotating Channel with Hall and Ion-Slip Effect Using Effective Thermal Conductivity Model. *Entropy* **2019**, *21*, 52. [CrossRef]
45. Alharbi, S.O.; Dawar, A.; Shah, Z.; Khan, W.; Idrees, M.; Islam, S.; Khan, I. Entropy Generation in MHD Eyring–Powell Fluid Flow over an Unsteady Oscillatory Porous Stretching Surface under the Impact of Thermal Radiation and Heat Source/Sink. *Appl. Sci.* **2018**, *8*, 2588. [CrossRef]
46. Liao, S.J. On Homotopy Analysis Method for Nonlinear Problems. *Appl. Math. Comput.* **2004**, *147*, 499–513. [CrossRef]
47. Nasir, N.; Shah, Z.; Islam, S.; Bonyah, E.; Gul, T. Darcy Forchheimer nanofluid thin film flow of SWCNTs and heat transfer analysis over an unsteady stretching sheet. *AIP Adv.* **2019**, *9*, 015223. [CrossRef]
48. Tlili, I.; Khan, W.A.; Khan, I. Multiple slips effects on MHD SA-Al2O3 and SA-Cu non-Newtonian nanofluids flow over a stretching cylinder in porous medium with radiation and chemical reaction. *Results Phys.* **2018**, *8*, 213–221. [CrossRef]
49. Khan, N.S.; Shah, Z.; Islam, S.; Khan, I.; Alkanhal, T.A.; Tlili, I. Entropy Generation in MHD Mixed Convection Non-Newtonian Second-Grade Nanoliquid Thin Film Flow through a Porous Medium with Chemical Reaction and Stratification. *Entropy* **2019**, *21*, 139. [CrossRef]
50. Fiza, M.; Islam, S.; Ullah, H.; Shah, Z.; Chohan, F. An Asymptotic Method with Applications to Nonlinear Coupled Partial Differential Equations. *Punjab Univ. J. Math.* **2018**, *50*, 139–151.

© 2019 by the authors. Licensee MDPI, Basel, Switzerland. This article is an open access article distributed under the terms and conditions of the Creative Commons Attribution (CC BY) license (http://creativecommons.org/licenses/by/4.0/).

Article

Peristaltic Blood Flow of Couple Stress Fluid Suspended with Nanoparticles under the Influence of Chemical Reaction and Activation Energy

Rahmat Ellahi [1,2,*], Ahmed Zeeshan [2], Farooq Hussain [2,3] and A. Asadollahi [4]

1. Center for Modeling & Computer Simulation, Research Institute, King Fahd University of Petroleum & Minerals, Dhahran 31261, Saudi Arabia
2. Department of Mathematics & Statistics, FBAS, IIUI, Islamabad 44000, Pakistan; ahmad.zeeshan@iiu.edu.pk (A.Z.); farooq.hussain@buitms.edu.pk (F.H.)
3. Department of Mathematics, (FABS), BUITEMS, Quetta 87300, Pakistan
4. Department of Mechanical Engineering & Energy Processes, Southern Illinois University, Carbondale, IL 62901, USA; arash.asadollahi@siu.edu
* Correspondence: rellahi@alumni.ucr.edu

Received: 28 January 2019; Accepted: 19 February 2019; Published: 21 February 2019

Abstract: The present study gives a remedy for the malign tissues, cells, or clogged arteries of the heart by means of permeating a slim tube (i.e., catheter) in the body. The tiny size gold particles drift in free space of catheters having flexible walls with couple stress fluid. To improve the efficiency of curing and speed up the process, activation energy has been added to the process. The modified Arrhenius function and Buongiorno model, respectively, moderate the inclusion of activation energy and nanoparticles of gold. The effects of chemical reaction and activation energy on peristaltic transport of nanofluids are also taken into account. It is found that the golden particles encapsulate large molecules to transport essential drugs efficiently to the effected part of the organ.

Keywords: chemical reaction; activation energy; peristalsis; couple stress fluid; nanoparticle; Keller-box method

1. Introduction

In any living organism peristaltic motion is mainly caused by the contraction and expansion of some flexible organs. This applies a pressure force to drive fluids, for example, blood in veins, urine to bladder, and transport of medicines to desired locations are a few common biological examples. The rapid developments in nano-science have noticeably revolutionized almost every field of life, particularly in medical sciences. The advent of nano-technology in medicines has brought miraculous changes by reshaping the primitive methods of treatment. Nowadays, in developed countries operations are preferably performed without involving any prunes and cuts, which was once thought to be very complex and menacing for cancer treatment, brain tumors, lithotripsy, etc. Regardless of many other uses of nanofluids in industrial and practical settings, the primary objective of nanoparticles is the enhancement of heat transfer [1]. It is mainly due to their high conductivity. In addition to the size and type of nanoparticles, other factors, such as temperature, volume fraction, and thermal conductivity are also very important to maximize the thermal conductivity. In pursuit of attaining such enhancement in the system, with the passage of time many useful models based on the physical properties of the matter have been developed. On the said topic, scholars have made full use of these models in their analyses, experiments, and conditions, which have been discussed here very briefly. For instance, the investigation of Tripathi and Beg [2] explains the application of peristaltic micropumps and novel drug delivery systems in pharmacological engineering. They formulated their study with the help of the Buongiornio nanofluid model and treated blood as Newtonian fluid. El-dabe et al. [3] have explained the significant contribution of nanofluids in peristalsis. They produced their results for flexible wall properties, lubrication, MHD, and porosity. As generally it is believed that the motion of blood is likely to be non-Newtonian, therefore, Swarnalathamma and Krishana [4] have studied the physiological flow of the blood in the micro circulatory system by taking account of the particle size effect. They considered couple stress fluid for the given peristaltic analysis, which is further affected by magnetic fields. The effects of channel inclination are studied by Shit and Roy [5]. The couple stress fluid influenced by constant application of magnetic fields is used as the base fluid. Jamalabad et al. [6] reported the effects of biomagnetic blood flow through a stenosis artery by means of non-Newtonian flow of a Carreau-Yasuda fluid model. They carried out a numerical simulation of an unsteady blood flow problem. Hosseini et al. [7] have presented the thermal conductivity of a nanofluid model. To perform this investigation, the nanofluid model is considered as the function of thermal conductivity of nanoparticles, base fluid, and interfacial shell properties by considering temperature as the most effective of parameters involved in the study. The most noteworthy contributions on the matter can be seen in the list of references [8–19]. Furthermore, activation energy has a key role in industries, in particular, effectively aggravating slow chemical reactions in chemistry laboratories to improve the efficiency of various mechanisms by adding activation energy to respective physical and mechanical processes. Mustafa et al. [20] have proposed a chemical and activation energy MHD-effected mix convection flow of nanofluids. In this study the flow over the vertical sheet expands due to high temperature and causes the fluid motion is analyzed numerically. A few of the latest works related to this present work have been listed in [21–28].

In view of the existing literature, one can feel the application of nanotechnology in medical science opens a new dimension for researchers to turn their attention towards the effective role of chemical reaction and activation energy [29–31], since nanoparticles help in treating different diseases by means of the peristaltic movement of blood. Such biological transport of blood helps to deliver drugs or medicine effectively to the damaged tissue or organ. As a matter of fact, this effort is devoted to inspecting the simultaneous effects of chemical reaction and activation energy for the peristaltic flow of couple stress nanofluids in a single model, which is yet not available in literature, and could have dual applications in expediting the treatment process.

2. Formulism

The inner tube is of a rigid configuration, while the outer tube is flexible in nature as shown in Figure 1. The sinusoidal waves travel with a constant speed through its walls, due to the stress caused by an unsteady movement of heated nanofluid through the space between both tubes. The general form of equations governing the two-dimensional flows are given as:

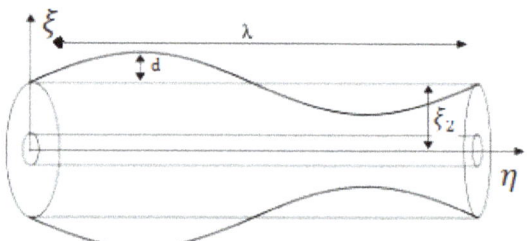

Figure 1. Configuration of coaxial tubes.

Conservation of mass

$$\nabla . \vec{V} = 0. \tag{1}$$

Conservation of momentum

$$\rho_f \left(\vec{V} \cdot \vec{\nabla} \right) \vec{V} = -\vec{\nabla} P^* + \mu \nabla^2 \vec{V} + \left[\varphi \rho_p + (1-\varphi) \{ \rho_f (1 - \beta_T (v - v_w)) \} \right] g - \gamma_1 \nabla^4 \vec{V}. \tag{2}$$

Thermal energy

$$(\rho c)_f \left(\vec{V} \cdot \vec{\nabla} \right) v = k \nabla^2 v + (\rho c)_p \left[D_b \vec{\nabla} \varphi \cdot \vec{\nabla} v + \frac{D_T}{v_w} \vec{\nabla} v \cdot \vec{\nabla} v \right]. \tag{3}$$

Concentration of nanoparticles

$$\rho_p \left(\vec{V} \cdot \vec{\nabla} \right) \varphi = D_b \nabla^2 \varphi + \frac{D_T}{v_w} \nabla^2 v - k_r^2 \left(\frac{v}{v_w} \right)^n (\varphi - \varphi_w) \exp \left(\frac{-E_a}{kv} \right). \tag{4}$$

One can easily identify that the last term in the momentum equation describes the velocity of couple stress fluid involving a constant associated with the couple stress fluid γ_1. The last term in Equation (4) on the right side is known as "Arrhenius term", which shows the effects of chemical reaction and activation energy incorporated to a nanofluid. The radial and axial velocity components of nanofluids are respectively defined by $u(\xi, \eta)$ and $w(\xi, \eta)$ in two concentric tubes, such that there is no rotation about their axes. A peristaltic flow of a heated couple stress fluid carrying the gold nanoparticles (GNPs) through these coaxial tubes due to the contraction and expansion of flexible walls of the outer tube is assumed. If the two-dimensional peristaltic motion of the concerned nanofluid is denoted by $[u(\xi, \eta) \; 0 \; w(\xi, \eta)]$, then the Equations (1)–(4) in the component's form will take the following form:

$$\frac{u}{\xi} + \frac{\partial u}{\partial \xi} + \frac{\partial w}{\partial \eta} = 0, \tag{5}$$

$$\rho_f \left(\frac{\partial u}{\partial t} + u \frac{\partial u}{\partial \xi} + w \frac{\partial u}{\partial \eta} \right)$$
$$= -\gamma_1 \left[\frac{\partial^4 u}{\partial \xi^4} + \frac{\partial^4 u}{\partial \eta^4} + \frac{\partial^4 u}{\partial \xi^2 \partial \eta^2} + \frac{\partial^4 u}{\partial \eta^2 \partial \xi^2} + 2 \frac{\partial^3 u}{\xi \partial \xi^3} + \frac{\partial^3 u}{\xi \partial \xi \partial \eta^2} + \frac{\partial^3 u}{\xi \partial \eta^2 \partial \xi} \right. \tag{6}$$
$$\left. + \frac{\partial^2 u}{\xi^2 \partial \xi^2} + \frac{\partial u}{\xi^3 \partial \xi} \right] - \frac{\partial P}{\partial \xi} + \mu \left(\frac{\partial^2 u}{\partial \xi^2} + \frac{\partial u}{\xi \partial \xi} + \frac{\partial^2 u}{\partial \eta^2} \right),$$

$$\rho_f \left(\frac{\partial w}{\partial t} + u \frac{\partial w}{\partial \xi} + w \frac{\partial w}{\partial \eta} \right)$$
$$= -\gamma_1 \left[\frac{\partial^4 w}{\partial \xi^4} + \frac{\partial^4 w}{\partial \eta^4} + \frac{\partial^4 w}{\partial \xi^2 \partial \eta^2} + \frac{\partial^4 w}{\partial \eta^2 \partial \xi^2} + \frac{\partial^3 w}{\xi \partial \xi^3} + \frac{\partial^3 w}{\xi \partial \xi \partial \eta^2} + \frac{\partial^3 w}{\xi \partial \eta^2 \partial \xi} \right.$$
$$\left. - \frac{\partial^2 w}{\xi^2 \partial \xi^2} + \frac{\partial w}{\xi^3 \partial \xi} \right] + \left[\varphi \rho_p + (1-\varphi) \{ \rho_f (1 - \beta_T (v - v_w)) \} \right] g - \frac{\partial P}{\partial \eta} \tag{7}$$
$$+ \mu \left(\frac{\partial^2 w}{\partial \xi^2} + \frac{1}{\xi} \frac{\partial w}{\partial \xi} + \frac{\partial^2 w}{\partial \eta^2} \right),$$

$$(\rho c)_f \left(\frac{\partial v}{\partial t} + u \frac{\partial v}{\partial \xi} + w \frac{\partial v}{\partial \eta} \right)$$
$$= (\rho c)_p \left[D_b \left\{ \frac{\partial \varphi}{\partial \xi} \frac{\partial v}{\partial \xi} + \frac{\partial \varphi}{\partial \eta} \frac{\partial v}{\partial \eta} \right\} + \frac{D_T}{v_w} \left\{ \left(\frac{\partial v}{\partial \xi} \right)^2 + \left(\frac{\partial v}{\partial \eta} \right)^2 \right\} \right] \tag{8}$$
$$+ k \left(\frac{\partial^2 v}{\partial \xi^2} + \frac{1}{\xi} \frac{\partial v}{\partial \xi} + \frac{\partial^2 v}{\partial \eta^2} \right),$$

$$\left(\frac{\partial \varphi}{\partial t} + u \frac{\partial \varphi}{\partial \xi} + w \frac{\partial \varphi}{\partial \eta} \right)$$
$$= D_b \left(\frac{\partial^2 \varphi}{\partial \xi^2} + \frac{1}{\xi} \frac{\partial \varphi}{\partial \xi} + \frac{\partial^2 \varphi}{\partial \eta^2} \right) + \frac{D_T}{v_w} \left(\frac{\partial^2 v}{\partial \xi^2} + \frac{1}{\xi} \frac{\partial v}{\partial \xi} + \frac{\partial^2 v}{\partial \eta^2} \right) \tag{9}$$
$$- k_r^2 \left(\frac{v}{v_w} \right)^n (\varphi - \varphi_w) \exp \left(\frac{-E_a}{kv} \right).$$

The corresponding boundary at the extreme wall.
At the rigid wall:

$$\left. \begin{array}{l} (i). \; w(\xi) = 0, \\ (ii). \; v(\xi) = v_m, \\ (iii). \; \varphi(\xi) = \varphi_m. \end{array} \right\} ; \text{When } \xi = \xi_1. \tag{10}$$

At the flexible wall:

$$\left.\begin{array}{l}(i).\ w(\xi) = 0,\\ (ii).\ v(\xi) = v_w,\\ (iii).\ \varphi(\xi) = \varphi_w.\end{array}\right\};\ \text{When}\ \xi = \xi_2. \tag{11}$$

As the unsteady peristaltic flow of nanofluids in the laboratory frame (ξ, η) is considered, thus a wave frame (ξ^*, η^*), which moves corresponding to the wave that travels on the flexible and parallel walls of the outer tube, is taken into account. Let "c" be the constant velocity of the wave frame, such that:

$$\begin{array}{l}\xi^* = \xi;\quad \eta^* = \eta - ct;\quad u^* = u;\quad w^* = w - c;\\ v^*(\xi^*, \eta^*) = v(\xi, \eta, t);\quad \varphi^*(\xi^*, \eta^*) = \varphi(\xi, \eta, t).\end{array} \tag{12}$$

In view of the transformation given in Equation (10), the governing Equations (6)–(9) in wave frame can be written as:

$$\begin{aligned}\rho_f&\left[\left(\frac{-\delta c^2}{\lambda}\right)\frac{\partial \bar{u}}{\partial \eta} + \left(\frac{\delta^2 c^2}{\xi_2}\right)\bar{u}\frac{\partial \bar{u}}{\partial \xi} + \left(\frac{\delta c^2}{\lambda}\right)\bar{w}\frac{\partial \bar{u}}{\partial \eta}\right]\\ &= -\gamma_1\left[\left(\frac{\delta c}{\xi_2^4}\right)\frac{\partial^4 \bar{u}}{\partial \xi^4} + \left(\frac{\delta c}{\lambda^4}\right)\frac{\partial^4 \bar{u}}{\partial \eta^4} + \left(\frac{\delta c}{\lambda^2 \xi_2^2}\right)\frac{\partial^4 \bar{u}}{\partial \xi^2 \partial \eta^2} + \left(\frac{\delta c}{\lambda^2 \xi_2^2}\right)\frac{\partial^4 \bar{u}}{\partial \eta^2 \partial \xi^2}\right.\\ &\quad + \left(\frac{2\delta c}{\xi_2^4}\right)\frac{\partial^3 \bar{u}}{\xi \partial \xi^3} + \left(\frac{\delta c}{\lambda^2 \xi_2^2}\right)\frac{\partial^3 \bar{u}}{\xi \partial \xi \partial \eta^2} + \left(\frac{\delta c}{\lambda^2 \xi_2^2}\right)\frac{\partial^3 \bar{u}}{\xi \partial \eta^2 \partial \xi} + \left(\frac{\delta c}{\xi_2^4}\right)\frac{\partial^2 \bar{u}}{\xi^2 \partial \xi^2}\\ &\quad \left. + \left(\frac{\delta c}{\xi_2^4}\right)\frac{\partial \bar{u}}{\xi^3 \partial \xi}\right] - \left(\frac{\mu \lambda c}{\xi_2^3}\right)\frac{\partial \bar{P}}{\partial \xi} + \mu\left[\left(\frac{\delta c}{\xi_2^2}\right)\frac{\partial^2 \bar{u}}{\partial \xi^2} + \left(\frac{\delta c}{\xi_2^2}\right)\frac{\partial \bar{u}}{\xi \partial \xi} + \left(\frac{\delta c}{\lambda^2}\right)\frac{\partial^2 \bar{u}}{\partial \eta^2}\right],\end{aligned} \tag{13}$$

$$\begin{aligned}\rho_f&\left[-c\frac{\partial(w^*+c)}{\partial \eta^*} + u^*\frac{\partial(w^*+c)}{\partial \xi^*} + (w^*+c)\frac{\partial(w^*+c)}{\partial \eta^*}\right]\\ &= -\gamma_1\left[\frac{\partial^4(w^*+c)}{\partial \xi^{*4}} + \frac{\partial^4(w^*+c)}{\partial \eta^{*4}} + \frac{\partial^4(w^*+c)}{\partial \xi^{*2}\partial \eta^{*2}} + \frac{\partial^4(w^*+c)}{\partial \eta^{*2}\partial \xi^{*2}}\right.\\ &\quad + \frac{2\partial^3(w^*+c)}{\xi^* \partial \xi^{*3}} + \frac{\partial^3(w^*+c)}{\xi^* \partial \xi^* \partial \eta^{*2}} + \frac{\partial^3(w^*+c)}{\xi^* \partial \eta^{*2} \partial \xi^*} + \frac{\partial^2(w^*+c)}{\xi^{*2} \partial \xi^{*2}}\\ &\quad \left. + \frac{\partial(w^*+c)}{\xi^{*3}\partial \xi^*}\right] + \left[\varphi^* \rho_p + (1-\varphi^*)\{\rho_f(1-\beta_T(v^*-v_w))\}\right]g - \frac{\partial P^*}{\partial \xi^*}\\ &\quad + \mu\left[\frac{\partial^2(w^*+c)}{\partial \xi^{*2}} + \frac{\partial(w^*+c)}{\xi^*\partial \xi^*} + \frac{\partial^2(w^*+c)}{\partial \eta^{*2}}\right].\end{aligned} \tag{14}$$

$$\begin{aligned}u^*\frac{\partial v^*}{\partial \xi^*} - c\frac{\partial v^*}{\partial \eta^*} + (w^*+c)\frac{\partial v^*}{\partial \eta^*} &= \frac{k}{(\rho c)_f}\left(\frac{\partial^2 v^*}{\partial \xi^{*2}} + \frac{1}{\xi^*}\frac{\partial v^*}{\partial \xi^*} + \frac{\partial^2 v^*}{\partial \eta^{*2}}\right) + \frac{(\rho c)_p}{(\rho c)_f}\\ &\quad \left[D_b(\varphi_m-\varphi_w)\left\{\frac{\partial \varphi^*}{\partial \xi^*}\frac{\partial v^*}{\partial \xi^*} + \frac{\partial \varphi^*}{\partial \eta^*}\frac{\partial v^*}{\partial \eta^*}\right\} + \frac{D_T(v_m-v_w)}{v_w}\left\{\left(\frac{\partial v^*}{\partial \xi^*}\right)^2 + \left(\frac{\partial \varphi^*}{\partial \eta^*}\right)^2\right\}\right].\end{aligned} \tag{15}$$

$$\begin{aligned}\left[u^*\frac{\partial \varphi^*}{\partial \xi^*} + w^*\frac{\partial \varphi^*}{\partial \eta^*}\right] &= \frac{D_T}{v_w}\left(\frac{\partial^2 v^*}{\partial \xi^{*2}} + \frac{1}{\xi^*}\frac{\partial v^*}{\partial \xi^*} + \frac{\partial^2 v^*}{\partial \eta^{*2}}\right) - k_r^2\left(\frac{v^*}{v_w}\right)^n(\varphi^* - \varphi_w)\exp\left(\frac{-E_a}{kv^*}\right)\\ &\quad + D_b\left(\frac{\partial^2 \varphi^*}{\partial \xi^{*2}} + \frac{1}{\xi^*}\frac{\partial \varphi^*}{\partial \xi^*} + \frac{\partial^2 \varphi^*}{\partial \eta^{*2}}\right).\end{aligned} \tag{16}$$

3. Results

Dealing with an unsteady peristaltic transport of couple stress fluid suspended with heated golden nano-sized particles ends up with a system of ordinary differential equations. These differential equations were mutually intermingled with each other, involving a nonlinear composition. Therefore, for such complex geometry, an exact solution was not possible. This means one has to turn to a numerical scheme suitable for tackling the said issue. In order to achieve the desired goal, first we have to make the entire system a non-dimensional form, by using the following transformations along with Oberbeck-Boussinesq approximation and long wave length assumption:

$$\begin{aligned}
&\frac{\bar{\zeta}^*}{\zeta_2} = \bar{\zeta}; \quad \frac{\eta^*}{\lambda} = \bar{\eta}; \quad \frac{u^*}{c\delta} = \bar{u}; \quad \frac{w^*}{c} = \bar{w}; \quad \frac{\zeta_2}{\lambda} = \delta; \quad \frac{\zeta_2^2 P^*}{c\lambda\mu} = \bar{P}; \quad \frac{(\rho c)_p}{(\rho c)_f} = \tau; \\
&\frac{k}{(\rho c)_f} = \alpha; \quad \sqrt{\frac{\mu}{\gamma_1}}\zeta_2 = \gamma; \quad 1 + \bar{\epsilon}\cos(2\pi\bar{\eta}) = R_2; \quad E^* = \frac{E_a}{k\,v_w}; \\
&\frac{\zeta_2^2 (v_m - v_w)(1 - \varphi_w)g\,\rho_f\,\beta_T}{c\mu} = G_r; \quad D_b(\varphi_m - \varphi_w) = N_b; \quad A^* = \frac{k_r^2}{D_b}; \\
&\frac{v^* - v_w}{v_m - v_w} = \bar{v}; \quad \frac{\varphi^* - \varphi_w}{\varphi_m - \varphi_w} = \bar{\varphi}; \quad \frac{D_T(v_m - v_w)}{v_w} = N_t; \quad \frac{d}{\zeta_2} = \bar{\epsilon}; \\
&\frac{\zeta_2^2 (\varphi_m - \varphi_w)(\rho_p - \rho_f)g}{c\mu} = B_r; \quad \frac{\zeta_1}{\zeta_2} = R_1; \quad \beta^* = \frac{(v_m - v_w)}{v_w}.
\end{aligned} \quad (17)$$

Equations (13)–(16) in dimensionless form can be obtained as:

$$\frac{d\bar{P}}{d\bar{\eta}} = \frac{d^2\bar{w}}{d\bar{\zeta}^2} + \frac{1}{\bar{\zeta}}\frac{d\bar{w}}{d\bar{\zeta}} - \frac{1}{\gamma^2}\left[\frac{d^4\bar{w}}{d\bar{\zeta}^4} + \frac{2}{\bar{\zeta}}\frac{d^3\bar{w}}{d\bar{\zeta}^3} + \frac{d^2\bar{w}}{\bar{\zeta}^2 d\bar{\zeta}^2} + \frac{d\bar{w}}{\bar{\zeta}^3 d\bar{\zeta}}\right] + B_r\bar{\varphi} + G_r\,\bar{v}, \quad (18)$$

$$\alpha\left(\frac{\partial^2\bar{v}}{\partial\bar{\zeta}^2} + \frac{1}{\bar{\zeta}}\frac{\partial\bar{v}}{\partial\bar{\zeta}}\right) + \tau\left\{N_b\left(\frac{\partial\bar{\varphi}}{\partial\bar{\zeta}}\right)\left(\frac{\partial\bar{v}}{\partial\bar{\zeta}}\right) + N_t\left(\frac{\partial\bar{v}}{\partial\bar{\zeta}}\right)^2\right\} = 0, \quad (19)$$

$$N_b\left(\frac{\partial^2\bar{\varphi}}{\partial\bar{\zeta}^2} + \frac{1}{\bar{\zeta}}\frac{\partial\bar{\varphi}}{\partial\bar{\zeta}}\right) + N_t\left(\frac{\partial^2\bar{v}}{\partial\bar{\zeta}^2} + \frac{1}{\bar{\zeta}}\frac{\partial\bar{v}}{\partial\bar{\zeta}}\right) - \{A^*(\beta^*\bar{v} + 1)^n N_b\}\bar{\varphi}\exp\left(\frac{-E^*}{\beta^*\bar{v} + 1}\right) = 0. \quad (20)$$

Also, the corresponding boundary conditions in dimensionless form are as follows.
At the rigid wall:

$$\left.\begin{aligned}
&(i).\ \bar{w}(\bar{\zeta}) = -1, \\
&(ii).\ \bar{v}(\bar{\zeta}) = 1, \\
&(iii).\ \bar{\varphi}(\bar{\zeta}) = 1.
\end{aligned}\right\}; \text{ When } \bar{\zeta} = R_1. \quad (21)$$

At the flexible wall:

$$\left.\begin{aligned}
&(i).\ \bar{w}(\bar{\zeta}) = -1, \\
&(ii).\ \bar{v}(\bar{\zeta}) = 0, \\
&(iii).\ \bar{\varphi}(\bar{\zeta}) = 0.
\end{aligned}\right\}; \text{ When } \bar{\zeta} = R_2. \quad (22)$$

Finally, to obtain reliable solutions of Equations (18)–(20) subject to corresponding boundary conditions given in Equations (21) and (22), the most efficient numerical approach, Keller-box scheme, [32] is utilized. This method is much faster and more flexible to use as compared to other methods. It has been extensively used and tested on boundary layer flows. By means of said method, the solution can be attained by using four steps: (i) First reduce the system of equations to a first order system; (ii) then write the difference equations by means of central differences; (iii) now linearize the resulting nonlinear equation by Newton's method, if needed; and (iv) finally the block-tridiagonal-elimination technique is used to solve the linear system.

4. Discussion

This graphical section is relevant to the effectively contributing parameters, which influence axial velocity of couple stress fluid, temperature of nanofluid, and concentration of nano sized Hafnium particles, respectively. The involved parameters have a greater impression on the flow, namely, couple stress parameter γ, Brownian motion N_b and thermophoresis parameters N_t, thermophoresis diffusion G_r, and Brownian parameter B_r emerging due to the presence of heat and metallic particles. Moreover, a modified Arrhenius mathematical term yields some additional parameters, such as reaction rate A^*, activation energy E^*, temperature difference parameter β^*, and the fitted rate constant n, assuming the contribution of peristaltic pressure to be constant. To make this more systematic, the main portion is further divided into four subsections.

4.1. Axial Velocity

Axial velocity is spotted in Figures 2–4 for couple stress parameter, Brownian diffusion constant, and Grashof number. Axial velocity, as shown in Figure 2, accelerates in response to an increases in couple stress parameter. This is mainly due to the decrease in friction, which arises from the particle (i.e., base-fluid particles) additives, which constitute a size-dependent effect in couple stress fluids. In addition to the preceding remark, the rotational field of fluid particles is minimal as well. The peristaltic motion of outer walls of the tube also contributes by rapidly pushing the fluid in the axial direction, as B_r gets numerically variated in Figure 3. Figure 4 displays a different picture of the velocity of the fluid for the case of thermophoresis diffusion constant. The diagram basically describes the influence of buoyancy in terms of Grashof number G_r. As one can see from Equation (17), the buoyancy effects are mainly due to gravity and temperature difference. Therefore, increase in G_r attenuates the fluid's momentum by aggravating buoyant force. This brings a vivid decline in the velocity of the fluid. Furthermore, the relation defining Gr suggests that if $Gr > 0$, then this physically describes the heating of the nanofluid, while a reverse case can be expected for $Gr < 0$.

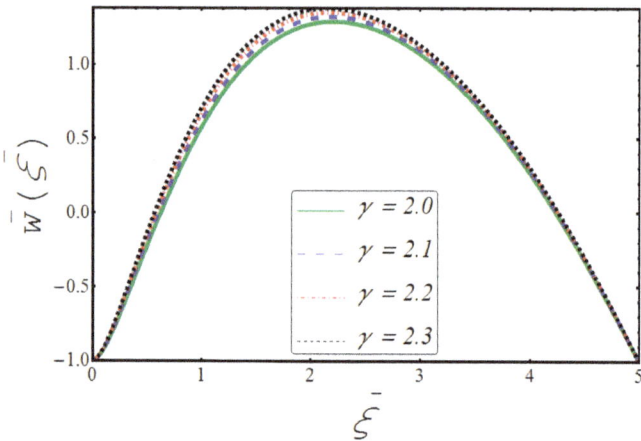

Figure 2. Axial velocity influenced by γ.

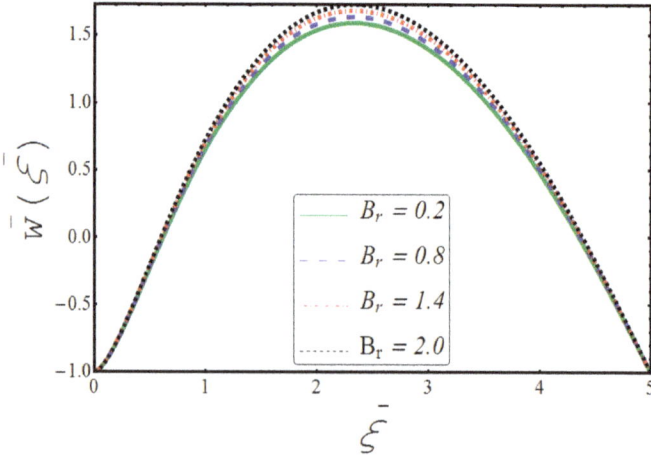

Figure 3. Axial velocity influenced by B_r.

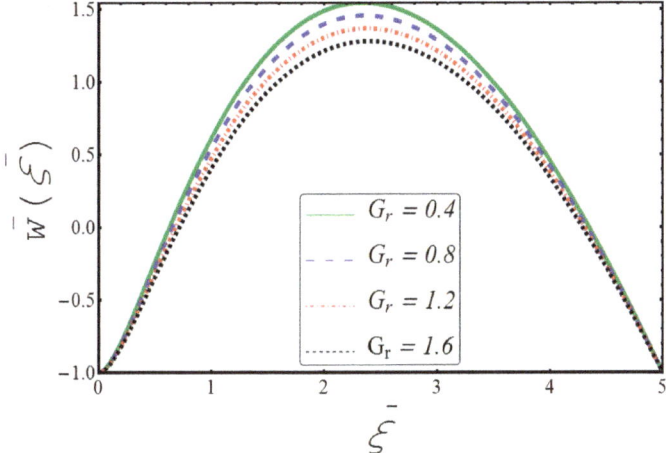

Figure 4. Axial velocity influenced by G_r.

4.2. Thermal Distribution

The temperature distribution of the nanofluid in the presence of additional chemical reaction and activation energy are portrayed in Figures 5 and 6. The variation of the Brownian motion parameter has noticeable effects on the nanofluid temperature, as the Brownian motion is generated due the collision of nanoparticles, driving the particles to a random motion. The collision of the particles, whether mutual or with the fluid molecules, is enhanced by the inward contraction of the flexible walls. Due to this factor Brownian motion parameter, N_b accumulates some additional thermal energy in the fluid, as shown in Figure 5. The nanoparticles were further thermally charged by the increase in N_t. It is important to keep in mind that the thermophoresis forces become stronger in the response of larger values of N_t, which finally result in higher temperature, as seen in Figure 6. Sometimes, such variations are credited to the thermal boundary layer thickness as well. Obviously, this increase in fluid temperature is due to increase in the random motion of nanoparticles when the above-mentioned parameters are increased.

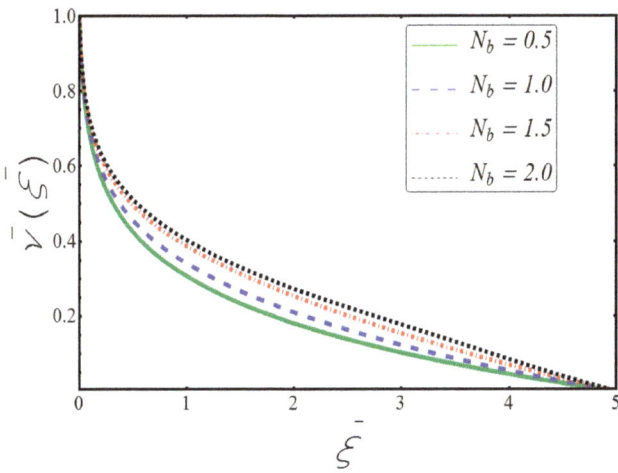

Figure 5. Temperature influenced by N_b.

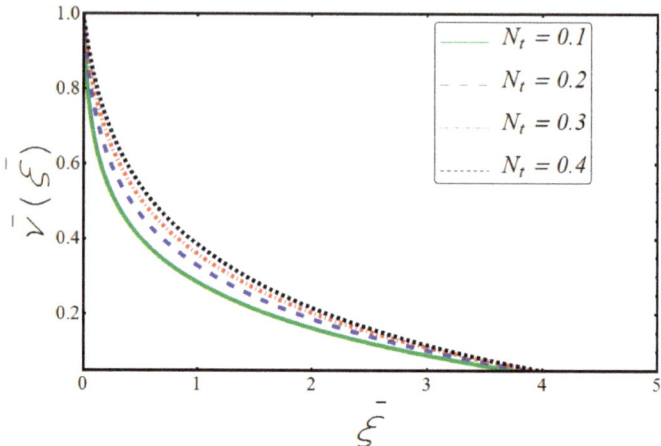

Figure 6. Temperature influenced by N_t.

4.3. Nanoparticle Concentration Profile

The concentration of golden particles is observed in Figures 7 and 8, when the Brownian motion parameter and thermophoresis parameter, respectively, are given higher numeric values. The random motion of the nanoparticles is seen to be faster in response to increase in the values of said parameters, which makes diffusion of nanoparticles rapid and fast. Therefore, rising curves show an increase in the concentration of nanoparticles. Moreover, this contribution of Brownian motion identifies the quick movement of hotter gold particle, from the region of higher temperature to lower temperature. The thermophoresis forces also bring positive effects on the golden particles by making the concentration strong against the higher numerical variation in N_t, as is noticeable in Figure 8. With the same trend of influence, an onward surge of activation energy again gives a rise to the golden solution. One can see in Figure 9 that the boundary layer thickness of the particles gets depreciated when E_a is further motivated to transport the required drug or medicine to the desired target. The Arrhenius equation, which gives the mathematical description of the introduction of activation energy into any system, clearly reveals that the reduction in heat and acceleration of E_a returns a low reaction rate constant. In the process, this slows down the chemical reaction and results in higher concentration of the particles, which confirms the accumulation of gold nanoparticles at the location of the malign tissue or organ to cured. Finally, the surge in concentration of gold particles is evidenced by the decline in Figures 10–12. The temperature difference ratio brings a remarkable decline in concentration of the heated nanoparticles. As the difference between the ambient fluid temperature and wall temperature widens, the concentration boundary layer thickness expands. This thickness resists the increase in particle concentration displayed in Figure 10. Similarly, retardation can be witnessed for reaction rate and fitted rate constant. It can be conceived that the rise in these parameters and constants sharpens the chemical reaction, which motivates the concentration gradient at the wall of the inner tube. Hence, a vivid reduction in the concentration of the particles occurs, as is seen in Figures 11 and 12).

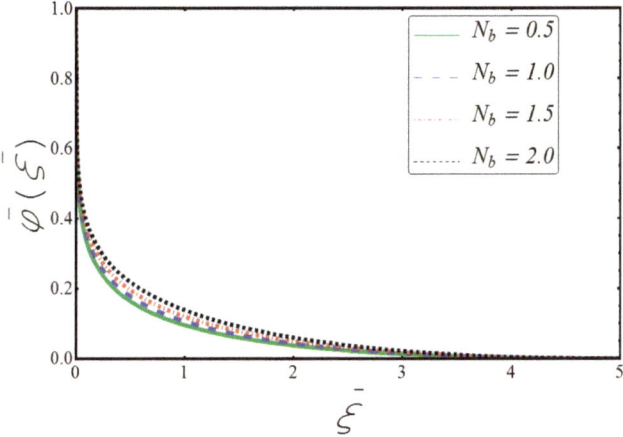

Figure 7. Concentration under the influence of N_b.

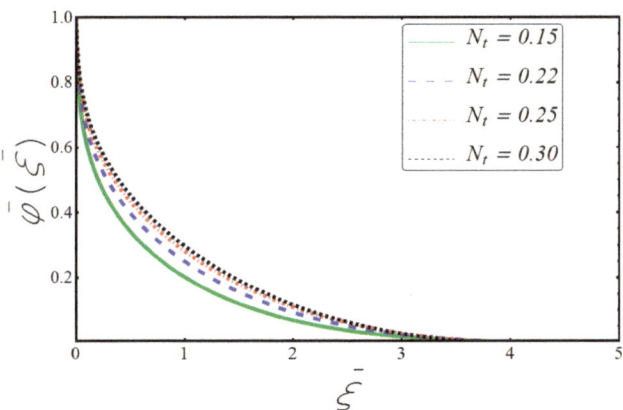

Figure 8. Concentration under the influence of N_t.

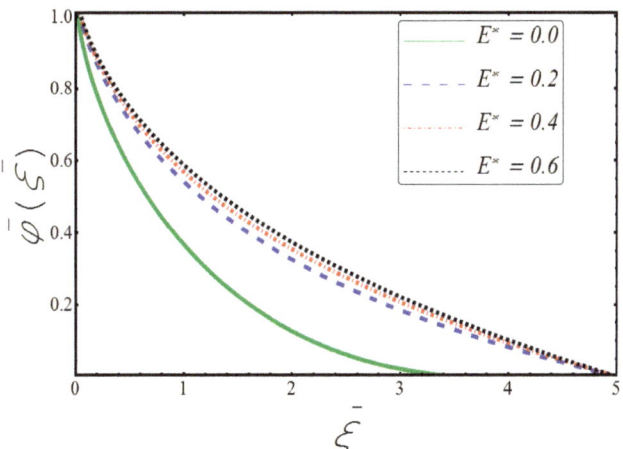

Figure 9. Concentration under the influence of E^*.

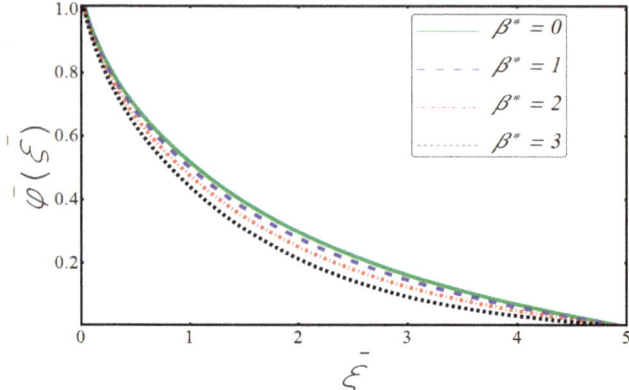

Figure 10. Concentration under the influence of β^*.

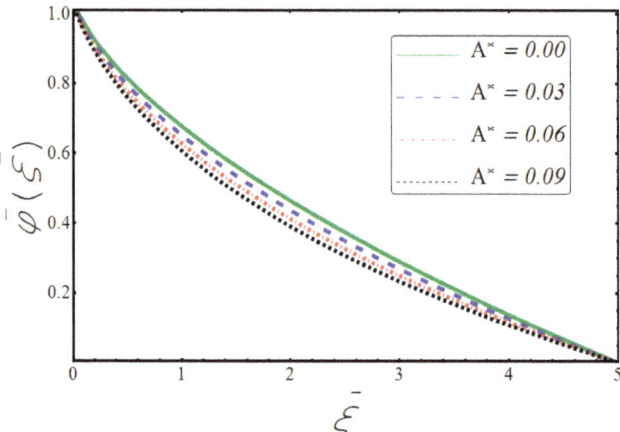

Figure 11. Concentration under the influence of A^*.

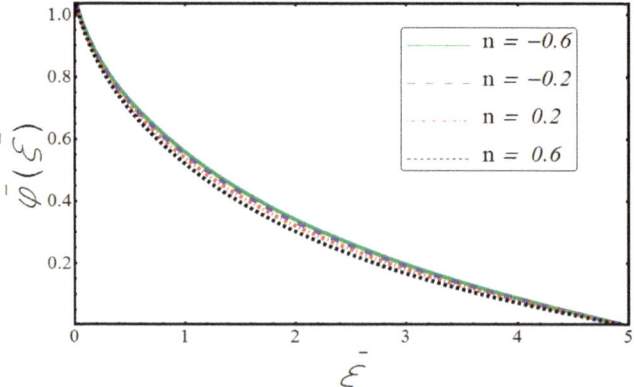

Figure 12. Concentration under the influence of n.

4.4. Trapping Phenomenon/Streamline Configuration

Finally, the most significant phenomena relevant to any peristaltic motion in a living organism is known as "Trapping". Essentially, this is the appearance of a round closed bolus, which is identified as the hallow cavity, transporting the required medication to the desired tissues or organs, as shown in Figures 13–19. In Figures 13 and 14, one can easily notice that the fluids face less resistance when traveling through the coaxial space, as the contours reduce in size and configuration. In contrast, the couple stress fluid results in shrinking the streamlines and generates the circulating boluses, as depicted in Figure 15. Isotherms of the Brownian motion parameter keep binding closer together, which allows the bolus to expand, as established in Figures 16 and 17, whereas the thermophoresis parameter provides extra potential for isotherms to compress the bolus inwards. Hence, the bolus keeps getting smaller. In the last two graphs, contours are sketched in order to see how concentration is influenced by the reaction rate constant and thermophoresis parameter. One can see in Figure 18 that the bolus bulges out as the reaction rate constant gets stronger, whereas a reverse trend is observed for the thermophoresis parameter in Figure 19.

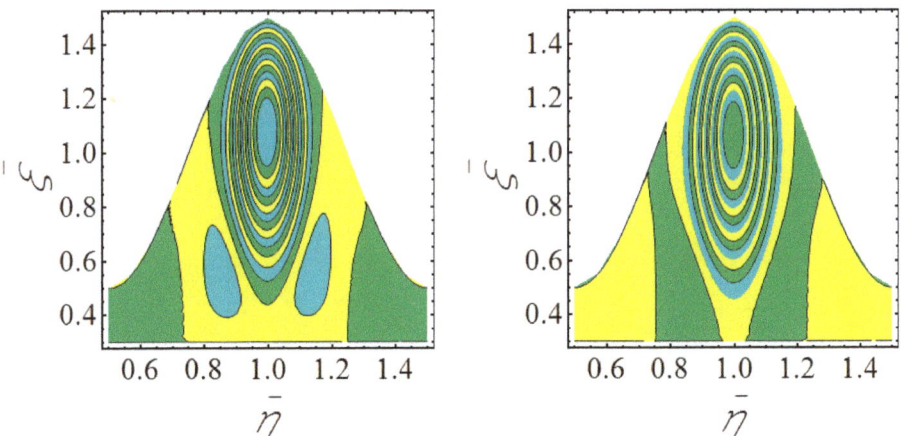

Figure 13. Stream lines for Brownian diffusion constant. (**a**): For $B_r = 0.2$; (**b**): For $B_r = 0.2$.

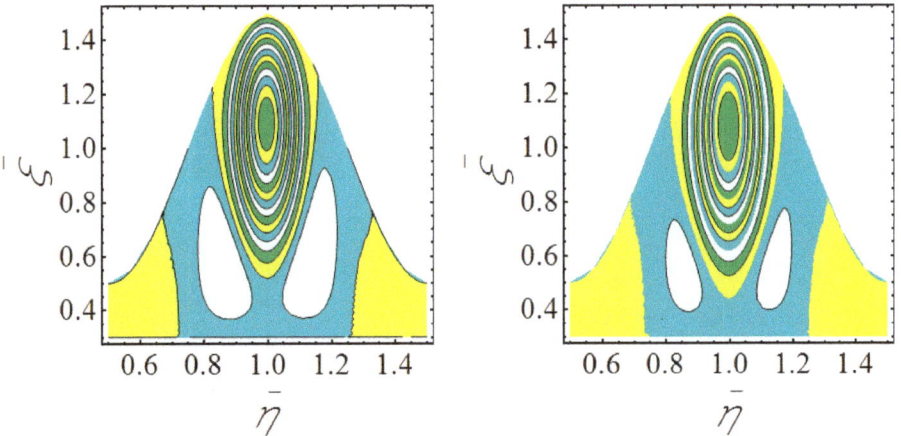

Figure 14. Stream lines for Grashof number. (**a**): For $G_r = 0.1$; (**b**): For $G_r = 0.3$.

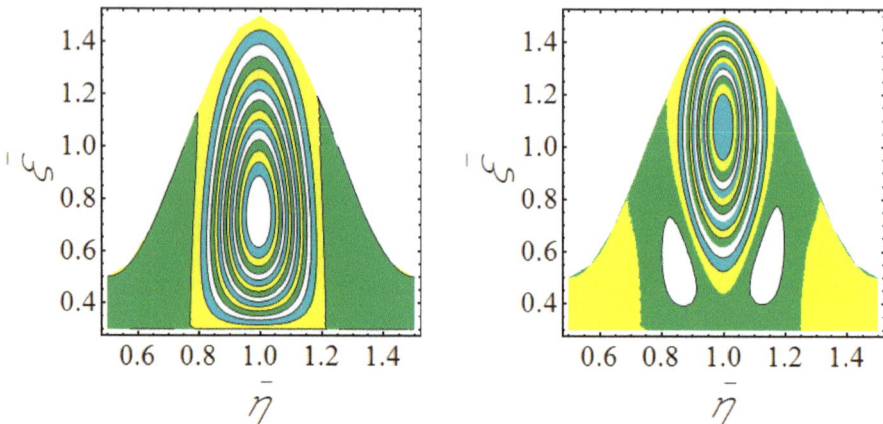

Figure 15. Stream lines for couple stress parameter. (**a**): For $\gamma = 1.0$; (**b**): For $\gamma = 2.0$.

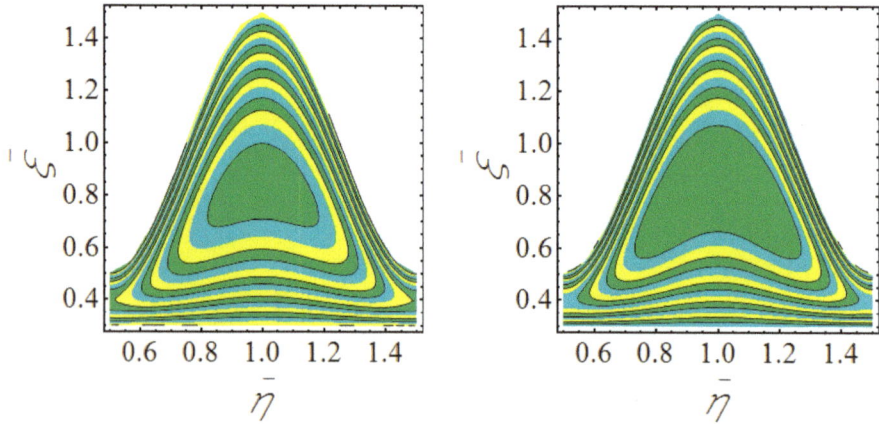

Figure 16. Isotherms for Brownian motion parameter. (**a**): For $N_b = 1.0$; (**b**): For $N_b = 1.5$.

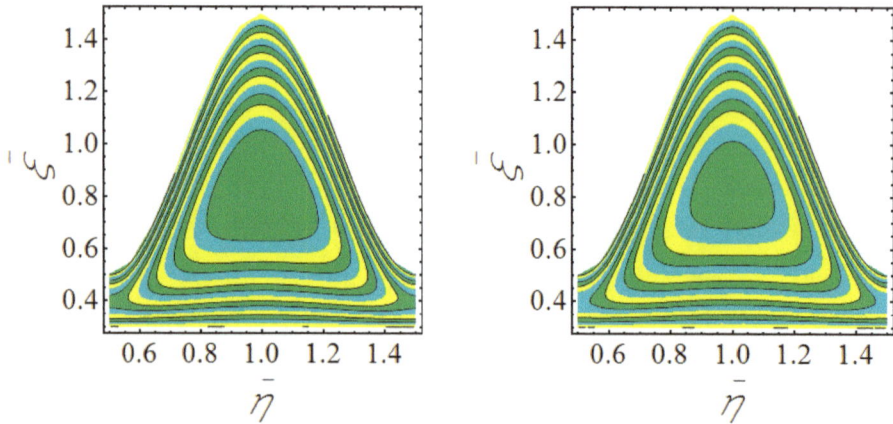

Figure 17. Isotherms for Thermophoresis parameter. (**a**): For $N_t = 0.2$; (**b**): For $N_t = 0.5$.

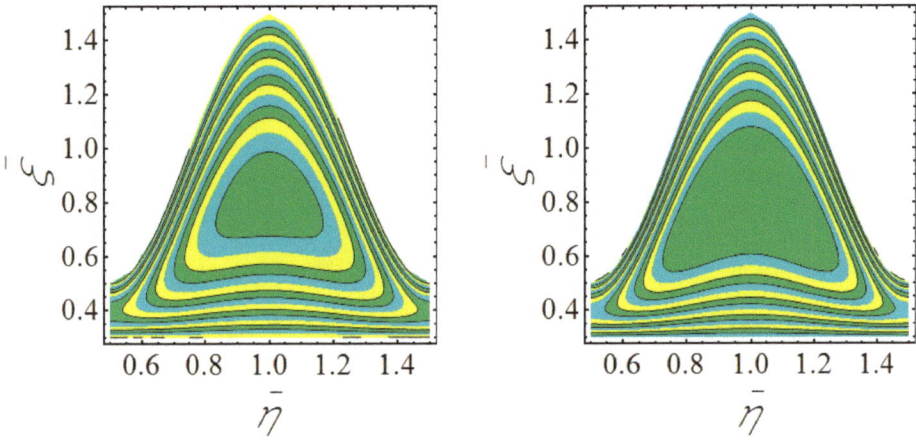

Figure 18. Contour plot for reaction rate constant. (**a**): For $A^* = 0.5$; (**b**): For $A^* = 1.0$.

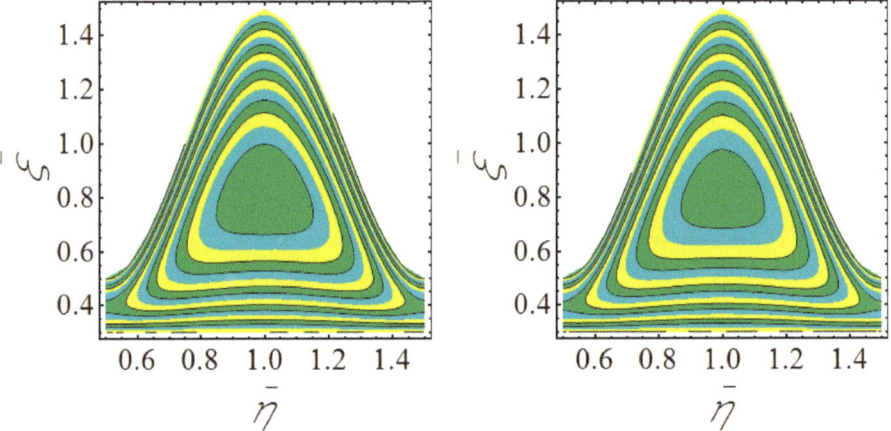

Figure 19. Contour plot for Thermophoresis parameter. (**a**): For $N_t = 0.2$; (**b**): For $N_t = 0.5$.

5. Conclusions

A numerical investigation is carried out for the peristaltic flow of nanofluids between the gap of two coaxial tubes with different configurations and structures. The nanofluid is composed of gold particles, while the couple stress fluid serves as the solvent. To enhance the mutual interaction of gold particles, or the interaction of molecules with the base fluid, additional effects of chemical reaction and activation energy have also been taken into consideration. The performed study reveals very informative results. Such results include that axial velocity is fully supported by the couple stress parameter and Brownian diffusion constant, in contrast to the Grashof number. The temperature of the nanofluid remains high for both involved parameters, which are thermophoresis and Brownian motion parameter. Looking at the graphs of concentrations of the metallic particles, it is inferred that activation energy, thermophoresis, and Brownian motion parameters cause an increase in the concentration of particles, whereas temperature ratio, reaction rate, and fitted rate constants do not support the increase. In the final portion of the graphical study, the number and size of the circulating boluses are depicted. One can easily notice that boluses get enlarged in response to the Brownian motion parameter, couples stress parameter, and reaction rate constant. However, a reverse trend is observed for the Grashof number, thermophoresis parameter, and Brownian diffusion constant. The key finding can be summarized as:

- Strong buoyant force results in retarded axial velocity for the thermophoresis parameter.
- Peristaltic movement of the outer tube enhances the Brownian motion and raises the temperature of the nanofluid.
- Activation energy entering the process maximizes the concentration boundary layer thickness.
- The reaction rate constant increases concentration at the catheter, which decreases the concentration of nanoparticles.
- The thermophoresis parameter shrinks the size of the bolus by strengthening isotherms and closed paths of concentration lines.
- The couple stress parameter and reaction rate constant give freedom to the bolus to swell by binding the stream lines closer to each another.

Author Contributions: Supervision, R.E.; conceptualization, A.Z.; investigation, F.H.; methodology, A.A.

Funding: This research received no external funding.

Acknowledgments: R. Ellahi gratefully thanks Prof. Sadiq M. Sait, the Director Office of Research Chair Professors, King Fahd University of Petroleum and Minerals, Dhahran, Saudi Arabia, to honor him with the Chair Professor at KFUPM. F. Hussain is also HEC (Higher Education Commission Pakistan) to provide him the title of indigenous scholar for the pursuance of his Ph.D. studies.

Conflicts of Interest: The authors declare no conflict of interest.

Nomenclature

V	Nanofluid velocity
G	Gravitational acceleration
u	Radial velocity Component (fixed frame)
w	Axial velocity component (Fixed frame)
u^*	Radial velocity component (wave frame)
w^*	Axial velocity component (Wave frame)
$\bar{u}$	Dimensionless radial velocity component
$\bar{w}$	Dimensionless lateral velocity component
d	Amplitude of peristaltic wave
t	Time
k_r	Rate of reaction
c	Propagating velocity of wave
N_t	Thermophoresis parameter
k	Thermal conductivity
N_b	Brownian motion parameter
G_r	Grashof number
D_t	Thermophoretic diffusion coefficient
D_b	Brownian motion coefficient
d	Amplitude of peristaltic wave
t	Time
R_2	Dimensionless radius of outer tube
R_1	Dimensionless radius of inner tube
P^*	Dimensional pressure
B_r	Brownian diffusion constant
A^*	Reaction rate constant
E^*	Activation energy (Dimensionless)
E_a	Activation energy (Dimensional)
n	Fitted rate constant

Greek Symbols

ξ	Radial direction of the flow (Fixed frame)
η	Axial direction of the flow (Fixed frame)
ξ^*	Radial direction of the flow (Wave frame)
η^*	Axial direction of the flow (Wave frame)
$\bar{\xi}$	Radial direction of the flow (Dimensionless)
$\bar{\eta}$	Axial direction of the flow (Dimensionless)
ξ_1	Radius of inner tube (Dimensional)
ξ_2	Radius of outer tube (Dimensional)
$\vec{\varphi}$	Nanoparticle concentration (Fixed frame)
$\vec{\nu}$	Nanofluid temperature (Fixed frame)
φ^*	Nanoparticle concentration (Wave frame)
ν^*	Nanofluid temperature (Wave frame)
$\bar{\varphi}$	Nanoparticle concentration (Dimensionless)
$\bar{\nu}$	Nanofluid temperature (Dimensionless)
γ_1	Couple stress fluid's constant
γ	Couple stress parameter
τ	A ratio defined as $\frac{(\bar{\rho}c)_f}{(\bar{\rho}c)_p}$
β^*	Temperature ratio
ρ_p	Density of nanoparticle at reference temperature
ρ_f	Density of nanofluid at reference temperature
$(\rho c)_f$	Heat capacity of base fluid
$(\rho c)_p$	Heat capacity of particle
μ	Dynamic Viscosity
ν	Kinematic viscosity
λ	Wavelength
α	Ratio defined as $\frac{k}{(\rho c)_f}$
$\bar{\epsilon}$	A constant ratio
β_T	Volumetric coefficient of expansion
φ_w	Reference concentration
ν_w	Reference temperature
φ_m	Mass concentration
ν_m	Fluid temperature

Subscripts

f	Base fluid
p	Particle

References

1. Karimipour, A.; Orazio, A.D.; Shadloo, M.S. The effects of different nano particles of Al2O3 and Ag on the MHD nano fluid flow and heat transfer in a microchannel including slip velocity and temperature jump. *Physica E* **2017**, *86*, 146–153. [CrossRef]
2. Tripathi, D.; Beg, O.A. A study on peristaltic flow of nanofluids: Application in drug delivery systems. *Int. J. Heat Mass Transf.* **2014**, *70*, 61–70. [CrossRef]
3. Nabil, T.M.; El-dabe, N.T.M.; Moatimid, G.M.; Hassan, M.A.; Godh, W.A. Wall properties of peristaltic MHD Nanofluid flow through porous channel. *Fluid Mech. Res. Int.* **2018**, *2*, 19. [CrossRef]
4. Swarnalathamma, B.V.; Krishna, M.V. Peristaltic hemodynamic flow of couple stress fluid through a porous medium under the influence of magnetic field with slip effect. *AIP Conf. Proc.* **2016**, *1728*, 020603. [CrossRef]
5. Shit, G.C.; Roy, M. Hydromagnetic effect on inclined peristaltic flow of a couple stress fluid, Hydromagnetic effect on inclined peristaltic flow of a couple stress fluid. *Alex. Eng. J.* **2014**, *53*, 949–958. [CrossRef]
6. Jamalabadi, M.Y.A.; Daqiqshirazi, M.; Nasiri, H.; Safaei, M.R.; Nguyen, T.K. Modeling and analysis of biomagnetic blood Carreau fluid flow through a stenosis artery with magnetic heat transfer: A transient study. *PLoS ONE* **2018**, *13*, e0192138. [CrossRef]

7. Hosseini, S.M.; Safaei, M.R.; Goodarzi, M.; Alrashed, A.A.A.A.; Nguyen, T.K. New temperature, interfacial shell dependent dimensionless model for thermal conductivity of nanofluids. *Int. J. Heat Mass Transf.* **2017**, *114*, 207–210. [CrossRef]
8. Mekheimer, K.; Hasona, W.; Abo-Elkhair, R.E.; Zaher, A. Peristaltic blood flow with gold nanoparticles as a third grade nanofluid in catheter. *Appl. Cancer Ther. Phys. Lett. A* **2018**, *382*, 85–93.
9. Nasiri, H.; Jamalabadi, M.Y.A.; Sadeghi, R.; Safaei, M.R.; Nguyen, T.K.; Shadloo, M.S. A smoothed particle hydrodynamics approach for numerical simulation of nano-fluid flows. *J. Anal. Calorim.* **2018**, 1–9. [CrossRef]
10. Shadloo, M.S.; Kimiaeifar, A. Application of homotopy perturbation method to find an analytical solution for magnetohydrodynamic flows of viscoelastic fluids in converging/diverging channels. *Proc. Mech. Eng. Part C J. Mech. Eng.* **2011**, *225*, 347–353. [CrossRef]
11. Safaei, M.R.; Ahmadi, G.; Goodarzi, M.S.; Shadloo, M.S.; Goshayeshi, H.R.; Dahari, M. Heat transfer and pressure drop in fully developed turbulent flows of graphene nanoplatelets–silver/water nanofluids. *Fluids* **2016**, *1*, 20. [CrossRef]
12. Mohyud-Din, S.T.; Khan, U.; Ahmed, N.; Hassan, S.M. Magnetohydrodynamic flow and heat transfer of nanofluids in stretchable convergent/divergent channels. *Appl. Sci.* **2015**, *5*, 1639–1664. [CrossRef]
13. Chamkha, A.J.; Selimefendigil, F. Forced convection of pulsating nanofluid flow over a backward facing step with various particle shapes. *Energies* **2018**, *11*, 3068. [CrossRef]
14. Xu, Z.; Kleinstreuer, C. Direct nano-drug delivery for tumor targeting subject to shear-augmented diffusion in blood flow. *Med. Biol. Eng. Comput.* **2018**, *56*, 1949–1958. [CrossRef] [PubMed]
15. Rashidi, M.M.; Nasiri, M.; Shadloo, M.S.; Yang, Z. Entropy generation in a circular tube heat exchanger using nanofluids: Effects of different modeling approaches. *Heat Transf. Eng.* **2017**, *38*, 853–866. [CrossRef]
16. Jamalabadi, M.Y.A.; Safaei, M.R.; Alrashed, A.A.A.A.; Nguyen, T.K.; Filho, E.P.B. Entropy generation in thermal radiative loading of structures with distinct heaters. *Entropy* **2017**, *19*, 506. [CrossRef]
17. Sadiq, M.A. MHD stagnation point flow of nanofluid on a plate with anisotropic slip. *Symmetry* **2019**, *11*, 132. [CrossRef]
18. Jawad, M.; Shah, Z.; Islam, S.; Majdoubi, J.; Tlili, I.; Khan, W.; Khan, I. Impact of nonlinear thermal radiation and the viscous dissipation effect on the unsteady three-dimensional rotating flow of single-wall carbon nanotubes with aqueous suspensions. *Symmetry* **2019**, *11*, 207. [CrossRef]
19. Akbarzadeh, O.; Zabidi, N.A.M.; Wahab, Y.A.; Hamizi, N.A.; Chowdhury, Z.Z.; Merican, Z.M.A.; Rahman, M.A.; Akhter, S.; Rasouli, E.; Johan, M.R. Effect of cobalt catalyst confinement in carbon nanotubes support on fischer-tropsch synthesis performance. *Symmetry* **2018**, *10*, 572. [CrossRef]
20. Mustafa, M.; Khan, J.A.; Hayat, T.; Alsaedi, A. Buoyancy effects on the MHD nanofluid flow past a vertical surface with chemical reaction and activation energy. *Int. J. Heat Mass Transf.* **2017**, *108*, 1340–1346. [CrossRef]
21. Xu, H.; Fan, T.; Pop, I. Analysis of mixed convection flow of a nanofluid in a vertical channel with the Buongiorno mathematical model. *Int. Commun. Heat Mass Transf.* **2013**, *44*, 15–22. [CrossRef]
22. Sheikholeslami, M.; Bhatti, M.M. Active method for nanofluid heat transfer enhancement by means of EHD. *Int. J. Heat Mass Transf.* **2017**, *109*, 115–122. [CrossRef]
23. Bhatti, M.M.; Rashidi, M.M. Effects of thermo-diffusion and thermal radiation on Williamson nanofluid over a porous shrinking/stretching sheet. *J. Mol. Liq.* **2016**, *221*, 567–573. [CrossRef]
24. Marin, M. An approach of a heat-flux dependent theory for micropolar porous media. *Meccanica* **2016**, *51*, 127–1133. [CrossRef]
25. Rashidi, S.; Esfahani, J.A.; Ellahi, R. Convective heat transfer and particle motion in an obstructed duct with two side by side obstacles by means of DPM model. *Appl. Sci.* **2017**, *7*, 431. [CrossRef]
26. Zeeshan, A.; Ijaz, N.; Abbas, T.; Ellahi, R. The sustainable characteristic of Bio-bi-phase flow of peristaltic transport of MHD Jeffery fluid in human body. *Sustainability* **2018**, *10*, 2671. [CrossRef]
27. Hussain, F.; Ellahi, R.; Zeeshan, A. Mathematical models of electro magnetohydrodynamic multiphase flows synthesis with nanosized hafnium particles. *Appl. Sci.* **2018**, *8*, 275. [CrossRef]
28. Haq, R.U.; Soomro, F.A.; Hammouch, Z.; Rehman, S. Heat exchange within the partially heated C-shape cavity filled with the water based SWCNTs. *Int. J. Heat Mass Transf.* **2018**, *127*, 506–514. [CrossRef]
29. Kumar, R.V.M.S.S.K.; Kumar, G.V.; Raju, C.S.K.; Shehzad, S.A.; Varma, S.V.K. Analysis of Arrhenius activation energy in magnetohydrodynamic Carreau fluid flow through improved theory of heat diffusion and binary chemical reaction. *J. Phys. Commun.* **2018**, *2*, 35–49. [CrossRef]

30. Anuradha, S.; Yegammai, M. MHD radiative boundary layer flow of nanofluid past a vertical plate with effects of binary chemical reaction and activation energy. *J. Pure Appl. Math.* **2017**, *13*, 6377–6392.
31. Pal, D.; Talukdar, B. Perturbation analysis of unsteady magnetohydrodynamic convective heat and mass transfer in a boundary layer slip flow past a vertical permeable plate with thermal radiation and chemical reaction. *Commun. Nonlinear Sci. Numer. Simul.* **2010**, *15*, 1813–1830. [CrossRef]
32. Hossain, A., Md.; Subba, R.; Gorla, R. Natural convection flow of non-Newtonian power-law fluid from a slotted vertical isothermal surface. *Int. J. Numer. Methods Heat Fluid Flow* **2009**, *19*, 835–846. [CrossRef]

© 2019 by the authors. Licensee MDPI, Basel, Switzerland. This article is an open access article distributed under the terms and conditions of the Creative Commons Attribution (CC BY) license (http://creativecommons.org/licenses/by/4.0/).

Article

A Numerical Simulation of Silver–Water Nanofluid Flow with Impacts of Newtonian Heating and Homogeneous–Heterogeneous Reactions Past a Nonlinear Stretched Cylinder

Muhammad Suleman [1,2], Muhammad Ramzan [3,4,*], Shafiq Ahmad [5], Dianchen Lu [1], Taseer Muhammad [6] and Jae Dong Chung [4]

1. Department of Mathematics, Faculty of Science, Jiangsu University, Zhenjiang 212013, China; suleman@ujs.edu.cn (M.S.); dclu@ujs.edu.cn (D.L.)
2. Department of Mathematics, Comsats University, Islamabad 44000, Pakistan
3. Department of Computer Science, Bahria University, Islamabad Campus, Islamabad 44000, Pakistan
4. Department of Mechanical Engineering, Sejong University, Seoul 143-747, Korea; jdchung@sejong.ac.kr
5. Department of Mathematics, Quaid-i-Azam University, Islamabad 44000, Pakistan; ashafiq@math.qau.edu.pk
6. Department of Mathematics, Government College Women University, Sialkot 51310, Pakistan; taseer_qau@yahoo.com
* Correspondence: mramzan@bahria.edu.pk; Tel.: +92-300-5122700

Received: 29 December 2018; Accepted: 29 January 2019; Published: 24 February 2019

Abstract: The aim of the present study is to address the impacts of Newtonian heating and homogeneous–heterogeneous (h-h) reactions on the flow of Ag–H2O nanofluid over a cylinder which is stretched in a nonlinear way. The additional effects of magnetohydrodynamics (MHD) and nonlinear thermal radiation are also added features of the problem under consideration. The Shooting technique is betrothed to obtain the numerical solution of the problem which is comprised of highly nonlinear system ordinary differential equations. The sketches of different parameters versus the involved distributions are given with requisite deliberations. The obtained numerical results are matched with an earlier published work and an excellent agreement exists between both. From our obtained results, it is gathered that the temperature profile is enriched with augmented values radiation and curvature parameters. Additionally, the concentration field is a declining function of the strength of h-h reactions.

Keywords: Newtonian heating; nonlinear thermal radiation; nonlinear stretching cylinder; homogeneous/heterogeneous reactions; nanofluid

1. Introduction

In copious engineering processes, the role of poor thermal conductivity of certain base fluids is considered to be a big hurdle to shape a refined product. To overcome such snag, numerous practices such as clogging, abrasion, and pressure loss were engaged but the results were not very promising. The novel idea of nanofluid, which is an engineered amalgamation of metallic particles with size (<100 nm) and some traditional fluids like ethylene glycol, presented by Choi [1], has revolutionized the modern world. Many heat transfer applications [2] such as domestic refrigerators, microelectronics, hybrid-power engines, and fuel cells possess numerous characteristics that make them valuable because of nanofluids. In all aforementioned applications, enriched thermal conductivity is observed whenever some metallic particles are added to the ordinary base fluid [3]. The idea of a nanofluid with multiple dimensions has been conversed by many researchers and scientists in the last two decades. Amongst these, Li et al. [4] used a mixture of H_2O–CuO for the enhancement of the solidification rate. The finite element method is engaged to obtain the numerical solution of the problem. Sheikholeslami [5] pondered over the influence of radiation and magnetohydrodynamic on the Al_2O_3–H_2O mixture past a spongy semi-annulus. The numerical solution of the problem is witnessed by employing the Control Volume Finite Element Method (CVFEM). The flow of Casson nanofluid with inserted multi-walled carbon nanotubes past a swirling cylinder was deliberated by Ramzan et al. [6] using bvp4c MATLAB software. The said problem is pondered with impacts of entropy generation and melting heat transfer. The flow of micropolar nanofluid with binary chemical reaction, double stratification, and activation energy is also studied by Ramzan et al. [7]. The flow of viscoelastic nanofluid with analysis of entropy generation past an exponential stretched surface was discussed by Suleman et al. [8]. Farooq et al. [9] deliberated the flow of Newtonian fluid with the amalgamation of nanoparticles by utilizing the BVPh 2.0 technique and many therein [10–15].

The role of MHD is vital in abundant fluid flows and has many applications like medicine, aerospace, nuclear reactors, MHD generators, petroleum processes, and astrophysics. The explorations highlighting the effects of MHD may include a study by Azam et al. [16] who found the numerical solution for the time-dependent MHD Cross nanofluid flow under the influence of nonlinear thermal radiation and zero mass flux conditions. The flow of MHD nanofluid with thermal diffusion and heat generation past a permeable medium over an oscillating vertical plate is perceived by Sheikholeslami et al. [17]. Makinde and Animasaun [18] examined the MHD nanofluid flow past the upper surface of a paraboloid of revolution with impacts of quartic autocatalysis chemical reaction and nonlinear thermal radiation. Lu et al. [19] debated the rotating flow of 3D MHD Maxwell fluid with a non-Fourier heat flux and binary chemical reaction using a BVP-4c MATLAB built-in technique. The analytical technique is engaged to find the multiple solutions for MHD Jeffery-Hamel flow using the KKL nanofluid model by Rana et al. [20]. The effect of generalized Fourier and Fick laws combined with temperature-dependent thermal conductivity on 3D MHD second-grade nanofluid is considered by Ramzan et al. [21]. Yuan et al. [22] established the numerical solution of MHD nanofluid flow past a baffled U-shaped enclosure by engaging the KKL technique. The 3D flow of MHD nanofluid with varied nanoparticles including Fe_3O_4, Cu, Al_2O_3 and TiO_2 and water as the base fluid past an exponential stretched surface was discussed by Jusoh et al. [23]. Some recent investigations highlighting the importance of MHD may also be found in References [24–27].

The Newtonian heating or the conjugate convective flow is termed as a direct proportionate amid the local temperature and heat transfer rate. The role of Newtonian heating is pivotal in many processes such as heat exchanger's designing, convective flows where heat is taken from the solar radiators, and conjugate heat transfer around the fins etc. The four distinct categories of the heat transfer viz. (i) conjugate boundary condition, (ii) constant or prescribed surface heat flux, (iii) Newtonian heating, and (iv) constant or prescribed surface temperature, are given by Merkin [28]. The flow of Micropolar fluid with Newtonian heating and chemical reaction past a stretching/shrinking sheet was studied by Kamran and Wiwatanapataphee [29]. Mehmood et al. [30] examined the Oldroyd-B nanofluid flow with a transverse magnetic field and Newtonian heating. The flow of Tangent hyperbolic nanofluid under the influences of Newtonian heating, MHD and bi-convection was examined by Shafiq et al. [31]. The flow of nanofluid which is an amalgamation of the base fluid (Sodium alginate) and nanoparticles (Silver, Titanium oxide, Copper and Aluminum oxide) with effects of radiation and Newtonian heating past an isothermal vertical plate are scrutinized by Khan et al. [32]. Many researchers have undertaken the impact of Newtonian heating owing to its wide-ranging practical applications [33–36].

A close review of the literature specifies that copious research is done on the subject of nanofluids with linear/nonlinear stretching sheets in comparison to the flows past the curved stretched surfaces. This topic gets even more limited if we talk about the flows of nanofluids past cylinders stretched in a nonlinear way. So, our prime goal here is to ponder the nanofluid flow past a nonlinear stretched cylinder with nonlinear thermal radiation, h-h reactions, and Newtonian heating. The nanofluid used here is the mixture of water and silver. The numerical simulations are conducted for the proposed problem using the Runge–Kutta method by shooting technique. To corroborate the presented results, a comparison with an already published article is done and an excellent correlation between the two results is found.

2. Mathematical Modeling

Here, we assume a silver–water nanofluid incompressible flow with impacts of h-h reactions, nonlinear thermal radiation and Newtonian heating over a horizontal cylinder which is stretched in a nonlinear way. The magnetic field $B = B_0 x^{(n-1)/2}$ is applied in the radial direction. Owing to our assumption of a small Reynolds number, the induced magnetic field is overlooked (Figure 1).

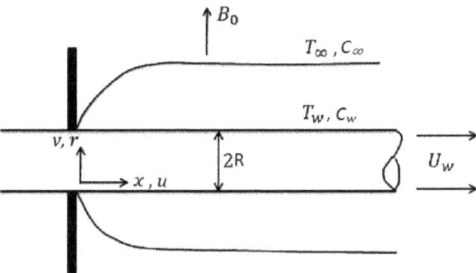

Figure 1. Diagram of the flow geometry.

The homogeneous reaction for cubic autocatalysis can be written as:

$$A_1 + 2B_1 \rightarrow 3B_1, \text{ rate } = k_c a b^2, \tag{1}$$

$$A_1 \rightarrow B_1, \text{ rate } = k_s a. \tag{2}$$

The reaction rate dies out in the outer boundary layer. The system of boundary layer equations of the subject model is given as:

$$\frac{\partial(ru)}{\partial x} + \frac{\partial(rv)}{\partial r} = 0, \tag{3}$$

$$u\frac{\partial u}{\partial x} + v\frac{\partial u}{\partial r} = \nu_{nf}\left(\frac{\partial^2 u}{\partial r^2} + \frac{1}{r}\frac{\partial u}{\partial r}\right) - \frac{\sigma_{nf} B^2(x)}{\rho_{nf}} u, \tag{4}$$

$$u\frac{\partial T}{\partial x} + v\frac{\partial T}{\partial r} = \alpha_{nf}\left(\frac{\partial^2 T}{\partial r^2} + \frac{1}{r}\frac{\partial T}{\partial r}\right) - \frac{1}{(\rho c_p)_{nf}}\frac{\partial q_r}{\partial r}, \tag{5}$$

$$u\frac{\partial a}{\partial x} + v\frac{\partial a}{\partial r} = D_A\left(\frac{\partial^2 a}{\partial r^2} + \frac{1}{r}\frac{\partial a}{\partial r}\right) - k_1 a b^2, \tag{6}$$

$$u\frac{\partial b}{\partial x} + v\frac{\partial b}{\partial r} = D_B\left(\frac{\partial^2 b}{\partial r^2} + \frac{1}{r}\frac{\partial b}{\partial r}\right) + k_1 a b^2, \tag{7}$$

Accompanied by the conditions:

$$u|_{r=R} = U_w(x),\ v|_{r=R} = 0,\ \frac{\partial T}{\partial r}\Big|_{r=R} = h_s T,\ D_A \frac{\partial a}{\partial r}\Big|_{r=R} = k_s a,\ D_B \frac{\partial b}{\partial r}\Big|_{r=R} = -k_s a, \tag{8}$$
$$u|_{r\to\infty} \to 0,\ T|_{r\to\infty} \to T_\infty,\ a|_{r\to\infty} \to a_0,\ b|_{r\to\infty} \to 0,$$

where $U_w(x) = U_0 x^n$, and $q_r = \frac{4\sigma^*}{3k^*}\frac{\partial T^4}{\partial r} = \frac{16\sigma^* T^3}{3k^*}\frac{\partial T}{\partial r}$. The numerical values of specific heat, density, and thermal conductivity of silver (Ag) and water (H$_2$O) are given in Table 1.

Table 1. Thermo-physical characteristics of the base fluid and nanoparticles [1,37].

Physical Properties	Water	Ag
C_p (J/kg K)	4179	235.0
ρ (kg/m^3)	997.1	10,500.0
K (W/mK)	0.61300	429.0

With the following characteristics:

$$\alpha_{nf} = \frac{k_{nf}}{(\rho C_p)_{nf}}, \tag{9}$$

$$\frac{(\rho C_p)_{nf}}{(\rho C_p)_f} = (1 - \phi) + \phi \frac{(\rho C_p)_s}{(\rho C_p)_f}, \tag{10}$$

$$\frac{\mu_{nf}}{\mu_f} = (1.005 + 0.497\phi - 0.1149\phi^2), \tag{11}$$

$$\frac{\rho_{nf}}{\rho_f} = 1 - \phi + \phi \frac{\rho_s}{\rho_f}, \tag{12}$$

$$\frac{k_{nf}}{k_f} = (0.9692\phi + 0.9508), \tag{13}$$

The use of under-mentioned similarity transformations

$$\eta = \frac{r^2 - R^2}{2R}\sqrt{\frac{U_w}{\nu_f x}}, \quad \psi = \sqrt{U_w \nu_f x} R f(\eta), \quad \theta = \frac{T - T_\infty}{T_\infty} \tag{14}$$
$$a = a_0 h, b = a_0 g.$$

Satisfy the Equation (3) and convert the Equations (4)–(7) in non-dimensional form

$$(1 + 2\gamma\eta)f''' + 2\gamma f'' + (1.005 + 0.497\phi - 0.1149\phi^2)(1 - \phi + \phi\frac{\rho_s}{\rho_f})\left(\left(\frac{n+1}{2}\right)ff'' - nf'^2\right) \tag{15}$$
$$- (1.005 + 0.497\phi - 0.1149\phi^2)Mf' = 0,$$

$$(1 + 2\gamma\eta)\left(\frac{k_{nf}}{k_f} + (1 + (N_r - 1)\theta)^3\right)\theta'' + \left(\frac{1}{2K^*}\frac{k_{nf}}{k_f} + (1 + (N_r - 1)\theta)^3\right)\gamma\theta' + \frac{Pr}{4K^*}$$
$$\left(1 - \phi + \phi\frac{(\rho C_p)_s}{(\rho C_p)_f}\right)\left(\frac{n+1}{2}f\theta' - nf'\theta\right) + 2(1 + 2\gamma\eta)(N_r - 1)(1 + (N_r - 1)\theta)^2\theta'^2 = 0, \tag{16}$$

$$\frac{1}{Sc}(1 + 2\gamma\eta)h'' + 2\gamma h') + fh' - \frac{2K}{n+1}hg^2 = 0, \tag{17}$$

$$\frac{1}{Sc}(1 + 2\gamma\eta)g'' + 2\gamma g') + fg' - \frac{2K}{n+1}hg^2 = 0, \tag{18}$$

Supported by the boundary conditions

$$f(\eta) = 0, \ f'(\eta) = 1, \ \theta'(\eta) = -\lambda(1 + \theta(\eta)), \ h'(\eta) = K_s h(\eta), \ \delta g'(\eta) = -K_s h(\eta) \text{ as } \eta = 0, \tag{19}$$
$$\theta(\eta) \to 0, \ f'(\eta) \to 0, \ h(\eta) \to 1, \ g(\eta) \to 0, \text{ at } \eta \to \infty,$$

where $\lambda = h_s \left(\nu_f x/U_w\right)^{1/2}$, $M = \frac{\sigma_f B_0^2}{U_0 \rho_f}$, $K^* = \frac{16\sigma^* T_\infty^3}{3k^* k_f}$, $\gamma = \left(\frac{x\nu_f}{R^2 U_w}\right)^{1/2}$, $N_r = T_w/T_\infty$, $K = \frac{a_0^2 x k_1}{U_w}$, $K_s = \frac{k_s}{D_A}\sqrt{\frac{x\nu_f}{U_w}}$ and $S_c = \frac{\nu_f}{D_A}$.

Here, it is expected that A_1 and B_1 are equivalent. From this assumption, it is inferring that D_A and D_B (diffusion coefficients) are equal i.e., $\delta = 1$, and on account of this supposition, we have

$$g(\eta) + h(\eta) = 1. \tag{20}$$

Equations (17) and (18) after the use of Equation (20) and the relevant boundary conditions take the shape

$$\frac{1}{S_c}((1+2\gamma\eta)h'' + 2\gamma h') + fh' - \frac{2k_1}{n+1}h(1-h)^2 = 0, \tag{21}$$

$$h'(0) = k_2 h(0), h(\infty) \to 1. \tag{22}$$

The physical quantities like Skin friction factor and Local Nusselt number in non-dimensional form are labelled as

$$C_f = \frac{2\tau_w}{\rho_f u_w^2}, Nu_x = \frac{xq_w}{k_f(T_f - T_\infty)}, \tag{23}$$

With τ_w and q_w given by

$$\tau_w = \mu_{nf}\frac{\partial u}{\partial r}\bigg|_{r=R}, q_w = -\left(\frac{\partial T}{\partial r}\right)_{r=R} k_{nf} + (q_r)_{r=R}. \tag{24}$$

Equation (23), after the use of Equations (14) and (24), takes the form

$$\begin{aligned}Re_x^{1/2}C_f &= \left(\frac{1}{1.005+0.497\phi-0.1149\phi^2}\right)f''(0),\\ Re_x^{-1/2}Nu_x &= \frac{k_{nf}}{k_f}\lambda\left(1 + \frac{1}{\theta(0)}\right).\end{aligned} \tag{25}$$

3. Numerical Scheme

The numerical solution of Equations (15), (16) and (21) supported by the boundary conditions (19) and (22) is found by the Shooting scheme. In the calculation of the numerical solution of the problem, the second and third order differential equations are transformed to first order with the help of new parameters. The selection of the initial guess estimate is pivotal in the Shooting scheme as it needs to satisfy the equation and the boundary conditions asymptotically. We have selected the tolerance as 10^{-7} for this specific problem. The first order system obtained in this regard is appended below:

$$f(\eta) = y(1), \tag{26}$$

$$f'(\eta) = y(2), \tag{27}$$

$$f''(\eta) = y(3), \tag{28}$$

$$f'''(\eta) = F(3) = -\frac{1}{(1+2\gamma\eta)}\left[\begin{array}{c}(1.005+0.497\phi-0.1149\phi^2)(1-\phi+\phi\frac{\rho_s}{\rho_f})\left(\left(\frac{n+1}{2}\right)ff'' - nf'^2\right)\\ +2\gamma f'' - (1.005+0.497\phi-0.1149\phi^2)Mf'\end{array}\right], \tag{29}$$

$$\theta(\eta) = y(4), \tag{30}$$

$$\theta'(\eta) = y(5), \tag{31}$$

$$\theta''(\eta) = F(5) = -\frac{1}{(1+2\gamma\eta)\left(\frac{k_{nf}}{k_f}+(1+(N_r-1)\theta)^3\right)}\left[\begin{array}{c}\left(\frac{1}{2K^*}\frac{k_{nf}}{k_f}+(1+(N_r-1)\theta)^3\right)\gamma\theta'\\ +\frac{Pr}{4K^*}\left(1-\phi+\phi\frac{(\rho C_p)_s}{(\rho C_p)_f}\right)\left(\begin{array}{c}\frac{n+1}{2}f\theta'\\ -nf'\theta\end{array}\right)\\ +2(1+2\gamma\eta)(N_r-1)(1+(N_r-1)\theta)^2\theta'^2\end{array}\right], \tag{32}$$

$$h(\eta) = y(6), \tag{33}$$

$$h'(\eta) = y(7), \tag{34}$$

$$h''(\eta) = F(7) = -\frac{1}{(1+2\gamma\eta)}\left[2\gamma h' + S_c fh' - S_c\frac{2k_1}{n+1}h(1-h)^2\right], \tag{35}$$

and the boundary condition becomes

$$y_0(1) = 0, \; y_0(2) = 1, \; y_0(5) = -\lambda(1 + y_0(4)), \; y_0(7) = k_2 y_0(6),$$
$$y_{\inf}(2) \to 0, \; y_{\inf}(4) \to 0, \; y_{\inf}(6) \to 0. \tag{36}$$

We have chosen $\eta_\infty = 6$, that guarantees every numerical solution's asymptotic value accurately.

Here, Table 2 depicts the comparative estimates of the present model with Qasim et al. [38] in the limiting case. Both results are found in an excellent correlation.

Table 2. Nusselt number $(Re_x^{-1/2} Nu_x)$ for numerous estimates of γ and Pr with $M = 0$, $\phi = 0.0$, $\lambda = 0.0$.

γ	Pr	$Re_x^{-1/2} Nu_x$	
		Qasim et al. [38]	Present Result
0.0	0.72	1.23664	1.236651
	1.0	1.00000	1.000000
	6.7	0.33330	0.333310
	10	0.26876	0.268770
1.0	0.72	0.87018	0.870190
	1.0	0.74406	0.744070
	6.7	0.29661	0.296620
	10	0.24217	0.242180

4. Results and Discussion

In this section, we have plotted the Figures 2–11 that exhibit the influences of various parameters like volume fraction ($0.0 \leq \phi \leq 0.3$), magnetic parameter ($1.0 \leq M \leq 4.0$), nonlinearity exponent ($1.0 \leq n \leq 5.0$), curvature parameter ($0.0 \leq \gamma \leq 0.4$), conjugate parameter ($0.4 \leq \lambda \leq 0.7$), radiation parameter ($0.7 \leq K^* \leq 1.0$), strength of homogeneous reaction ($0.1 \leq \kappa_1 \leq 1.8$), strength of heterogeneous reaction ($0.1 \leq \kappa_2 \leq 1.8$) and Schmidt number ($3.0 \leq S_c \leq 4.5$) on involved distributions. Figures 2–5 are drawn to portray the influence of ϕ, M, n and γ on axial velocity. The impact of ϕ is discussed in Figure 2. For growing values of ϕ, the axial velocity also augments. In Figure 3, the influence of M versus the velocity field is debated. It is perceived that the velocity field deteriorates for escalated values of M. The Lorentz force is enforced by the strong magnetic field that hinders the fluid's velocity. Figure 4 illustrates the impression of n on the axial velocity. Reduced velocity is witnessed for larger values of n. This is because higher values of n create more collision between the particles of the fluid that hinder the fluid flow and feeble velocity if perceived. In Figures 5 and 6, the behavior of the velocity and temperature fields for increasing values of γ is given. It is seen that the velocity and temperature of the fluid augment for growing estimates of γ. Larger values of γ mean a smaller radius, comparatively minimum contact region of the cylinder with the fluid and increased heat transport. That is why augmented velocity and temperature are witnessed. Figure 7 is drawn to depict the relation between the temperature of the fluid and the λ. It is detected that temperature enhances for improved values of λ. In fact, the sturdier heat transfer process is observed for larger values of λ as more heat is moved from the cylinder to the fluid. Remembering that $\lambda = 0$ means the insulated walls and $\lambda \to \infty$ constant wall temperature. Figures 8 and 9 are sketched to elaborate the influences of K^* and M on the temperature field. It is clearly perceived that temperature enhances when both K^* and M increase. The decrease in the mean absorption coefficient represents an enriched heat transfer rate and ultimately temperature is enhanced. Similarly, a stronger Lorentz force hinders the movement of the fluid, thus causing more collision between the molecules of the fluid that turns into the improved temperature. In Figures 10 and 11, the behavior of the concentration profile versus h-h reactions is depicted. The concentration diminishes for growing values of h-h reactions.

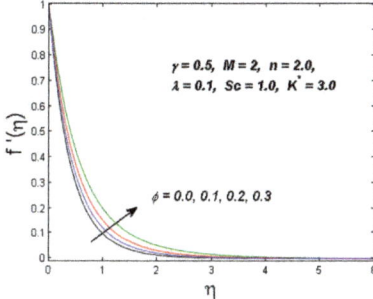

Figure 2. Diagram of axial velocity versus ϕ.

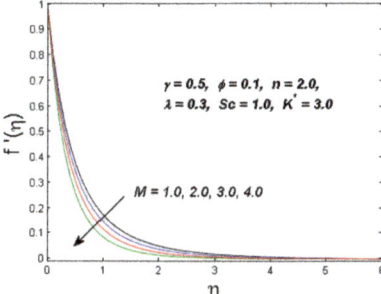

Figure 3. Diagram of axial velocity versus M.

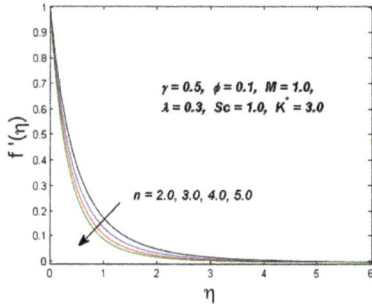

Figure 4. Diagram of axial velocity versus n.

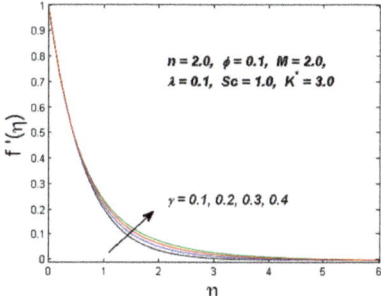

Figure 5. Diagram of axial velocity versus γ.

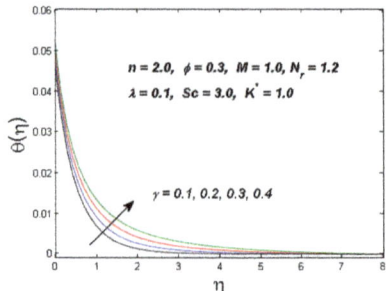

Figure 6. Diagram of the temperature field versus.

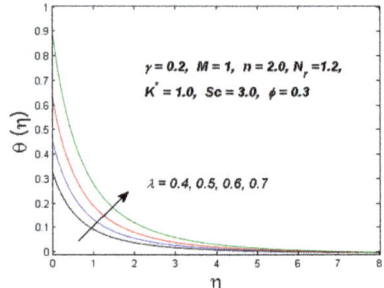

Figure 7. Diagram of the temperature field versus λ.

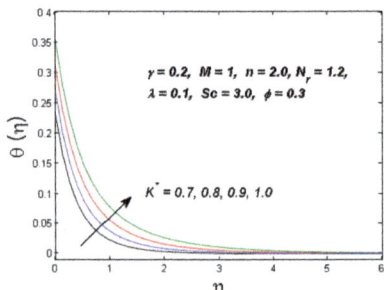

Figure 8. Diagram of the temperature field versus K^*.

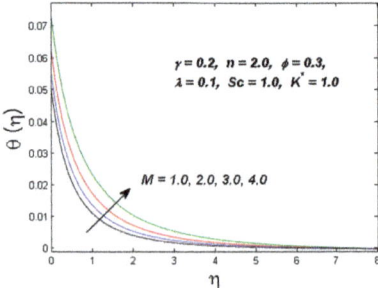

Figure 9. Diagram of the temperature field versus M.

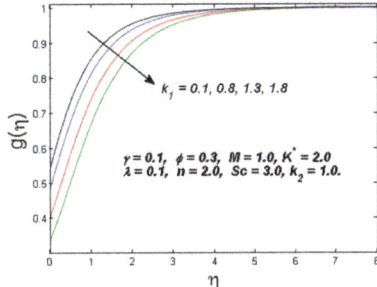

Figure 10. Diagram of concentration field versus k_1.

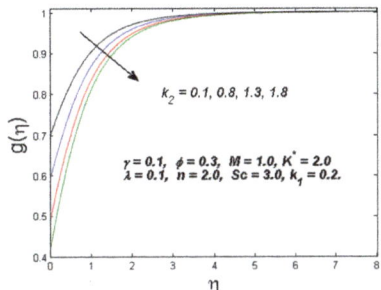

Figure 11. Diagram of concentration field versus k_2.

Table 3 depicts the numerical values of skin friction ($-\text{Re}_x^{1/2}C_f$) and local Nusselt number ($\text{Re}_x^{-1/2}Nu_x$) for numerous estimates of thee parameters. It is perceived that skin friction and Nusselt number upsurge for growing values of solid volume friction ϕ and M (magnetic parameter). It is also observed that for escalated values of curvature parameter γ and nonlinearity parameter n, both skin friction and local Nusselt number show opposite behavior. Moreover, Skin friction and local Nusselt number exhibit a constant trend for values of the temperature ratio parameter N_r and radiation parameter K^*.

Table 3. Numerical values of $-\text{Re}_x^{1/2}C_f$ and $\text{Re}_x^{-1/2}Nu_x$ for Ag–water with Pr = 6.2.

n	γ	ϕ	M	N_r	K^*	$-\text{Re}_x^{1/2}C_f$	$\text{Re}_x^{-1/2}Nu_x$
1.0	0.1	0.1	1.0	1.2	1.0	1.92700	0.32149
2.0	0.1	0.1	1.0	1.2	1.0	2.45720	0.30963
3.0	0.1	0.1	1.0	1.2	1.0	2.89050	0.30305
2.0	1.0	0.1	1.0	1.2	1.0	2.81410	0.28200
2.0	2.0	0.1	1.0	1.2	1.0	3.18730	0.26410
2.0	3.0	0.1	1.0	1.2	1.0	3.54330	0.25369
2.0	1.0	0.1	1.0	1.2	1.0	2.81410	0.28200
2.0	1.0	0.2	1.0	1.2	1.0	3.42830	0.31250
2.0	1.0	0.3	1.0	1.2	1.0	4.04270	0.34319
2.0	1.0	0.1	1.0	1.2	1.0	2.81410	0.28200
2.0	1.0	1.0	2.0	1.2	1.0	2.99960	0.28387
2.0	1.0	1.0	3.0	1.2	1.0	3.17070	0.28553
2.0	1.0	1.0	1.0	0.1	1.0	2.81410	0.28200
2.0	1.0	1.0	1.0	0.7	1.0	2.81410	0.28200
2.0	1.0	1.0	1.0	1.0	1.0	2.81410	0.28200
2.0	1.0	1.0	1.0	1.2	1.0	2.81410	0.28200
2.0	1.0	1.0	1.0	1.2	2.0	2.81410	0.28200
2.0	1.0	1.0	1.0	1.2	3.0	2.81410	0.28200

5. Final Comments

In the present exploration, we have pondered over the nanofluid flow (with base fluid water and nanoparticles as silver) past a nonlinear stretching cylinder with impacts of h-h reactions and nonlinear thermal radiation. Additional effects of Newtonian heating and magnetohydrodynamics have also been taken into account. A numerical solution of the dimensionless mathematical model is achieved via the shooting scheme. The core outcomes of the present study are as follows:

- The temperature profile is a growing function of radiation and magnetic parameters.
- For larger values of the curvature parameter, augmented velocity is observed.
- The concentration of the fluid decreases for growing values of homogeneous–heterogeneous reactions.
- For escalated values of the magnetic parameter, velocity and temperature distributions show the opposite trend.
- The skin friction and local Nusselt number show opposite behavior for curvature and nonlinearity parameters.

Author Contributions: Data curation, J.D.C.; Formal analysis, S.A.; Funding acquisition, M.S.; Investigation, D.L.; Methodology, T.M.; Resources, J.D.C.; Software, M.S.; Supervision, M.R.; Validation, S.A.; Visualization, T.M.; Writing—original draft, M.R.; Writing—review & editing, D.L.

Funding: This research was funded by National Natural Science Foundation China (No. 11571140 and 11671077), Faculty of Science, Jiangsu University, Zhenjiang, China.

Acknowledgments: This research was funded by National Natural Science Foundation China (No. 11571140 and 11671077), Faculty of Science, Jiangsu University, Zhenjiang, China.

Conflicts of Interest: Authors have no conflict of interest regarding this publication.

Nomenclature

U	along x-axis fluid velocity [m/s]
V	along r-axis fluid velocity [m/s]
U_w	Stretching velocity [m/s]
U_e	Free stream velocity [m/s]
M	Magnetic parameter
ρ_f, ρ_s	Density of fluid and solid particle respectively [kg/m^3]
ϕ	nanofluid volume fraction
τ_w	surface shear stress [N/m^2]
T_∞	Ambient temperature [K]
λ	conjugate parameter
σ_{nf}	electric conductivity of fluid and nanofluid respectively [S/m]
γ	curvature parameter
k_f	thermal conductivities of fluid [J/mKs^{n-2}]
k_s	thermal conductivities of nanomaterial [J/mKs^{n-2}]
δ	ratio of mass diffusion coefficients
q_r	radiative heat flux [kg/m^2]
h_f	convective heat transfer coefficient
C_f	Skin friction coefficient
Nu_x	Nusselt number
T	Temperature [K]
T_f	convective fluid temperature [K]
h_s	heat transfer coefficient
μ_{nf}	Nanofluid dynamic viscosity [kg/ms^{n-2}]
Re_x	local Reynolds number
k^*	mean absorption coefficient
σ^*	Stefan-Boltzmann constant

D_A, D_B	diffusion coefficients [m^2/s]
q_w	surface heat flux [W/m^2]
S_c	Schmidt number
k_c, k_s	Rate constants
K^*	Radiation parameter
μ_f	Fluid dynamic viscosity [kg/ms^{n-2}]
α_{nf}	Nanofluid thermal diffusivity [m^2/s^{n-2}]
k_1	Strength of homogeneous reaction
ν_{nf}	nanofluid kinematic viscosity [m^2/s]
k_2	strength of heterogeneous reaction
R	radius of cylinder [m]
B	Magnetic field strength [A/m]
N_r	temperature ratio parameter
A_1, B_1	concentrations of chemical species a, b
n	Nonlinearity exponent

References

1. Choi, S.U.S. Enhancing conductivity of fluids with nanoparticles, ASME Fluid Eng. *Division* **1995**, *231*, 99–105.
2. Minkowycz, W.J.; Sparrow, E.M.; Abraham, J.P. (Eds.) *Nanoparticle Heat Transfer and Fluid Flow*; CRC Press: Boca Raton, FL, USA, 2012; Volume 4.
3. Kakac, S.; Pramuanjaroenkij, A. Review of convective heat transfer enhancement with nanofluids. *Int. J. Heat Mass Transf.* **2009**, *52*, 3187–3196. [CrossRef]
4. Li, Z.; Sheikholeslami, M.; Shafee, A.; Ramzan, M.; Kandasamy, R.; Al-Mdallal, Q.M. Influence of adding nanoparticles on solidification in a heat storage system considering radiation effect. *J. Mol. Liq.* **2019**, *273*, 589–605. [CrossRef]
5. Sheikholeslami, M.; Shafee, A.; Ramzan, M.; Li, Z. Investigation of Lorentz forces and radiation impacts on nanofluid treatment in a porous semi annulus via Darcy law. *J. Mol. Liq.* **2018**, *272*, 8–14. [CrossRef]
6. Lu, D.; Ramzan, M.; Ahmad, S.; Chung, J.D.; Farooq, U. Upshot of binary chemical reaction and activation energy on carbon nanotubes with Cattaneo-Christov heat flux and buoyancy effects. *Phys. Fluids* **2017**, *29*, 123103. [CrossRef]
7. Ramzan, M.; Ullah, N.; Chung, J.D.; Lu, D.; Farooq, U. Buoyancy effects on the radiative magneto Micropolar nanofluid flow with double stratification, activation energy and binary chemical reaction. *Sci. Rep.* **2017**, *7*, 12901. [CrossRef] [PubMed]
8. Suleman, M.; Ramzan, M.; Zulfiqar, M.; Bilal, M.; Shafee, A. Entropy analysis of 3D non-Newtonian MHD nanofluid flow with nonlinear thermal radiation past over exponential stretched surface. *Entropy* **2018**, *20*, 930. [CrossRef]
9. Farooq, U.; Lu, D.C.; Ahmed, S.; Ramzan, M.; Chung, J.D.; Chandio, F.A. Computational analysis for mixed convective flows of viscous fluids with nanoparticles. *J. Therm. Sci. Eng. Appl.* **2019**, *11*, 021013. [CrossRef]
10. Ramzan, M.; Sheikholeslami, M.; Saeed, M.; Chung, J.D. On the convective heat and zero nanoparticle mass flux conditions in the flow of 3D MHD Couple Stress nanofluid over an exponentially stretched surface. *Sci. Rep.* **2019**, *9*, 562. [CrossRef] [PubMed]
11. Ramzan, M.; Bilal, M.; Chung, J.D.; Mann, A.B. On MHD radiative Jeffery nanofluid flow with convective heat and mass boundary conditions. *Neural Comput. Appl.* **2019**, *30*, 2739–2748. [CrossRef]
12. Ramzan, M.; Sheikholeslami, M.; Chung, J.D.; Shafee, A. Melting heat transfer and entropy optimization owing to carbon nanotubes suspended Casson nanoliquid flow past a swirling cylinder—A numerical treatment. *AIP Adv.* **2018**, *8*, 115130. [CrossRef]
13. Lu, D.; Ramzan, M.; Ullah, N.; Chung, J.D.; Farooq, U. A numerical treatment of radiative nanofluid 3D flow containing gyrotactic microorganism with anisotropic slip, binary chemical reaction and activation energy. *Sci. Rep.* **2017**, *7*, 17008. [CrossRef] [PubMed]

14. Ramzan, M.; Bilal, M.; Kanwal, S.; Chung, J.D. Effects of variable thermal conductivity and non-linear thermal radiation past an Eyring Powell nanofluid flow with chemical reaction. *Commun. Theor. Phys.* **2017**, *67*, 723. [CrossRef]
15. Muhammad, T.; Lu, D.C.; Mahanthesh, B.; Eid, M.R.; Ramzan, M.; Dar, A. Significance of Darcy-Forchheimer porous medium in nanofluid through carbon nanotubes. *Commun. Theor. Phys.* **2018**, *70*, 361. [CrossRef]
16. Azam, M.; Shakoor, A.; Rasool, H.F.; Khan, M. Numerical simulation for solar energy aspects on unsteady convective flow of MHD Cross nanofluid: A revised approach. *Int. J. Heat Mass Transf.* **2019**, *131*, 495–505. [CrossRef]
17. Sheikholeslami, M.; Kataria, H.R.; Mittal, A.S. Effect of thermal diffusion and heat-generation on MHD nanofluid flow past an oscillating vertical plate through porous medium. *J. Mol. Liq.* **2018**, *257*, 12–25. [CrossRef]
18. Makinde, O.D.; Animasaun, I.L. Bioconvection in MHD nanofluid flow with nonlinear thermal radiation and quartic autocatalysis chemical reaction past an upper surface of a paraboloid of revolution. *Int. J. Therm. Sci.* **2016**, *109*, 159–171. [CrossRef]
19. Lu, D.; Ramzan, M.; Bilal, M.; Chung, J.D.; Farooq, U. A numerical investigation of 3D MHD rotating flow with binary chemical reaction, activation energy and non-Fourier heat flux. *Commun. Theor. Phys.* **2018**, *70*, 089. [CrossRef]
20. Rana, P.; Shukla, N.; Gupta, Y.; Pop, I. Analytical prediction of multiple solutions for MHD Jeffery–Hamel flow and heat transfer utilizing KKL nanofluid model. *Phys. Lett. A* **2019**, *383*, 176–185. [CrossRef]
21. Ramzan, M.; Bilal, M.; Chung, J.D.; Lu, D.; Farooq, U. Impact of generalized Fourier's and Fick's laws on MHD 3D second grade nanofluid flow with variable thermal conductivity and convective heat and mass conditions. *Phys. Fluids* **2017**, *29*, 093102. [CrossRef]
22. Yuan, M.; Mohebbi, R.; Rashidi, M.M.; Yang, Z.; Sheremet, M.A. Numerical study of MHD nanofluid natural convection in a baffled U-shaped enclosure. *Int. J. Heat Mass Transf.* **2019**, *130*, 123–134.
23. Rahimah, J.; Nazar, R.; Pop, I. Magnetohydrodynamic boundary layer flow and heat transfer of nanofluids past a bidirectional exponential permeable stretching/shrinking sheet with viscous dissipation effect. *J. Heat Transf.* **2019**, *141*, 012406.
24. Benos, L.; Sarris, I.E. Analytical study of the magnetohydrodynamic natural convection of a nanofluid filled horizontal shallow cavity with internal heat generation. *Int. J. Heat Mass Transf.* **2019**, *130*, 862–873. [CrossRef]
25. Sajjadi, H.; Delouei, A.A.; Izadi, M.; Mohebbi, R. Investigation of MHD natural convection in a porous media by double MRT lattice Boltzmann method utilizing MWCNT–Fe_3O_4/water hybrid nanofluid. *Int. J. Heat Mass Transf.* **2019**, *132*, 1087–1104. [CrossRef]
26. Zhao, G.; Wang, Z.; Jian, Y. Heat transfer of the MHD nanofluid in porous microtubes under the electrokinetic effects. *Int. J. Heat Mass Transf.* **2019**, *130*, 821–830. [CrossRef]
27. Ibrahim, M.G.; Hasona, W.M.; ElShekhipy, A.A. Concentration-dependent viscosity and thermal radiation effects on MHD peristaltic motion of Synovial Nanofluid: Applications to rheumatoid arthritis treatment. *Comput. Methods Programs Biomed.* **2019**, *170*, 39–52. [CrossRef] [PubMed]
28. Merkin, J.H. Natural-convection boundary-layer flow on a vertical surface with Newtonian heating. *Int. J. Heat Fluid Flow* **1994**, *15*, 392–398. [CrossRef]
29. Kamran, M.; Wiwatanapataphee, B. Chemical reaction and Newtonian heating effects on steady convection flow of a micropolar fluid with second order slip at the boundary. *Eur. J. Mech. B/Fluids* **2018**, *71*, 138–150. [CrossRef]
30. Mehmood, R.; Rana, S.; Nadeem, S. Transverse thermopherotic MHD Oldroyd-B fluid with Newtonian heating. *Results Phys.* **2018**, *8*, 686–693. [CrossRef]
31. Shafiq, A.; Hammouch, Z.; Sindhu, T.N. Bioconvective MHD flow of tangent hyperbolic nanofluid with Newtonian heating. *Int. J. Mech. Sci.* **2017**, *133*, 759–766. [CrossRef]
32. Khan, A.; Khan, D.; Khan, I.; Ali, F.; Karim, F.U.; Imran, M. MHD flow of Sodium Alginate-based Casson type nanofluid passing through a porous medium with Newtonian heating. *Sci. Rep.* **2018**, *8*, 8645. [CrossRef] [PubMed]
33. El-Hakiem, M.A.; Ramzan, M.; Chung, J.D. A numerical study of magnetohydrodynamic stagnation point flow of nanofluid with Newtonian heating. *J. Comput. Theor. Nanosci.* **2016**, *13*, 8419–8426. [CrossRef]

34. Ramzan, M.; Yousaf, F. Boundary layer flow of three-dimensional viscoelastic nanofluid past a bi-directional stretching sheet with Newtonian heating. *AIP Adv.* **2015**, *5*, 057132. [CrossRef]
35. Shehzad, S.A.; Hussain, T.; Hayat, T.; Ramzan, M.; Alsaedi, A. Boundary layer flow of third grade nanofluid with Newtonian heating and viscous dissipation. *J. Cent. South Univ.* **2015**, *22*, 360–367. [CrossRef]
36. Ramzan, M. Influence of Newtonian heating on three dimensional MHD flow of couple stress nanofluid with viscous dissipation and Joule heating. *PLoS ONE* **2015**, *10*, e0124699. [CrossRef] [PubMed]
37. Upreti, H.; Pandey, A.K.; Kumar, M. MHD flow of Ag-water nanofluid over a flat porous plate with viscous-Ohmic dissipation, suction/injection and heat generation/absorption. *Alex. Eng. J.* **2018**, *57*, 1839–1847. [CrossRef]
38. Qasim, M.; Khan, Z.H.; Khan, W.A.; Shah, I.A. MHD boundary layer slip flow and heat transfer of ferrofluid along a stretching cylinder with prescribed heat flux. *PLoS ONE* **2014**, *9*, e83930. [CrossRef] [PubMed]

© 2019 by the authors. Licensee MDPI, Basel, Switzerland. This article is an open access article distributed under the terms and conditions of the Creative Commons Attribution (CC BY) license (http://creativecommons.org/licenses/by/4.0/).

Article

MHD Flow and Heat Transfer over Vertical Stretching Sheet with Heat Sink or Source Effect

Ibrahim M. Alarifi [1], Ahmed G. Abokhalil [2,3], M. Osman [1,4], Liaquat Ali Lund [5], Mossaad Ben Ayed [6,7], Hafedh Belmabrouk [8,9] and Iskander Tlili [1,*]

1. Department of Mechanical and Industrial Engineering, College of Engineering, Majmaah University, Al-Majmaah 11952, Saudi Arabia; i.alarifi@mu.edu.sa (I.M.A.); m.othman@mu.edu.sa (M.O.)
2. Department of Electrical Engineering, College of Engineering, Majmaah University, Al-Majmaah 11952, Saudi Arabia; a.abokhalil@mu.edu.sa
3. Electrical Engineering Department, Assiut University, Assiut 71515, Egypt
4. Mechanical Design Department, Faculty of Engineering Mataria, Helwan University, Cairo El-Mataria 11724, Egypt
5. Sindh Agriculture University, Tandojam Sindh 70060, Pakistan; balochliaqatali@gmail.com
6. Computer Science Department, College of Science and Humanities at Alghat, Majmaah University, Al-Majmaah 11952, Saudi Arabia; mm.ayed@mu.edu.sa
7. Computer and Embedded System Laboratory, Sfax University, Sfax 3011, Tunisia
8. Electronics and Microelectronics Laboratory, Faculty of Science of Monastir, University of Monastir, Monastir 5019, Tunisia; ha.belmabrouk@mu.edu.sa
9. Department of Physics, College of Science at Zulfi, Majmaah University, Al Zulfi 11932, Saudi Arabia
* Correspondence: l.tlili@mu.edu.sa; Tel.: +96-650-910-7698

Received: 4 January 2019; Accepted: 15 February 2019; Published: 26 February 2019

Abstract: A steady laminar flow over a vertical stretching sheet with the existence of viscous dissipation, heat source/sink, and magnetic fields has been numerically inspected through a shooting scheme based Runge—Kutta–Fehlberg-integration algorithm. The governing equation and boundary layer balance are expressed and then converted into a nonlinear normal system of differential equations using suitable transformations. The impact of the physical parameters on the dimensionless velocity, temperature, the local Nusselt, and skin friction coefficient are described. Results show good agreement with recent researches. Findings reveal that the Nusselt number at the sheet surface augments, since the Hartmann number, stretching velocity ratio A, and Hartmann number Ha increase. Nevertheless, it reduces with respect to the heat generation/absorption coefficient δ.

Keywords: steady laminar flow; nanofluid; heat source/sink; magnetic field; stretching sheet

1. Introduction

The steady laminar flow and heat transfer of a viscous fluid over a vertical stretching sheet with the existence of heat source/sink and magnetic fields has gained significant interest because of its various usages in engineering procedures like geothermal energy extraction, glass fiber, and plasma studies, etc. Several researchers investigated numerically MHD mixed convective stagnation point flow lengthways a perpendicular widening piece in the presence of a heat source/sink in order to evaluate the impacts of relevant physical parameters especially Hartmann number, Stretching velocity ratio and Biot number on velocity and temperature profiles besides to skin friction and heat transfer properties. Using the Runge-Kutta-Fehlberg methods joined with shooting technique [1–4]. P.R. Sharma et al. [3] analyzed numerically the impacts of an external magnetic field. Tarek M. A. El-Mistikawy [5] focused on the flow resulting from a linearly stretching sheet with a transverse magnetic field. A. Mohammadeina et al. [1] investigated the impacts of thermal radiation and magnetic field on flow of CuO-water nanofluid past a stretching sheet characterized by forced-convection with a stagnation point with suction/injection. S.S. Ghadikolaei et al. [4] investigated the impact of thermal radiation and Joule heating. Wubshet Ibrahim [2] studied numerically the melting heat transfer and magneto hydrodynamic stagnation point flow of a nanofluid pasta extending piece.

Homotopy technique and shooting method have been widely used in many studies to obtain exact and wide-ranging analytic solution [6–9]. Arif Hussain et al. [6] focused on the thermal and physical properties of features of MHD hyperbolic refraction of fluid flow over a non-linear widening sheet taking into account convective boundary conditions and viscous dissipation. Tasawar Hayat et al. [9] studied convective stream of ferrofluid owing to nonlinear widening curved slip while M. N. Tufail et al. [7] analyzed numerically the heat transmission ended an unsteady widening sheet for an MHD Casson fluid through viscous dissipation impacts using the homotopy method. M. Y. Malik et al. [10] examined numerically an MHD flow of the Carreau fluid over a stretching sheet with inconstant thickness with the Keller box method. T.M. Agbaje et al. [11] proposed the spectral perturbation technique, which is a sequence development method which spreads the usage of the typical trepidation methods. When joined with the Chebyshev pseudo-spectral process, the SPM can provide higher order approximate mathematical resolutions for intricate increases faced in perturbation patterns. Siti Khuzaimah Soida et al. [12] analyzed numerically a stable MHD flow through a centrifugally widening or lessening floppy using the boundary value problem solver in Matlab software.

The temperature profile behavior was investigated by several researchers. U.S. Mahabaleshwar et al. [13] studied an MHD couple stress liquid caused by a perforated sheet experiencing lined widening with radiation. It was found that the temperature rises as the heat source/sink NI parameter and Chandrasekhar number Q augment [13]. M. Ferdows et al. [14] studied a steady two-dimensional free convective flow of a viscous incompressible fluid lengthways a perpendicular stretching sheet. The temperature profiles increase as the combined effect of porous diffusivity and magnetic field R, perturbation parameter increase. The temperature profile augments with the Eckert number Ec [15] and Biot number [6,13] and the magnetic field parameter, radiation and viscous dissipation [7,8,14–17]. In fact, when variable thickness exists, the radiative heat transfer and viscous dissipation enhance the nanofluid temperature [18]. When the unsteadiness of the stretching sheet extends, temperature rises [19]. The fluid temperature increases with increasing of Brownian motion, thermophoresis [20,21], temperature ratio [1] and heat source/sink parameter [3]. In the other hand, temperature drops when the Casson fluid parameter augments [7] and when the Prandtl number increases [13,20] and as the power law increases [6]. G.S. Seth et al. [19] reported that temperature is reduced when the stretching sheet is nonlinear. The fluid temperature declines with higher values of velocity ratio parameter, suction/injection parameter [1,16], Hartmann number Ha, Prandtl number Pr, thermal stratification parameter [18] or mixed convection parameter k [3]. Whereas, Kai-Long Hsiao [15] found that the temperature impact is improved in parallel with the rise of Prandtl number.

The impacts of significant factors on the solitary velocity for stretching sheet circumstance were discussed in many studies. Decrease in the velocity profile has been reported verses the increase of the Casson fluid parameter [7], non-linearity restriction, flow comportment index, the magnetic field, power law index and Weissenberg number [1,6–8]. However, the impacts of power law index and magnetic field parameter are more significant than nonlinearity parameter and Weissenberg number [6]. The dimensionless velocity is reduced when values of volume fraction and dimensionless velocity slip parameters augment [22]. It declines also when mass concentration parameter and permeability parameter augment [4]. Velocity profile is reduced by the combined effect of porous diffusivity [14], with the increase of Hartmann number and suction/injection parameter [23]. Greater values of suction and stretching parameter m and wall thickness parameter reduce velocity profile while lead to the rise of the volume fraction and porosity parameter [17]. The fluid velocity is improved in parallel with increase of velocity ratio, curvature parameter, electric field parameter and material parameter [4].

M. Ferdows et al. [14] reported that velocity profile is enhanced with the increase of the proportion of the flow velocity constraint to stretching sheet indicator and viscosity ratio. G.S. Seth et al. [24] reported that nanofluid velocity is reduced when the stretching sheet is nonlinear whereas when the unsteadiness of the stretching sheet augments velocity is decreased. M. N. Tufail et al. [7] found that as the unsteadiness parameter growths, the velocity decreases close the sheet and increases distant to the sheet. The higher of the slight order derivative causes the quicker velocity of viscoelastic fluids close the platter [25–32]. The temperature and velocity profile behavior was investigated by several researchers with the effect of heat generation/absorption on MHD flow though rare of them trait the effect of an external magnetic field and heat generation/absorption on mixed convictive flow lengthways a perpendicular extending piece [33–37].

An inclusive analysis of the works about nanofluids is presented by Wang et al. [19,20]. The inactivity point flow of a nanofluid near a stretching sheet has been explored by Khan and Pop [21], Mustafa et al. [22], Nazar et al. [23], and Ibrahim et al. [24]. Nadeem and Haq [29] inspected MHD three-dimensional flow of nanofluids past a shrinking sheet with thermal radiation. They employed Boungiorno model and considered the influences of thermophoresis and Brownian motion on the local Nusselt ad Sherwood numbers.

This study includes a numerically solution of MHD boundary layer flow over a vertical stretching sheet with company of heat sink/ source and magnetic fields effect. The governing equations for the problematic have been clearly described with some appropriate changes and then explained numerically via shooting arrangement based RKFI procedure. In this work the effects of relevant physical indicators essentially Hartmann number, Stretching velocity ratio and Biot number on velocity and temperature distributions in addition to the skin friction and heat transfer properties have been investigated. Comparison of the results with research results [3,26–30] shows a good agreement as shown in Table 1.

Table 1. Numerical results of Nusselt number—$\theta(0)$ for diverse Prandtl number when $A = 1$, $A = 0$, $Ha = 0$, and $\delta = 0$.

Pr	Ramchandran et al. [26]	Hassanien and Gorla [30]	Lok et al. [27]	Ishak et al. [28]	Ali et al. [29]	Sharma et al. [3]	Present
1	-	-	-	0.8708	0.8708	0.8707	0.87078
10	-	1.9446	-	1.9446	1.9448	1.94463	1.94465
20	2.4576	-	2.4577	2.4576	2.4579	2.4576	2.4577
40	3.1011	-	3.1023	3.1011	3.1017	3.1011	3.1015
60	3.5514	-	3.556	3.5514	3.5524	3.55142	3.55148
80	3.9055	-	3.9195	3.9095	3.9108	3.90949	3.90919
100	4.2116	4.2337	4.2289	4.2116	4.2133	4.21163	4.21135

2. Problem Description

The present study deals with a viscous incompressible fluid that was analyzed through steady laminar flow two-dimensional condition lengthways on an upright stretching sheet that was positioned in the x track and the y axis is perpendicular to the plan of this sheet.

u is the velocity component in x direction.

v is the velocity component in y direction.

Taking into consideration that mutually c then a are positive constants; $u = u_e(x) = ax$ represents the unrestricted stream velocity, while $u = u_w(x) = cx$ represents the velocity when there is stretching on the sheet.

When a heat source/sink is present, H_0 is an outside magnetic field that is practically perpendicular to the sheet.

The principal equations of continuity, momentum, and energy are written as

$$\frac{\partial u}{\partial x} + \frac{\partial v}{\partial y} = 0, \tag{1}$$

$$u\frac{\partial u}{\partial x} + v\frac{\partial u}{\partial y} = -\frac{1}{\rho}\frac{dp}{dx} + v\frac{\partial^2 u}{\partial y^2} - \frac{\sigma \mu_e^2 H_0^2}{\rho}u + g\beta(T - T_\infty), \tag{2}$$

$$u\frac{\partial T}{\partial x} + v\frac{\partial T}{\partial y} = \frac{k}{\rho C_p}\frac{\partial^2 T}{\partial y^2}P + \frac{Q}{\rho C_p}(T - T_\infty), \tag{3}$$

where

σ is the electrical conductivity,
μ_e is the magnetic permeability,
T_∞ is the temperature of free stream,
g is acceleration due to gravity,
β is the volumetric coefficient of thermal expansion,
k is the thermal conductivity,
$v(=\mu/\rho)$ is the kinematic viscosity, and
$T_w = T_\infty + bx$ is the temperature of the sheet.

When the superficial is heated $b > 0$ so that cfr4 whereas for cooled superficial $b < 0$ and $T_w \langle T_\infty$. The boundary conditions are

$$\begin{array}{l} v = 0,\ u = u_w(x) = cx,\ -k\frac{\partial T}{\partial y} = h_f(T_f - T)\ at\ y = 0, \\ u = u_e(x) = ax,\ T = T_\infty asy \to \infty, \end{array} \tag{4}$$

The forces will be in equilibrium because of the presence of the hydrostatic and magnetic pressure gradient, as mentioned below

$$-\frac{1}{\rho}\frac{dp}{dx} = u_e\frac{du_e}{dx} + \frac{\sigma \mu_e^2 H_0^2}{\rho}u_e, \tag{5}$$

Therefore, the momentum equation turns into

$$u\frac{\partial u}{\partial x} + v\frac{\partial u}{\partial y} = u_e\frac{du_e}{dx} - \frac{\sigma \mu_e^2 H_0^2}{\rho}(u - u_e) + v\frac{\partial^2 u}{\partial y^2} + g\beta(T - T_\infty). \tag{6}$$

3. Scheme Analysis

The subsequent change and dimensionless quantities are used into equations, while taking into account the mentioned boundary condition (4) within the solution of Equations (1) to (3), Equation (7), obtained as:

$$\eta = \sqrt{\frac{a}{v}}y,\ \psi = x\sqrt{av}f(\eta),\ \theta(\eta) = \frac{T - T_\infty}{T_w - T_\infty} and u = \frac{\partial \psi}{\partial y},\ v = -\frac{\partial \psi}{\partial x} \tag{7}$$

Therefore, the equation of continuity is approved.
The momentum and energy equation becomes

$$f''' + ff'' - (f')^2 + 1 + H_a^2(1 - f') + \lambda\theta = 0 \tag{8}$$

$$\theta'' + P_r(f\theta' - f'\theta + \delta\theta) = 0 \tag{9}$$

where major represents the derivative according to η,
$Ha\left(= \mu_e H_0 \sqrt{\frac{\sigma}{\rho a}}\right)$ is the Hartmann number,
$\lambda\left(= \frac{Gr_x}{Re_x^{\frac{4}{2}}}\right)$ is mixed convection parameter,

$Gr_x \left(= g\beta(T_w - T_\infty)\frac{x^3}{v^2}\right)$ is the local Grashof number,
$Re_x \left(= u_e(x)\frac{x}{v}\right)$ and $P_r\left(\frac{v}{x}\right)$, and
$\delta\left(= \frac{Q}{\rho a C_p}\right)$ represent the factor of heat generation/absorption.
The related boundary circumstances are limited to

$$f = 0, \ f' = c/a = A, \ \theta' = B(\theta - 1) \text{ at } \eta = 0 \qquad (10)$$
$$f' = 1, \ \theta = 0 \text{ as } \eta \to \infty$$

where $A\left(= \frac{c}{a}\right)$ represent the rate of velocity

Also, $B = \frac{h_f}{k}\sqrt{\frac{v}{a}}$.

Therefore, the Nusselt number and Skin friction are expressed by

$$C_f = \frac{2\tau_w}{\rho u_e^2} = Re_x^{-1/2} f''(0), \ Nu_x = \frac{xq_w}{k(T_w - T_\infty)} = -Re_x^{1/2}\theta'(0). \qquad (11)$$

Additionally, respectively, the wall shear stress τ_w and the heat flux q_w are illustrated, as below:

$$\tau_w = \mu \left(\frac{\partial u}{\partial y}\right)_{y=0}, \ q_w = -k\left(\frac{\partial T}{\partial y}\right)_{y=0}$$

It is intricate to get the locked procedure explanations because calculations (8) and (9) are significantly nonlinear. Consequently, schemes are obtained by replacing

$$f = f_1, f' = f_2, f'' = f_3, f''' = f_3', \theta' = f_5, \theta'' = f_5'$$

Thus, the scheme of calculations becomes

$$f_1' = f_2, f_2' = f_3, f_3' = f_2^2 - f_1 f_3 - 1 + Ha^2(f_2 - 1) - \lambda f_4$$
$$f_4' = f_5, f_5' = P_r(f_2 f_4 - f_1 f_5 - \delta f_4)$$

Depending on the next conditions

$$f_1(0) = 0, f_2(0) = A, f_3(0) = s_1, f_4(0) = 1, f_5(0) = s_2 \text{ and } f_2(\infty) = 1, f_4(\infty) = 0$$

In order to obtain stage-by-stage integration and scheming, which are determined by relying on MATLAB software; the Runge–Kutta fourth order method with the shooting technique is actually used.

4. Stability Analysis

When there exists more than one solution in any fluid flow problem, the stability of solution is necessary to perform in that problem. In order to perform stability analysis, we adopt the algorithm of Merkin [33], Weidman et al. [36], and Rosca and Pop [37].

Step 1: To convert the governing Equations (2) and (3) of fluid flow in unsteady form, we have

$$\frac{\partial u}{\partial t} + u\frac{\partial u}{\partial x} + v\frac{\partial u}{\partial y} = -\frac{1}{\rho}\frac{dP}{dx} + v\frac{\partial^2 u}{\partial y^2} - g\beta(T - T_\infty) - \frac{\sigma \mu_e H_0^2 u}{\rho} \qquad (12)$$

$$\frac{\partial T}{\partial t} + u\frac{\partial T}{\partial x} + v\frac{\partial T}{\partial y} = \frac{k}{\rho C_p}\frac{\partial^2 T}{\partial y^2} + \frac{k}{\rho C_p}(T - T_\infty) \qquad (13)$$

Step 2: To introduce a new non-dimensional time variable $\tau = a.t$, and all other similarity variables are also a function of τ, can be written as,

$$\psi = x\sqrt{av}f(\eta,\tau); \eta = y\sqrt{\frac{a}{v}}; \theta(\eta,\tau) = \frac{(T-T_\infty)}{(T_w-T_\infty)}; \tag{14}$$

Step 3: By applying Equation (14) on Equations (12) and (13), we have

$$\frac{\partial^3 f(\eta,\tau)}{\partial \eta^3} + f\frac{\partial^2 f(\eta,\tau)}{\partial \eta^2} - \left(\frac{\partial f(\eta,\tau)}{\partial \eta}\right)^2 + 1 + Ha^2\left(1 - \frac{\partial f(\eta,\tau)}{\partial \eta}\right) - \lambda\theta(\eta,\tau) \\ - \frac{\partial^2 f(\eta,\tau)}{\partial \tau \partial \eta} = 0 \tag{15}$$

$$\frac{1}{Pr}\frac{\partial^2 \theta(\eta,\tau)}{\partial \eta^2} + f(\eta,\tau)\frac{\partial \theta(\eta,\tau)}{\partial \eta} - \frac{\partial f(\eta,\tau)}{\partial \eta}\theta(\eta,\tau) + \delta\theta(\eta,\tau) - \frac{\partial \theta(\eta,\tau)}{\partial \tau} = 0 \tag{16}$$

and the related boundary conditions are

$$f(0,\tau) = 0; \frac{\partial f(0,\tau)}{\partial \eta} = A; \frac{\partial \theta(0,\tau)}{\partial \eta} = B(\theta(0,\tau) - 1), \\ \frac{\partial f(\eta,\tau)}{\partial \eta} \to 1; \theta(\eta,\tau) \to 0 \text{ as } \eta \to \infty \tag{17}$$

Step 4: To check the stability of steady flow solutions $f(\eta) = f_0(\eta)$ and $\theta(\eta) = \theta_0(\eta)$ will satisfy the basic model by introducing the following functions

$$f(\eta,\tau) = f_0(\eta) + e^{-\tau} F(\eta,\tau); \theta(\eta,\tau) = \theta_0(\eta) + e^{-\tau} G(\eta,\tau) \tag{18}$$

Here, $F(\eta,\tau)$, and $G(\eta,\tau)$ are small relative to $f_0(\eta)$, and $\theta_0(\eta)$. The unknown eigenvalue is γ, which is to be found out.

Step 5: By putting Equation (18) into Equations (15) and (16) and keeping $\tau = 0$, we have

$$F_0''' + f_0(\eta)F_0'' + F_0 f_0'' - 2f_0' F_0' - \lambda G_0 - Ha^2 F_0' + \gamma F_0' = 0 \tag{19}$$

$$\frac{1}{Pr}G_0'' + f_0 G_0' + F_0 \theta_0' - f_0' G_0 - F_0' \theta_0 + \delta G_0 + \gamma G_0 = 0 \tag{20}$$

Along with boundary conditions

$$F_0(0) = 0, F_0'(0) = 0, G_0'(0) = BG_0 \\ F_0'(\eta) \to 0, G_0(\eta) \to 0, \text{ as } \eta \to \infty \tag{21}$$

Step 6: To relax one boundary condition into an initial condition, as suggested by Weidman et al. [36] and Harris et al. [34]. In this problem, we relaxed $G_0(\eta) \to 0$, as $\eta \to \infty$ into $G_0'(0) = 1$. We have to solve Equation (19) and (20) with boundary and relaxed initial condition in order to find the values of smallest eigenvalue γ.

It is worth mentioning that the negative values of γ indicate the growth of disturbance and the flow becomes unstable. On the other hand, if the values of γ are positive, which means that the flow is stable and shows an initial decay. The values of smallest eigenvalue are given in Table 2, which indicate that only first (second) solution is stable (unstable).

Table 2. Smallest eigenvalue γ when $\lambda = -0.2$, $Pr = 1$, $A < 0$ (for Shrinking surface) and $A > 0$ (for Stretching surface).

ε	Ha	γ	
		1st Solution	2nd Solution
0.5	0.3	0.97533	−0.09572
	0.5	0.65753	−0.06946
−0.5	0	1.34857	−0.75392
-	0.5	1.02349	−0.58327

The bvp4c solver function has performed stability analysis. According to Rahman et al. [35], "this collocation formula and the collocation polynomial provides a C^1 continuous solution that is fourth order accurate uniformly in [a,b]. Mesh selection and error control are based on the residual of the continuous solution". As we know, only the first solution is the stable and only the stable solution has physical meaning. In these regards, the various effect of different physical parameters on velocity and temperature profiles have been demonstrated for the first solution only. Finally, from Figure 1a,b, we draw some graphs in order to show the existence of multiple solutions for the opposing flow case.

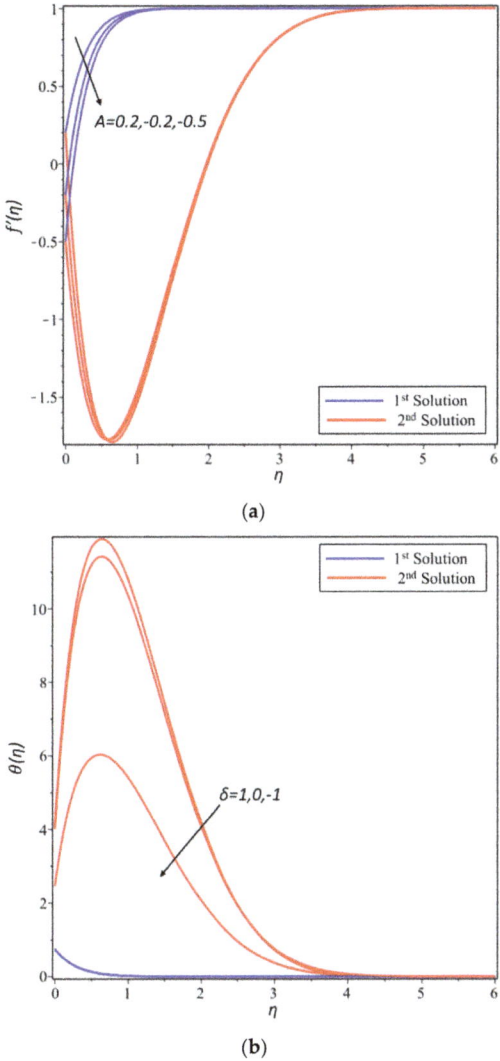

Figure 1. (a) The existence of multiple solutions for the opposing flow case different parameter A; and, (b) the existence of multiple solutions for the opposing flow case different parameter δ.

5. Results and Discussion

A steady laminar flow above a vertical stretching sheet with the existence of viscous dissipation, heat sink/source, and magnetic fields has been numerically explored by using of RKFI process through shooting scheme. In this work, the effects of pertinent physical parameters are essentially Hartmann number, Stretching velocity ratio, and Biot number on temperature and velocity distributions; also, the skin friction and heat transfer properties have been examined.

The effect of Hartmann number on dimensionless velocity for mutually opposing and assisting flow are shown in Figure 2a,b respectively, with supplementary parameters are usually the regularly of velocity ratio parameter $A = 1$, Heat generation/absorption coefficient $\delta = 1$, Prandtl number $Pr = 1$, and Biot number $B_i = 1$. It is perceived that, for assisting flow ($\lambda > 0$), the dimensionless velocity is the maximum at the superficial of the vertical stretching sheet and it increasingly reduces to the minimum value $f' = 1$ as it changes to gone after the superficial, while for opposing flow ($\lambda < 0$), the dimensionless velocity is the lowest at the surface of the vertical stretching sheet and gradually increases to the maximum value $f' = 1$, as it changes away from the superficial, this effect is mathematically obvious in Equation (10). It is additional observed that the velocity profile decreases with the Hartmann number for the assisting flow, whereas, it increases with the Hartmann number for the opposing flow. Consequently, the hydrodynamic boundary layer thickness also depends upon Ha. It is well known that the Hartmann number represents the proportion of electromagnetic force to the viscous force, thus, in the case of assisting flow, increasing the Hartmann number means that electromagnetic force was enhanced when compared to viscous force, which in turn Lorentz force augments, then opposes the flow, and then reduces the velocity profile. Nonetheless, in the situation of opposing flow, there is a reverse effect of the Hartmann number on dimensionless velocity. It should be pointed out that, in the case of assisting flow, ($\lambda > 0$) means the heating of the fluid, therefore the thermal buoyancy forces were enhanced. It can be interpreted on this fact that the highest value of dimensionless velocity is near the stretched surface; however, for opposing flow ($\lambda > 0$), which means that the fluid is consequently cooled; the thermal buoyancy forces decreases and then we realize the lowest value of dimensionless velocity near the stretched surface. It is worthwhile to note that the velocity profile increases with the mixed convection parameter λ for mutually case opposing and assisting flow due to an increasing of the thermal buoyancy forces. It can be seen that, for buoyancy opposed ($\lambda < 0$, opposing), the velocity profile will be significantly affected.

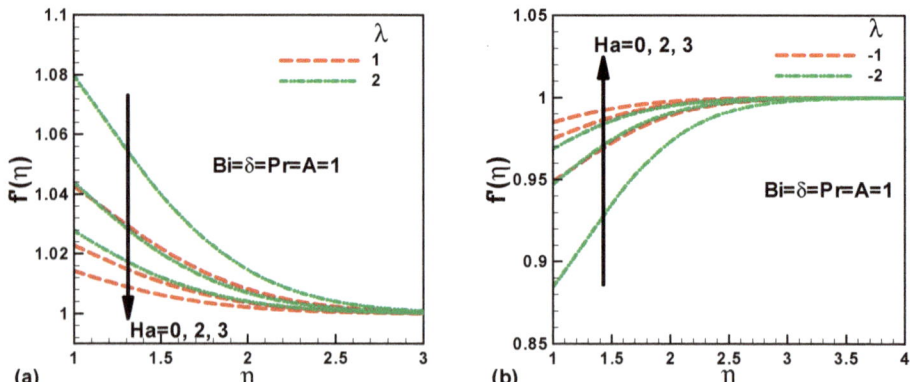

Figure 2. Variation of dimensionless velocity with Hartmann number for (**a**) assisting flow and (**b**) opposing flow.

Figure 3a,b, respectively, show the effects of stretching velocity ratio A when stretching in the flow and in the opposite direction, with further parameters sustaining the constant Heat generation/absorption coefficient $\delta = -1$, Prandtl number $Pr = 1$, Biot number $B_i = 1$, and Hartmann number $Ha = 1$. It can be seen that the velocity profile increases with stretching velocity ratio when stretching in the flow direction. Whereas, the velocity profile decreases with stretching velocity ratio when stretching in the opposite direction, this can be attributed to the significant enhancement in pressure on the sheet. Furthermore, it is remarked that the velocity profile augment with mixed convection parameter λ for both stretching in the flow direction and stretching in the opposite direction, as proven in Figure 1. The physical reason behind this is that, by augmenting the mixed convection parameter, the thermal buoyancy forces rise and help to push the flow in y direction, which in turn increases the velocity profile.

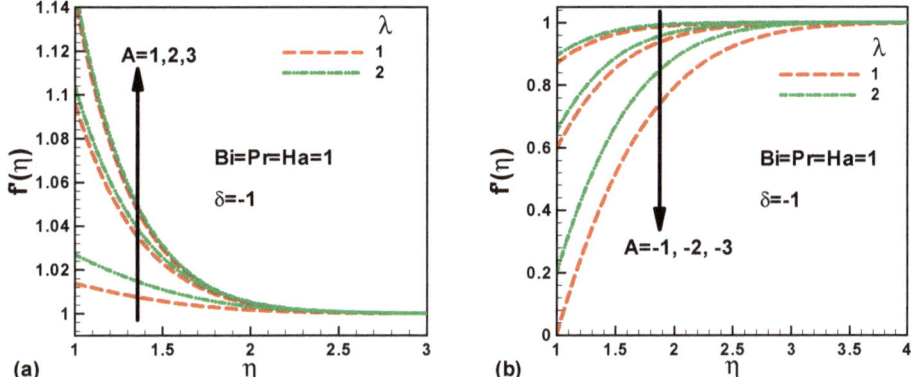

Figure 3. Variation of dimensionless velocity with stretching velocity ratio when (**a**) stretching in the flow direction and (**b**) stretching in the opposite direction.

The effects of Biot number in the presence of heat source on the dimensionless temperature for assisting and opposing flow are shown in Figure 4a,b, respectively at $P_r = H_a = A = 1$. It is clear that the temperature profile is greater at the stretching sheet surface and then exponentially lessens along the streamwise path up to the zero value for both assisting and opposing flow; this effect is proved and designated by the choice of boundary conditions and it is mathematically noticeable in Equation (10). It is worthwhile to note that the dimensionless temperature augment with both Biot number Bi and Heat generation/absorption coefficient δ for both case assisting and opposing flow; therefore, the thermal boundary layer thickness increases. It is well known that the Biot number signifies the proportion of heat convection to heat conduction; therefore increasing the Biot number leads to more heat will be released to the fluid flow, which in turn augments the temperature profile. Similarly, augmenting heat source leads to applying more heat to the fluid flow and it results in enhancing the dimensionless temperature.

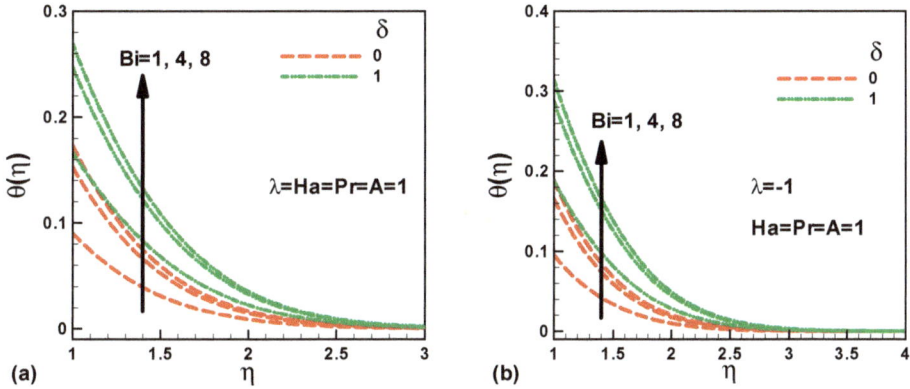

Figure 4. Variation of dimensionless temperature with Biot number in the presence of heat source for (**a**) assisting flow and (**b**) opposing flow.

Figure 5a,b, respectively, illustrate the result of Biot number in the company of heat sink on the dimensionless temperature for assisting and opposing flow at $P_r = H_a = A = 1$. As expected, the temperature profile satisfies the boundary conditions, starting with a higher value at the surface of the sheet and then meaningfully declining to zero value when η increases. Furthermore, it is remarked that dimensionless temperature increases with Biot number, even in the company of heat sink, as shown in Figure 3 in the circumstance of heat basis, so it can be concluded that the temperature profile increase with biot number independent of Heat generation/absorption coefficient δ. It should be pointed out that, for positive values of Heat generation/absorption coefficient, δ acts as a heat source, but for a negative value of Heat generation/absorption coefficient, δ acts as sink source, this signifies that the dimensionless temperature reduced in case of a negative value of δ (heat sink) when compared to a positive value of δ (heat source), as presented in Figure 3. Similarly, the thermal boundary layer thickness decreases.

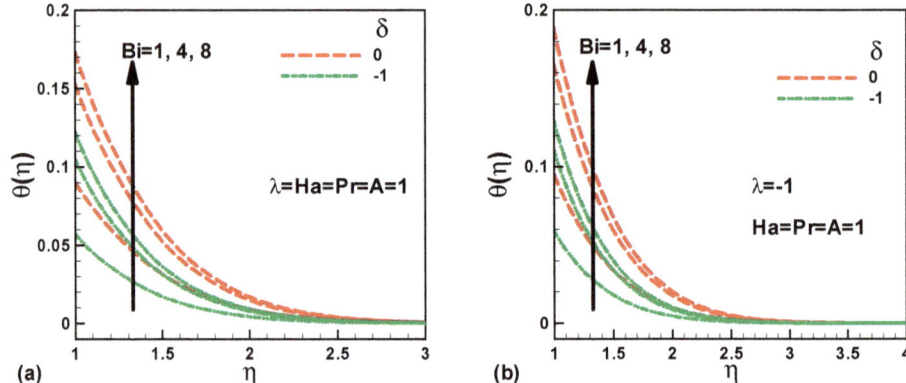

Figure 5. Variation of dimensionless temperature with Biot number in the presence of heat sink for (a) assisting flow and (b) opposing flow.

Figure 6a,b, respectively, display the effects of stretching velocity ratio and mixed convection parameter on dimensionless skin friction for both assisting and opposing flow. It is found that the skin friction increases with the mixed convection parameter and Hartmann number Ha, although it drops with the stretching velocity ratio for mutually opposing and assisting flow. This can be attributed to the result of velocity profile in the boundary layer and consequently disturbs the boundary layer thickness, as shown by Figures 1 and 2. On the other hand, the decreasing of skin friction with stretching velocity ratio can be associated to the augmentation of the velocity; therefore, the velocity boundary layer increases. It is well recognized that augmenting velocity means growing the Reynold number which in turn leads to lessening viscous force regarding inertial force, consequently the dropping in viscous force will reduce skin friction. It is valuable to mention that the effect of assisting flows on dimensionless skin friction is slightly more pronounced than that of opposing flow, because the pressure near to the surface is greater than not near to the surface. As observed, the skin friction continually and significantly increases with the Hartmann number, since it represents the ratio of electromagnetic force to the viscous force, therefore the magnetic field will increase and accordingly the Lorentz force will oppose and push the flow to the surface, which in turn augments the skin friction.

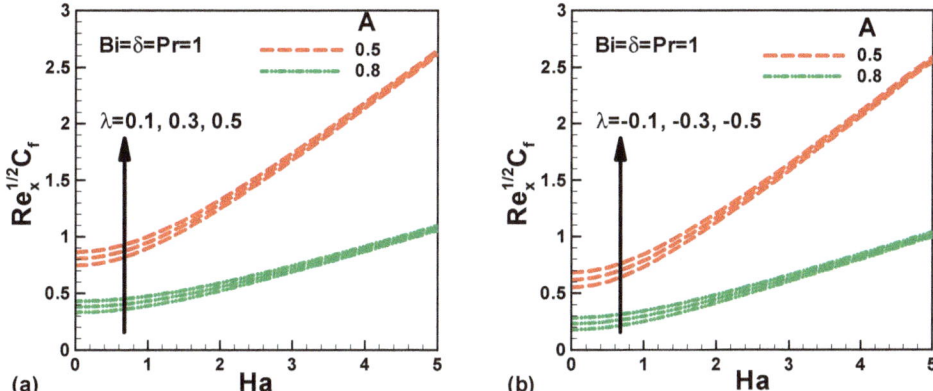

Figure 6. Variation of dimensionless skin friction with several parameters in the presence of heat source for (**a**) assisting flow and (**b**) opposing flow.

The effects of Biot number Bi, Heat generation/absorption coefficient δ, and mixed convection parameter λ on the dimensionless heat transfer rate for both assisting and opposing flow are illustrated, respectively, in Figure 7a,b. It is indicates that the dimensionless heat transfer rate increase with the Biot number and mixed convection parameter; whereas, it decreases with heat generation/absorption coefficient for both assisting and opposing flow. Physically, this can be attributed to the increase of temperature gradient with respect to the Biot number and mixed convection parameter, as presented by Figures 3 and 4. However, in an unexpected and perplexing result, it can be seen that the dimensionless heat transfer rate decrease with heat generation/absorption coefficient δ, it is well known that Nusselt numbers represent the proportion of convection to conduction heat transfer and in both situation assisting and opposing flow the effect of heat generation/absorption coefficient is more pronounced in conduction more than the convection heat transfer. It is clear that the effect of the mixed convection parameter on the heat transfer rate is slightly perceptible. Finally, the Nusselt number at the sheet surface augments, because the Hartmann number, stretching velocity ratio A, Hartmann number Ha, and mixed convection parameter λ increase. Though, it declines with respect to heat generation/absorption coefficient δ.

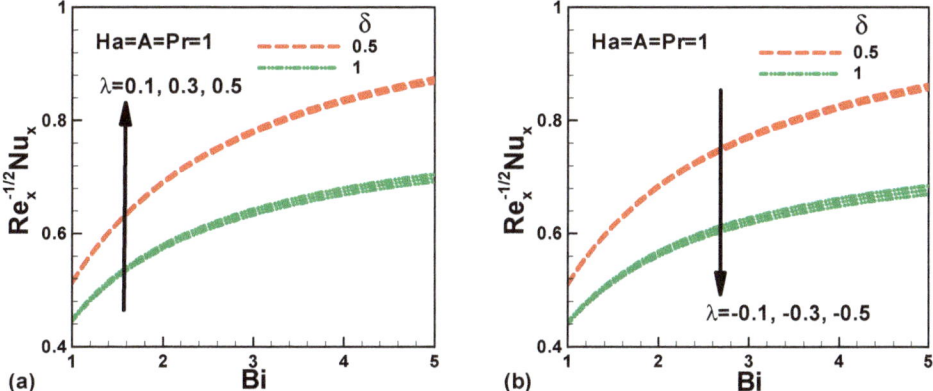

Figure 7. Variation of dimensionless heat transfer rate with several parameters in the presence of heat source for (**a**) assisting flow and (**b**) opposing flow.

6. Conclusions

A steady laminar flow over a vertical stretching sheet with the existence of viscous dissipation, heat source/sink, and magnetic fields has been mathematically explored through shooting arrangement based RKFI procedure. A comparison of the results with research results [3,26–29] demonstrates good agreement. The influence of relevant involved indicators on temperature and velocity, as well as the Nusselt number and skin friction coefficient are perceived. The main results that are enclosed in this work are concluded, as follows:

1. It is perceived that, for assisting flow ($\lambda > 0$), the dimensionless velocity is the maximum at the superficial of the vertical stretching sheet and it gradually lessens to the minimum value $f' = 1$, as it transfers to gone after the superficial, while for opposing flow ($\lambda < 0$), the dimensionless velocity is the lowest at the superficial of the vertical stretching sheet and gradually increases to the maximum value $f' = 1$.
2. The velocity profile augments with mixed convection parameter λ for both stretching in the flow direction and stretching in the opposite direction.
3. The temperature profile is greater at the stretching sheet superficial then exponentially lessens along the streamwise path up to the zero value for both assisting and opposing flow.
4. Skin friction increase with a mixed convection parameter and Hartmann number Ha, though it declines by stretching speed ratio for together opposing and assisting flow.
5. The effect of mixed convection parameter on the heat transfer rate is slightly perceptible. The Nusselt number at the sheet surface augments, because the Hartmann number, Hartmann number Ha, stretching velocity ratio A, and mixed convection parameter λ increase. Though, it declines according to heat generation/absorption coefficient δ.

Author Contributions: All authors contribute equally in this work in all parts and in all steps.

Acknowledgments: Ibrahim M. Alarifi would like to thank Deanship of Scientific Research at Majmaah University for supporting this work under the Project Number No. 1440-50.

Conflicts of Interest: The authors declare no conflict of interest.

References

1. Mohammadein, S.A.; Raslan, K.; Abdel-Wahed, M.S.; Abedel-Aal, E.M. KKL-model of MHD CuO-nanofluid flow over a stagnation point stretching sheet with nonlinear thermal radiation and suction/injection. *Results Phys.* **2018**, *10*, 194–199. [CrossRef]
2. Ibrahim, W. Magnetohydrodynamic (MHD) boundary layer stagnation point flow and heat transfer of a nanofluid past a stretching sheet with melting. *Propuls. Power Res.* **2017**, *6*, 214–222. [CrossRef]
3. Sharma, P.R.; Sinha, S.; Yadav, R.S.; Filippov, A.N. MHD mixed convective stagnation point flow along a vertical stretching sheet with heat source/sink. *Int. J. Heat Mass Transf.* **2018**, *117*, 780–786. [CrossRef]
4. SGhadikolaei, S.; Hosseinzadeh, K.H.; Ganji, D.D. MHD raviative boundary layer analysis of micropolar dusty fluid with graphene oxide (Go)-engine oil nanoparticles in a porous medium over a stretching sheet with joule heating effect. *Powder Technol.* **2018**. [CrossRef]
5. El-Mistikawy, T.M.A. MHD flow due to a linearly stretching sheet with induced magnetic field. *Acta Mech.* **2016**, *227*, 3049–3053. [CrossRef]
6. Hussain, A.; Malik, M.Y.; Salahuddin, T.; Rubab, A.; Khan, M. Effects of viscous dissipation on MHD tangent hyperbolic fluid over a nonlinear stretching sheet with convective boundary conditions. *Results Phys.* **2017**, *7*, 3502–3509. [CrossRef]
7. Tufail, M.N.; Butt, A.S.; Ali, A. Computational modeling of an MHD flow of a non-newtonian fluid over an unsteady stretching sheet with viscous dissipation effects. *J. Appl. Mech. Tech. Phys.* **2016**, *57*, 900–907. [CrossRef]
8. Mabood, F.; Khan, W.A.; Ismail, A.I.M. MHD flow over exponential radiating stretching sheet using homotopy analysis method. *J. King Saud Univ. Eng. Sci.* **2017**, *29*, 68–74. [CrossRef]

9. Hayat, T.; Qayyum, S.; Alsaedi, A.; Ahmad, B. Magnetohydrodynamic (MHD) nonlinear convective flow of Walters-B nanofluid over a nonlinear stretching sheet with variable thickness. *Int. J. Heat Mass Transf.* **2017**, *110*, 506–514. [CrossRef]
10. Malik, M.Y.; Khan, M.; Salahuddin, T. Study of an MHD flow of the carreau fluid flow over a stretching sheet with a variable thickness by using an implicit finite difference scheme. *J. Appl. Mech. Tech. Phys.* **2017**, *58*, 1033–1039. [CrossRef]
11. Agbaje, T.M.; Mondal, S.; Makukula, Z.G.; Motsa, S.S.; Sibanda, P. A new numerical approach to MHD stagnation point flow and heat transfer towards a stretching sheet. *Ain Shams Eng. J.* **2018**, *9*, 233–243. [CrossRef]
12. Soid, S.K.; Ishak, A.; Pop, I. MHD flow and heat transfer over a radially stretching/shrinking disk. *Chin. J. Phys.* **2018**, *56*, 58–66. [CrossRef]
13. Mahabaleshwar, U.S.; Sarris, I.E.; Hill, A.A.; Lorenzini, G.; Pop, I. An MHD couple stress fluid due to a perforated sheet undergoing linear stretching with heat transfer. *Int. J. Heat Mass Transf.* **2017**, *105*, 157–167. [CrossRef]
14. Ferdows, M.; Khalequ, T.S.; Tzirtzilakis, E.E.; Sun, S. Effects of Radiation and Thermal Conductivity on MHD Boundary Layer Flow with Heat Transfer along a Vertical Stretching Sheet in a Porous Medium. *J. Eng. Thermophys.* **2017**, *26*, 96–106. [CrossRef]
15. Hsiao, K.-L. Micropolar nanofluid flow with MHD and viscous dissipation effects towards a stretching sheet with multimedia feature. *Int. J. Heat Mass Transf.* **2017**, *112*, 983–990. [CrossRef]
16. Daniel, Y.S.; Aziz, Z.A.; Ismail, Z.; Salah, F. Thermal stratification effects on MHD radiative flow of nanofluid over nonlinear stretching sheet with variable thickness. *J. Comput. Des. Eng.* **2018**, *5*, 232–242. [CrossRef]
17. Hayat, T.; Rashid, M.; Alsaedi, A. MHD convective flow of magnetite-Fe_3O_4 nanoparticles by curved stretching sheet. *Results Phys.* **2017**, *7*, 3107–3115. [CrossRef]
18. Daniel, Y.S.; Aziz, Z.A.; Ismail, Z.; Salah, F. Thermal radiation on unsteady electrical MHD flow of nanofluid over stretching sheet with chemical reaction. *J. King Saud Univ. Sci.* **2017**, in press. [CrossRef]
19. Seth, G.S.; Singha, A.K.; Mandal, M.S.; Banerjee, A.; Bhattacharyya, K. MHD stagnation-point flow and heat transfer past a non-isothermal shrinking/stretching sheet in porous medium with heat sink or source effect. *Int. J. Mech. Sci.* **2017**, *134*, 98–111. [CrossRef]
20. Khan, M.; Hussain, A.; Malik, M.Y.; Salahuddin, T.; Khan, F. Boundary layer flow of MHD tangent hyperbolic nanofluid over a stretching sheet: A numerical investigation. *Results Phys.* **2017**, *7*, 2837–2844. [CrossRef]
21. Madhu, M.; Kishan, N. MHD boundary-layer flow of a non-newtonian nanofluid past a stretching sheet with a heat source/sink. *J. Appl. Mech. Tech. Phys.* **2016**, *57*, 908–915. [CrossRef]
22. Babu, M.J.; Sandeep, N. Three-dimensional MHD slip flow of nanofluids over a slendering stretching sheet with thermophoresis and Brownian motion effects. *Adv. Powder Technol.* **2016**, *27*, 2039–2050. [CrossRef]
23. Khan, I.; Malik, M.Y.; Hussain, A.; Salahuddin, T. Effect of homogenous-heterogeneous reactions on MHD Prandtl fluid flow over a stretching sheet. *Results Phys.* **2017**, *7*, 4226–4231. [CrossRef]
24. Seth, G.S.; Mishra, M.K. Analysis of transient flow of MHD nanofluid past a non-linear stretching sheet considering Navier's slip boundary condition. *Adv. Powder Technol.* **2017**, *28*, 375–384. [CrossRef]
25. Chen, X.; Ye, Y.; Zhang, X.; Zheng, L. Lie-group similarity solution and analysis for fractional viscoelastic MHD fluid over a stretching sheet. *Comput. Math. Appl.* **2018**, *75*, 3002–3011. [CrossRef]
26. Ramchandran, N.; Chen, T.S.; Armaly, B.F. Mixed convection in stagnation flows adjacent to vertical surfaces. *J. Heat Transf.* **1988**, *110*, 373–377. [CrossRef]
27. Lok, Y.Y.; Amin, N.; Pop, I. Unsteady mixed convection flow of a micropolar fluid near the stagnation point on a vertical surface. *Int. J. Therm. Sci.* **2006**, *45*, 1149–1152. [CrossRef]
28. Ishak, A.; Nazar, R.; Pop, I. Dual solutions in mixed convection flow near a stagnation point on a vertical porous plate. *Int. J. Therm. Sci.* **2008**, *47*, 417–422. [CrossRef]
29. Ali, F.M.; Nazar, R.; Arifin, N.M.; Pop, I. Mixed convection stagnation point flow on vertical stretching sheet with external magnetic field. *Appl. Math. Mech. Engl. Ed.* **2014**, *35*, 155–166. [CrossRef]
30. Hassanien, I.A.; Gorla, R. Nonsimilar Solutions for Natural Convection in Micropolar Fluids on a Vertical Plate. *Int. J. Fluid Mech. Res.* **2003**, *30*, 4–14. [CrossRef]
31. Li, J.; Zheng, L.; Liu, L. MHD viscoelastic flow and heat transfer over a vertical stretching sheet with Cattaneo-Christov heat flux effects. *J. Mol. Liq.* **2016**, *221*, 19–25. [CrossRef]

32. Freidoonimehr, N.; Rahimi, A.B. Exact-solution of entropy generation for MHD nanofluid flow induced by a stretching/shrinking sheet with transpiration: Dual solution. *Adv. Powder Technol.* **2017**, *28*, 671–685. [CrossRef]
33. Merkin, J.H. On dual solutions occurring in mixed convection in a porous medium. *J. Eng. Math.* **1986**, *20*, 171–179. [CrossRef]
34. Harris, S.D.; Ingham, D.B.; Pop, I. Mixed convection boundary-layer flow near the stagnation point on a vertical surface in a porous medium: Brinkman model with slip. *Transp. Porous Media* **2009**, *77*, 267–285. [CrossRef]
35. Rahman, M.M.; Roşca, A.V.; Pop, I. Boundary layer flow of a nanofluid past a permeable exponentially shrinking/stretching surface with second order slip using Buongiorno's model. *Int. J. Heat Mass Transf.* **2014**, *77*, 1133–1143. [CrossRef]
36. Weidman, P.D.; Kubitschek, D.G.; Davis, A.M.J. The effect of transpiration on self-similar boundary layer flow over moving surfaces. *Int. J. Eng. Sci.* **2006**, *44*, 730–737. [CrossRef]
37. Roşca, A.V.; Pop, I. Flow and heat transfer over a vertical permeable stretching/shrinking sheet with a second order slip. *Int. J. Heat Mass Transf.* **2013**, *60*, 355–364. [CrossRef]

© 2019 by the authors. Licensee MDPI, Basel, Switzerland. This article is an open access article distributed under the terms and conditions of the Creative Commons Attribution (CC BY) license (http://creativecommons.org/licenses/by/4.0/).

Article

Fractional Order Forced Convection Carbon Nanotube Nanofluid Flow Passing Over a Thin Needle

Taza Gul [1,2], Muhammad Altaf Khan [1], Waqas Noman [1], Ilyas Khan [3,*], Tawfeeq Abdullah Alkanhal [4] and Iskander Tlili [5]

1. Department of Mathematics, City University of Science and Information Technology, Peshawar 25000, Pakistan; tazagul@cusit.edu.pk (T.G.); makhan@cusit.edu.pk (M.A.K.); waqarnoman55@yahoo.com (W.N.)
2. Department of Mathematics, Govt. Superior Science College Peshawar, Khyber Pakhtunkhwa, Pakistan
3. Faculty of Mathematics and Statistics, Ton Duc Thang University, Ho Chi Minh City 72915, Vietnam
4. Department of Mechatronics and System Engineering, College of Engineering, Majmaah University, Majmaah 11952, Saudi Arabia; t.alkanhal@mu.edu.sa
5. Energy and Thermal Systems Laboratory, National Engineering School of Monastir, Street Ibn El Jazzar, 5019 Monastir, Tunisia; iskander.tlili@enim.rnu.tn
* Correspondence: ilyaskhan@tdt.edu.vn

Received: 7 January 2019; Accepted: 2 February 2019; Published: 2 March 2019

Abstract: In the fields of fluid dynamics and mechanical engineering, most nanofluids are generally not linear in character, and the fractional order model is the most suitable model for representing such phenomena rather than other traditional approaches. The forced convection fractional order boundary layer flow comprising single-wall carbon nanotubes (SWCNTs) and multiple-wall carbon nanotubes (MWCNTs) with variable wall temperatures passing over a needle was examined. The numerical solutions for the similarity equations were obtained for the integer and fractional values by applying the Adams-type predictor corrector method. A comparison of the SWCNTs and MWCNTs for the classical and fractional schemes was investigated. The classical and fractional order impact of the physical parameters such as skin fraction and Nusselt number are presented physically and numerically. It was observed that the impact of the physical parameters over the momentum and thermal boundary layers in the classical model were limited; however, while utilizing the fractional model, the impact of the parameters varied at different intervals.

Keywords: SWCNT/MWCNT nanofluid; thin needle; classical and fractional order problems; APCM technique

1. Introduction

This study is concerned with the enhancement of heat transfer through nanofluid, which will play a dynamic role in the field of chemical sciences and the energy sector. The enhancement of heat transfer through nanofluid was studied by many scientists in the field of geometry under diverse conditions. Sparrow and Gregg [1] scrutinized the removal of humidity, using centrifugal force procedures on a cooled rotary disk. The energy obtaining and cooling behavior of the devices mainly depend on the heat transfer liquid used, and the lower thermal efficiency of these fluids can create harsh restrictions for device performance. The limitations and low thermal efficiency of these liquids delay the device performance and compression of heat exchangers. Choi [2] explored the idea of nanofluids by utilizing small nanosized (10–50 nm) particles in base fluids. The anticipated factors influencing the performance of nanofluids during heat transfer were: (i) thermal properties; (ii) chemical stability; (iii) compatibility with the base fluid; (iv) toxicity; (v) accessibility; and (vi) cost. Possible nanomaterials include metals, metal oxides, and carbon materials. Carbon materials play a significant role in enhancing the thermal efficiency of base fluids, and carbon nanotubes (CNTs) are the renowned family of carbons that have been used for thermal and cooling applications in recent studies. Carbon nanotubes are further divided into two classes: single-walled carbon nanotubes (SWCNTs) and multiple-walled carbon nanotubes (MWCNTs). Single-walled carbon nanotubes are created by packaging a layer of carbon one-atom thick, while MWCNTs contain multiple rolled layers of carbon. Carbon nanotubes nanofluids have many important applications in industries such as aerospace, electronics, optics, and energy conservation, as reported by Volder et al. [3] and Terrones [4]. The higher thermal conductivity (2000–6000 W/mK) of carbon nanotubes make them more valuable for the augmentation of heat transfer devices. Ellahi et al. [5] investigated CNTs' nanofluid flow along a vertical cone under the influence of a variable wall temperature, and a comparison between SWCNTs and MWCNTs was made in their study. Gohar et al. [6] have studied SWCNTs/MWCNTs' nanofluid flow over a non-linear stretching disc. The high thermal efficiency of CNTs increased the heat flux and thermal efficiency of the base liquids as the heat fluxed, compared to other nanofluids as reported by Murshed et al. [7,8]. Various thermal conductivity models have been proposed by researchers for nanofluid flow problems. The appropriate and frequently-used thermal conductivity models for CNTs were reported by Xue [9]. The flow problem which passes over a thin needle under the effect of convection has been considered by many scholars Narain and Uberoi [10,11] and Chen [12]. Wang [13] and Grosan and Pop [14] have deliberated the mixed convection boundary layer flows over an upright thin needle including an intense heat source at the tip of the needle.

This study was carried out considering water-based CNT nanofluid flow over a thin needle. Further, the variable surface temperature with forced convection comprising single walled carbon nanotubes (SWCNT) and multi walled carbon nanotube (MWCNT) water-based nanofluid past over a thin needle was investigated in classical and fractional models, respectively.

The integer order derivatives or the classical model of fluid dynamics investigate the flow behavior at the integer steps, while the fractional order derivatives of the same fluid flow explore the natural phenomena to expose the internal behavior of the fluid flow by taking the fractional values among the integers. However, the idea of fractional calculus has been conventional for approximately three hundred years [15–17].

In fluid mechanics, most fluids are not generally linear in characteristic and the fractional order model is more appropriate for the illustration of such a kind of spectacle, rather than traditional methods. Caputo [18] introduced the idea of fractional derivatives from the modified Darcy's law using the concept of unsteadiness. This idea was further modified by other researchers [19–21] through the introduction of a variety of new fractional derivatives and their applications. Agarwal et al. [22] studied neural network models using ynchronization and impulsive Caputo fractional differential equations. Khan et al. [23] examined the fractional order solution of the Phi-4 equations using the GO/G expansion technique. Hameed et al. [24] examined the fractional order second grade fluid peristaltic transport in a vertical cylinder. A variety of numerical techniques have been used to find solutions to the classical models [25–30], and these techniques have been further combined to find solutions for fractional order problems.

The aim of this study is to analyze the force convectional CNT nanofluid flow passing over a thin needle including the elastic heat flux. The FDE-12 method was used for the solution for the fractional order non-linear differential equations. It is the execution of the predictor corrector method of the Adams–Bash Forth–Moulton technique derived by Diethelm and Freed [31]. Diethelm et al. [32] found the convergence and validity of this method for the solution of fractional order differential equations. The solution for classical and fractional order mathematical models containing (SWCNT/MWCNT) water-based nanofluids was obtained through the solution for fractional order systems, which was solved by the Adams-type predictor corrector method as used in References [33,34]. The range of the parameters in this study were selected as per the investigation by Gul et al. [35] using the BVP 2.0 package and the Optimal Homotopy Analysis Method OHAM technique. They used the 20th order approximation for the selected range of the parameters and obtained the minimum square residual error. The important outcomes were presented physically and numerically.

2. Mathematical Formulation

The axisymmetric boundary layer comprising SWCNT and MWCNT nanofluids' flow over the surface of a thin needle, including position-dependent wall temperature at the ambient temperature T_∞ was considered. The radius of the thin needle is defined as. The surface temperature, T_w, of the thin needle is considered heavier than Ambient temperature T_∞, ($T_w > T_\infty$). The external flow velocity of the nanofluid is considered to be $u_e(x)$. The momentum and thermal boundary layer equations were derived in the axial and radial coordinates and all the assumption are imposed as [14]:

$$\frac{\partial \tilde{r}\tilde{u}}{\partial \tilde{x}} + \frac{\partial \tilde{r}\tilde{v}}{\partial \tilde{r}} = 0, \qquad (1)$$

$$\left(\tilde{u}\frac{\partial \tilde{u}}{\partial \tilde{x}} + \tilde{v}\frac{\partial \tilde{u}}{\partial \tilde{r}}\right) = \tilde{u}_e \frac{d\tilde{u}_e}{d\tilde{x}} + v_{nf}\frac{1}{\tilde{r}}\frac{\partial}{\partial \tilde{r}}\left(\tilde{r}\frac{\partial \tilde{u}}{\partial \tilde{r}}\right), \qquad (2)$$

$$\left(\tilde{u}\frac{\partial \tilde{T}}{\partial \tilde{x}} + \tilde{v}\frac{\partial \tilde{T}}{\partial \tilde{r}}\right) = \alpha_{nf}\frac{1}{\tilde{r}}\frac{\partial}{\partial \tilde{r}}\left(\tilde{r}\frac{\partial \tilde{T}}{\partial \tilde{r}}\right). \qquad (3)$$

The physical conditions satisfy [14] and are defined as:

$$\begin{array}{l}\tilde{u} = 0, \tilde{v} = 0, \tilde{T} = T_w \text{ at } \tilde{r} = R(\tilde{x}),\\ \tilde{u} = \tilde{u}_e(\tilde{x}), \tilde{T} = T_\infty, \text{at } \tilde{r} \to \infty.\end{array} \qquad (4)$$

The velocity components are represented by $\tilde{u}, \tilde{v}$ towards the axial and radial $(\tilde{x}, \tilde{r})$ directions, respectively.

ρ_{nf} is the density of the nanofluids, μ_{nf} is the dynamic viscosity of the nanofluids such that $v_{nf} = \frac{\mu_{nf}}{\rho_{nf}}$ is the kinematic viscosity of the nanofluid, ϕ is the solid particle volume fraction, k_{nf} is the thermal conductivity, and $(\rho C_p)_{nf}$ is the specific heat capacity of the nanofluids such that $\alpha_{nf} = \frac{k_{nf}}{(\rho C_p)_{nf}}$. The thermophysical properties for the CNT nanofluids were presented and satisfy Xue [9]:

$$\rho_{nf} = \rho_f - \phi\rho_f + \phi\rho_s, \mu_{nf} = \mu_f(1-\phi)^{-2.5}, (\rho C_p)_{nf} = (\rho C_p)_f - \phi(\rho C_p)_f + \phi(\rho C_p)_s$$
$$\frac{k_{nf}}{k_f} = \frac{1-\phi+2\left(\frac{k_{CNT}}{k_{CNT}-k_f}\right)\ln\left(\frac{k_{CNT}+k_f}{2k_f}\right)\phi}{1-\phi+2\left(\frac{k_f}{k_{CNT}-k_f}\right)\ln\left(\frac{k_{CNT}+k_f}{2k_f}\right)\phi}. \qquad (5)$$

To bring the basic Equations (1)–(3) into a dimensionless form, under boundary limitations, as per Equation (4), we adopted the scaling transformations as [14]:

$$\begin{array}{l}x = \tilde{x}/L, r = (\tilde{r}/L)\text{Re}^{\frac{1}{2}}, R(x) = (\tilde{R}(\tilde{x})/L)\text{Re}^{\frac{1}{2}}, u = \tilde{u}/U_\infty,\\ v = (\tilde{v}/U_\infty)\text{Re}^{\frac{1}{2}}, u_e(x) = \tilde{u}_e(\tilde{x})/U_\infty, T = (\tilde{T}-T_\infty)/\Delta T.\end{array} \qquad (6)$$

Here, $\text{Re} = \frac{U_\infty L}{v_f}$ is the Reynolds number, L is the characteristic length of the needle, $R(x)$ is the dimensionless radial coordinate, r is the dimensionless radius of the needle, U_∞ is the characteristic velocity, ΔT is the characteristic temperature, and x is the dimensionless axial coordinate. Bringing Equation (6) into the basic Equations (1)–(4) cuts into the following non-linear differential form as:

$$\frac{\partial ru}{\partial x} + \frac{\partial rv}{\partial r} = 0, \qquad (7)$$

$$\left(u\frac{\partial u}{\partial x}+v\frac{\partial u}{\partial r}\right)=u_e\frac{du_e}{dx}+\frac{v_{nf}}{v_f}\frac{1}{r}\frac{\partial}{\partial r}\left(r\frac{\partial u}{\partial r}\right), \tag{8}$$

$$\Pr\left(u\frac{\partial T}{\partial x}+v\frac{\partial T}{\partial r}\right)=\frac{\alpha_{nf}}{\alpha_f}\frac{1}{r}\frac{\partial}{\partial r}\left(r\frac{\partial T}{\partial r}\right). \tag{9}$$

The suitable boundary conditions are:

$$u=0, v=0, T=T_w(x) \text{ at } r=R(x),$$
$$u=u_e(x), T=0, \text{ at } r\to\infty. \tag{10}$$

Next, the similarity variables are:

$$u_e(x)=x^m, T_w(x)=x^n, \psi=x\,f(\eta), \eta=x^{m-1}r^2, T(x)=x^n\Theta(\eta). \tag{11}$$

Here, $u_e(x)$ is the dimensionless velocity of the external flow, ψ is used to demonstrate the stream function and satisfy the continuity Equation (7). The velocity components derived from the stream function ψ are defined as: $u=\frac{1}{r}\frac{\partial \psi}{\partial r}, v=-\frac{1}{r}\frac{\partial \psi}{\partial x}$. Putting $\eta=a$ into Equation (11) describes the size of the needle: $r=R(x)=\sqrt{ax^{(1-m)}}$, along the surface. Using Equation (11) in the basic Equations (7)–(10), the continuity equation is satisfied characteristically, and the rest of the equations are transformed as:

$$\frac{8}{(1-\phi)^{2.5}\left(1-\phi+\phi\frac{\rho_{CNT}}{\rho_f}\right)}(\eta f'')'+4ff''+m(1-4(f')^2)=0, \tag{12}$$

$$\frac{2\left(\frac{k_{nf}}{k_f}\right)}{\Pr\left(1-\phi+\phi\frac{(\rho C_p)_{CNT}}{(\rho C_p)_f}\right)}(\eta\Theta')'+f\Theta'-nf'\Theta=0. \tag{13}$$

The suitable boundary conditions are:

$$f(a)=0, f'(a)=0, \Theta(a)=1,$$
$$f'(\infty)=\tfrac{1}{2}, \Theta(\infty)=0. \tag{14}$$

The skin friction coefficient and the local Nusselt number satisfy [14]:

$$\operatorname{Re}_x^{\frac{1}{2}}C_f=4a^{\frac{1}{2}}(1-\phi)^{-2.5}f''(a), \quad \operatorname{Re}_x^{-\frac{1}{2}}Nu_x=\left[-2a^{\frac{1}{2}}\frac{K_{nf}}{K_f}\right]\Theta'(a). \tag{15}$$

Here, $\operatorname{Re}_x=\frac{u_e(x)\,x}{v_f}$, is the local Reynolds number.

3. Preliminaries on the Caputo Fractional Derivatives

The useful definition of Caputo fractional order derivatives and their properties are presented below.

Definition 1. *Let $a>0$, $t>a$; $a, \alpha, t \in \Re$. The Caputo fractional derivative of order α of the function $f \in C^n$ is given by:*

$${}^C_aD^\alpha_t f(t)=\frac{1}{\Gamma(n-\alpha)}\int_a^t \frac{f^{(n)}(\xi)}{(t-\xi)^{\alpha+1-n}}d\xi, \quad n-1<\alpha<n\in N. \tag{16}$$

Property 1. Let $f(t)$, $g(t) : [a, b] \to \Re$ be such that ${}^C_aD^\alpha_t f(t)$ and ${}^C_aD^\alpha_t g(t)$ exist almost everywhere, and let $c_1, c_2 \in \Re$. Then ${}^C_aD^\alpha_t\{c_1 f(t) + c_2 g(t)\}$ exists almost everywhere and

$${}^C_aD^\alpha_t\{c_1 f(t) + c_2 g(t)\} = c_1 {}^C_aD^\alpha_t f(t) + c_2 {}^C_aD^\alpha_t g(t). \tag{17}$$

Property 2. The function $f(t) \equiv c$ is constant and therefore, the fractional derivative is zero: ${}^C_aD^\alpha_t c = 0$. The general description of the fractional differential equation was assumed including the Caputo concept:

$${}^C_aD^\alpha_t x(t) = f(t, x(t)), \quad \alpha \in (0,1). \tag{18}$$

With the initial conditions $x_0 = x(t_0)$.

4. Solution Methodology

The following variables were selected for the momentum and thermal boundary layer (12, 13) to reduce the system into the first order differential equations as:

$$y_1 = \eta, y_2 = f, y_3 = f', y_4 = f'', y_5 = \Theta, y_6 = \Theta'. \tag{19}$$

The Caputo fractional order derivative applied to the first order ODE system was obtained from (12, 13) with the efforts of the proposed variables given in Equation (19).

The fractional order system was obtained from Reference [32]:

$$\begin{pmatrix} D^\alpha_\eta y_1 \\ D^\alpha_\eta y_2 \\ D^\alpha_\eta y_3 \\ D^\alpha_\eta y_4 \\ D^\alpha_\eta y_5 \\ D^\alpha_\eta y_6 \end{pmatrix} = \begin{pmatrix} 1 \\ y_3 \\ y_4 \\ \frac{-(1-\phi)^{2.5}\left(1-\phi+\phi\frac{\rho_{CNT}}{\rho_f}\right)}{8y_1}\left(y_4 + 4y_1y_3 + m(1-4(y_3)^2)\right) \\ y_6 \\ \frac{-\Pr\left(1-\phi+\phi\frac{(\rho C_p)_{CNT}}{(\rho C_p)_f}\right)}{2\left(\frac{k_{nf}}{k_f}\right)}(y_6 + y_2y_6 - ny_3y_5) \end{pmatrix} \begin{pmatrix} y_1 \\ y_2 \\ y_3 \\ y_4 \\ y_5 \\ y_6 \end{pmatrix} = \begin{pmatrix} 0 \\ 0 \\ 0 \\ u_1 \\ 1 \\ u_2 \end{pmatrix}. \tag{20}$$

Equation (20) represents a matrix system of fractional order equations of an initial value problem. Considering ($\alpha = 1$), we have an integer order model or a classical model.

5. Results and Discussion

The two-dimensional forced conventional boundary layer SWCNT/MWCNT nanofluid flow for the enhancement of heat transmission over a thin needle was examined. A comparison of the influence of the physical constraints was studied for the integer and fractional order values

The fractional order system was solved numerically through the Adams-type predictor corrector method.

The geometry of the problem is displayed in Figure 1. The influence of the constant m versus velocity field $f'(\eta)$ is shown in Figures 2 and 3, for the classical and fractional order values, respectively. The larger values of the parameter m cause lower velocity. Physically, the rising values of m enhance the non-linearity to generate a friction force to decline the radial velocity. This decline is comparatively fast in the fractional order scheme. Due to the high thermophysical properties, the decline effect is comparatively rapid using the SWCNTs. The impact of ϕ over the $f'(\eta)$ for the integer and fractional order values is displayed in Figures 4 and 5, respectively. The larger value of ϕ causes a decrease in the velocity, and this effect is clearly larger when using the SWCNTs when compared to the MWCNTs. In fact, the larger amount of ϕ enhances the efficiency of the frictional force, and as a result, the viscous forces become strong enough to stop the fluid motion. Again, the decline effect is stronger using the fractional values. Figures 6 and 7 indicate the influence of the various values of the nanoparticle volume fraction versus the temperature field. The larger value of ϕ raises the temperature profile, and this effect is comparatively strong by means of the SWCNTs. In fact, the thermal conductivity of SWCNTs is high and provides rapid thermal efficiency to enhance the temperature field.

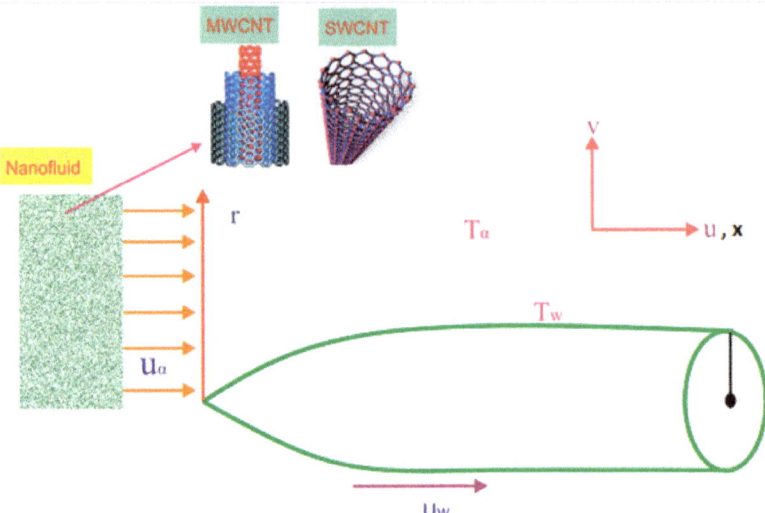

Figure 1. The geometry of the problem.

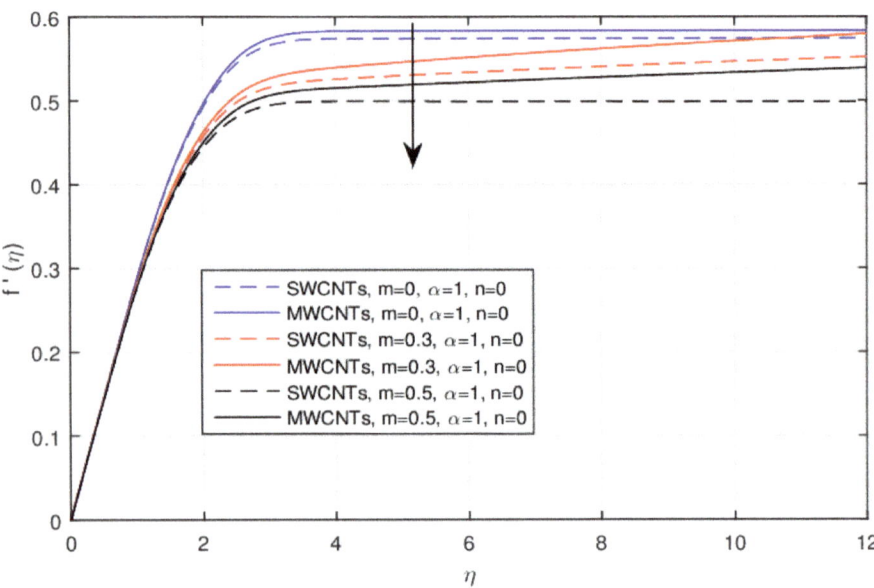

Figure 2. The impact of m over the $f'(\eta)$ for the integer values.

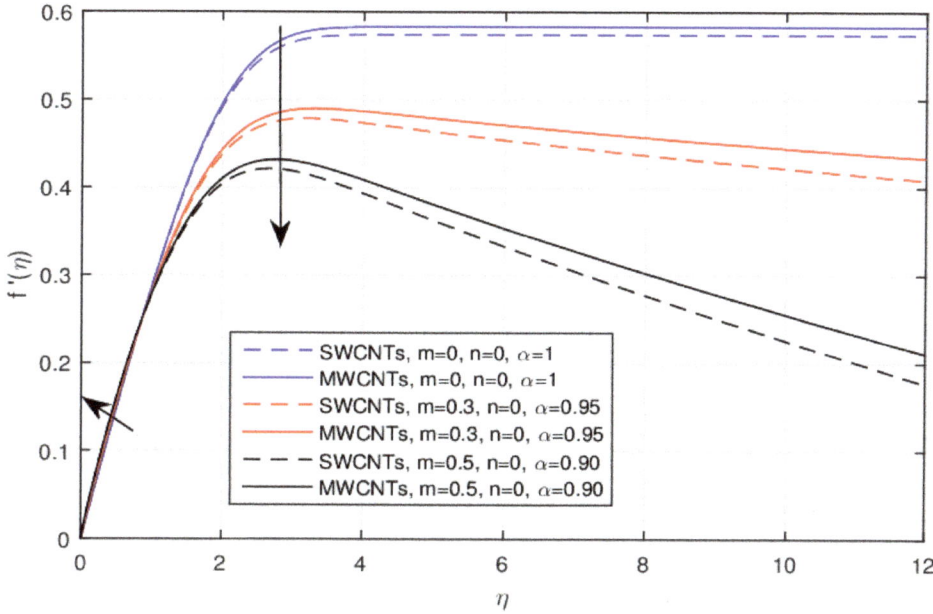

Figure 3. The impact of m over the $f'(\eta)$ for the fractional values.

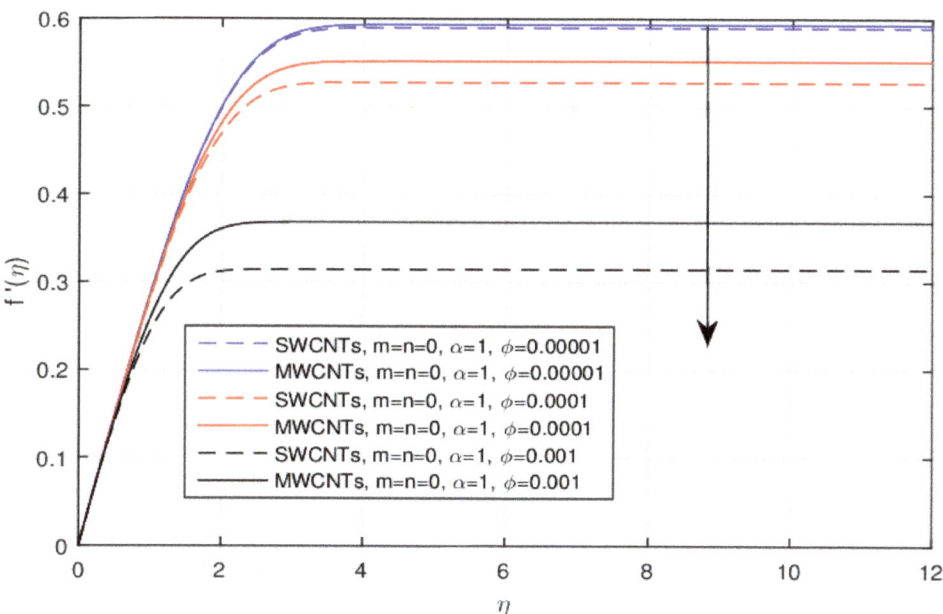

Figure 4. The impact of ϕ over the $f'(\eta)$ for the integer values.

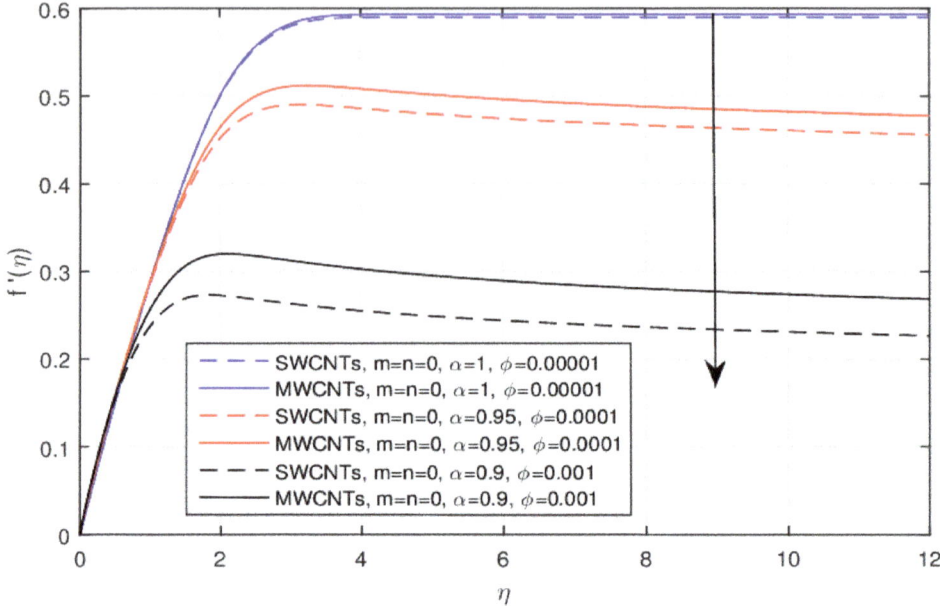

Figure 5. The impact of ϕ over the $f'(\eta)$ for the fractional values.

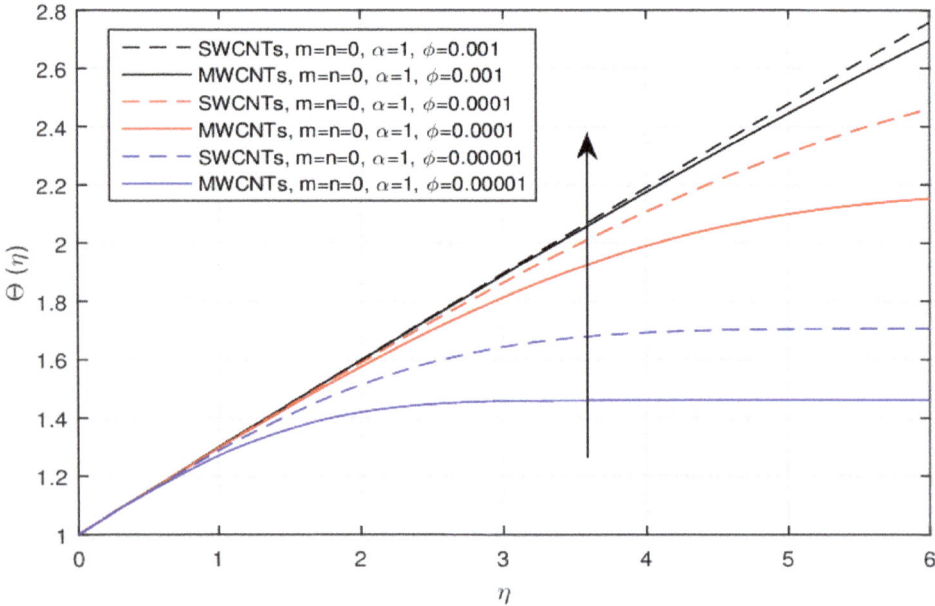

Figure 6. The impact of ϕ over the $\Theta(\eta)$ for the integer values.

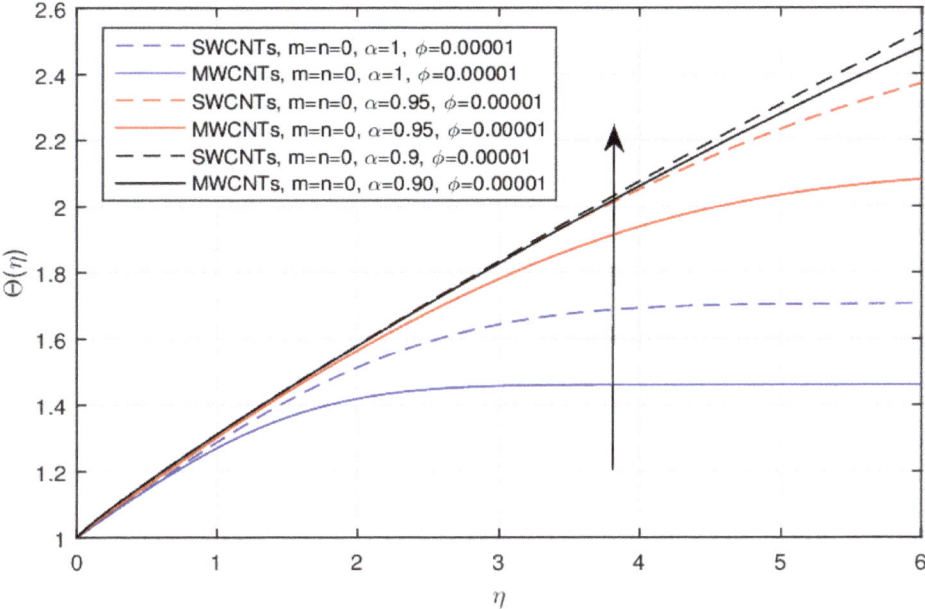

Figure 7. The impact of ϕ over the $\Theta(\eta)$ for the fractional values.

The impact of the wall temperature profile parameter n over the $\Theta(\eta)$ for the integer and fractional values are shown in Figures 8 and 9, respectively. The smaller values of n are enhancing the cooling effect, and as a result, the temperature field declines for the integer values and this effect is reversed for the fractional order values. The performance of the parameter n decreases the temperature field near the surface of the needle for the fractional values $\alpha = 1, 0.95, 0.90$, and this effect changes to increase the temperature profile after the critical point, as shown in Figure 9. The impact of the Prandtl number Pr over the temperature profile $\Theta(\eta)$ for the integer and fractional values is displayed in Figures 10 and 11, respectively. The rising values of Pr causes lower values compared to the classical model, as usually shown in the literature, but using the fractional model for the same values as the Prandtl number, the temperature profile near the needle surface increases and declines after the point of inflection.

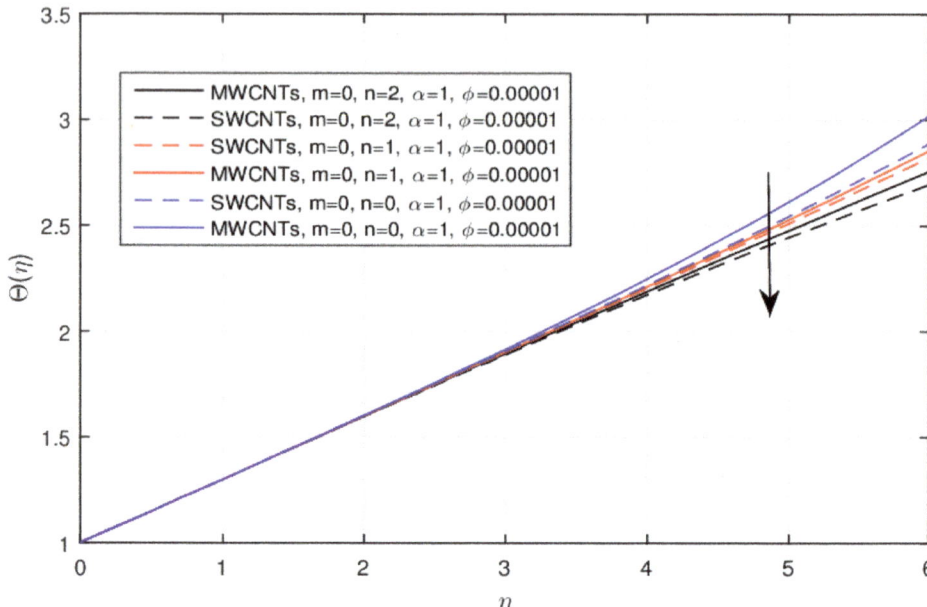

Figure 8. The impact of n over the $\Theta(\eta)$ for the integer values.

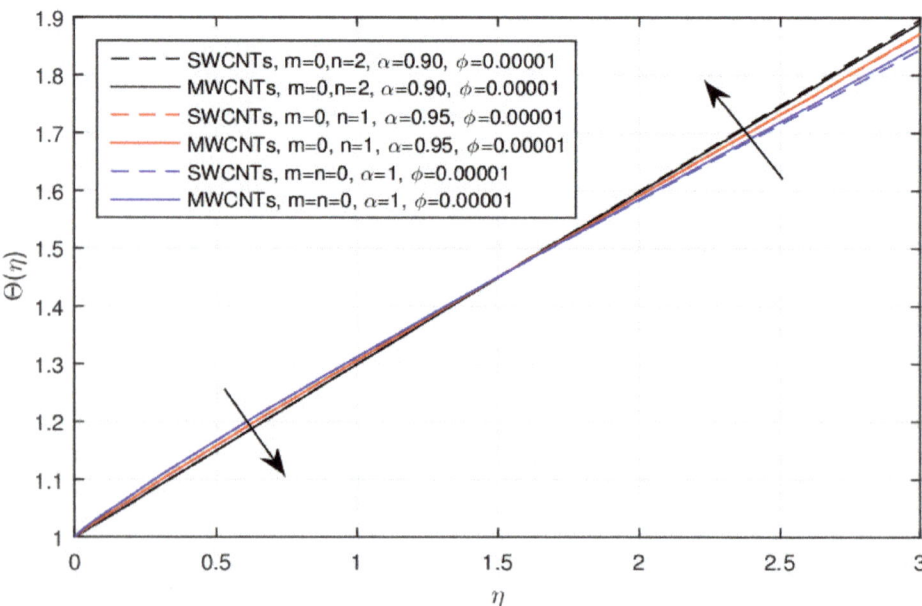

Figure 9. The impact of n over the $\Theta(\eta)$ for the fractional values.

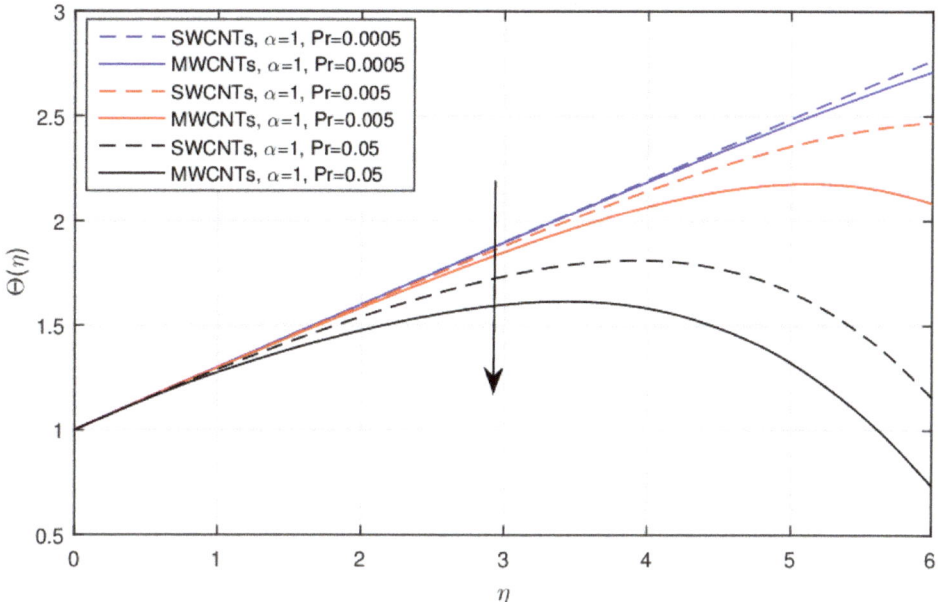

Figure 10. The impact of Pr over the $\Theta(\eta)$ for the integer values.

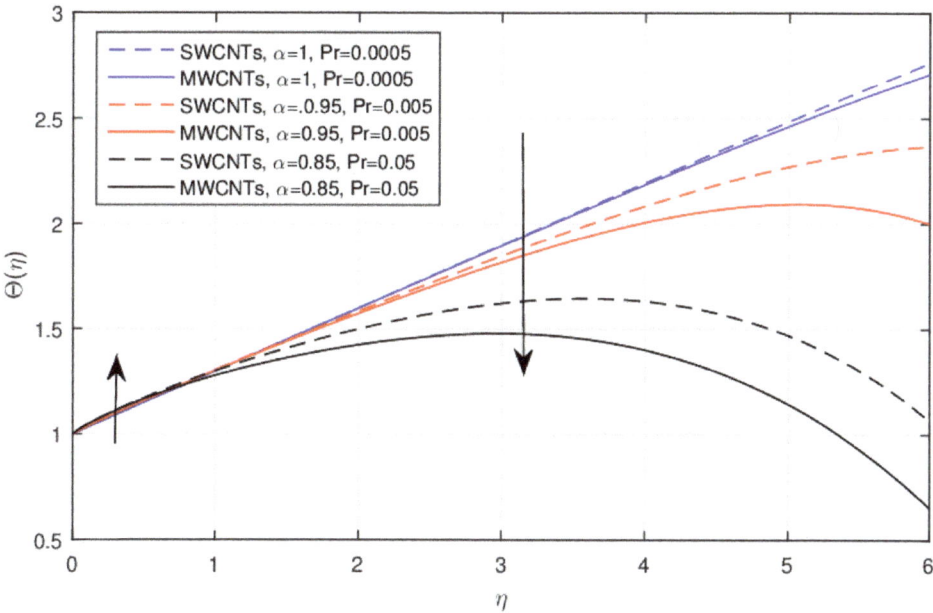

Figure 11. The impact of Pr over the $\Theta(\eta)$ for the fractional values.

The thermophysical properties of the base fluid and SWCNTs/MWCNTs are shown in Table 1. Skin friction and the Nusselt number are the physical parameters of interest under the influence of classical and fractional values. The classical and fractional model outputs for the skin friction and Nusselt number are displayed in Tables 2 and 3, respectively. Both tables specify the decline in the numerical values using the fractional order model. The two types of CNTs were compared in these tables for the fractional order values, and it was observed that the impact of the SWCNTs and MWCNTs varies using the fractional model, which is completely different from the classical model, where identical outputs occur in all cases.

Table 1. The thermo physical properties of carbon nanotubes (CNTs) and the base fluid water.

Physical Properties		Density $\rho(Kg/m^3)$	Thermal Conduct $k(Wm^{-1}/k^{-1})$	Specific Heat $c_p(Kg^{-1}/k^{-1})$
Base fluid Water		997	0.613	4179
Nanoparticles	SWCNT	2600	6600	425
	MWCNT	1600	3000	796

Table 2. The classical and fractional order comparison for the skin fraction comprising (SWCNTs/MWCNTs). When $\alpha = 1, 0.95, 0.90, \Pr = 0.0005, \phi = 0.0001, m = 0.01, n = 0$.

$\alpha=1,\eta.$	$f''(a)$ SWCNTs	$f''(a)$ MWCNTs	$\alpha=0.95,\eta.$	$f''(a)$ SWCNTs	$f''(a)$ MWCNTs	$\alpha=0.90,\eta.$	$f''(a)$ SWCNTs	$f''(a)$ MWCNTs
0.1	0.0988	0.0988	0.1	0.0985	0.0985	0.1	0.0982	0.0982
0.2	0.0967	0.0967	0.2	0.0960	0.0961	0.2	0.0953	0.0954
0.3	0.0938	0.0939	0.3	0.0927	0.0928	0.3	0.0915	0.0916
0.4	0.0901	0.0903	0.4	0.0886	0.0888	0.4	0.0869	0.0871
0.5	0.0856	0.0859	0.5	0.0837	0.0839	0.5	0.0815	0.0818
0.6	0.0804	0.0807	0.6	0.0781	0.0784	0.6	0.0755	0.0758
0.7	0.0745	0.0749	0.7	0.0717	0.0722	0.7	0.0688	0.0693
0.8	0.0679	0.0683	0.8	0.0648	0.0653	0.8	0.0615	0.0621
0.9	0.0606	0.0612	0.9	0.0572	0.0579	0.9	0.0538	0.0545
1.0	0.0527	0.0534	1.0	0.0492	0.0499	1.0	0.0456	0.0464

Table 3. The classical and fractional order comparison for the Nusselt number (SWCNTs/MWCNTs). When $\alpha = 1, 0.95, 0.90, \Pr = 0.0005, \phi = 0.0001, m = 0.0, n = 1$.

$\alpha=1,\eta.$	$\Theta'(a)$ SWCNTs	$\Theta'(a)$ MWCNTs	$\alpha=0.95,\eta.$	$\Theta'(a)$ SWCNTs	$\Theta'(a)$ MWCNTs	$\alpha=0.90,\eta.$	$\Theta'(a)$ SWCNTs	$\Theta'(a)$ MWCNTs
0.1	0.3000	0.3000	0.1	0.3000	0.3000	0.1	0.3000	0.3000
0.2	0.3000	0.2999	0.2	0.3000	0.2999	0.2	0.3000	0.2999
0.3	0.3000	0.2999	0.3	0.3000	0.2999	0.3	0.2999	0.2999
0.4	0.2999	0.2998	0.4	0.2999	0.2998	0.4	0.2999	0.2998
0.5	0.2999	0.2998	0.5	0.2999	0.2997	0.5	0.2999	0.2997
0.6	0.2999	0.2997	0.6	0.2999	0.2997	0.6	0.2998	0.2996
0.7	0.2998	0.2996	0.7	0.2998	0.2996	0.7	0.2998	0.2995
0.8	0.2998	0.2995	0.8	0.2998	0.2995	0.8	0.2998	0.2994
0.9	0.2998	0.2994	0.9	0.2997	0.2994	0.9	0.2997	0.2993
1.0	0.2997	0.2993	1.0	0.2997	0.2992	1.0	0.2997	0.2992

6. Conclusions

The SWCNT and MWCNT water-based nanofluids' flow over a thin needle was analyzed for the enhancement of temperature. Classical and fractional models were used to investigate the impact of the physical parameters and for similar values for the boundary conditions. The non-linear system was solved through the FDE-12 method. The classical and fractional results were obtained for $\alpha = 1$, and $\alpha = 0.95, 0.90$, respectively. The impact of the physical parameters over the velocity and temperature profiles in the classical model were limited, but utilizing the fractional model, the impact of the parameters varied for different intervals. It was observed that the fractional order model specifies the accuracy of the physical parameters more precisely considering the small interval of the derivative between 0 and 1, which have important applications, such as for a fractional order PID controller which may provide a more effective way to improve the system control routine; similarly, non-Fickian transport and anomalous diffusion in porous media, polymer flows, or very high gradients of concentration or heat are important application areas of the fractional order derivative in the field of engineering.

The main findings of this study are:

- Greater values of Pr cause decreases in the thickness of the thermal boundary layer when using the classical model, but by means of the fractional model for the same values of the Prandtl number, the thermal boundary layer near the needle surface increases and decreases after the critical point.
- Lower values of n lead to a decrease in the temperature profile using the classical model values, and this effect is upturned for the fractional order values $\alpha = 0.95, 0.90$ near the wall and change to an upsurge in the thermal boundary layer after the point of inflection.

Author Contributions: T.G.; conceptualization, M.A.K. and W.N.; methodology, I.K.; software, T.A.A.; and I.T.; validation, T.G., M.A.K. and W.N.; formal analysis, I.K.; investigation, T.A.A.; I.K.; and I.T.; writing—original draft preparation, T.G.; M.A.K. and W.N.; writing—review and editing.

Funding: This research received no external funding.

Acknowledgments: Authors acknowledge anonymous referees for their valuable suggestions.

Conflicts of Interest: The authors declare that they have no conflict of interest.

References

1. Sparrow, E.M.; Gregg, J.L. A theory of rotating condensation. *J. Heat Transf.* **1959**, *81*, 113–120.
2. Choi, S.U.S. Enhancing thermal conductivity of fluids with nanoparticles, developments and applications of non-Newtonian flows. *FED-231/MD* **1995**, *66*, 99–105.
3. Volder, M.D.; Tawfick, S.; Baughman, R.; Hart, A. Carbon nanotubes: Present and future commercial applications. *Science* **2013**, *339*, 535–539. [CrossRef] [PubMed]
4. Terrones, M. Science and technology of the twenty-first century: Synthesis, properties, and applications of carbon nanotubes. *Annu. Rev. Mater. Res.* **2003**, *33*, 491–501. [CrossRef]
5. Ellahi, R.; Hassan, M.; Zeeshan, A. Study of Natural Convection MHD Nanofluid by Means of Single and Multi-Walled Carbon Nanotubes Suspended in a Salt-Water Solution. *IEEE Trans. Nanotechnol.* **2015**, *14*, 726–734. [CrossRef]
6. Taza, G.; Waris, K.; Muhammad, S.; Muhammad, A.K.; Ebenezer, B. MWCNTs/SWCNTs Nanofluid Thin Film Flow over a Nonlinear Extending Disc: OHAM Solution. *J. Therm. Sci.* **2018**. [CrossRef]
7. Murshed, S.M.S.; Nieto de Castro, C.A.; Loureno, M.J.V.; Lopes, M.L.M.; Santos, F.J.V. A review of boiling and convective heat transfer with nanofluids. *Renew. Sustain. Energy Rev.* **2011**, *15*, 2342–2354. [CrossRef]
8. Murshed, S.M.S.; Leong, K.C.; Yang, C. Thermophysical and electro kinetic properties of nanofluids a critical review. *Appl. Therm. Eng.* **2008**, *28*, 2109–2125. [CrossRef]
9. Xue, Q. Model for thermal conductivity of carbon nanotube-based composites. *Phys. B Condens. Matter.* **2005**, *368*, 302–307. [CrossRef]
10. Narain, J.P.; Uberoi, M.S. Combined Forced and Free-Convection Heat Transfer From Vertical Thin Needles in a Uniform Stream. *Phys. Fluids* **1972**, *15*, 1879–1882. [CrossRef]
11. Narain, J.P.; Uberoi, M.S. Combined Forced and Free- Convection Over Thin Needles. *Int. J. Heat Mass Transf.* **1973**, *16*, 1505–1512. [CrossRef]
12. Chen, J.L.S. Mixed Convection Flow About Slender Bodies of Revolution. *ASME J. Heat Transf.* **1987**, *109*, 1033–1036. [CrossRef]
13. Wang, C.Y. Mixed convection on a vertical needle with heated tip. *Phys. Fluids A* **1990**, *2*, 622–625. [CrossRef]
14. Grosan, T.; Pop, I. Forced convection boundary layer flow past nonisothermal thin needles in nanofluids. *J. Heat Transf.* **2011**, *133*, 1–4. [CrossRef]
15. Oldham, K.B.; Spanier, J. *The Fractional Calculus*; Academic Press: New York, NY, USA, 1974.
16. Benson, D.; Wheatcraft, S.W.; Meerschaert, M.M. The fractional-order governing equation of Lévy motion. *Water Resour. Res.* **2000**, *36*, 1413–1423. [CrossRef]
17. Benson, D.; Wheatcraft, S.W.; Meerschaert, M.M. Application of a fractional advection–dispersion equation. *Water Resour. Res.* **2000**, *36*, 1403–1412. [CrossRef]

18. Caputo, M. Models of flux in porous media with memory. *Water Resour. Res.* **2000**, *36*, 693–705. [CrossRef]
19. El Amin, M.F.; Radwan, A.G.; Sun, S. Analytical solution for fractional derivative gas-flow equation in porous media. *Results Phys.* **2017**, *7*, 2432–2438. [CrossRef]
20. Atangana, A.; Alqahtani, R.T. Numerical approximation of the space-time Caputo-Fabrizio fractional derivative and application to groundwater pollution equation. *Adv. Differ. Equ.* **2016**, *156*, 1–13. [CrossRef]
21. Alkahtani, B.S.T.; Koca, I.; Atangan, A. A novel approach of variable order derivative: Theory and Methods. *J. Nonlinear Sci. Appl.* **2016**, *9*, 4867–4876. [CrossRef]
22. Agarwal, R.; Hristova, S.; O'Regan, D. Global Mittag-Leffler Synchronization for Neural Networks Modeled by Impulsive Caputo Fractional Differential Equations with Distributed Delays. *Symmetry* **2018**, *10*, 473. [CrossRef]
23. Khan, U.; Ellahi, R.; Khan, R.; Mohyud-Din, S.T. extracting new solitary wave solutions of Benny–Luke equation and Phi-4 equation of fractional order by using (G0/G)-expansion method. *Opt. Quant. Electron.* **2017**, *49*, 362–376. [CrossRef]
24. Hameed, M.; Ambreen, A.K.; Ellahi, R.; Raza, M. Study of magnetic and heat transfer on the peristaltic transport of a fractional second grade fluid in a vertical tube. *Eng. Sci. Technol. Int. J.* **2015**, *18*, 496–502. [CrossRef]
25. Shirvan, K.M.; Ellahi, R.; Sheikholeslami, T.F.; Behzadmehr, A. Numerical investigation of heat and mass transfer flow under the influence of silicon carbide by means of plasmaenhanced chemical vapor deposition vertical reactor. *Neural Comput. Appl.* **2018**, *30*, 3721–3731. [CrossRef]
26. Barikbin, Z.; Ellahi, R.; Abbasbandy, S. The Ritz-Galerkin method for MHD Couette ow of non-Newtonian fluid. *Int. J. Ind. Math.* **2014**, *6*, 235–243.
27. Hayat, T.; Saif, R.S.; Ellahi, R.; Muhammad, T.; Ahmad, B. Numerical study of boundary-layer flow due to a nonlinear curved stretching sheet with convective heat and mass conditions. *Results Phys.* **2017**, *7*, 2601–2606. [CrossRef]
28. Hayat, T.; Saif, R.S.; Ellahi, R.; Muhammad, T.; Ahmad, B. Numerical study for Darcy-Forchheimer flow due to a curved stretching surface with Cattaneo-Christov heat flux and homogeneous heterogeneous reactions. *Results Phys.* **2017**, *7*, 2886–2892. [CrossRef]
29. Javeed, S.; Baleanu, D.; Waheed, A.; Khan, M.S.; Affan, H. Analysis of Homotopy Perturbation Method for Solving Fractional Order Differential Equations. *Mathematics* **2019**, *7*, 40. [CrossRef]
30. Srivastava, H.M.; El-Sayed, A.M.A.; Gaafar, F.M. A Class of Nonlinear Boundary Value Problems for an Arbitrary Fractional-Order Differential Equation with the Riemann-Stieltjes Functional Integral and Infinite-Point Boundary Conditions. *Symmetry* **2018**, *10*, 508. [CrossRef]
31. Diethelm, K.; Freed, A.D. The Frac PECE subroutine for the numerical solution of differential equations of fractional order. In *Forschung und Wissenschaftliches Rechnen*; Heinzel, S., Plesser, T., Eds.; Gessellschaft fur Wissenschaftliche Datenverarbeitung: Gottingen, Germany, 1999; pp. 57–71.
32. Diethelm, K.; Ford, N.J.; Freed, A.D. Detailed error analysis for a fractional Adams method. *Numer. Algorithms* **2004**, *36*, 31–52. [CrossRef]
33. Saifullah Khan, M.A.; Farooq, M. A fractional model for the dynamics of TB virus. *Chaos Solitons Fractals* **2018**, *116*, 63–71.
34. Gul, T.; Khan, M.A.; Khan, A.; Shuaib, M. Fractional-order three-dimensional thin-film nanofluid flow on an inclined rotating disk. *Eur. Phys. J. Plus* **2018**, *133*, 500–5011. [CrossRef]
35. Gul, T.; Haleem, I.; Ullah, I.; Khan, M.A.; Bonyah, E.; Khan, I.; Shuaib, M. The study of the entropy generation in a thin film flow with variable fluid properties past over a stretching sheet. *Adv. Mech. Eng.* **2018**, *10*, 1–15. [CrossRef]

© 2019 by the authors. Licensee MDPI, Basel, Switzerland. This article is an open access article distributed under the terms and conditions of the Creative Commons Attribution (CC BY) license (http://creativecommons.org/licenses/by/4.0/).

Article

Cattaneo–Christov Heat Flux Model for Three-Dimensional Rotating Flow of SWCNT and MWCNT Nanofluid with Darcy–Forchheimer Porous Medium Induced by a Linearly Stretchable Surface

Zahir Shah [1], Asifa Tassaddiq [2], Saeed Islam [1], A.M. Alklaibi [3] and Ilyas Khan [4],*

1. Department of Mathematics, Abdul Wali Khan University, Mardan 23200, Pakistan; zahir1987@yahoo.com (Z.S.); saeedislam@awkum.edu.pk (S.I.)
2. College of Computer and Information Sciences, Majmaah University, Al-Majmaah 11952, Saudi Arabia; a.tassaddiq@mu.edu.sa
3. Department of Mechanical and Industrial Engineering, College of Engineering, Majmaah University, P.O. Box 66 Majmaah 11952, Saudi Arabia; a.alklaibi@mu.edu.sa
4. Faculty of Mathematics and Statistics, Ton Duc Thang University, Ho Chi Minh City 72915, Vietnam
* Correspondence: ilyaskhan@tdt.edu.vn

Received: 2 January 2019; Accepted: 14 February 2019; Published: 6 March 2019

Abstract: In this paper we investigated the 3-D Magnetohydrodynamic (MHD) rotational nanofluid flow through a stretching surface. Carbon nanotubes (SWCNTs and MWCNTs) were used as nano-sized constituents, and water was used as a base fluid. The Cattaneo–Christov heat flux model was used for heat transport phenomenon. This arrangement had remarkable visual and electronic properties, such as strong elasticity, high updraft stability, and natural durability. The heat interchanging phenomenon was affected by updraft emission. The effects of nanoparticles such as Brownian motion and thermophoresis were also included in the study. By considering the conservation of mass, motion quantity, heat transfer, and nanoparticles concentration the whole phenomenon was modeled. The modeled equations were highly non-linear and were solved using homotopy analysis method (HAM). The effects of different parameters are described in tables and their impact on different state variables are displayed in graphs. Physical quantities like Sherwood number, Nusselt number, and skin friction are presented through tables with the variations of different physical parameters.

Keywords: SWCNTs; MWCNTs; stretched surface; rotating system; nanofluid; MHD; thermal radiation; HAM

1. Introduction

Heat transfer phenomenon is important in manufacturing and life science applications, for example in freezing electronics, atomic power plant refrigeration, tissue heat transfer, energy production, etc. Fluids that flow on a stretched surface are more significant among researchers in fields such as manufacturing and commercial processes, for instance in making and withdrawing polymers and gum pieces, crystal and fiber production, food manufacturing, condensed fluid layers, etc. Considering these applications, heat transfer is an essential subject for further investigation in order to develop solutions to stretched surface fluid film problems. The flow of a liquefied sheet was initially considered to obtain a viscid stream and was further extended to stretched surface for non-Newtonian liquids. Choi [1] examined the enhancement of thermal conductivity in nanoparticles deferrals. For the enhancement of thermal conductivity and heat transfer, Hsiao [2,3] performed a successful survey using the Carreau-Nanofluid and Maxwell models, and obtained some interesting results. Ramasubramaniam et al. [4] treated a homogeneous carbon nanotube composite for electrical purposes. Xue [5] work as presenting a CNT model for grounded compounds. Nasir et al. [6] deliberate the nanofluid tinny liquid flow of SWCNTs using an optimal approach. Ellahi et al. [7] presented the usual transmitting nanofluids based on CNTs. Shah et al. [8,9] investigated nanofluid flow in a rotating frame with microstructural and inertial properties with Hall effects in parallel plates. Hayat and others [10] examined Darcy–Forchheimer flow carbon nanotube flow due to a revolving disk. Recently, scholars have been working on finding a rotational flow close to the flexible or non-expandable geometries due its wide array of uses in rotating-generator systems, food handling, spinning devices, disc cleaners, gas transformer designs, etc. Wang [11] presented a perturbation solution for rotating liquid flow through an elastic sheet. The magnetic flux features of rotating flow above a flexible surface was premeditated by Takhar et al. [12]. Shah et al. [13–15] studied nanofluid and heat transfer with radiative and electrical properties using an optimal approach. Rosali et al. [16] presented a numerical survey for flow with rotation over porous surface with exponential contraction. Hayat et al. [17] used the non-Fourier heat fluctuation hypothesis to get a three-dimensional turning stream of the Jeffrey substances. Mustafa. [18] discussed non-linear aspects of rotating nano-fluid flow through the flexible plane. Sheikholeslami et al. [19] inspected the consistent magnetic and radiated effect on water-based nanofluid in a permeable enclosure. Hsiao [20,21] researched the microploar nanofluid stream with MHD on a stretching surface. Khan et al. [22,23] examined nanofluid of micrpoler fluid with the Darcy–Forchheimer and irregular heat generation/absorption between two plates.

The classical Fourier law of conduction [24] is one best model for explanation of the temperature transmission process under numerous relevant conditions. Cattaneo [25] successfully extended the Fourier model in combining significant properties of the temperature reduction period. Cattaneo's work produced a hyperbolic energy equation for the temperature field which allows heat to be transferred by transmission of heat waves with finite velocity. Heat transfer has many practical applications, from the flow of nanofluids to the simulation of skin burns (see Tibullo and Zampoli [26]). Christov [27] discussed the Cattaneo–Maxwell model for finite heat conduction. Straughan and Ciarletta [28] demonstrated the rareness of the solution of the Cattaneo–Christov equation. Straughan et al. [29] presented the heat transfer analysis for this model with a brief discussion of a solution for the model. Recently, Han et al. [30] deliberated the sliding stream with temperature transmission through the Maxwell fluids for the Christov–Cattaneo model. A numerical comparative survey was also presented for the validation of their described results. The current exploration of nanofluid with entropy analysis can be studied in References [31–36].

This paper is based on the features of the Christov–Cattaneo heat flux in rotating nano liquid. A three-dimensional nanofluid flow is considered over a stretching surface with carbon nanotubes (CNTs). An effective thermal conductivity model was used in the enhancement of heat transfer. The problem was modeled from the schematic diagram with concentration. These modeled equations were transformed into a system of non-linear ordinary differential equations. The modeled equations were coupled and highly non-linear and were tackled by an analytical and numerical approach. Homotopy analysis method (HAM) (a high-precision analytical technique proposed by Liao [37]) was used for the solution of the reduced system. Many researcher [38–46] used HAM due to it to it excellent results. Various parameters are presented via graphs. Different physical parameters (thermal relaxation time, skin friction, etc.) with the variations of other physical constraints are presented via graphs and discussed in detail.

2. Effective Thermal Conductivity Models Available in the Literature

Maxwell's [47] proposed a thermal conductivity model as

$$\frac{k_{nf}}{k_f} = 1 + \frac{3\left(\varsigma - 1\right)\psi}{\left(\varsigma + 2\right) - \left(\varsigma - 1\right)\psi}. \tag{1}$$

where $\varsigma = \frac{k_{nf}}{k_f}$ and ψ is a volumetric fraction. Also, Jeffery [48] proposed the following model:

$$\frac{k_{nf}}{k_f} = 1 + 3\chi\psi + \left(3\chi^2 + \frac{3\chi^2}{4} + \frac{9\chi^3}{16}\left(\frac{\varsigma+2}{2\varsigma+3}\right) + \ldots\ldots\right)\psi^2. \qquad (2)$$

where $\chi = \frac{(\varsigma-1)}{(\varsigma+2)}$. After a little modification, Davis [49] presented a model defined as:

$$\frac{k_{nf}}{k_f} = 1 + \frac{3(\varsigma-1)\psi}{(\varsigma+2) - (\varsigma-1)\psi}\left\{\psi + \psi(\varsigma)\psi^2 + O(\psi^3)\right\}. \qquad (3)$$

This model gives a good approximation of thermal conductivity even for a very small capacity and is independent of the atom's form.

Hamilton and Crosser [50] presented a particle-form-based model defined as:

$$\frac{k_{nf}}{k_f} = \frac{\varsigma + (\hbar - 1) - (\varsigma - 1)(\hbar - 1)\psi}{\varsigma + (\hbar - 1) + (1 - \varsigma)\psi}. \qquad (4)$$

Here $\hbar$ denotes the particle form used. The main limitation of the models discussed above is that they can only be used for rotating or circular components and cannot be used for CNTs, especially for their spatial distribution. To overcome this deficiency, Xue [5] presented a model of very large axel relation and used it for the spatial distribution of CNTs. This model has a mathematical description given as:

$$\frac{k_{nf}}{k_f} = \frac{1 - \psi + 2\left(\frac{k_{nf}}{k_{nf}-k_f}\ln\frac{k_{nf}+k_f}{2k_f}\right)\psi}{1 - \psi + 2\left(\frac{k_f}{k_{nf}-k_f}\ln\frac{k_{nf}+k_f}{2k_f}\right)\psi}. \qquad (5)$$

In the present work we implement the Xue [5] model to calculate thermal conductivity.

3. Formulation of the Problem

A three-dimensional rotational flow of CNTs was carried through a linear flexible surface. The temperature distribution was deliberated by the Xue model [5]. The compact fluid describing the Darcy–Forchheimer relationship saturates the permeable area. The stretching surface was adjusted in the Cartesian plane that plates associated in the xy plane. We assumed only the positive values of liquid for z. Surface is extended in the x-direction with a positive rate c. In addition, the liquid is uniformly rotated at a continuous uniform speed ω around the z-axis. Surface temperature is due to convective heating, which is provided by the high temperature of the fluid T_f. The coefficient of this heat transfer is h_f. The relevant equations after applying assumptions are [13–18]:

$$\vec{u}_x + \vec{v}_y + \vec{w}_z = 0 \qquad (6)$$

$$u\vec{u}_x + v\vec{u}_y + w\vec{u}_z - 2\omega v = v_{nf}\vec{u}_{zz} - \frac{v_{nf}}{K^*}u - Fu^2 \qquad (7)$$

$$u\vec{v}_x + v\vec{v}_y + w\vec{v}_z - 2\omega u = v_{nf}\vec{v}_{zz} - \frac{v_{nf}}{K^*}v - Fv^2 \qquad (8)$$

$$\rho c_p(uT_x + vT_y + wT_z) = -\nabla \cdot \vartheta \qquad (9)$$

$$\vartheta + \lambda_2(\vartheta_t + V.\nabla\vartheta - \vartheta.\nabla V + (\nabla.V)\vartheta) = -k\nabla T \qquad (10)$$

$$uT_x + vT_y + wT_z = \frac{k}{\rho c_p}T_{zz} - \lambda_2\left[\begin{array}{c} u^2T_{xx} + v^2T_{yy} + w^2T_{zz} + 2uvT_{xy} + 2vwT_{yz} \\ +2uwT_{xz} + \left(u\vec{u}_x + v\vec{u}_y + w\vec{u}_z\right)T_x \\ +\left(u\vec{v}_x + v\vec{v}_y + w\vec{v}_z\right)T_y + \left(u\vec{w}_x + v\vec{w}_y + w\vec{w}_z\right)T_z \end{array}\right] \qquad (11)$$

$$u\,C_x + v\,C_y + w\,C_z = D_B\,C_{zz} + \frac{D_T}{T_0}\,T_{zz}. \tag{12}$$

The related boundary conditions are:

$$\begin{aligned}
& u = u_w(x) = cx, \quad v = 0, \quad w = 0, \quad -k_{nf}T_z = h_f\left(T_f - T\right), \\
& -k_{nf}C_z = h_f\left(C_f - C\right) \text{ at } z = 0 \\
& u \to 0, \quad v \to 0, \quad T \to T_\infty, \quad C \to C_\infty \text{ as } z \to \infty
\end{aligned} \tag{13}$$

K is the permeability, $F = \frac{C_b}{xK^{*1/2}}$ is the irregular inertial coefficient of the permeable medium, C_b represents drag constant and T_∞ represents the ambient fluid temperature. The basic mathematical features of CNTs are [5]:

$$\mu_{nf} = \frac{\mu_f}{(1-\phi)^{2.5}},\ v_{nf} = \frac{\mu_{nf}}{\rho_{nf}},\ \alpha_{nf} = \frac{k_{nf}}{(\rho c_p)_{nf}},\ \rho_{nf} = \rho_f(1-\phi) + \rho_{CNT}\phi,$$

$$(\rho c_p)_{nf} = (\rho c_p)_f(1-\phi) + (\rho c_p)_{CNT}\phi,\ \frac{k_{nf}}{k_f} = \frac{(1-\phi) + 2\phi\frac{k_{CNT}}{k_{CNT}-k_f}\ln\frac{k_{CNT}+k_f}{2k_f}}{(1-\phi) + 2\phi\frac{k_f}{k_{CNT}-k_f}\ln\frac{k_{CNT}+k_f}{2k_f}} \tag{14}$$

Transformations are taken as follows:

$$\begin{aligned}
& u = cxf'(\eta), \quad v = cxg(\eta), \quad w = -(cv_f)^{1/2}f(\eta), \\
& \Theta(\eta) = \frac{T-T_\infty}{T_f - T_\infty}, \quad \Phi(\eta) = \frac{C-C_\infty}{C_f - C_\infty}, \quad \eta = \left(\frac{c}{v_f}\right)^{1/2}z.
\end{aligned} \tag{15}$$

Now Equation (6) is identically satisfied and Equations (7), (8), (11)–(13) were reduced to

$$\frac{1}{(1-\phi)^{2.5}\left(1 - \phi + \frac{\rho_{CNT}}{\rho_f}\phi\right)}\left(f''' - \lambda f'\right) + ff'' + 2Kg - (1 + F_r)f'^2 = 0 \tag{16}$$

$$\frac{1}{(1-\phi)^{2.5}\left(1 - \phi + \frac{\rho_{CNT}}{\rho_f}\phi\right)}\left(g'' - \lambda g\right) + fg' - f'g - 2Kf' - F_r g^2 = 0 \tag{17}$$

$$\frac{k_{nf}}{k_f}\Theta'' + Pr\left[(1-\phi) + \frac{(\rho c_p)_{nf}}{(\rho c_p)_f}\right]\left[\Theta'\left(\Phi'\right)Nb + (\Theta')^2 Nt + f\Theta'\right] = 0 \tag{18}$$

$$\Phi'' + Scf\Phi' + \frac{Nt}{Nb}\Theta'' = 0. \tag{19}$$

$$\begin{aligned}
& f = 0,\ f' = 1,\ g = 0,\ \Theta' = -\frac{k_f}{k_{nf}}\gamma(1-\Theta),\ \Phi' = \frac{k_f}{k_{nf}}\gamma(1-\Phi) \text{ at } \eta = 0 \\
& f' \to 0,\ g \to 0,\ \Theta \to 0,\ \Phi \to 0,\ \text{when } \eta \to \infty
\end{aligned} \tag{20}$$

The dimensionless parameters are defined as

$$\begin{aligned}
& \lambda = \frac{v_f}{cK^*},\ F_r = \frac{C_b}{K^{*1/2}},\ K = \frac{\omega}{c},\ Pr = \frac{v_f}{\alpha_f},\ \gamma = \frac{h_f}{k_f}\sqrt{\frac{v_f}{c}},\ Sc = \frac{v_f}{D_B}, \\
& Pr = \frac{(\rho c_p)_f}{k_f},\ Nb = \frac{\tau D_B(C_f - C_\infty)}{v_f},\ Nt = \frac{\tau D_T(T_f - T_\infty)}{T_0 v_f}.
\end{aligned} \tag{21}$$

where λ represents porosity K rotation parameter, F_r signifies coefficient of is the inertia, Pr, signifies Prandtl number, Nb is the parameter of Brownian motion, Sc signifies Schmidt number, and γ is Biot number and Nt is the thermophoresis parameter which are defined in Equation (21).

4. Results and Discussion

3-D Magnetohydrodynamic (MHD) rotational nanofluid flow through a stretching surface is modeled. The Cattaneo–Christov heat flux model was used for heat transport phenomenon. By considering the conservation of mass, motion quantity, heat transfer, and nanoparticles concentration the whole phenomenon was modeled. The modeled equations were solved using homotopy analysis method (HAM). Figure 1 Show the geometry of the flow pattern.

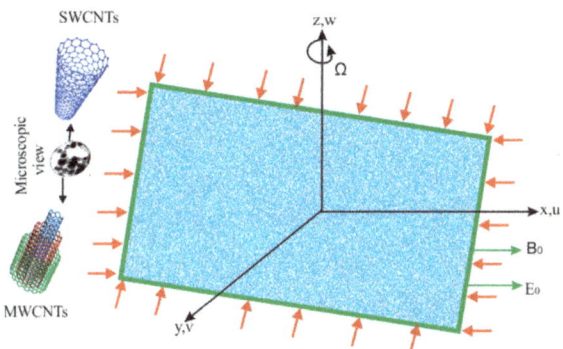

Figure 1. Schematic physical geometry.

Figure 2a–d presents the impact of λ on $f'(\eta)$, $g(\eta)$ & $\Theta(\eta)$ and Biot number γ on $\Theta(\eta)$. Figure 2a displays the deviation of $f'(\eta)$ for different numbers of λ. It was observed that greater porosity parameter λ values indicate a decline in velocity field $f'(\eta)$. Figure 2b reflects the $g(\eta)$ for dissimilar values of the permeability constraint λ. It is detected that for greater permeability constraint λ, the velocity field $g(\eta)$ increased. Figure 2c shows the impact of permeability parameters λ on $\Theta(\eta)$. It is observed that $\Theta(\eta)$ enhanced by increasing the permeability constraint λ for SWCNTs and MWCNTs. Figure 2d illustrates that a greater Biot number γ yields stronger convection, which results in a greater temperature field $\Theta(\eta)$ and hotter sheet wideness.

Figure 3a–d present the impact of K on $f'(\eta)$, $g(\eta)$, $\Theta(\eta)$, and Biot number γ on $\Phi(\eta)$. Figure 3a shows in what way the K affects $f'(\eta)$. A rise in K produced a lesser velocity field $f'(\eta)$ and a smaller momentum sheet wideness of the SWCNTs and MWCNTs. Greater rotational parameter K values resulted in greater rotational rates than tensile rates. Thus, a greater turning effect relates to inferior velocity field $f'(\eta)$ and smaller momentum sheet wideness. Figure 3b describes $g(\eta)$ for K. Larger values of the rotation parameter K, caused a decrease in the velocity field $g(\eta)$. Figure 3c illustrates $\Theta(\eta)$ variations for dissimilar values of K. Greater rotational parameter K decreases the temperature field $\Theta(\eta)$ and supplementary thermal layer width. Figure 3d demonstrates the concentration distribution $\Phi(\eta)$ for varying Biot numbers γ. Higher values of γ indicate enhancement in $\Phi(\eta)$.

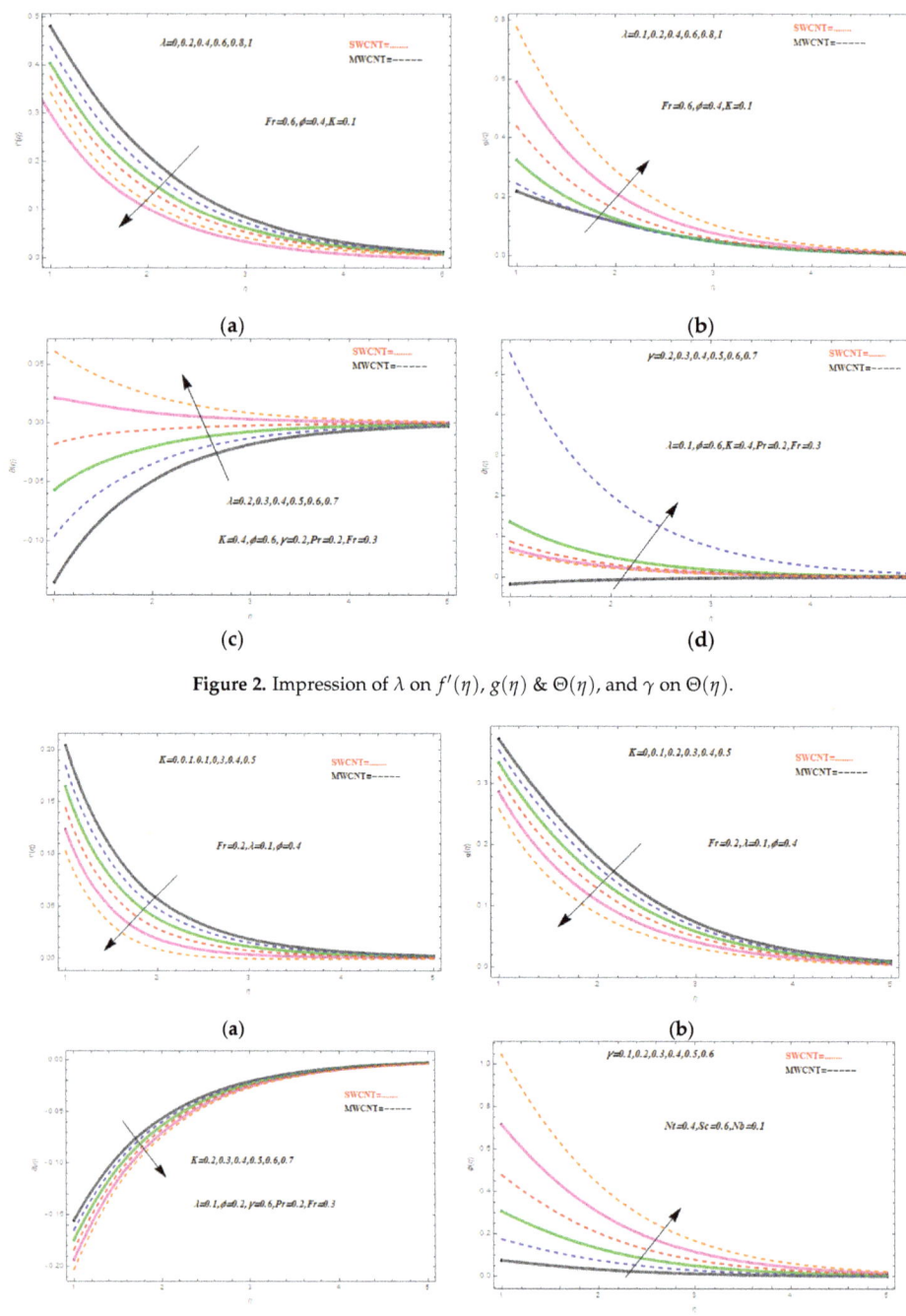

Figure 2. Impression of λ on $f'(\eta)$, $g(\eta)$ & $\Theta(\eta)$, and γ on $\Theta(\eta)$.

Figure 3. Impression of K on $f'(\eta)$, $g(\eta)$ & $\Theta(\eta)$, and γ on $\Phi(\eta)$.

The influences of inertia coefficient F_r on $f'(\eta)$, $g(\eta)$, $\Theta(\eta)$, and Prandtl number P_r on $\Theta(\eta)$ are shown in Figure 4a–d. Figure 4a shows the inertia coefficient F_r on $f'(\eta)$. It is observed that greater values of inertia coefficients F_r resulted in the decline of $f'(\eta)$. Figure 4b depicts the effect of inertia coefficient F_r over the velocity field $g(\eta)$. For greater inertia coefficients F_r of SWCNTs and MWCNTs, there is an increase in the velocity field $g(\eta)$. The influence of the inertia constant F_r on $\Theta(\eta)$ is shown in Figure 4c. Greater rates of inertia factor F_r resulted in powerful temperature field $\Theta(\eta)$ and additional thermal layer thicknesses for SWCNTs and MWCNTs. Figure 4d shows that greater Prandtl number P_r resulted in the decline of the temperature field $\Theta(\eta)$ of SWCNTs and MWCNTs.

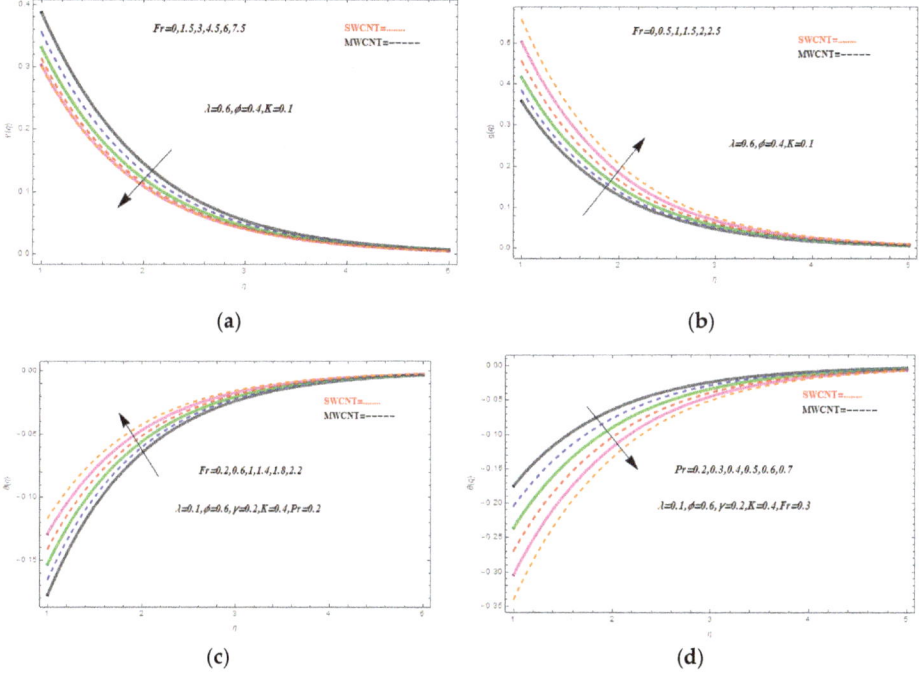

Figure 4. Impression of F_r on $f'(\eta)$, $g(\eta)$, $\Theta(\eta)$, and P_r on $\Theta(\eta)$.

The influences of nanoparticle capacity fraction ϕ on $f'(\eta)$, $g(\eta)$ & $\Theta(\eta)$, Sc on $\Phi(\eta)$ are shown in Figure 5a–d. Figure 5a demonstration the modification in $f'(\eta)$ of the changing nanoparticle capacity fraction ϕ. It was noted that with the rise of the nanoparticle capacity fraction ϕ, an increase $f'(\eta)$ is observed. Results of nanoparticle capacity fraction ϕ on the $g(\eta)$ is shown in Figure 5b. The higher values of the nanoparticle capacity fraction ϕ caused a decreases $g(\eta)$. Figure 5c represents $\Theta(\eta)$ for different nanoparticles volume fraction ϕ. It is observed that greater nanoparticle capacity fraction ϕ resulted in the decline of the temperature field $\Theta(\eta)$. Figure 5d displays the consequence of Sc on $\Phi(\eta)$ of the nanoparticles. It is noticed that an increase in Sc caused a decline in $\Phi(\eta)$.

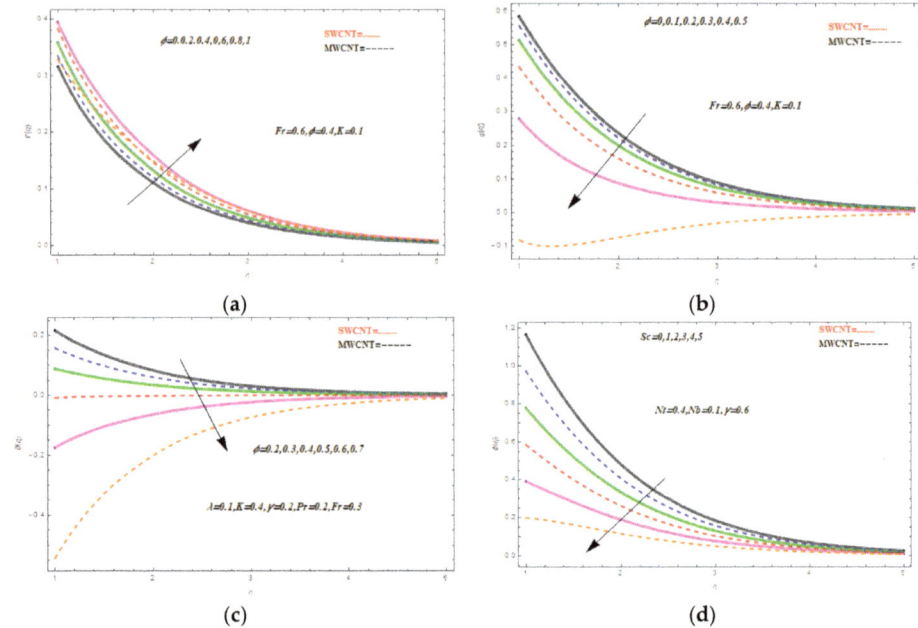

Figure 5. Impression of ϕ on $f'(\eta)$, $g(\eta)$, $\Theta(\eta)$ and Sc on $\Phi(\eta)$.

Figure 6a depicts the concentration distribution $\Phi(\eta)$ for dissimilar values of thermophoretic parameter Nt for SWCNTs and MWCNTs. Higher values of Nt designate the augmentation in $\Phi(\eta)$. Figure 6b depicts the concentration distribution $\Phi(\eta)$ for the varying Brownian motion parameter Nb of SWCNTs and MWCNTs. We noted that greater values of Nb show a reduction in $\Phi(\eta)$ and the connected boundary film thickness.

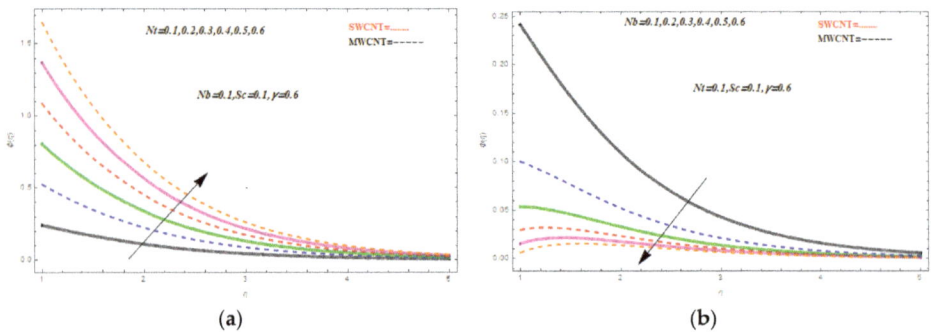

Figure 6. Impression of Nt on $\Theta(\eta)$ and Nb. on $\Phi(\eta)$.

4.1. Table Discussion

Physical values of skin friction for dissimilar values of SWCNTs and MWCNTs in the case of different parameters for C_{fx} and C_{fy} are calculated numerically in Table 1. It was perceived that amassed values of F_r, λ and γ increasing C_{fx} and C_{fy} for SWCNTs nanofluid. Similar results were obtained for MWCNTs. The higher value of K reduces C_{fx} and C_{fy} for SWCNT nanofluid, while for MWCNTs the result was opposite. Physical values for the heat and mass fluxes for dissimilar parameters at $Pr = 7.0$ are calculated in Table 2. Greater values of F_r and K augmented the heat flux as well as the mass flux for both SWCNTs and MWCNTs. The higher value of Nt and Nb reduced the heat flux as well as the mass flux while increasing γ decreased it for both SWCNTs and MWCNTs.

Table 1. Variation in skin friction.

F_r	K	λ	γ	$-C_{fx}$		$-C_{fy}$	
				SWCNT	MWCNT	SWCNT	MWCNT
0.0	0.1	0.1	0.1	1.08621	1.03466	1.24014	1.14536
0.1	–	–	–	1.15376	1.12687	1.27592	1.19880
0.3	–	–	–	1.19736	1.14350	1.34626	1.24368
0.1	0.0	–	–	1.19574	1.04576	1.19574	1.13667
–	0.3	–	–	0.98487	0.23571	1.44073	1.17462
–	0.5	–	–	0.85351	0.02547	1.61115	1.32445
–	0.1	0.0	–	0.96939	0.23419	1.10121	1.03462
–	–	0.3	–	1.54123	1.45483	1.23261	1.100344
–	–	0.5	–	1.68412	1.52445	1.69793	1.37352
–	–	0.1	0.1	1.69222	1.57245	1.72350	1.22359
–	–	–	0.3	1.73432	1.62355	1.88193	1.31912
–	–	–	0.5	1.92365	1.81199	1.90843	1.59331

Table 2. Variation in Nusselt number and Sherwood Number at $Pr = 7.0$.

F_r	K	Nt	Nb	γ	$-Nu_x$		$-Sh_x$	
					SWCNT	MWCNT	SWCNT	MWCNT
0.0	0.1	0.1	0.1	0.1	0.116964	0.231567	0.120208	0.243362
0.1	–	–	–	–	0.116964	0.231390	0.120212	0.243021
0.3	–	–	–	–	0.116965	0.233321	0.120212	0.243204
0.1	0.0	–	–	–	0.116965	0.134136	0.120202	0.243204
–	0.3	–	–	–	0.116966	0.134342	0.120196	0.243192
–	0.5	–	–	–	0.116967	0.134351	0.128786	0.243145
–	0.1	0.3	–	–	0.116953	0.261532	0.13737	0.243145
–	–	0.5	–	–	0.116943	0.261531	0.15026	0.950126
–	–	0.8	–	–	0.116926	0.261530	0.118472	0.950139
–	–	0.1	0.5	–	0.118451	0.156382	0.116457	0.936457
–	–	–	1.0	–	0.120623	0.234521	0.114607	0.934607
–	–	–	1.5	–	0.122502	0.267373	0.11678	0.932678
–	–	–	0.1	0.5	0.116945	0.234536	0.116352	0.566352
–	–	–	–	1.0	0.116928	0.198342	0.116209	0.566209
–	–	–	–	1.5	0.116895	0.162455	0.116198	0.526198

5. Conclusions

Three-dimensional MHD rotational flow of nanofluid over a stretching surface with Cattaneo–Christov heat flux was numerically investigated. Nanofluid is formed as a suspension of SWCNTs and MWCNTs. The modeled equations under different physical parameters were analyzed via graphs for SWNTs. The following main points were concluded from this work.

- It was observed that greater porosity parameter λ values reduced $f'(\eta)$ while increasing $g(\eta)$ and also increasing temperature field $\Theta(\eta)$ for SWCNTs and MWCNTs.
- Greater Biot number γ yields stronger convection, which results in a greater temperature field $\Theta(\eta)$ and hotter sheet wideness.
- Greater rotational parameter K values resulted in greater rotational rates than tensile rates. The higher value of K increased the fluid velocity.
- It was observed that greater values of inertia coefficients F_r resulted in the decline of the velocity field.
- The higher value of the nanoparticle capacity fraction ϕ increased the velocity field and decreased temperature.
- Higher values of Nt indicated enhancement in $\Phi(\eta)$.
- Greater values of Nb showed reduction in $\Phi(\eta)$.
- Larger values of Sc demoted nanoparticle concentration $\Phi(\eta)$.
- It was perceived that increasing values of F_r, λ, and γ increased C_{fx} and C_{fy} for SWCNT nanofluid. Similar results were obtained for MWCNTs.
- The higher values of Nt and Nb reduced the heat flux as well as the mass flux, while increasing γ decreased it for both SWCNTs and MWCNTs.

Author Contributions: Z.S. and S.I. modeled the problem and wrote the manuscript. A.T. and I.K. thoroughly checked the mathematical modeling and English corrections. Z.S. and A.M.A. solved the problem using Mathematica software. I.K., S.A. and A.T. contributed to the results and discussions. All authors finalized the manuscript after its internal evaluation.

Funding: This research received no external funding.

Conflicts of Interest: The authors declare no conflict of interest.

References

1. Choi, S.U.S.; Zhang, Z.G.; Yu, W.; Lockwood, F.E.; Grulke, E.A. Anomalous thermal conductivity enhancement in nanotube suspensions. *Appl. Phys. Lett.* **2001**, *79*, 2252–2254. [CrossRef]
2. Hsiao, K. To promote radiation electrical MHD activation energy thermal extrusion manufacturing system efficiency by using Carreau-Nanofluid with parameters control method. *Energy* **2017**, *130*, 486–499. [CrossRef]
3. Hsiao, K. Combined Electrical MHD Heat Transfer Thermal Extrusion System Using Maxwell Fluid with Radiative and Viscous Dissipation Effects. *Appl. Therm. Eng.* **2017**, *112*, 1281–1288. [CrossRef]
4. Shah, Z.; Dawar, A.; Islam, S.; Khan, I.; Ching, D.L.C. Darcy-Forchheimer Flow of Radiative Carbon Nanotubes with Microstructure and Inertial Characteristics in the Rotating Frame. *Case Stud. Therm. Eng.* **2018**, *12*, 823–832. [CrossRef]
5. Xue, Q.Z. Model for thermal conductivity of carbon nanotube-based composites. *Phys. B Condens. Matter* **2005**, *368*, 302–307. [CrossRef]
6. Nasir, N.; Shah, Z.; Islam, S.; Bonyah, E.; Gul, T. Darcy Forchheimer nanofluid thin film flow of SWCNTs and heat transfer analysis over an unsteady stretching sheet. *AIP Adv.* **2019**, *9*, 015223. [CrossRef]
7. Ellahi, R.; Hassan, M.; Zeeshan, A. Study of natural convection MHD nanofluid by means of single and multi-walled carbon nanotubes suspended in a salt-water solution. *IEEE Trans. Nanotechnol.* **2015**, *14*, 726–734. [CrossRef]
8. Shah, Z.; Islam, S.; Ayaz, H.; Khan, S. The electrical MHD and hall current impact on micropolar nanofluid flow between rotating parallel plates. *Results Phys.* **2018**, *9*, 1201–1214. [CrossRef]
9. Shah, Z.; Islam, S. Radiative heat and mass transfer analysis of micropolar nanofluid flow of Casson fluid between two rotating parallel plates with effects of Hall current. *ASME J. Heat Transf.* **2019**, *141*, 022401. [CrossRef]
10. Hayat, T.; Haider, F.; Muhammad, T.; Alsaedi, A. On Darcy-Forchheimer flow of carbon nanotubes due to a rotating disk. *Int. J. Heat Mass Transf.* **2017**, *112*, 248–254. [CrossRef]

11. Wang, C.Y. Stretching a surface in a rotating fluid. *Zeitschrift für angewandte Mathematik und Physik ZAMP* **1988**, *39*, 177–185. [CrossRef]
12. Takhar, H.S.; Chamkha, A.J.; Nath, G. Flow and heat transfer on a stretching surface in a rotating fluid with a magnetic field. *Int. J. Therm. Sci.* **2003**, *42*, 23–31. [CrossRef]
13. Shah, Z.; Bonyah, E.; Islam, S.; Gul, T. Impact of thermal radiation on electrical mhd rotating flow of carbon nanotubes over a stretching sheet. *Aip Adv.* **2019**, *9*, 015115. [CrossRef]
14. Shah, Z.; Dawar, A.; Islam, S.; Khan, I.; Ching, D.L.C.; Khan, Z.A. Cattaneo-Christov model for Electrical MagnetiteMicropoler Casson Ferrofluid over a stretching/shrinking sheet using effective thermal conductivity model. *Case Stud. Therm. Eng.* **2018**. [CrossRef]
15. Shah, Z.; Gul, T.; Khan, A.M.; Ali, I.; Islam, S. Effects of hall current on steady three dimensional non-newtonian nanofluid in a rotating frame with brownian motion and thermophoresis effects. *J. Eng. Technol.* **2017**, *6*, 280–296.
16. Rosali, H.; Ishak, A.; Nazar, R.; Pop, I. Rotating flow over an exponentially shrinking sheet with suction. *J. Mol. Liquids* **2015**, *211*, 965–969. [CrossRef]
17. Hayat, T.; Qayyum, S.; Imtiaz, M.; Alsaedi, A. Three-dimensional rotating flow of Jeffrey fluid for Cattaneo-Christov heat flux model. *AIP Adv.* **2016**, *6*, 025012. [CrossRef]
18. Mustafa, M.; Mushtaq, A.; Hayat, T.; Alsaedi, A. Rotating flow of magnetite-water nanofluid over a stretching surface inspired by non-linear thermal radiation. *PLoS ONE* **2016**, *11*, e0149304. [CrossRef] [PubMed]
19. Sheikholeslami, M.; Shah, Z.; Shafi, A.; Khan, I.; Itili, I. Uniform magnetic force impact on water based nanofluid thermal behavior in a porous enclosure with ellipse shaped obstacle. *Sci. Rep.* **2019**, *9*, 1196. [CrossRef] [PubMed]
20. Hsiao, K. Micropolar Nanofluid Flow with MHD and Viscous Dissipation Effects Towards a Stretching Sheet with Multimedia Feature. *Int. J. Heat Mass Transf.* **2017**, *112*, 983–990. [CrossRef]
21. Hsiao, K. Stagnation Electrical MHD Nanofluid Mixed Convection with Slip Boundary on a Stretching Sheet. *Appl. Therm. Eng.* **2016**, *98*, 850–861. [CrossRef]
22. Khan, A.; Shah, Z.; Islam, S.; Khan, S.; Khan, W.; Khan, Z.A. Darcy–Forchheimer flow of micropolar nanofluid between two plates in the rotating frame with non-uniform heat generation/absorption. *Adv. Mech. Eng.* **2018**, *10*, 1–16. [CrossRef]
23. Khan, A.; Shah, Z.; Islam, S.; Dawar, A.; Bonyah, E.; Ullah, H.; Khan, Z.A. Darcy-Forchheimer flow of MHD CNTs nanofluid radiative thermal behaviour andconvective non uniform heat source/sink in the rotating frame with microstructureand inertial characteristics. *AIP Adv.* **2018**, *8*, 125024. [CrossRef]
24. Fourier, J. Theorie Analytique de la Chaleur, par M. Fourier; Chez Firmin Didot, père et Fils. 1822. Available online: https://gallica.bnf.fr/ark:/12148/bpt6k1045508v.texteImage (accessed on 1 January 2019).
25. Cattaneo, C. Sulla conduzione del calore. *Atti Sem. Mat. Fis. Univ. Modena* **1948**, *3*, 83–101.
26. Tibullo, V.; Zampoli, V. A uniqueness result for the Cattaneo–Christov heat conduction model applied to incompressible fluids. *Mech. Res. Commun.* **2011**, *38*, 77–79. [CrossRef]
27. Christov, C.I. On frame indifferent formulation of the Maxwell–Cattaneo model of finite-speed heat conduction. *Mech. Res. Commun.* **2009**, *36*, 481–486. [CrossRef]
28. Straughan, B. Thermal convection with the Cattaneo–Christov model. *Int. J. Heat Mass Transf.* **2010**, *53*, 95–98. [CrossRef]
29. Ciarletta, M.; Straughan, B. Uniqueness and structural stability for the Cattaneo–Christov equations. *Mech. Res. Commun.* **2010**, *37*, 445–447. [CrossRef]
30. Han, S.; Zheng, L.; Li, C.; Zhang, X. Coupled flow and heat transfer in viscoelastic fluid with Cattaneo–Christov heat flux model. *Appl. Math. Lett.* **2014**, *38*, 87–93. [CrossRef]
31. Feroz, N.; Shah, Z.; Islam, S.; Alzahrani, E.O.; Khan, W. Entropy Generation of Carbon Nanotubes Flow in a Rotating Channel with Hall and Ion-Slip Effect Using Effective Thermal Conductivity Model. *Entropy* **2019**, *21*, 52. [CrossRef]
32. Alharbi, S.O.; Dawar, A.; Shah, Z.; Khan, W.; Idrees, M.; Islam, S.; Khan, I. Entropy Generation in MHD Eyring–Powell Fluid Flow over an Unsteady Oscillatory Porous Stretching Surface under the Impact of Thermal Radiation and Heat Source/Sink. *Appl. Sci.* **2018**, *8*, 2588. [CrossRef]
33. Khan, N.S.; Shah, Z.; Islam, S.; Khan, I.; Alkanhal, T.A.; Tlili, I. Entropy Generation in MHD Mixed Convection Non-Newtonian Second-Grade Nanoliquid Thin Film Flow through a Porous Medium with Chemical Reaction and Stratification. *Entropy* **2019**, *21*, 139. [CrossRef]

34. Ishaq, M.; Ali, G.; Shah, Z.; Islam, S.; Muhammad, S. Entropy Generation on Nanofluid Thin Film Flow of Eyring–Powell Fluid with Thermal Radiation and MHD Effect on an Unsteady Porous Stretching Sheet. *Entropy* **2018**, *20*, 412. [CrossRef]
35. Dawar, A.; Shah, Z.; Khan, W.; Idrees, M.; Islam, S. Unsteady squeezing flow of MHD CNTS nanofluid in rotating channels with Entropy generation and viscous Dissipation. *Adv. Mech. Eng.* **2019**, *10*, 1–18. [CrossRef]
36. Sheikholeslami, M.; Shah, Z.; Tassaddiq, A.; Shafee, A.; Khan, I. Application of Electric Field for Augmentation of Ferrofluid Heat Transfer in an Enclosure including Double Moving Walls. *IEEE Access* **2019**. [CrossRef]
37. Liao, S. Notes on the homotopy analysis method: Some definitions and theorems. *Commun. Nonlinear Sci. Numer. Simul.* **2009**, *14*, 983–997. [CrossRef]
38. Shah, Z.; Bonyah, E.; Islam, S.; Khan, W.; Ishaq, M. Radiative MHD thin film flow of Williamson fluid over an unsteady permeable stretching. *Heliyon* **2018**, *4*, e00825. [CrossRef] [PubMed]
39. Jawad, M.; Shah, Z.; Islam, S.; Islam, S.; Bonyah, E.; Khan, Z.A. Darcy-Forchheimer flow of MHD nanofluid thin film flow with Joule dissipation and Navier's partial slip. *J. Phys. Commun.* **2018**. [CrossRef]
40. Khan, N.; Zuhra, S.; Shah, Z.; Bonyah, E.; Khan, W.; Islam, S. Slip flow of Eyring-Powell nanoliquid film containing graphene nanoparticles. *AIP Adv.* **2018**, *8*, 115302. [CrossRef]
41. Fiza, M.; Islam1, S.; Ullah, H.; Shah, Z.; Chohan, F. An Asymptotic Method with Applications to Nonlinear Coupled Partial Differential Equations. *Punjab Univ. J. Math.* **2018**, *50*, 139–151.
42. Ali, A.; Sulaiman, M.; Islam, S.; Shah, Z.; Bonyah, E. Three-dimensional magnetohydrodynamic (MHD) flow of Maxwell nanofl uid containing gyrotactic micro-organisms with heat source/sink. *AIP Adv.* **2018**, *8*, 085303. [CrossRef]
43. Dawar, A.; Shah, Z.; Idrees, M.; Khan, W.; Islam, S.; Gul, T. Impact of Thermal Radiation and Heat Source/Sink on Eyring–Powell Fluid Flow over an Unsteady Oscillatory Porous Stretching Surface. *Math. Comput. Appl.* **2018**, *23*, 20. [CrossRef]
44. Hammed, K.; Haneef, M.; Shah, Z.; Islam, I.; Khan, W.; Asif, S.M. The Combined Magneto hydrodynamic and electric field effect on an unsteady Maxwell nanofluid Flow over a Stretching Surface under the Influence of Variable Heat and Thermal Radiation. *Appl. Sci.* **2018**, *8*, 160. [CrossRef]
45. Khan, A.S.; Nie, Y.; Shah, Z.; Dawar, A.; Khan, W.; Islam, S. Three-Dimensional Nanofluid Flow with Heat and Mass Transfer Analysis over a Linear Stretching Surface with Convective Boundary Conditions. *Appl. Sci.* **2018**, *8*, 2244. [CrossRef]
46. Maxwell, J.C. *Electricity and Magnetism*, 3rd ed.; Clarendon: Oxford, UK, 1904.
47. Jawad, M.; Shah, Z.; Islam, S.; Majdoubi, J.; Tlili, I.; Khan, W.; Khan, I. Impact of Nonlinear Thermal Radiation and the Viscous Dissipation Effect on the Unsteady Three-Dimensional Rotating Flow of Single-Wall Carbon Nanotubes with Aqueous Suspensions. *Symmetry* **2019**, *11*, 207. [CrossRef]
48. Jeffrey, D.J. Conduction through a random suspension of spheres. *Proc. R. Soc. Lond. A* **1973**, *335*, 355–367. [CrossRef]
49. Davis, R.H. The effective thermal conductivity of a composite material with spherical inclusions. *Int. J. Thermophys.* **1986**, *7*, 609–620. [CrossRef]
50. Hamilton, R.L.; Crosser, O.K. Thermal conductivity of heterogeneous two-component systems. *Ind. Eng. Chem. Fundam.* **1962**, *1*, 187–191. [CrossRef]

© 2019 by the authors. Licensee MDPI, Basel, Switzerland. This article is an open access article distributed under the terms and conditions of the Creative Commons Attribution (CC BY) license (http://creativecommons.org/licenses/by/4.0/).

Article

Analytical Study of the Head-On Collision Process between Hydroelastic Solitary Waves in the Presence of a Uniform Current

Muhammad Mubashir Bhatti [1,2] **and Dong Qiang Lu** [1,2,*]

1 Shanghai Institute of Applied Mathematics and Mechanics, Shanghai University, Yanchang Road, Shanghai 200072, China; muhammad09@shu.edu.cn
2 Shanghai Key Laboratory of Mechanics in Energy Engineering, Yanchang Road, Shanghai 200072, China
* Correspondence: dqlu@shu.edu.cn

Received: 7 January 2019; Accepted: 1 March 2019; Published: 6 March 2019

Abstract: The present study discusses an analytical simulation of the head-on collision between a pair of hydroelastic solitary waves propagating in the opposite directions in the presence of a uniform current. An infinite thin elastic plate is floating on the surface of water. The mathematical modeling of the thin elastic plate is based on the Euler–Bernoulli beam model. The resulting kinematic and dynamic boundary conditions are highly nonlinear, which are solved analytically with the help of a singular perturbation method. The Poincaré–Lighthill–Kuo method is applied to obtain the solution of the nonlinear partial differential equations. The resulting solutions are presented separately for the left- and right-going waves. The behavior of all the emerging parameters are presented mathematically and discussed graphically for the phase shift, maximum run-up amplitude, distortion profile, wave speed, and solitary wave profile. It is found that the presence of a current strongly affects the wavelength and wave speed of both solitary waves. A graphical comparison with pure-gravity waves is also presented as a particular case of our study.

Keywords: nonlinear hydroelastic waves; uniform current; thin elastic plate; solitary waves; PLK method

1. Introduction

The interaction between a deformable body and a moving fluid has received remarkable attention due to its numerous applications in offshore, polar engineering and industrial problems. Some applications in transportation systems can be observed in the cold region, where the ice sheet is treated as runways and roads, while air-cushion vehicles are very helpful in breaking the ice. These kinds of problems involve various mathematical challenges and present significant difficulties in the mathematical modeling of wave motion and ice deformation. Most of the previous theoretical and numerical results based on linear wave theories are not applicable to large amplitude waves. Hydroelasticity is associated with the deformation of elastic bodies due to hydrodynamics excitations, and together, these excitations are a result of body deformation. In hydroelastic problems, the elastic body and the fluid motion are coupled, which indicates that the deformation of the elastic body relies on the hydrodynamic forces and vice versa. Hydroelastic problems are difficult to analyze numerically and theoretically because, on the surface of the elastic body, hydrodynamic forces actively depend on the accelerations of the surface displacements.

In the past few years, various theoretical and numerical studies have been presented with the help of the Kirchhoff–Love plate theory to examine hydroelastic wave problems. For instance, Xia and Shen [1] analyzed the nonlinear interaction between hydroelastic solitary waves covered with ice. They used a simple perturbation method to obtain the solution for the nonlinear equations. They found that the wavelength, shape and celerity of nonlinear solitary waves depend on the wave amplitude. The wave speed is less than the wave speed in an open water region. Milewski et al. [2] discussed hydroelastic waves in deep water using a numerical method. They used a nonlinear model for an elastic plate and particularly discussed the dynamics of unforced and forced waves. Vanden–Broeck and Părău [3] investigated the generalized form of hydroelastic periodic and solitary waves in a two-dimensional channel. They derived weakly nonlinear solutions using a perturbation scheme, and fully nonlinear solutions were obtained with the help of a numerical method. Deike et al. [4] experimentally examined the behavior of nonlinear and linear waves propagating beneath an elastic sheet in the presence of flexural waves and surface tension. By using an optical method to derive a full space–time wave field, Deike et al. [4] observed that nonlinear shift occurs due to tension in a sheet by transverse motion of the fundamental mode of an elastic plate. They further noticed that the separation between associated timescales is satisfactory at each scale of a turbulent cascade which coincides with theoretical results. Wang and Lu [5] studied nonlinear hydroelastic waves traveling under an infinite elastic plate on the surface of deep water through the homotopy analysis method.

In the studies mentioned above, less attention has been given to hydroelastic waves in the presence of a uniform current. There are different reasons, i.e., thermal, wind, and tidal effects and the rotation of the earth, why ocean currents are often produced. According to engineering applications, it is essential to determine the behavior of current when it is required to perform refraction calculations, examine the water particle acceleration and velocities for force calculations on ocean structures [6] and calculate the wave height from subsurface pressure recordings. The presence of a current influences the wave speed and affects the observed wave period and the relationship between wavelengths. Physically, when the wave travels from one region to another region in the presence of a current, not only will the wavelength and wave speed change but also, probably, current-induced refraction will occur; furthermore, the wave height will be affected. Schulkes et al. [7] analyzed hydroelastic waves in the presence of a uniform current using linear potential flow theory. Bhattacharjee and Sahoo [8] addressed the interaction of flexural gravity waves with the wave current. They also used a linear approach to discuss the physical features of a floating elastic plate under the impact of a current. Later, Bhattacharjee and Sahoo [9] examined the effect of a uniform current on flexural-gravity waves that occur due to an initial disturbance at a point. Mohanty et al. [10] explored the simultaneous effects of compressive forces and a current on time-dependent hydroelastic waves with both single- and double-layer fluids propagating through a finite and infinite depth in a two-dimensional channel. They presented the asymptotic results for the Green function and the deflection of the elastic plate using the stationary phase method. Lu and Yeung [11] examined the unsteady flexural-gravity waves that occur due to the interaction of a fixed concentrated line load with the impact of a uniform current. They observed that the flexural-gravity wave motion depends on the ratio of the current speed to the group or phase speeds.

In recent decades, various authors have investigated the collision between solitary waves using different methodologies from different geometrical aspects [12–15]. Gardner et al. [16] introduced the inverse scattering transform method to determine the exact solution of the Korteweg–de Vries (KdV) equation and discussed various engrossing characteristics of the collision between solitary waves. According to this technique, one can easily obtain the solution for overtaking solitary waves, but this technique is not suitable for determining the solution of a head-on collision process between solitary waves. When two solitary waves come close to each other, they collide and transfer their positions and energies with each other. After separating, they regain their original shapes and positions. During this entire process of interaction, both solitary waves are very stable and preserve their identities. The features of solitary waves such as striking and colliding, can only be maintained in a conservative system. Su and Mirie [17] studied the head-on collision between two solitary waves with the help of the Poincaré–Lighthill–Kuo (PLK) method. Later, Mirie and Su [18] again numerically studied the head-on collision between solitary waves and observed that after the collision of solitary waves, they recovered their original amplitudes and positions; however, a difference of less than 2% was observed. Dai [19] investigated solitary waves at the interface of a two-layer fluid and considered a rigid bottom and surface of the channel. Mirie and Su [20] examined the head-on collision between internal solitary waves using a perturbation method. With the third-order solution, they observed that the amplitude and energy of the wave train diminish with time. Later, Mirie and Su [21] considered a similar mathematical modeling [20] with a different asymptotic expansion and derived a modified form of the KdV solution. They concluded that the collision process is inelastic, and a dispersive wave train occurs behind each emerging solitary wave. Recently, Ozden and Demiray [22] explored the work of Su and Mirie [17] with a different asymptotic assumption of the trajectory functions. The order of the trajectory functions considered by Su and Mirie [17] is ε, where ε is the perturbation parameter related to the wave amplitude. Ozden and Demiray [22] considered the order of trajectory functions to be ε^2 with a similar definition for ε. Marin and Öchsner [23] discussed the initial boundary value problem for a dipolar medium using the Green–Naghdi thermoelastic theory. Some more relevant studies on the head-on collision in single and two-layer fluids can be found in Refs. [24–28].

According to the previously published results, less attention has been given to hydroelastic solitary waves, and no attempt has been made to analyze the head-on collision mechanism between hydroelastic solitary waves in the presence of a uniform current. Recently, Bhatti and Lu [29] examined the head-on collision between two hydroelastic solitary waves using the Euler–Bernoulli beam model in the presence of compression. Therefore, the present study aims to discuss the head-on collision between two hydroelastic solitary waves under uniform current and surface tension effects. We apply a singular perturbation method to obtain the analytic results for the highly nonlinear coupled partial differential equations. The PLK method is the most appropriate technique to determine the collision properties, i.e., the head-on collision, wave speed, phase shift, distortion profile, and maximum run-up amplitude. The resulting series solutions are presented up to the third-order approximation. A graphical comparison with previously published results is also presented.

2. Mathematical Formulation

Consider a pair of nonlinear hydroelastic solitary waves propagating in the opposite directions through a finite channel. A Cartesian coordinate is selected to formulate the mathematical model, i.e., the x-axis is proposed to lie along the horizontal direction, and the z-axis is considered to lie along the vertical direction as shown in Figure 1. A thin elastic plate is floating on the surface of water at $z = H(x,t)$, and the horizontal bottom is located at $z = 0$. Let U_c be the intensity of an underlying uniform current moving from left to right ($U_c > 0$). An opposing current is defined as that moving from the right to left ($U_c < 0$). The normal velocity of the governing fluid is taken as zero. The fluid is supposed to be incompressible, homogenous and inviscid, and the motion be irrotational. The velocity field in terms of potential function $\phi(x,z,t)$ satisfies

$$\nabla^2 \phi = 0, \qquad (0 < z < H). \qquad (1)$$

The bottom boundary condition at $z = 0$ is written as

$$\frac{\partial \phi}{\partial z} = 0. \qquad (2)$$

The kinematic boundary condition at the water–plate interface ($z = H(x,t)$) is defined as [11,29]

$$\frac{\partial H}{\partial t} + U_c \frac{\partial H}{\partial x} + \nabla \phi \cdot \nabla H = \frac{\partial \phi}{\partial z}. \qquad (3)$$

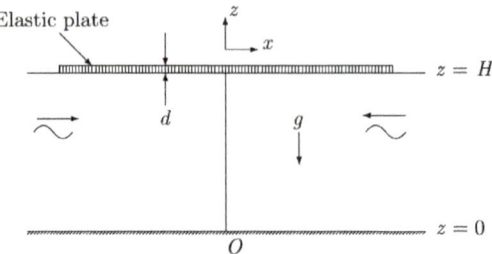

Figure 1. Schematic diagram.

The dynamic boundary condition reads [11,29]

$$\frac{\partial \phi}{\partial t} + \frac{1}{2}(U_c + |\nabla \phi|)^2 + gH - \frac{T}{\rho}\frac{\partial^2 H}{\partial x^2} + \frac{P_e}{\rho} = B_c(t). \tag{4}$$

In the above equation, U_c is a uniform current, g the gravitational acceleration, ρ the density of the fluid, and $B_c(t)$ the Bernoulli constant which is considered to be zero. T is coefficient of surface tension of the fluid. The expression for pressure P_e consists on the Euler–Bernoulli beam theory which can be written as

$$P_e = D\frac{\partial^4 H}{\partial x^4} + M\frac{\partial^2 H}{\partial t^2}, \tag{5}$$

where $D = Ed^3/[12(1-\nu^2)]$ is the flexural rigidity of the plate, $M = \rho_e d$, E Young's modulus, d constant thickness, ν Poisson's ratio, and ρ_e the uniform mass density of the elastic plate.

With the help of Equations (1) and (2), the potential function $\phi(x,z,t)$ can be describe using the Taylor series expansion at $z = 0$, we have [17]

$$\phi(x,z,t) = \sum_{i=0}^{\infty}(-1)^i \frac{z^{2i}}{(2i)!}\nabla^{2i}\Phi, \tag{6}$$

where

$$\phi(x,0,t) = \Phi(x,t). \tag{7}$$

Using Equation (6), the kinematic and dynamic boundary condition can be obtained as

$$\frac{\partial H}{\partial t} + U_c \frac{\partial H}{\partial x} + \nabla \cdot \left[\sum_{i=0}^{\infty}(-1)^i \frac{H^{2i+1}}{(2i+1)!}\nabla^{2i}(\nabla \Phi)\right] = 0, \tag{8}$$

$$\frac{\partial \Phi}{\partial t} + \frac{1}{2}(U_c + |\nabla \phi|)^2 + gH + \frac{P_e}{\rho} - \frac{T}{\rho}\frac{\partial^2 H}{\partial x^2} \\ + \sum_{i=1}^{\infty}(-1)^i \frac{H^{2i}}{(2i)!}\left[\nabla^{2i}\Phi_t + U_c \nabla^{2j+1}\Phi + \frac{1}{2}\sum_{j=0}^{2i}(-1)^j C_j^{2i} \nabla^{j+1}\Phi \ast \nabla^{2i-j+1}\Phi\right], \tag{9}$$

where

$$C_j^{2i} = \binom{2i}{j} = \frac{2i!}{j!(2i-j)!}, \tag{10}$$

is a binomial coefficient. The asterisk in Equation (9) indicates an inner vector product for the multiplication of even j and odd i.

Equations (8) and (9) can be simplified in the following form as

$$\frac{\partial H}{\partial t} + U_c \frac{\partial H}{\partial x} + \frac{\partial}{\partial x}\left[HU + \sum_{i=1}^{\infty}(-1)^i \frac{H^{2i+1}}{(2i+1)!} \frac{\partial^{2i} U}{\partial x^{2i}}\right] = 0, \tag{11}$$

$$\frac{\partial U}{\partial t} + U_c \frac{\partial U}{\partial x} + \frac{\partial}{\partial x}\left[gH - \frac{T}{\rho}\frac{\partial^2 H}{\partial x^2} + \frac{U^2}{2} + \frac{P_e}{\rho} + \sum_{i=1}^{\infty}(-1)^i \frac{H^{2i}}{(2i)!}\left(\frac{\partial^{2i} U}{\partial t \partial x^{2i-1}}\right.\right.$$
$$\left.\left. + U_c \frac{\partial^{2i} U}{\partial x^{2i}} + \frac{1}{2}\sum_{j=0}^{\infty}(-1)^j C_j^{2i} \frac{\partial^{2i-j} U}{\partial x^{2i-j}} \frac{\partial^j U}{\partial x^j}\right)\right], \tag{12}$$

where $U = \partial \Phi / \partial x$ is the tangential velocity at the bottom of the channel.

3. Solution Methodology

We will employ the PLK method in the ensuing section. Let us introduce the following coordinate transformations in the wave frame, we have

$$\xi_0 = \varepsilon^\delta k(x - Ct), \qquad \eta_0 = \varepsilon^\delta \bar{k}(x + \bar{C}t), \tag{13}$$

where k and $\bar{k}$ are the wave numbers of order unity for the right- and left-going waves, respectively; ε with $0 < \varepsilon \ll 1$ is a dimensionless parameter that represents the order of magnitude of the wave amplitude. C and $\bar{C}$ are the wave speeds of the right- and left-going solitary waves. Using the method of strained coordinates, we introduce the following transformation of wave frame coordinates with phase functions:

$$\xi_0 = \xi - \varepsilon k \theta(\xi, \eta), \qquad \eta_0 = \eta - \varepsilon \bar{k} \varphi(\xi, \eta), \tag{14}$$

where $\theta(\xi,\eta)$ and $\varphi(\xi,\eta)$ are the phase functions to be deduced in the perturbation analysis. The purpose of these functions is to obtain the asymptotic approximations which acquiesce us to analyze the phase changes due to a collision.

According to Ursell's theory of shallow water waves, we consider the scaling of the horizontal wavelength as $\delta = 1/2$. Although other values of δ are also possible, i.e., $0, 1/4, 1/8, 1$, these values have some restriction in our present case. However, these values are used by different authors [21,24,25] to derive various forms of the KdV equation in a two-layer fluid model but fail to give a KdV equation in our case. Later, Dai et al. [27] used $\delta = 0$ to discuss the head-on collision among solitary waves propagating in a compressible Mooney–Rivlin elastic rod. The value of δ plays a significant role and mainly depends upon physical and mathematical assumptions of the governing problem.

Let

$$H = H_0(1 + \zeta), \tag{15}$$

where ζ is the non-dimensional elevation of the plate–fluid interface, and H_0 is the undisturbed depth of the fluid. Considering the linear part of Equations (11) and (12), and assuming the linear solutions of U and ζ take form of $\exp^{i(kx - \omega t)}$, where ω is the wave frequency, then we have the phase speed for the right-going wave as

$$C_c = \frac{c_0}{\chi_2}\left(F + \sqrt{F^2 + \chi_1 \chi_2}\right), \tag{16}$$

and for the left-going wave it reads

$$\bar{C}_c = \frac{c_0}{\chi_2}\left(-F + \sqrt{F^2 + \chi_1 \chi_2}\right), \tag{17}$$

where

$$\chi_1 = 1 + \Gamma(kH_0)^4 + \tau(kH_0)^2 - F^2, \tag{18}$$

$$\chi_2 = 1 + \sigma(kH_0)^2, \tag{19}$$

$$c_0 = \sqrt{gH_0}, \quad \Gamma = \frac{D}{\rho g H_0^4}, \quad \sigma = \frac{M}{\rho H_0}, \quad \tau = \frac{T}{\rho g H_0^2}, \quad F = \frac{U_c}{c_0}. \tag{20}$$

The value c_0 is the phase speed for pure gravity waves in shallow water of finite depth. The hydroelastic phase speed can be reduced for pure-gravity waves [17] by taking $D \to 0$, $M \to 0$, $T \to 0$, and $U_c \to 0$.

For convenience, let us introduce a column vector **T** defined as

$$\mathbf{T} = \begin{pmatrix} \zeta \\ u \end{pmatrix}. \tag{21}$$

New variables are introduced in the form of the following power series:

$$\theta(\xi, \eta) = \theta_0(\eta) + \varepsilon \theta_1(\xi, \eta) + \varepsilon^2 \theta_2(\xi, \eta) + \ldots \tag{22}$$

$$\varphi(\xi, \eta) = \varphi_0(\xi) + \varepsilon \varphi_1(\xi, \eta) + \varepsilon^2 \varphi_2(\xi, \eta) + \ldots \tag{23}$$

$$C = C_c \left(1 + \varepsilon a c_1 + \varepsilon^2 a^2 c_2 + \ldots\right), \tag{24}$$

$$\bar{C} = \bar{C}_c \left(1 + \varepsilon b \bar{c}_1 + \varepsilon^2 b^2 \bar{c}_2 + \ldots\right), \tag{25}$$

$$\mathbf{T} = \varepsilon \mathbf{T}_1(\xi, \eta) + \varepsilon^2 \mathbf{T}_2(\xi, \eta) + \varepsilon^3 \mathbf{T}_3(\xi, \eta) + \ldots, \tag{26}$$

where c_n and $\bar{c}_n$ (i.e., $n = 1, 2, \ldots$) are beneficial for removing the secular terms during the solution procedure; a and b are the amplitude factors.

4. Perturbation Analysis

Substituting Equation (22) into the resulting nonlinear partial differential equations, we obtain a set of coupled differential equations with coefficients in the form of $\varepsilon^{\frac{3}{2}}, \varepsilon^{\frac{5}{2}}, \varepsilon^{\frac{7}{2}}, \ldots$, which are expressed in a sequence as follows.

4.1. Coefficients of $\varepsilon^{3/2}$

The system of first-order equations reduces to the following form as

$$k\mathbf{N}\frac{\partial \mathbf{T}_1}{\partial \xi} + \bar{k}\bar{\mathbf{N}}\frac{\partial \mathbf{T}_1}{\partial \eta} = 0, \tag{27}$$

where

$$\mathbf{N} = \begin{pmatrix} C_c \beta_- & 1 \\ c_0^2 & C_c \beta_- \end{pmatrix}, \quad \bar{\mathbf{N}} = \begin{pmatrix} \bar{C}_c \beta_+ & 1 \\ c_0^2 & \bar{C}_c \beta_+ \end{pmatrix}, \tag{28}$$

$$\beta_- = -1 + F\sqrt{\chi}, \quad \beta_+ = 1 + F\sqrt{\bar{\chi}}, \quad \chi = \frac{c_0^2}{C_c^2}, \quad \bar{\chi} = \frac{c_0^2}{\bar{C}_c^2}. \tag{29}$$

We define a matrix system to determine the solutions of the first order equations and higher order equations. Su and Mirie [17] introduced a new form of transformation to obtain the solutions, but the transformation fails to provide a solution for the hydroelastic wave speed (see Equation (16)). Zhu and Dai [24] successfully used a similar methodology for a two-layer fluid model, and Dai et al. [27] used a matrix system to examine the solutions for a single-layer fluid model.

The right and left characteristic vectors of $\mathbf{N}$ and $\overline{\mathbf{N}}$ are

$$\mathbf{R} = \begin{pmatrix} 1 \\ -C_c \beta_- \end{pmatrix}, \quad \overline{\mathbf{R}} = \begin{pmatrix} 1 \\ -\overline{C}_c \beta_+ \end{pmatrix}, \tag{30}$$

$$\mathbf{L} = \begin{pmatrix} 1, & -\dfrac{1}{C_c \beta_-} \end{pmatrix}, \quad \overline{\mathbf{L}} = \begin{pmatrix} 1, & -\dfrac{1}{\overline{C}_c \beta_+} \end{pmatrix}. \tag{31}$$

The right and left characteristic vectors in Equations (30) and (31) are introduced to determine the solution at each order. The right characteristic vectors are helpful for obtaining the solution at each order of approximation, whereas the left characteristic vectors are beneficial for solving the coupled equations at a higher order. In higher-order approximations, the resulting equations are more complex and highly coupled. It is impossible to solve the equations directly. Therefore, left characteristic vectors are beneficial for making the coupled equations into one equation at each order of approximation.

Let us consider the solution of Equation (27) in the following form

$$\mathbf{T}_1 = aA(\xi)\mathbf{R} + bB(\eta)\overline{\mathbf{R}}, \tag{32}$$

where $A(\xi)$ and $B(\eta)$ are arbitrary functions to be determined in the next order.

The first-order solution can be written as

$$\zeta_1 = aA(\xi) + bB(\eta), \tag{33}$$

$$U_1 = -\left(\overline{C}_c \beta_+ bB(\eta) + C_c \beta_- aA(\xi)\right). \tag{34}$$

In the above equation, by taking $F = 0$, the present results reduce to results similar to those obtained by Bhatti and Lu [29] for hydroelastic solitary waves.

4.2. Coefficients of $\varepsilon^{5/2}$

The system of second-order equations reduces to the following form as

$$\begin{aligned} & \mathbf{N}k\dfrac{\partial \mathbf{T}_2}{\partial \xi} + aC_c k\left(\mathbf{E}_1 A' + \mathbf{E}_2 AA' + \mathbf{E}_3 A''' + \mathbf{E}_4 A'\right) \\ & + \overline{\mathbf{N}}\overline{k}\dfrac{\partial \mathbf{T}_2}{\partial \eta} + b\overline{C}_c \overline{k}\left(\overline{\mathbf{E}}_1 B' + \overline{\mathbf{E}}_2 BB' + \overline{\mathbf{E}}_3 B''' + \overline{\mathbf{E}}_4 B'\right) = 0, \end{aligned} \tag{35}$$

where $\mathbf{E}_n$ and $\overline{\mathbf{E}}_n$ i.e., ($n = 1, 2, 3, 4$) are presented in Appendix A.

Let us assume a general solution of the following form

$$\mathbf{T}_2 = X(\xi, \eta)\mathbf{R} + Y(\xi, \eta)\overline{\mathbf{R}}. \tag{36}$$

Using Equation (36) in Equation (35), and multiplying it by $\mathbf{L}$ and $\overline{\mathbf{L}}$, we obtain

$$\begin{aligned} & \mathbf{L}\overline{\mathbf{N}}\mathbf{R}\overline{k}\dfrac{\partial X}{\partial \eta} + aC_c k\left(\mathbf{LE}_1 A' + \mathbf{LE}_2 AA' + \mathbf{LE}_3 A''' + \mathbf{LE}_4 A'\right) \\ & + b\overline{C}_c \overline{k}\left(\mathbf{L}\overline{\mathbf{E}}_1 B' + \mathbf{L}\overline{\mathbf{E}}_2 BB' + \mathbf{L}\overline{\mathbf{E}}_3 B''' + \mathbf{L}\overline{\mathbf{E}}_4 B'\right) = 0, \end{aligned} \tag{37}$$

$$\bar{\mathbf{L}}\bar{\mathbf{N}}\bar{\mathbf{R}}k\frac{\partial Y}{\partial \zeta} + aC_c k \left(\bar{\mathbf{L}}\bar{\mathbf{E}}_1 A' + \bar{\mathbf{L}}\bar{\mathbf{E}}_2 AA' + \bar{\mathbf{L}}\bar{\mathbf{E}}_3 A''' + \bar{\mathbf{L}}\bar{\mathbf{E}}_4 A'\right) \\ + b\bar{C}_c \bar{k} \left(\bar{\mathbf{L}}\bar{\mathbf{E}}_1 B' + \bar{\mathbf{L}}\bar{\mathbf{E}}_2 BB' + \bar{\mathbf{L}}\bar{\mathbf{E}}_3 B''' + \bar{\mathbf{L}}\bar{\mathbf{E}}_4 B'\right) = 0, \quad (38)$$

where $\mathbf{L}\bar{\mathbf{N}}\mathbf{R}$, $\bar{\mathbf{L}}\bar{\mathbf{N}}\bar{\mathbf{R}}$, $\mathbf{LE}_n$ and $\bar{\mathbf{L}}\bar{\mathbf{E}}_n$ ($n = 1, 2, 3, 4$) represent the inner products $\mathbf{L} \cdot \mathbf{N} \cdot \mathbf{R}$, $\bar{\mathbf{L}} \cdot \mathbf{N} \cdot \bar{\mathbf{R}}$, $\mathbf{L} \cdot \mathbf{E}_n$ and $\mathbf{L} \cdot \bar{\mathbf{E}}_n$, respectively.

The above equation is further divided into three parts namely, (i) secular terms, (ii) local terms and (iii) non-local terms.

4.2.1. Secular Terms

Secular terms in Equation (37) are those terms which do not depend on η. The other terms cannot be treated as secular terms, because if we integrate these terms with respect to η, then these terms become unbounded in space and time and show a secular behavior. The secular terms are

$$\mathbf{LE}_1 A' + \mathbf{LE}_2 AA' + \mathbf{LE}_3 A''' = 0. \quad (39)$$

Let

$$c_1 = \frac{1}{2}, \quad k^2 H_0^2 = 3a, \quad (40)$$

then Equation (39) reduces to the following form

$$\gamma A''' - 3\beta_- AA' - A' = 0, \quad (41)$$

where

$$\gamma = -\frac{1}{\beta_-}\left(\beta_-^2 + 3\alpha\right), \quad \alpha = \sigma - \chi\tau. \quad (42)$$

The solution of the above KdV equation is found as

$$A = -\frac{1}{\beta_-}\mathrm{sech}^2 \frac{\zeta}{2\sqrt{\gamma}}. \quad (43)$$

Similarly,

$$\bar{\mathbf{L}}\bar{\mathbf{E}}_1 B' + \bar{\mathbf{L}}\bar{\mathbf{E}}_2 BB' + \bar{\mathbf{L}}\bar{\mathbf{E}}_3 B''' = 0. \quad (44)$$

Let

$$\bar{c}_1 = \frac{1}{2}, \quad \bar{k}^2 H_0^2 = 3b. \quad (45)$$

Then Equation (44) reduces to the following form

$$\bar{\gamma} B''' + 3\beta_+ BB' - B' = 0, \quad (46)$$

$$B = \frac{1}{\beta_+}\mathrm{sech}^2 \frac{\eta}{2\sqrt{\bar{\gamma}}}, \quad (47)$$

where

$$\bar{\gamma} = \frac{1}{\beta_+}\left(\beta_+^2 + 3\alpha\right). \quad (48)$$

We have obtained a third-order KdV equation for the hydroelastic wave profile. A third-order KdV profile was also presented by Bhatti and Lu [29] in the absence of a uniform current, i.e., $F = 0$. Furthermore, the stiffness of the plate appears in the third-order approximation.

4.2.2. Non-Local Terms

The non-local terms are not secular. However, these terms are helpful for determining the phase shifts. Therefore, we will leave these terms as they are. The non-local terms appearing in Equation (37) are

$$\mathbf{LE}_4 = 0. \tag{49}$$

It follows that

$$\theta_0 = \frac{b}{\bar{k}Y} \int_{-\infty}^{\eta} B d\eta_1, \tag{50}$$

where

$$Y = \frac{2\beta_+ \beta_- - \sqrt{\chi\bar{\chi}} - S\beta_-^2}{\beta_-(2\beta_+ + S\beta_-)}, \quad S = \frac{C_c}{\bar{C}_c}. \tag{51}$$

Similarly, we have

$$\overline{\mathbf{LE}_4} = 0. \tag{52}$$

It follows that

$$\varphi_0 = \frac{a}{k\bar{Y}} \int_{+\infty}^{\zeta} A d\zeta_1, \tag{53}$$

where

$$\bar{Y} = \frac{2\beta_- \beta_+ - \sqrt{\chi\bar{\chi}} - \bar{S}\beta_+^2}{\beta_+(2\beta_- + \bar{S}\beta_+)}, \quad \bar{S} = \frac{\bar{C}_c}{C_c}. \tag{54}$$

4.2.3. Local Terms

The local terms are those terms that are helpful in examining the wave speed for the left- and right-going solitary waves. The local terms appearing in Equation (37) can be written in the following form

$$\mathbf{L\bar{N}R}\bar{k}\frac{\partial X}{\partial \eta} + b\overline{C}_c\bar{k}\left(\mathbf{L\bar{E}}_1 B' + \mathbf{L\bar{E}}_2 BB' + \mathbf{L\bar{E}}_3 B''' + \mathbf{L\bar{E}}_4 B'\right) = 0. \tag{55}$$

Integrating the above equation with respect to η, we obtain the resulting equation after simplification

$$X(\zeta, \eta) = C_1 b^2 B + \frac{C_2 b^2}{2} B^2 + \frac{C_3 b^2}{\bar{\gamma}} \left(B - \frac{3\beta_+}{2} B^2 \right) + C_4 ab AB + a^2 A_1(\zeta), \tag{56}$$

where C_n ($n = 1 \ldots 4$) are presented in Appendix B.
Similarly

$$\mathbf{\bar{L}\bar{N}\bar{R}}k\frac{\partial Y}{\partial \zeta} + aC_c k \left(\mathbf{\bar{L}\bar{E}}_1 A' + \mathbf{\bar{L}\bar{E}}_2 AA' + \mathbf{\bar{L}\bar{E}}_3 A''' + \mathbf{\bar{L}\bar{E}}_4 A'\right). \tag{57}$$

Integrating the above equation with respect to ξ, we get the resulting equation after simplification

$$Y(\xi,\eta) = \overline{C}_1 a^2 A + \frac{a^2 \overline{C}_2}{2} A^2 + \frac{\overline{C}_3 a^2}{\gamma}\left(A + \frac{3\beta_-}{2}A^2\right) + \overline{C}_4 ab AB + b^2 B_1(\eta), \qquad (58)$$

where $\overline{C}_n$ ($n = 1\ldots 4$) are presented in Appendix C. The unknown arbitrary functions $A_1(\xi)$ and $B_1(\eta)$ will be determined in the next order.

4.3. Coefficients of $\varepsilon^{7/2}$

In the third-order system, we obtain the following equation

$$\begin{aligned}
& \mathbf{N}k\frac{\partial \mathbf{T}_3}{\partial \xi} + aC_c k \left(\mathbf{F}_1 A' + \mathbf{F}_2 AA' + \mathbf{F}_3 A'A^2 + \mathbf{F}_4\left(A'A_1 + AA_1'\right) + \mathbf{F}_5 A_1' + \mathbf{F}_6 A_1'''\right. \\
& \left. +\mathbf{F}_7 A' + \overline{\mathbf{F}}_8 A'\right) + \overline{\mathbf{N}}k\frac{\partial \mathbf{T}_3}{\partial \eta} + b\overline{C}_c \bar{k}\left(\overline{\mathbf{F}}_1 B' + \overline{\mathbf{F}}_2 BB' + \overline{\mathbf{F}}_3 B'B^2\right. \\
& \left. +\overline{\mathbf{F}}_4\left(B'B_1 + B_1'B\right) + \overline{\mathbf{F}}_5 B_1' + \overline{\mathbf{F}}_6 B_1''' + \overline{\mathbf{F}}_7 B' + \overline{\mathbf{F}}_8 B'\right) = 0,
\end{aligned} \qquad (59)$$

where $\mathbf{F}_n$ and $\overline{\mathbf{F}}_n$ i.e., ($n = 1\ldots 8$) are presented in Appendix A.

Let us assume a general solution of the following form

$$\mathbf{T}_3 = X_1(\xi,\eta)\mathbf{R} + Y_1(\xi,\eta)\overline{\mathbf{R}}. \qquad (60)$$

Using Equation (60) in Equation (59), and multiplying it by $\mathbf{L}$ and $\overline{\mathbf{L}}$, we obtain

$$\begin{aligned}
& \mathbf{L}\overline{\mathbf{N}}\mathbf{R}\bar{k}\frac{\partial X_1}{\partial \eta} + aC_c k\left[(\mathbf{LF}_1 + \mathbf{LF}_2 A + \mathbf{LF}_3 A^2 + \mathbf{LF}_7 + \mathbf{L\overline{F}}_8)A' + \mathbf{LF}_5 A_1'\right. \\
& \left. +\mathbf{LF}_6 A_1''' + \mathbf{LF}_4\left(A'A_1 + AA_1'\right)\right] + b\overline{C}_c \bar{k}\left[(\mathbf{LF}_8 + \mathbf{L\overline{F}}_1 + \mathbf{L\overline{F}}_7 + \mathbf{L\overline{F}}_2 B\right. \\
& \left. +\mathbf{L\overline{F}}_3 B^2)B' + \mathbf{L\overline{F}}_4\left(B'B_1 + B_1'B\right) + \mathbf{L\overline{F}}_5 B_1' + \mathbf{L\overline{F}}_6 B_1'''\right] = 0,
\end{aligned} \qquad (61)$$

$$\begin{aligned}
& \overline{\mathbf{L}}\mathbf{N}\mathbf{R}k\frac{\partial Y_1}{\partial \xi} + aC_c k\left[(\overline{\mathbf{L}}\mathbf{F}_1 + \overline{\mathbf{L}}\mathbf{F}_2 A + \overline{\mathbf{L}}\mathbf{F}_3 A^2 + \overline{\mathbf{L}}\mathbf{F}_7 + \overline{\mathbf{L}}\overline{\mathbf{F}}_8)A' + \overline{\mathbf{L}}\mathbf{F}_5 A_1'\right. \\
& \left. +\overline{\mathbf{L}}\mathbf{F}_6 A_1''' + \overline{\mathbf{L}}\mathbf{F}_4\left(A'A_1 + AA_1'\right)\right] + b\overline{C}_c \bar{k}\left[(\overline{\mathbf{L}}\mathbf{F}_8 + \overline{\mathbf{L}}\overline{\mathbf{F}}_1 + \overline{\mathbf{L}}\overline{\mathbf{F}}_7 + \overline{\mathbf{L}}\overline{\mathbf{F}}_2 B\right. \\
& \left. +\overline{\mathbf{L}}\overline{\mathbf{F}}_3 B^2)B' + \overline{\mathbf{L}}\overline{\mathbf{F}}_4\left(B'B_1 + B_1'B\right) + \overline{\mathbf{L}}\overline{\mathbf{F}}_5 B_1' + \overline{\mathbf{L}}\overline{\mathbf{F}}_6 B_1'''\right] = 0,
\end{aligned} \qquad (62)$$

where $\mathbf{L}\overline{\mathbf{N}}\mathbf{R}$, $\overline{\mathbf{L}}\mathbf{N}\mathbf{R}$, $\mathbf{LF}_n$ and $\mathbf{L\overline{F}}_n$ ($n = 1\ldots 8$) represent the inner products $\mathbf{L}\cdot\overline{\mathbf{N}}\cdot\mathbf{R}$, $\overline{\mathbf{L}}\cdot\mathbf{N}\cdot\mathbf{R}$, $\mathbf{L}\cdot\mathbf{F}_n$ and $\mathbf{L}\cdot\overline{\mathbf{F}}_n$, respectively.

The above equation is further divided into the following three parts: (i) secular terms, (ii) local terms and (iii) non-local terms.

4.3.1. Secular Terms

The secular terms appearing in this order are found as

$$(\mathbf{LF}_1 + \mathbf{LF}_2 A + \mathbf{LF}_3 A^2)A' + \mathbf{LF}_4\left(A'A_1 + AA_1'\right) + \mathbf{LF}_5 A_1' + \mathbf{LF}_6 A_1'''. \qquad (63)$$

The above equation is simplified as

$$A_1''' - A_1' - 3\beta_-\left(A_1 A' + A_1' A\right) + \left(-2c_2 + C_6\right)A' + C_7 AA' + C_8 A^2 A'. \qquad (64)$$

Upon integrating the above equation we get

$$A_1'' - A_1 - 3\beta_- A_1 A + \left(-2c_2 + C_6\right)A + \frac{C_7}{2}A^2 + \frac{C_8}{3}A^3, \qquad (65)$$

where C_6, C_7 and C_8 are presented in Appendix B.

Let
$$c_2 = \frac{C_6}{2}. \tag{66}$$

Then the solution of Equation (64) can be written as
$$A_1 = C_9 A + C_{10} A^2, \tag{67}$$

where C_9 and C_{10} are presented in Appendix B.

Similarly, we have
$$(\overline{\mathbf{LF}}_1 + \overline{\mathbf{LF}}_2 B + \overline{\mathbf{LF}}_3 B^2) B' + \overline{\mathbf{LF}}_4 \left(B' B_1 + B'_1 B \right) + \overline{\mathbf{LF}}_5 B'_1 + \overline{\mathbf{LF}}_6 B'''_1. \tag{68}$$

The above equation is simplified as
$$B'''_1 - B'_1 + 3\beta_+ \left(B_1 B' + B'_1 B \right) - (2\bar{c}_2 + \overline{C}_6) B' - \overline{C}_7 A A' - \overline{C}_8 B^2 B'. \tag{69}$$

Upon integrating the above equation, we obtain
$$B''_1 - B_1 + 3\beta_+ B_1 B - (2\bar{c}_2 + \overline{C}_6) B - \frac{\overline{C}_7}{2} B^2 - \frac{\overline{C}_8}{3} B^3, \tag{70}$$

where $\overline{C}_6, \overline{C}_7$ and $\overline{C}_8$ are presented in Appendix C.

Let
$$\bar{c}_2 = -\frac{\overline{C}_6}{2}. \tag{71}$$

Then the solution of Equation (69) can be written as
$$B_1 = \overline{C}_9 B + \overline{C}_{10} B^2, \tag{72}$$

where $\overline{C}_9$ and $\overline{C}_{10}$ are presented in Appendix C.

This completes the solutions for Equations (56) and (58).

4.3.2. Non-Local Terms

The non-local terms appearing in this order are found as
$$\mathbf{LF}_7 A' = 0. \tag{73}$$

The above equation can be written as
$$\theta_1 = \bar{\theta}_{1,0} \int_{-\infty}^{\eta} B d\eta_1 + \bar{\theta}_{1,1} \int_{-\infty}^{\eta} B^2 d\eta_1, \tag{74}$$

where
$$\bar{\theta}_{1,0} = \frac{b}{\bar{\theta}_{1,2}} \left[C_{11} + a C_{12} A - \frac{C_{14} + a C_{15} A}{\beta_-} \right], \tag{75}$$

$$\bar{\theta}_{1,1} = \frac{b^2}{\bar{\theta}_{1,2}} \left[C_{13} - \frac{C_{16}}{\beta_-} \right], \tag{76}$$

$$\bar{\theta}_{1,2} = -\bar{k}\left[\frac{2\beta_+\beta_- - \sqrt{\chi\bar{\chi}} - S\beta_-^2}{\beta_-}\right], \qquad (77)$$

and C_n ($n = 11\ldots16$) are presented in Appendix B.

Similarly, we have

$$\overline{LF}_7 B' = 0. \qquad (78)$$

The above equation reduces to

$$\varphi_1 = \bar{\varphi}_{1,0}\int_{+\infty}^{\xi} A\,d\xi_1 + \bar{\varphi}_{1,1}\int_{+\infty}^{\xi} A^2\,d\xi_1, \qquad (79)$$

where

$$\bar{\varphi}_{1,0} = \frac{a}{\bar{\varphi}_{1,2}}\left[\overline{C}_{11} + b\overline{C}_{12}B - \frac{\overline{C}_{14} + b\overline{C}_{15}B}{\beta_-}\right], \qquad (80)$$

$$\bar{\varphi}_{1,1} = \frac{a^2}{\bar{\varphi}_{1,2}}\left[\overline{C}_{13} - \frac{\overline{C}_{16}}{\beta_-}\right], \qquad (81)$$

$$\bar{\varphi}_{1,2} = -k\left[\frac{2\beta_-\beta_+ - \sqrt{\chi\bar{\chi}} - \bar{S}\beta_+^2}{\beta_+}\right], \qquad (82)$$

and $\overline{C}_n$ ($n = 11\ldots16$) are presented in Appendix C.

In Equation (74), all the terms appearing are similar to the first-order phase shift and show a simple phase shift behavior except for the third term in $\bar{\theta}_{1,0}$. Few terms in $\bar{\theta}_{1,0}$ depend on ξ when $\eta \to +\infty$; therefore, the wave profile is different before and after the collision process (see Figures 6 and 7) because θ_1 enters into the argument of function $A(\xi)$. A similar behavior has been observed for the left-going wave.

4.3.3. Local Terms

The local terms are found as

$$\begin{aligned}\mathbf{L}\overline{\mathbf{N}}\mathbf{R}\bar{k}\frac{\partial X_1}{\partial \eta} + b\overline{C}_c\bar{k}\left[(\mathbf{LF}_8 + \mathbf{L}\overline{\mathbf{F}}_1 + \mathbf{L}\overline{\mathbf{F}}_7 + \mathbf{L}\overline{\mathbf{F}}_2 B + \mathbf{L}\overline{\mathbf{F}}_3 B^2)B'\right.\\ \left.+\mathbf{L}\overline{\mathbf{F}}_4\left(B'B_1 + B_1'B\right) + \mathbf{L}\overline{\mathbf{F}}_5 B_1' + \mathbf{L}\overline{\mathbf{F}}_6 B_1'''\right] + aC_c k \mathbf{L}\overline{\mathbf{F}}_8 A' = 0.\end{aligned} \qquad (83)$$

Integrating the above equation, we obtain

$$X_1 = \frac{1}{C_5}\left(C_{17}B + C_{18}B^2 + C_{19}B^3\right) + a^3 A_2(\xi), \qquad (84)$$

where C_{17}, C_{18} and C_{19} are presented in Appendix B.

Similarly, we have

$$\begin{aligned}\overline{\mathbf{L}}\mathbf{N}\overline{\mathbf{R}}k\frac{\partial Y_1}{\partial \xi} + aC_c k\left[(\overline{\mathbf{LF}}_1 + \overline{\mathbf{LF}}_2 A + \overline{\mathbf{LF}}_3 A^2 + \overline{\mathbf{LF}}_7 + \overline{\mathbf{LF}}_8)A' + \overline{\mathbf{LF}}_5 A_1'\right.\\ \left.+\overline{\mathbf{LF}}_6 A_1''' + \overline{\mathbf{LF}}_4\left(A'A_1 + AA_1'\right)\right] + b\overline{C}_c \bar{k}\overline{\mathbf{LF}}_8 = 0.\end{aligned} \qquad (85)$$

Integrating the above equation, we obtain

$$Y_1 = \frac{1}{\overline{C}_5}(\overline{C}_{17}A + \overline{C}_{18}A^2 + \overline{C}_{19}A^3) + b^3 B_2(\eta), \tag{86}$$

where $\overline{C}_{17}, \overline{C}_{18}$ and $\overline{C}_{19}$ are presented in Appendix C.

In the above equation, $A_2(\xi)$ and $B_2(\eta)$ are the undetermined functions. For further analysis, we will end our calculations here, and the solutions for $A_2(\xi)$ and $B_2(\eta)$ are neglected.

5. Analytical Results

The series solutions in the preceding section are summarized in the following form.

The interfacial elevation at the water–plate interface can be written with the help of Equations (32) and (36), and we have

$$\zeta = \varepsilon(aA + bB) + \varepsilon^2 [X(\xi, \eta) + Y(\xi, \eta)], \tag{87}$$

where $X(\xi, \eta)$ and $Y(\xi, \eta)$ are defined in Equations (56) and (58).

The distortion profile can be calculated with the help of Equation (87). Therefore, the functions that are products of $B(\eta)$ and $A(\xi)$ must be removed. For this purpose, by taking $B(\eta) = 0$ in Equation (87), the distortion profile at the water–plate interface can be written as

$$\zeta = a\varepsilon A + \varepsilon^2 a^2 \left[\overline{C}_1 A + \frac{\overline{C}_2}{2} A^2 + \frac{\overline{C}_3}{\gamma} \left(A + \frac{3\beta_-}{2} A^2 \right) + a^2 A_1(\xi) \right]. \tag{88}$$

The maximum run-up ζ_{max} during the collision process at the water–plate interface can be obtained by taking $A = B = 1$ in Equation (87), namely

$$\zeta_{max}\Big|_{A=B=1} = \zeta. \tag{89}$$

Following from Equations (32) and (36), the velocity at the bottom reads

$$U = -\varepsilon \left[\overline{C}_c \beta_+ bB + C_c \beta_- aA \right] - \varepsilon^2 \left[C_c \beta_- X(\xi, \eta) + \overline{C}_c \beta_+ Y(\xi, \eta) \right]. \tag{90}$$

Using Equations (40) and (65), the series solutions for the right- and left-going wave speeds are given by

$$C = C_c \left(1 + \frac{1}{2}\varepsilon a + \frac{C_6}{2}\varepsilon^2 a^2 + O(\varepsilon^3) \right), \tag{91}$$

$$\overline{C} = \overline{C}_c \left(1 + \frac{1}{2}\varepsilon b - \frac{\overline{C}_6}{2}\varepsilon^2 b^2 + O(\varepsilon^3) \right). \tag{92}$$

The phase shifts for the right- and left-going wave read

$$\theta = \theta_0 + \varepsilon \theta_1 + O(\varepsilon^2), \tag{93}$$

$$\varphi = \varphi_0 + \varepsilon \varphi_1 + O(\varepsilon^2), \tag{94}$$

where $\theta_0, \theta_1, \varphi_0$, and φ_1 are given in Equations (50), (53), (74) and (79), respectively.

6. Graphical Analysis

This section describes the graphical behaviors of all the physical parameters involved in the governing two-dimensional hydroelastic wave problem. To determine the results in a more significant manner, Figures 2–16 depict the water–plate interface, distortion profile, wave speed, phase shift, and maximum run-up amplitude during the collision process. We consider the physical parameters, unless otherwise stated, $E = 10^6 \, \text{N m}^{-2}$, $d = 0.05 \, \text{m}$, $F = 0.3 \, \text{m s}^{-1}$, $g = 9.8 \, \text{m s}^{-2}$, $\rho = 10^3 \, \text{kg m}^{-3}$, $T = 0.075 \, \text{N m}^{-1}$, $H_0 = 1 \, \text{m}$ and $\rho_e = 917 \, \text{kg m}^{-3}$ for the graphical results. It is worth mentioning here that by taking $\sigma = 0$, $\Gamma = 0$, $\tau = 0$, and $F = 0$ in Equations (3) and (4), the present results reduce to those obtained by Su and Mirie [17] for pure-gravity waves.

Figure 2 shows the wave profile for different values of Γ and σ. When the effects of the elastic plate are taken into account, then significantly changes in the wave profile are observed. The parameter Γ is directly proportional to the plate thickness d and Young's modulus E. When Γ and σ increase, the plate becomes significantly stiffer, which produces a reactive force that opposes the deformation of hydroelastic waves. Figure 3 is plotted to see the effect of the current on the solitary wave profile. We can see from this figure that when the current is moving from right to left, $F < 0$; then, the wavelength, amplitude and speed of the solitary waves are affected as shown in the region $x \in [-50, 0]$. However, when $F > 0$, a similar and converse behavior is found for the second solitary waves in the region $x \in [0, 50]$. Figure 4 shows that an increment in the surface tension parameter τ significantly diminishes the wave profile and that the wave profile before and after the collision process becomes narrower as the surface tension increases. Figure 5 shows a graphical comparison with previously published results presented by Su and Mirie [17]. We can observe that when $\Gamma = 0$, $\sigma = 0$, $\tau = 0$, and $F = 0$, our results are in excellent agreement with those of Su and Mirie [17] for pure gravity waves, which ensures the validity of the present results and the methodology used.

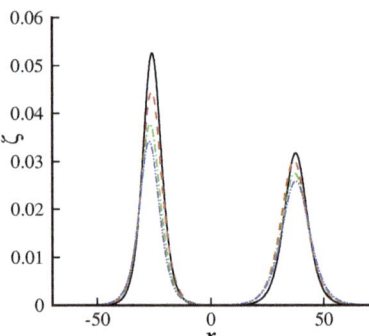

Figure 2. Head-on collision between two solitary waves for different values of Γ and σ. Solid line: $\Gamma = 0$, $\sigma = 0$; dashed line: $\Gamma = 0.07$, $\sigma = 0.5$; dot-dashed line: $\Gamma = 0.09$, $\sigma = 1.1$; dot-dot-dashed line: $\Gamma = 0.1$, $\sigma = 1.4$.

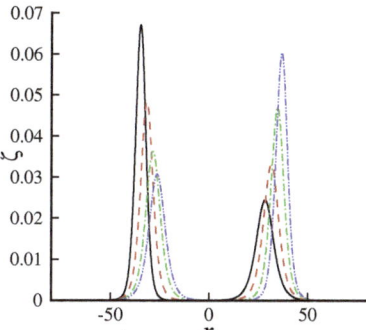

Figure 3. Head-on collision between two solitary waves for different values of F. Solid line: $F = -0.3$ (opposing current); dashed line: $F = -0.1$ (no current); dot-dashed line: $F = 0.3$ (following current); dot-dot-dashed line: $F = 0.5$ (following current).

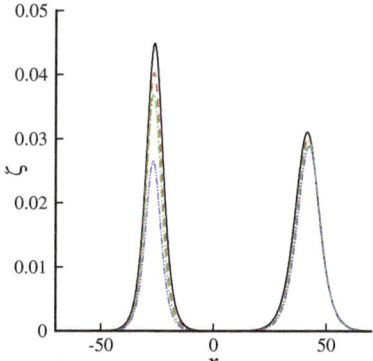

Figure 4. Head-on collision between two solitary waves for different values of τ. Solid line: $\tau = 0.5$; dashed line: $\tau = 0.7$; dot-dashed line: $\tau = 0.8$; dot-dot-dashed line: $\tau = 1$.

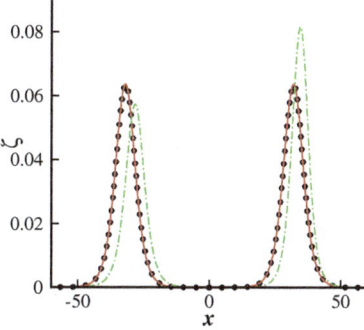

Figure 5. Comparison of head-on collision between two solitary waves with previously published results. Solid line (present results): $\Gamma = 0, \sigma = 0, \tau = 0, F = 0$; circles (Su and Mirie [17]): $\Gamma = 0, \sigma = 0, \tau = 0, F = 0$; dot-dashed line: $\Gamma = 0.11, \sigma = 0.09, \tau = 0.01, F = 0.3$.

Figures 6 and 7 show the distortion profile during the head-on collision process for $F > 0$ and $F < 0$. In both figures, we can see that before the collision process, the wave profile is similar for $F > 0$ and $F < 0$. We can observe from these figures that during the collision process, the wave profile tilts backward in the direction of wave propagation. However, the wave profile remains symmetric before the collision process. Further, we can see that the wave profile is less affected by the following current $F > 0$ compared with the opposing current $F < 0$. A similar behavior was also observed by Su and Mirie [17] for pure gravity waves and Bhatti and Lu [29] for nonlinear hydroelastic waves in the presence of a compressive force.

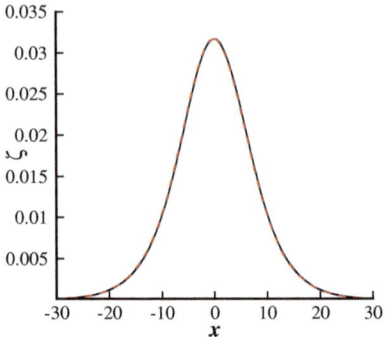

Figure 6. Distortion profile for $F > 0$ (following current). Solid line: before collision; dashed line: after collision.

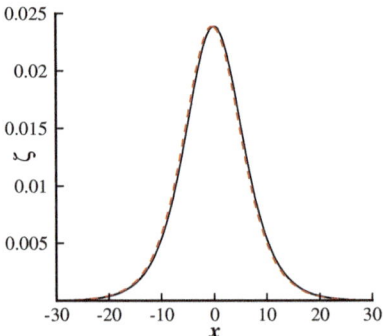

Figure 7. Distortion profile for $F < 0$ (opposing current). Solid line: before collision; dashed line: after collision.

Figures 8–10 are plotted for the phase shift profile against different values of Γ, σ, τ and F. It can be noted from Figure 8 that the initially phase shift profile increases due to the presence of the elastic plate, while its behavior becomes the opposite as the wave amplitude rises. Figure 9 shows that the phase shift profile increases for higher values of the following current ($F > 0$), whereas its behavior is converse for higher values of the opposing current ($F < 0$). Figure 10 shows the effects of the surface tension τ on the phase shift profile. From this figure, we observe that the surface tension results are uniform throughout the domain, whereas the phase shift remarkably decreases due to a strong influence of the surface tension.

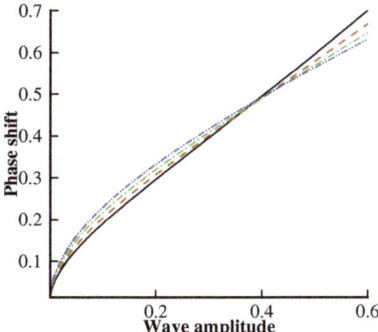

Figure 8. Phase shift *vs* wave amplitude. Solid line: $\Gamma = 0$, $\sigma = 0$; dashed line: $\Gamma = 0.11$, $\sigma = 0.09$; dot-dashed line: $\Gamma = 0.88$, $\sigma = 0.18$; dot-dot-dashed line: $\Gamma = 3$, $\sigma = 0.27$.

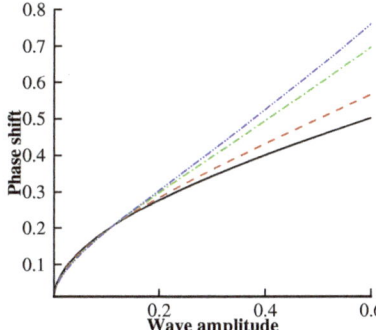

Figure 9. Phase shift *vs* wave amplitude. Solid line: $F = -0.3$ (opposing current); dashed line: $F = -0.1$ (opposing current); dot-dashed line: $F = 0.3$ (following current); dot-dot-dashed line: $F = 0.5$ (following current).

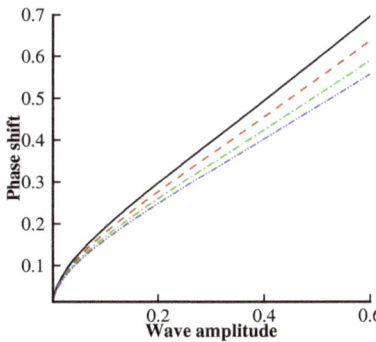

Figure 10. Phase shift *vs* wave amplitude. Solid line: $\tau = 0.1$; dashed line: $\tau = 0.5$; dot-dashed line: $\tau = 0.8$; dot-dot-dashed line: $\tau = 1$.

Figures 11–13 show the variation of the wave speed with multiple values of Γ, σ, τ and F. In Figure 11, we can observe that the wave speed decreases significantly when Γ increases. Furthermore, the behavior of the left-going wave speed will be opposite to that of the right-going one. Figure 12 is plotted to analyze the wave speed behavior for the left- and right-going solitary waves. In Figure 12, we can easily observe that the left-going wave speed tends to diminish with the following current $F > 0$, while the behavior is opposite for the right-going solitary wave. From Figure 13, we note that the surface tension strongly influences the wave speed. We can see that the wave speed decreases due to an increment in the surface tension parameter τ.

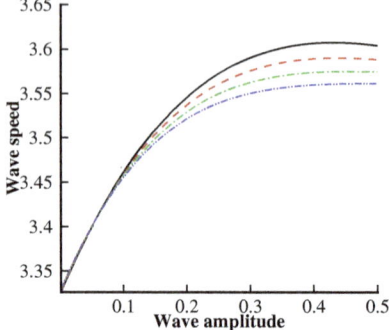

Figure 11. Wave speed *vs* wave amplitude. Solid line: $\Gamma = 0.3$; dashed line: $\Gamma = 0.4$; dot-dashed line: $\Gamma = 0.5$; dot-dot-dashed line: $\Gamma = 0.6$.

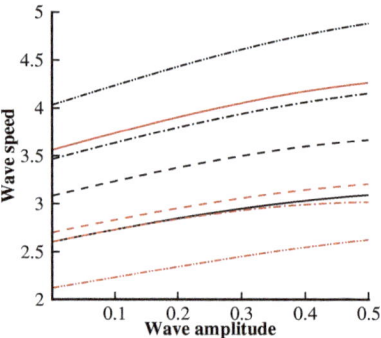

Figure 12. Wave speed *vs* wave amplitude. Solid line: $F = -0.4$ (opposing current); dashed line: $F = 0$ (no current); dot-dashed line: $F = 0.5$ (following current); dot-dot-dashed line: $F = 1$ (following current). Red line: left-going wave; black line: right-going wave.

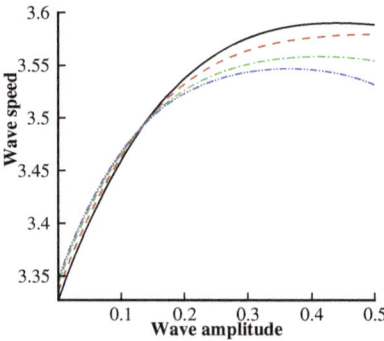

Figure 13. Wave speed *vs* wave amplitude. Solid line: $\tau = 0.1$; dashed line: $\tau = 0.3$; dot-dashed line: $\tau = 0.5$; dot-dot-dashed line: $\tau = 0.6$.

Figures 14–16 present the variations in the maximum run-up during the collision process. It can be noted from Figure 14 that the plate deflection creates an opposing force, which tends to resist the maximum run-up amplitude. Therefore, an increment of Γ or σ tends to diminish the maximum run-up amplitude during the collision process. However, in Figure 15, we can see that the maximum run-up amplitude increases for negative values of the current $F < 0$, whereas its behavior is similar for higher values of the current when $F > 0$. It can be observed from Figure 16 that the surface tension parameter τ significantly affects the wave speed compared with Γ, F and σ. An enhancement in the surface tension tends to reduce the maximum run-up amplitude.

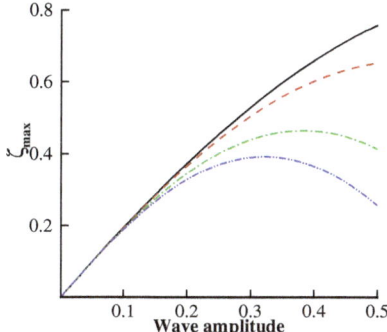

Figure 14. Maximum run-up *vs* wave amplitude. Solid line: $\Gamma = \sigma = 0$; dashed line: $\Gamma = 0.07, \sigma = 0.5$; dot-dashed line: $\Gamma = 0.09, \sigma = 1.1$; dot-dot-dashed line: $\Gamma = 0.1, \sigma = 1.4$.

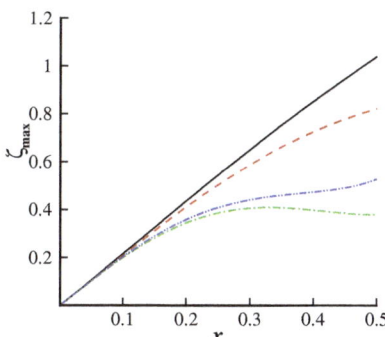

Figure 15. Maximum run-up *vs* wave amplitude. Solid line: $F = -0.5$ (opposing current); dashed line: $F = -0.4$ (opposing current); dot-dashed line: $F = 0.4$ (following current); dot-dot-dashed line: $F = 0.5$ (following current).

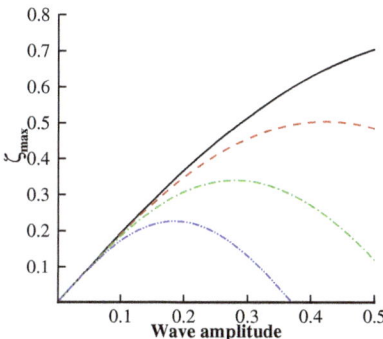

Figure 16. Maximum run-up *vs* wave amplitude. Solid line: $\tau = 0.5$; dashed line: $\tau = 1.0$; dot-dashed line: $\tau = 1.5$; dot-dot-dashed line: $\tau = 2$.

7. Concluding Remarks

In this article, we analytically studied the behavior of a current in the head-on collision process between a pair of hydroelastic solitary waves propagating in the opposite directions. The Poincaré–Lighthill–Kuo method is successfully applied to obtain the solution for the governing nonlinear partial differential equations. The resulting solutions are presented up to the third-order approximation for right- and left-going solitary waves. The impact of different essential parameters is discussed with the help of graphs and presented mathematically for the water–plate interface, wave speed, phase shift, distortion profile, and maximum run-up amplitude during the collision process.

A graphical comparison with previously published results is also presented, and it is found that the present results are in excellent agreement, which ensures that the results for the hydroelastic wave problem are correct. It is also found that the presence of a thin elastic plate significantly reduces the amplitude of the wave profile. Furthermore, we noted that the presence of a current not only affects the wavelength and wave amplitude but also produces a remarkable effect on the wave speed. The phase shift markedly decreases due to the more significant influence of the elastic plate. It is also noted that the phase shift tends to increase for the following current, whereas its behavior is converse for the opposing current. The maximum run-up amplitude increases due to the strong influence of the following and opposing currents. It is observed that very small tilting occurs in the distortion profile during the collision process for positive values of the current, whereas greater tilting in the wave profile is seen for negative current values.

Author Contributions: D.Q.L.; conceptualization, M.M.B.; methodology, validation, formal analysis and writing–original draft preparation, D.Q.L.; investigation, resources, supervision, project administration, funding acquisition and writing–review and editing.

Funding: This research received no external funding.

Acknowledgments: This research was sponsored by the National Natural Science Foundation of China under Grant No. 11872239.

Conflicts of Interest: The authors declare that they have no conflict of interest.

Appendix A

$$\mathbf{E}_1 = -ac_1 \begin{pmatrix} 1 \\ -C_c \beta_- \end{pmatrix}, \tag{A1}$$

$$\mathbf{E}_2 = -a\beta_- \begin{pmatrix} 2 \\ -C_c \beta_- \end{pmatrix}, \tag{A2}$$

$$\mathbf{E}_3 = 3a \begin{pmatrix} \dfrac{\beta_-}{6} \\ C_c \left(\dfrac{\beta_-^2}{2} + \alpha \right) \end{pmatrix}, \tag{A3}$$

$$\mathbf{E}_4 = \begin{pmatrix} -abk \left(\overline{S}\beta_+ + \beta_- \right) B + 2ak\overline{k} \left(\overline{S}\beta_+ - \beta_- \right) \dfrac{\partial \theta_0}{\partial \eta} \\ ak\overline{C}_c \left(b\beta_+\beta_- B + \overline{k}\dfrac{\partial \theta_0}{\partial \eta} \left(\sqrt{\overline{\chi}\overline{\chi}} - \beta_-\beta_+ \right) \right) \end{pmatrix}, \tag{A4}$$

136

$$\overline{\mathbf{E}}_1 = b\bar{c}_1 \begin{pmatrix} 1 \\ \overline{C}_c \beta_+ \end{pmatrix}, \qquad (A5)$$

$$\overline{\mathbf{E}}_2 = b\beta_+ \begin{pmatrix} 2 \\ \overline{C}_c \beta_+ \end{pmatrix}, \qquad (A6)$$

$$\overline{\mathbf{E}}_3 = 3b \begin{pmatrix} -\dfrac{\beta_+}{6} \\ \overline{C}_c \left(-\dfrac{\beta_+^2}{2} + \alpha \right) \end{pmatrix}, \qquad (A7)$$

$$\overline{\mathbf{E}}_4 = \begin{pmatrix} -ab\bar{k}\left(\beta_+ + S\beta_-\right)A - bk\bar{k}\left(\beta_+ - S\beta_-\right)\dfrac{\partial \varphi_0}{\partial \bar{\xi}} \\ b\bar{k}C_c \left(a\beta_+\beta_- A + k\dfrac{\partial \varphi_0}{\partial \bar{\xi}}\left(\sqrt{\chi\bar{\chi}} - \beta_-\beta_+\right)\right) \end{pmatrix}, \qquad (A8)$$

$$\mathbf{F}_1 = a^2 \begin{pmatrix} \dfrac{1}{2\gamma^2}\left(\overline{C}_3 \overline{S}\beta_+ - \overline{C}_3 \gamma + \overline{C}_1 \overline{S}\beta_+ \gamma - \gamma^2(\overline{C}_1 + 2c_2)\right) \\ C_c \left[c_1 \overline{C}_1 \overline{S}\beta_+ + c_2\beta_- + \dfrac{9\Gamma\chi}{\gamma^2} + \dfrac{3\overline{C}_3 \sigma}{\gamma^2} - \dfrac{3\overline{S}\overline{C}_3 \beta_+ \beta_-}{2\gamma^2} \right. \\ \left. -\dfrac{9\beta_-^2}{2\gamma^2} + \dfrac{3\overline{C}_1 \sigma}{\gamma} + \dfrac{\overline{S}\overline{C}_3 \beta_+}{2\gamma} + \dfrac{3\overline{S}\beta_+ \beta_- \overline{C}_1}{2\gamma} - \dfrac{3\beta_-}{4\gamma} \right. \\ \left. -\dfrac{3\overline{C}_3 \tau\chi}{\gamma^2} - \dfrac{3\overline{C}_1 \tau\chi}{\gamma} \right] \end{pmatrix}, \qquad (A9)$$

$$\mathbf{F}_2 = a^2 \begin{pmatrix} \left[-2\left\{(\overline{S}\beta_+ + \beta_-)\left(\overline{C}_1 + \dfrac{\overline{C}_3}{\gamma}\right)\right\} - c_1\left(\overline{C}_2 + \dfrac{3\overline{C}_3 \beta_-}{\gamma}\right) \right. \\ \left. + \dfrac{3\overline{S}\beta_+}{\gamma^2}\left(15\overline{C}_3 \beta_- + 4\overline{C}_2 \gamma + 3\overline{C}_1 \gamma\beta_-\right) \right] \\ C_c \left[2\overline{S}\beta_+\beta_- \left(\overline{C}_1 + \dfrac{\overline{C}_3}{\gamma}\right) + 3\left(\dfrac{\alpha - c_1 \overline{S}\beta_+ \beta_-}{\gamma^2}\right) \right. \\ \left. (15\overline{C}_3 \beta_- + 4\overline{C}_2 \gamma + 3\gamma\overline{C}_1 \beta_-) + c_1 \overline{S}\beta_+\left(\overline{C}_2 + \dfrac{3\overline{C}_3 \beta_-}{\gamma}\right) \right. \\ \left. + \dfrac{135\Gamma\beta_-}{\gamma^2} - \dfrac{45\beta_-^3}{8\gamma^2} + \dfrac{\beta_-^2}{4\gamma} \right] \end{pmatrix}, \qquad (A10)$$

$$\mathbf{F}_3 = a^2 \begin{pmatrix} -3c_1(\overline{S}\beta_+ + \beta_-)\left(\overline{C}_2 + \dfrac{3\overline{C}_3 \beta_-}{\gamma}\right) + 15c_1^2 \overline{S}\beta_+ \left(\dfrac{3\overline{C}_3 \beta_-^2}{\gamma^2} + \dfrac{\overline{C}_2 \beta_-}{\gamma}\right) \\ C_c \beta_- \left[\dfrac{45c_1(3\overline{C}_3 \beta_- + \overline{C}_2 \gamma)(2\alpha - \overline{S}\beta_-\beta_+)}{\gamma^2} + 3c_1 \overline{S}\beta_+ \right. \\ \left. \left(\overline{C}_2 + \dfrac{3\overline{C}_3 \beta_-}{\gamma}\right) + \dfrac{405\chi\Gamma\beta_-}{\gamma^2} + \dfrac{135c_1 \beta_-^3}{8\gamma^2} + \dfrac{27c_1 \beta_-^2}{2\gamma} \right] \end{pmatrix}, \qquad (A11)$$

$$\mathbf{F}_4 = \mathbf{E}_2, \quad \mathbf{F}_5 = \mathbf{E}_1, \quad \mathbf{F}_6 = \mathbf{E}_3, \tag{A12}$$

$$\mathbf{F}_7 = \begin{pmatrix} bC_{11}B + abC_{12}AB + b^2C_{13}B^2 + \bar{k}(\overline{S} - \beta_-)\dfrac{\partial \theta_1}{\partial \eta} \\ \overline{C}_c \left(bC_{14}B + abC_{15}AB + b^2C_{16}B^2 + \bar{k}\left(\sqrt{\overline{\chi}\chi} - \beta_-\beta_+\right)\dfrac{\partial \theta_1}{\partial \eta} \right) \end{pmatrix}, \tag{A13}$$

$$\mathbf{F}_8 = -k\overline{C}_c \dfrac{\partial \theta_1}{\partial \xi} \begin{pmatrix} 0 \\ S^2\chi + \beta_+^2 \end{pmatrix}, \tag{A14}$$

$$\overline{\mathbf{F}}_1 = b^2 \begin{pmatrix} \dfrac{1}{2}C_1 + \bar{c}_2 + \dfrac{3C_3 S\beta_-}{6\overline{\gamma}^2} + \dfrac{C_3}{2\overline{\gamma}} + \dfrac{3C_1 S\beta_-}{6\overline{\gamma}} \\ \overline{C}_c \left[-\bar{c}_2 \beta_+ - \dfrac{1}{2}C_1 S\beta_- + \dfrac{9\overline{\chi}\Gamma}{\overline{\gamma}^2} + \dfrac{3C_3\sigma}{\overline{\gamma}^2} - \dfrac{9\beta_+^2}{24\overline{\gamma}^2} - \dfrac{3C_3 S\beta_- \beta_+}{2\overline{\gamma}}{}^2 \right. \\ \left. + \dfrac{3c_1\sigma}{\overline{\gamma}} + \dfrac{3\beta_+}{4\overline{\gamma}} - \dfrac{C_3 S\beta_-}{2\overline{\gamma}} + \dfrac{3c_1 S\beta_- \beta_+}{2\overline{\gamma}} - 3\overline{\chi}\tau \left(\dfrac{C_3}{\overline{\gamma}^2} + \dfrac{C_1}{\overline{\gamma}} \right) \right] \end{pmatrix}, \tag{A15}$$

$$\overline{\mathbf{F}}_2 = b^2 \begin{pmatrix} \left[-2\left\{ (\beta_+ + S\beta_-)\left(C_1 + \dfrac{C_3}{\overline{\gamma}}\right)\right\} + \dfrac{1}{2}\left(C_2 - \dfrac{3\overline{C}_3 \beta_+}{\overline{\gamma}}\right) \right. \\ \left. + \dfrac{3S\beta_-}{6\overline{\gamma}^2}(15C_3\beta_+ + 4C_2\overline{\gamma} + 3C_1\overline{\gamma}\beta_+) \right] \\ \overline{C}_c \left[2S\beta_+\beta_- \left(C_1 + \dfrac{C_3}{\overline{\gamma}}\right) - 3\left(\dfrac{\alpha + c_1\beta_+ \beta_-}{\overline{\gamma}^2}\right) \right. \\ (15\overline{C}_3\beta_- - 4\overline{C}_2\overline{\gamma} + 3\overline{\gamma}\overline{C}_1\beta_-) - \bar{c}_1 S\beta_+ \left(C_2 + \dfrac{3C_3\beta_-}{\overline{\gamma}}\right) \\ \left. - \dfrac{135\Gamma S^2 \beta_+}{\overline{\gamma}^2} + \dfrac{45S^2 \beta_+^3}{8\overline{\gamma}^2} + \dfrac{\beta_+^2}{4\overline{\gamma}} \right] \end{pmatrix}, \tag{A16}$$

$$\overline{\mathbf{F}}_3 = b^2 \begin{pmatrix} -3c_1(\beta_+ + S\beta_-)\left(C_2 + \dfrac{3C_3\beta_-}{\overline{\gamma}}\right) + 15c_1^2 S\beta_- \left(\dfrac{3C_3\beta_+^2}{\overline{\gamma}^2} - \dfrac{C_2\beta_+}{\overline{\gamma}}\right) \\ \overline{C}_c \left[\dfrac{45c_1\beta_+ (3C_3\beta_- - C_2\overline{\gamma})(2\alpha + S\beta_-\beta_+)}{\overline{\gamma}^2} + 3c_1 S\beta_+ \beta_- \right. \\ \left. \left(C_2 - \dfrac{3C_3\beta_+}{\overline{\gamma}}\right) + \dfrac{405\Gamma\chi\beta_+^2}{\overline{\gamma}^2} - \dfrac{135c_1 S^2 \beta_+^4}{8\gamma^2} - \dfrac{27c_1\beta_+^3}{\overline{\gamma}} \right] \end{pmatrix}, \tag{A17}$$

$$\overline{\mathbf{F}}_4 = \overline{\mathbf{E}}_2, \quad \overline{\mathbf{F}}_5 = \overline{\mathbf{E}}_1, \quad \overline{\mathbf{F}}_6 = \overline{\mathbf{E}}_3, \tag{A18}$$

$$\overline{\mathbf{F}}_7 = \begin{pmatrix} a\overline{C}_{11}A + ab\overline{C}_{12}AB + a^2\overline{C}_{13}A^2 + k(\beta_+ + S\beta_-)\dfrac{\partial \varphi_1}{\partial \xi} \\ \overline{C}_c \left(a\overline{C}_{14}A + ab\overline{C}_{15}AB + ab\overline{C}_{16}A^2 + k\left(\sqrt{\overline{\chi}\chi} - \beta_-\beta_+\right)\dfrac{\partial \varphi_1}{\partial \xi} \right) \end{pmatrix}, \tag{A19}$$

$$\overline{\mathbf{F}}_8 = \bar{k}\dfrac{\partial \varphi_1}{\partial \eta} \begin{pmatrix} -2\beta_- \\ \overline{C}_c(\beta_-^2 - \overline{S}\sqrt{\overline{\chi}}) \end{pmatrix}, \tag{A20}$$

where C_n and $\overline{C}_n$ ($n = 11\ldots16$) are defined in Appendixs B and C.

Appendix B

$$C_1 = -\frac{\overline{S}\beta_+ + \beta_-}{2C_5\beta_-}, \tag{A21}$$

$$C_2 = \frac{\beta_+(\overline{S}\beta_+ + 2\beta_-)}{C_5\beta_-}, \tag{A22}$$

$$C_3 = \frac{3\overline{S}}{C_5\beta_-}\left(3\beta_+^2 + 6\alpha - S\beta_-\beta_+\right), \tag{A23}$$

$$C_4 = \frac{1}{C_5}\left(2\beta_+ + \overline{S}\beta_- + \frac{C_c\overline{Y}(\chi - \beta_-^2)}{\beta_-}\right), \tag{A24}$$

$$C_5 = 2\beta_+ - \frac{S}{\beta_-} - \frac{\sqrt{\chi}\sqrt{\overline{\chi}}}{\beta_-}, \tag{A25}$$

$$\begin{aligned}C_6 =\ & \frac{1}{2\gamma^2}(\overline{C}_3\overline{S}\beta_+ - \overline{C}_3\gamma + \overline{C}_1\overline{S}\beta_+\gamma - \gamma^2\overline{C}_1) - 2\beta_-A_2 - \frac{1}{\beta_-}\left[c_1\overline{C}_1\overline{S}\beta_+ + c_2\beta_- + \frac{9\Gamma\chi}{\gamma^2}\right.\\ & + \frac{3\overline{C}_3\sigma}{\gamma^2} - \frac{3\overline{SC}_3\beta_+\beta_-}{2\gamma^2} - \frac{9\beta_-^2}{2\gamma^2} + \frac{3\overline{C}_1\sigma}{\gamma} + \frac{\overline{SC}_3\beta_+}{2\gamma} + \frac{3\overline{S}\beta_+\beta_-\overline{C}_1}{2\gamma} - \frac{3\beta_-}{4\gamma} - \frac{3\overline{C}_3\tau\chi}{\gamma^2}\\ & \left. - \frac{3\overline{C}_1\tau\chi}{\gamma}\right],\end{aligned} \tag{A26}$$

$$\begin{aligned}C_7 =\ & \left[-2\left\{(\overline{S}\beta_+ + \beta_-)\left(\overline{C}_1 + \frac{\overline{C}_3}{\gamma}\right)\right\} - c_1\left(\overline{C}_2 + \frac{3\overline{C}_3\beta_-}{\gamma}\right) + (15\overline{C}_3\beta_- + 4\overline{C}_2\gamma)\right.\\ & + 3\overline{C}_1\gamma\beta_-)\frac{3\overline{S}\beta_+}{\gamma^2} - \frac{1}{\beta_-}\left[2\overline{S}\beta_+\beta_-\left(\overline{C}_1 + \frac{\overline{C}_3}{\gamma}\right) + 3\left(\frac{\alpha - c_1\overline{S}\beta_+\beta_-}{\gamma^2}\right)(15\overline{C}_3\beta_-\right.\\ & + 4\overline{C}_2\gamma + 3\gamma\overline{C}_1\beta_-) + c_1\overline{S}\beta_+\left(\overline{C}_2 + \frac{3\overline{C}_3\beta_-}{\gamma}\right) + \frac{135\Gamma\beta_-}{\gamma^2} - \frac{45\beta_-^3}{8\gamma^2} + \frac{\beta_-^2}{4\gamma}\right],\end{aligned} \tag{A27}$$

$$\begin{aligned}C_8 =\ & -3c_1(\overline{S}\beta_+ + \beta_-)\left(\overline{C}_2 + \frac{3\overline{C}_3\beta_-}{\gamma}\right) + 15c_1^2\overline{S}\beta_+\left(\frac{3\overline{C}_3\beta_-^2}{\gamma^2} + \frac{\overline{C}_2\beta_-}{\gamma}\right)\\ & - \left[\frac{45c_1(3\overline{C}_3\beta_- + \overline{C}_2\gamma)(2\alpha - \overline{S}\beta_-\beta_+)}{\gamma^2} + 3c_1\overline{S}\beta_+\left(\overline{C}_2 + \frac{3\overline{C}_3\beta_-}{\gamma}\right) + \frac{405\chi\Gamma\beta_-}{\gamma^2}\right.\\ & \left. + \frac{135c_1\beta_-^3}{8\gamma^2} + \frac{27c_1\beta_-^2}{2\gamma}\right],\end{aligned} \tag{A28}$$

$$C_9 = -\frac{4C_8 + 3C_7\beta_-}{9\beta_-^2}, \tag{A29}$$

$$C_{10} = \frac{2C_8}{3\beta_-}. \tag{A30}$$

$$\begin{aligned}C_{11} =& -c_1 a(C_4 + \overline{C}_4) - (b\beta_- + 2a\overline{S}\beta_+)\left(\overline{C}_1 + \frac{\overline{C}_3}{\gamma}\right) - 3c_1(b+a)\left(\frac{\overline{C}_4 \overline{S}\beta_+}{\overline{\gamma}} + \frac{C_4 \beta_-}{\overline{\gamma}}\right) \\ &+ c_1 a \overline{S} Y + \frac{c_1 a \beta_- Y}{\gamma} - \frac{c_1 b \overline{S} \beta_+ \overline{Y}}{\overline{\gamma}} + \frac{c_1 a \beta_-}{\overline{\gamma}} + \frac{b \overline{S} \beta_+ \overline{Y}}{\overline{\gamma}} + \frac{b\beta_- Y}{\overline{\gamma}} + \frac{a \beta_- Y}{\gamma},\end{aligned} \tag{A31}$$

$$\begin{aligned}C_{12} =& -2(C_4 + \overline{C}_4)\beta_- - 2(\overline{C}_4 \overline{S}\beta_+ + C_4 \beta_-) + \frac{3c_1 \overline{SC}_4 \beta_- \beta_+}{\gamma} + \frac{3c_1 C_4 \beta_-^2}{\gamma} \\ &- 2\overline{S}\beta_+ \left(\overline{C}_2 + \frac{3\overline{C}_3 \beta_-}{\gamma}\right) - 2\beta_- Y + \frac{9 c_1 \beta_-^2 Y}{\gamma},\end{aligned} \tag{A32}$$

$$\begin{aligned}C_{13} =& -2(C_4 + \overline{C}_4)\overline{S}\beta_+ - 2(\overline{C}_4 \overline{S}\beta_+ + C_4 \beta_-) + \frac{9 c_1 \overline{SC}_4 \beta_+^2}{\overline{\gamma}} + \frac{9 c_1 C_4 \beta_- \beta_+}{\overline{\gamma}} - \frac{3 \overline{S} \beta_+^2 \overline{Y}}{\overline{\gamma}} \\ &- 2\beta_- \left(\overline{C}_2 - \frac{3\overline{C}_3 \beta_+}{\overline{\gamma}}\right) + 2(C_4 + \overline{C}_4)\overline{S} Y \beta_+ - 2Y\left(\overline{S}\beta_+ + \beta_- + \overline{C}_4 \overline{S}\beta_+ + C_4 \beta_-\right) \\ &- \frac{3 c_1 \beta_+ \beta_- \overline{Y}}{\overline{\gamma}},\end{aligned} \tag{A33}$$

$$\begin{aligned}C_{14} =& c_1 a(\overline{C}_4 \overline{S}\beta_+ + C_4 \beta_-) + (b\beta_-^2 + a\beta_+^2 \overline{S}^2)\left(\overline{C}_1 + \frac{\overline{C}_3}{\overline{\gamma}}\right) + \sigma(3\overline{S}^2 b - 6b\overline{S} + 3a) \\ & \left(\frac{C_4}{\gamma} + \frac{\overline{C}_4}{\gamma}\right) + \frac{3 b \beta_-}{2\overline{\gamma}}(\overline{SC}_4 + C_4) + \frac{3 b \overline{S}\beta_+}{2\overline{\gamma}}(\overline{SC}_4 + C_4) + \frac{3 a \beta_-}{2\gamma}(\overline{SC}_4 + C_4) \\ &- a \overline{S}^2 Y \beta_+ \beta_- \left(\overline{C}_1 + \frac{\overline{C}_3}{\gamma}\right) + \frac{3 a \beta_-^2 Y}{\gamma} + \frac{3 a \beta_- \beta_+ \overline{S} Y}{2\gamma} + \frac{3 b \beta_+^2 \overline{S} Y}{2\overline{\gamma}} + \frac{3 b \beta_+ \beta_- \overline{S} Y}{\overline{\gamma}} \\ &+ \frac{3 b \beta_+ \beta_- \overline{S} Y}{2\overline{\gamma}} + \frac{3 a \beta_+ \beta_- \overline{S} Y}{2\gamma} - \frac{4 b \beta_+ \overline{S}^2 Y}{\overline{\gamma}} + \frac{4 a \beta_- \overline{S} Y}{\gamma} - \frac{7 b \overline{S}\beta_+ \overline{Y}}{2\overline{\gamma}} + \frac{3 b \overline{S}\beta_+ \beta_-}{2\overline{\gamma}} \\ &- \frac{3 a \overline{S} \beta_+ \beta_-}{2\overline{\gamma}} - 3 a b \chi \tau \left(\frac{b}{\overline{\gamma}} + \frac{a}{\gamma}\right)(C_4 + \overline{C}_4) - c_1 a \overline{S}\beta_- Y + a \overline{S}\sqrt{\chi\overline{\chi}} Y \left(\overline{C}_1 + \frac{\overline{C}_3}{\gamma}\right) \\ &- \frac{a^2 \sigma Y}{\gamma} - \frac{a^2 \chi \tau Y}{\gamma} - \tau \chi \left(\frac{3 b \overline{Y}}{\overline{\gamma}} + \frac{3 a Y}{\gamma}\right) - 3\sigma\left(\frac{b Y \overline{S}}{\overline{\gamma}} + \frac{a Y}{\gamma}\right) \\ &- 3\chi \tau\left(\frac{2 b \overline{Y}}{\overline{\gamma}} + \frac{b Y}{\overline{\gamma}} + \frac{a Y}{\gamma}\right) + 3\sigma\left(-\frac{2 b \overline{S} Y}{\overline{\gamma}} + \frac{b \overline{S}^2 Y}{\overline{\gamma}} + \frac{a Y}{\gamma}\right),\end{aligned} \tag{A34}$$

$$\begin{aligned}C_{15} =& 2\beta_-(\overline{C}_4 \overline{S}\beta_+ + C_4 \beta_-) + \overline{S}^2 \beta_+^2 \left(\overline{C}_2 + \frac{3\overline{C}_3 \beta_-}{\gamma}\right) + \frac{9\beta_- \alpha}{\gamma}(C_4 + \overline{C}_4) \\ &- \frac{3 a b^2 \overline{S}\beta_+ \beta_-^2}{2\gamma} + 2\beta_-^2 Y + \chi Y\left(\overline{C}_2 + \frac{3\overline{C}_3 \beta_-}{\gamma}\right) - \frac{9 \sigma \beta_- Y}{\gamma} - \frac{27 \beta_- \tau \chi Y}{\gamma} \\ &+ \frac{9 \overline{S}\beta_-^2 Y}{4\gamma} - \frac{9 \overline{S}\beta_-^3 Y}{\gamma} + \frac{9 \overline{S}\beta_-^2 \beta_+ Y}{\gamma} - \frac{3\beta_-^2}{\gamma}(\overline{SC}_4\beta_+ + C_4 \beta_-) - \overline{S}^2 \beta_+ Y \beta_- \\ &\left(\overline{C}_2 + \frac{\overline{C}_3}{\gamma}\right) + \frac{9\beta_-^3}{\gamma},\end{aligned} \tag{A35}$$

$$\begin{aligned}
C_{16} = \ & 2c_1\overline{S}\beta_+(\overline{C}_4\overline{S}\beta_+ + C_4\beta_-) + c_1\beta_-^2\left(C_2 - \frac{3C_3\beta_+}{\overline{\gamma}}\right) + \frac{3\sigma\overline{S}c_1\beta_+}{\overline{\gamma}}(\overline{S} - 2) \\
& (C_4 + \overline{C}_4) - \frac{9\overline{S}\beta_+^2\beta_-}{4\overline{\gamma}} + \frac{27\chi\tau\beta_+}{2\gamma}(C_4 + \overline{C}_4) - \frac{9\overline{S}\sigma\beta_+\overline{Y}}{2\overline{\gamma}} + \frac{9\chi\tau\overline{Y}}{2\overline{\gamma}} + Y \\
& (C_4 + \overline{C}_4) + 2\overline{S}\beta_+\beta_-Y + \frac{9\beta_+\tau\chi}{2\overline{\gamma}}(2\overline{Y} + Y) + \frac{9\sigma\beta_+\overline{S}}{2\overline{\gamma}}(2\overline{Y} - \overline{S}Y) - \frac{9\overline{S}\beta_+^2\overline{Y}}{4\overline{\gamma}} \\
& + \frac{81\overline{S}^2\beta_+^2\overline{Y}}{4\overline{\gamma}} + \frac{63\overline{S}\beta_+^2\overline{Y}}{4\overline{\gamma}} + \frac{18\overline{S}\beta_+\beta_-Y}{\overline{\gamma}},
\end{aligned} \qquad (A36)$$

$$C_{17} = b\left(\mathbf{L}\mathbf{F}_8 + \mathbf{L}\overline{\mathbf{F}}_1 + \mathbf{L}\overline{\mathbf{F}}_7 + \frac{\mathbf{L}\overline{\mathbf{F}}_6 C_8}{\gamma}\right) + Sa\mathbf{L}\overline{\mathbf{F}}_8 A', \qquad (A37)$$

$$C_{18} = b\left[\frac{\mathbf{L}\overline{\mathbf{F}}_2}{2} + \mathbf{L}\overline{\mathbf{F}}_4\overline{C}_9 + \mathbf{L}\overline{\mathbf{F}}_5\overline{C}_9 + \mathbf{L}\overline{\mathbf{F}}_6\left(\frac{4\overline{C}_{10}}{\gamma} - \frac{3\beta_+}{2\gamma}\right)\right], \qquad (A38)$$

$$C_{19} = b\left(\frac{\mathbf{L}\overline{\mathbf{F}}_3}{3} + \mathbf{L}\overline{\mathbf{F}}_4\overline{C}_{10} + \mathbf{L}\overline{\mathbf{F}}_5\overline{C}_{10} - \frac{5\mathbf{L}\overline{\mathbf{F}}_6\overline{C}_{10}\beta_+}{\gamma}\right). \qquad (A39)$$

Appendix C

$$\overline{C}_1 = \frac{\beta_+ + S\beta_-}{2\overline{C}_5\beta_+}, \qquad (A40)$$

$$\overline{C}_2 = \frac{\beta_-(2\beta_+ + S\beta_-)}{\overline{C}_5\beta_+}, \qquad (A41)$$

$$\overline{C}_3 = \frac{3S}{6\overline{C}_5\beta_+}\left(6\alpha - 3\beta_-^2 - \overline{S}\beta_+\beta_-\right), \qquad (A42)$$

$$\overline{C}_4 = \frac{1}{\overline{C}_5}\left(\overline{S}\beta_+ + 2\beta_- + \frac{\overline{C}_c Y(\chi - \beta_+^2)}{\beta_+}\right), \qquad (A43)$$

$$\overline{C}_5 = 2\beta_- - \frac{\overline{S}}{\beta_+} - \frac{\sqrt{\chi\overline{\chi}}}{\beta_+}. \qquad (A44)$$

$$\begin{aligned}
\overline{C}_6 = \ & \frac{1}{2}C_1 + \frac{3C_3S\beta_-}{6\overline{\gamma}^2} + \frac{C_3}{2\overline{\gamma}} + \frac{3C_1S\beta_-}{6\overline{\gamma}} \\
& - \frac{1}{\beta_+}\left[-\frac{1}{2}C_1S\beta_- + \frac{9\overline{\chi}\Gamma}{\overline{\gamma}^2} + \frac{3C_3\sigma}{\overline{\gamma}^2} - \frac{9\beta_+^2}{24\overline{\gamma}^2} - \frac{3C_3S\beta_-\beta_+^2}{2\overline{\gamma}}\right. \\
& \left. + \frac{3c_1\sigma}{\overline{\gamma}} + \frac{3\beta_+}{4\overline{\gamma}} - \frac{C_3S\beta_-}{2\overline{\gamma}} + \frac{3c_1S\beta_-\beta_+}{2\overline{\gamma}} - 3\overline{\chi}\tau\left(\frac{C_3}{\overline{\gamma}^2} + \frac{C_1}{\overline{\gamma}}\right)\right],
\end{aligned} \qquad (A45)$$

$$\overline{C}_7 = \left[-2\left\{(\beta_+ + S\beta_-)\left(C_1 + \frac{C_3}{\overline{\gamma}}\right)\right\} + \frac{1}{2}\left(C_2 - \frac{3\overline{C}_3\beta_+}{\overline{\gamma}}\right) + (15C_3\beta_+ + 4C_2\overline{\gamma})\right.$$
$$+3C_1\overline{\gamma}\beta_+)\frac{3S\beta_-}{6\overline{\gamma}^2}\right] - \frac{1}{\beta_+}\left[2S\beta_+\beta_-\left(C_1 + \frac{C_3}{\overline{\gamma}}\right) - 3\left(\frac{\alpha + c_1\beta_+\beta_-}{\overline{\gamma}^2}\right)(15\overline{C}_3\beta_- \quad \text{(A46)}$$
$$\left.-4\overline{C}_2\overline{\gamma} + 3\overline{\gamma}\overline{C}_1\beta_-) - \overline{c}_1 S\beta_+\left(C_2 + \frac{3C_3\beta_-}{\overline{\gamma}}\right) - \frac{135\Gamma S^2\beta_+}{\overline{\gamma}^2} + \frac{45S^2\beta_+^3}{8\overline{\gamma}^2} + \frac{\beta_+^2}{4\overline{\gamma}}\right],$$

$$\overline{C}_8 = -3c_1(\beta_+ + S\beta_-)\left(C_2 + \frac{3C_3\beta_-}{\overline{\gamma}}\right) + 15c_1^2 S\beta_-\left(\frac{3C_3\beta_+^2}{\overline{\gamma}^2} - \frac{C_2\beta_+}{\overline{\gamma}}\right) - \frac{1}{\beta_+}$$
$$\left[\frac{45c_1\beta_+(3C_3\beta_- - C_2\overline{\gamma})(2\alpha + S\beta_-\beta_+)}{\overline{\gamma}^2} + 3c_1 S\beta_+\beta_-\left(C_2 - \frac{3C_3\beta_+}{\overline{\gamma}}\right) + \frac{405\Gamma\chi\beta_+^2}{\overline{\gamma}^2}\right. \quad \text{(A47)}$$
$$\left.-\frac{135c_1 S^2\beta_+^4}{8\overline{\gamma}^2} - \frac{27c_1\beta_+^3}{\overline{\gamma}}\right],$$

$$\overline{C}_9 = -\frac{4\overline{C}_8 - 3\overline{C}_7\beta_+}{9\beta_+^2}, \quad \text{(A48)}$$

$$\overline{C}_{10} = \frac{2\overline{C}_8}{3\beta_+}, \quad \text{(A49)}$$

$$\overline{C}_{11} = c_1 b(C_4 + \overline{C}_4) - 2(bS\beta_- + a\beta_+)\left(\overline{C}_1 + \frac{\overline{C}_3}{\gamma}\right) + \left(\frac{b}{\overline{Y}} + \frac{3a}{2Y}\right)(\overline{C}_4\beta_+ + C_4\beta_- S)$$
$$-\frac{bS\overline{Y}}{2} + \frac{b\beta_+\overline{Y}}{6\overline{\gamma}} + \frac{a\beta_- YS}{2\gamma} + \frac{a\beta_+\overline{Y}}{2\gamma} + \frac{aS\beta_- Y}{\gamma}, \quad \text{(A50)}$$

$$\overline{C}_{12} = 2\beta_+(C_4 + \overline{C}_4) - 2(\overline{C}_4\beta_+ + C_4 S\beta_-) - \beta_- S\left(C_2 - \frac{3C_3\beta_+}{\overline{\gamma}}\right) - 2\beta_+\overline{Y} - \frac{3\beta_+^2\overline{Y}}{2\overline{\gamma}}$$
$$-\frac{\beta_+}{2\overline{\gamma}}(\overline{C}_4\beta_+ + C_4 S\beta_-), \quad \text{(A51)}$$

$$\overline{C}_{13} = (C_4 + \overline{C}_4) - (C_4 S\beta_- + \overline{C}_4\beta_+) - 2\beta_+\left(\overline{C}_2 + \frac{3\overline{C}_3\beta_-}{\gamma}\right) + \frac{3\beta_-}{4\gamma}(\beta_+\overline{Y} + 2S\beta_- Y)$$
$$+\frac{9\beta_-}{4\gamma}(\overline{C}_4\beta_+ + C_4 S\beta_-) + \beta_-(C_4 + \overline{C}_4)S\overline{Y} - (\beta_+ + S\beta_-) - \overline{Y}(\overline{C}_4\beta_+ + C_4 S\beta_-) \quad \text{(A52)}$$
$$+\frac{3SY\beta_-^2}{4\gamma},$$

$$\begin{aligned}
\overline{C}_{14} =& -c_1 b(\overline{C}_4\beta_+ + C_4 S\beta_-) + \left(b\beta_-^2 S^2 + a\beta_+^2\right)\left(\overline{C}_1 + \frac{\overline{C}_3}{\gamma}\right) + \frac{3a\beta_- YS^2}{\gamma} - \frac{3b\sigma\overline{Y}}{\overline{\gamma}} \\
& 3\sigma(\overline{C}_4 + C_4)\left[\frac{b}{\overline{\gamma}} + \frac{S^2 a}{\gamma} - \frac{2Sa}{\gamma}\right] - \frac{3\beta_+\beta_- S}{2}\left(\frac{b}{\overline{\gamma}} - \frac{a}{\gamma}\right) + \frac{3a\beta_+^2 YS^2}{\gamma} + 3b\beta_+^2 S^2 \\
& \left(\frac{\overline{Y}}{\overline{\gamma}} + \frac{1}{\gamma}\right) - 3\tau\chi(C_4 + \overline{C}_4)\left(\frac{b}{\overline{\gamma}} + \frac{a}{\gamma}\right) + \frac{bS\beta_+\overline{Y}}{2} + b\left(C_1 + \frac{C_3}{\overline{\gamma}}\right) - \frac{3b\tau\chi\overline{Y}}{\overline{\gamma}} \\
& -3\tau\chi\left(\frac{b\overline{Y}}{\overline{\gamma}} + \frac{a\overline{Y}}{\gamma} + \frac{2aY}{\gamma}\right) + 3\sigma\left(\frac{b}{\overline{\gamma}} + \frac{aS^2}{\gamma} - \frac{2aSY}{\gamma}\right) - 3\tau\sqrt{\chi\overline{\chi}}S\left(\frac{Y}{\overline{\gamma}} + \frac{Y}{\gamma}\right) \quad (\text{A53}) \\
& +3\left(\frac{b\overline{Y}}{\overline{\gamma}} + \frac{aSY}{\gamma}\right) + \frac{3}{2}S\left[\beta_+\overline{Y}\left(\frac{b}{\overline{\gamma}} + \frac{a}{\gamma}\right) - \frac{\beta_- Y}{\gamma}(Sa - b)\right] + \frac{3b\beta_+^2 \overline{Y}}{2\overline{\gamma}} \\
& +15 S\beta_+\overline{Y}\left(\frac{b}{\overline{\gamma}} + \frac{a}{\gamma}\right) + \frac{9b\beta_+\overline{Y}}{\overline{\gamma}} + \frac{3aS^2\beta_- Y}{\gamma} + (bc_1\beta_+ + 3c_1 a\beta_+ + aS\beta_-) \\
& \frac{1}{\gamma}(\beta_+ + S\beta_-) - bS^2\overline{Y}\beta_-^2\left(C_1 + \frac{C_3}{\overline{\gamma}}\right) + 3c_1 S^2\beta_+ \overline{Y}\beta_-\left(\frac{b}{\overline{\gamma}} + \frac{a}{\gamma}\right),
\end{aligned}$$

$$\begin{aligned}
\overline{C}_{15} =& 2\beta_+(\overline{C}_4\beta_+ + C_4 S\beta_-) + S\beta_+^2\left(C_2 - \frac{3C_3\beta_+}{\overline{\gamma}}\right) - \frac{9\beta_+\sigma}{\overline{Y}} + \frac{27\beta_+\chi\tau\overline{Y}}{\overline{\gamma}} + \frac{18\beta_+^2 \overline{Y}}{2\overline{\gamma}} \\
& (C_4 + \overline{C}_4) + \frac{3S\beta_+^2\beta_-}{2\overline{\gamma}} - \frac{9\beta_+\chi\tau}{\overline{\gamma}}(C_4 + \overline{C}_4) + 2\beta_+^2 \overline{Y} + \chi - S\beta_-^2 a^2 b\left(C_2 - \frac{3C_3\beta_+}{\overline{\gamma}}\right) \quad (\text{A54}) \\
& \overline{Y}\left(C_2 - \frac{3C_3\beta_+}{\overline{\gamma}}\right) + \frac{3\sigma\beta_+\overline{Y}}{\overline{Y}} - \frac{9\beta_+^2 \overline{Y}}{\overline{\gamma}}(1+S) - \frac{9\beta_+}{2\overline{\gamma}}(\overline{C}_4\beta_+^2 + C_4 S\beta_-) - \frac{27\beta_+^3 \overline{Y}}{2\overline{\gamma}},
\end{aligned}$$

$$\begin{aligned}
\overline{C}_{16} =& S\beta_-(\overline{C}_4\beta_+ + C_4 S\beta_-) + c_1\beta_+^2\left(\overline{C}_2 + \frac{3\overline{C}_3\beta_-}{\gamma}\right) + \frac{3c_1\sigma S}{\gamma}(S-2)(C_4 + \overline{C}_4) \\
& + \frac{9c_1 S\beta_+\beta_-^2}{2\gamma} + \frac{9c_1 S^2\chi\tau\beta_-}{\gamma}(C_4 + \overline{C}_4) + S^2\chi\overline{Y}(C_4 + \overline{C}_4) + 4S\beta_+\beta_-\overline{Y} \\
& + \frac{3b\sigma\beta_- YS}{2\gamma} - \frac{3S^2\chi\tau\beta_- Y}{2\gamma} - \frac{9\tau\chi S^2\beta_-}{2\gamma}(\overline{Y} + Y) + \frac{9S\sigma\beta_-}{2\gamma}(S\overline{Y} - 2Y) \quad (\text{A55}) \\
& + \frac{9\beta_-^2 YS^2}{4\gamma} + \frac{18\beta_+\beta_- S\overline{Y}}{\gamma} - \frac{63S\beta_-^2 Y}{2\gamma}(1+S) - S\overline{Y}\beta_-(\overline{C}_4\beta_+ + C_4 S\beta_-) + \frac{9\beta_-\beta_+}{4\gamma} \\
& (\overline{C}_4\beta_+ + S\beta_- C_4) - \frac{9S\beta_-^2}{2\gamma}(\overline{C}_4\beta_+ + S\beta_- C_4) + \frac{9\beta_+^2 S\beta_-\overline{Y}}{2\gamma} + \frac{9S^2\beta_-^2 Y}{4\gamma},
\end{aligned}$$

$$\overline{C}_{17} = a\left(\overline{\mathbf{L}F}_1 + \overline{\mathbf{L}F}_7 + \frac{\overline{\mathbf{L}F}_6 C_8}{\gamma} + \overline{\mathbf{L}F}_8\right) + \overline{S}b\overline{\mathbf{L}F}_8 B', \quad (\text{A56})$$

$$\overline{C}_{18} = a\left[\frac{\overline{\mathbf{L}F}_2}{2} + \overline{\mathbf{L}F}_4 C_9 + \overline{\mathbf{L}F}_5 C_9 + \overline{\mathbf{L}F}_6\left(\frac{4C_{10}}{\gamma} + \frac{3\beta_-}{2\gamma}\right)\right], \quad (\text{A57})$$

$$\overline{C}_{19} = a\left(\frac{\overline{\mathbf{L}F}_3}{3} + \overline{\mathbf{L}F}_4 C_{10} + \overline{\mathbf{L}F}_5 C_{10} + \frac{5\overline{\mathbf{L}F}_6 C_{10}\beta_-}{\gamma}\right). \quad (\text{A58})$$

References

1. Xia, X.; Shen, H.T. Nonlinear interaction of ice cover with shallow water waves in channels. *J. Fluid Mech.* **2002**, *467*, 259–268. [CrossRef]
2. Milewski, P.A.; Vanden-Broeck, J.M.; Wang, Z. Hydroelastic solitary waves in deep water. *J. Fluid Mech.* **2011**, *679*, 628–640. [CrossRef]
3. Vanden-Broeck, J.M.; Părău, E.I. Two-dimensional generalized solitary waves and periodic waves under an ice sheet. *Philos. Trans. R. Soc. A Math. Phys. Eng. Sci.* **2011**, *369*, 2957–2972. [CrossRef] [PubMed]
4. Deike, L.; Bacri, J.C.; Falcon, E. Nonlinear waves on the surface of a fluid covered by an elastic sheet. *J. Fluid Mech.* **2013**, *733*, 394–413. [CrossRef]
5. Wang, P.; Lu, D.Q. Analytic approximation to nonlinear hydroelastic waves traveling in a thin elastic plate floating on a fluid. *Sci. China Phys. Mech. Astron.* **2013**, *56*, 2170–2177. [CrossRef]
6. Hedges, T.S. Combinations of waves and currents: An introduction. *Proc. Inst. Civ. Eng.* **1987**, *82*, 567–585. [CrossRef]
7. Schulkes, R.M.S.M.; Hosking, R.J.; Sneyd, A.D. Waves due to a steadily moving source on a floating ice plate. part 2. *J. Fluid Mech.* **1987**, *180*, 297–318. [CrossRef]
8. Bhattacharjee, J.; Sahoo, T. Interaction of current and flexural gravity waves. *Ocean Eng.* **2007**, *34*, 1505–1515. [CrossRef]
9. Bhattacharjee, J.; Sahoo, T. Flexural gravity wave generation by initial disturbances in the presence of current. *J. Mar. Sci. Technol.* **2008**, *13*, 138–146. [CrossRef]
10. Mohanty, S.K.; Mondal, R.; Sahoo, T. Time dependent flexural gravity waves in the presence of current. *J. Fluids Struct.* **2014**, *45*, 28–49. [CrossRef]
11. Lu, D.Q.; Yeung, R.W. Hydroelastic waves generated by point loads in a current. *Int. J. Offshore Polar Eng.* **2015**, *25*, 8–12.
12. Fenton, J.D.; Rienecker, M.M. A fourier method for solving nonlinear water-wave problems: Application to solitary-wave interactions. *J. Fluid Mech.* **1982**, *118*, 411–443. [CrossRef]
13. Lin, P. A numerical study of solitary wave interaction with rectangular obstacles. *Coast. Eng.* **2004**, *51*, 35–51. [CrossRef]
14. Khan, U.; Ellahi, R.; Khan, R.; Mohyud-Din, S.T. Extracting new solitary wave solutions of benny–luke equation and phi-4 equation of fractional order by using (g'/g)-expansion method. *Opt. Quantum Electron.* **2017**, *49*, 362. [CrossRef]
15. Abdel-Gawad, H.; Tantawy, M. Mixed-type soliton propagations in two-layer-liquid (or in an elastic) medium with dispersive waveguides. *J. Mol. Liq.* **2017**, *241*, 870–874. [CrossRef]
16. Gardner, C.S.; Greene, J.M.; Kruskal, M.D.; Miura, R.M. Method for solving the Korteweg-de Vries equation. *Phys. Rev. Lett.* **1967**, *19*, 1095–1097. [CrossRef]
17. Su, C.H.; Mirie, R.M. On head-on collisions between two solitary waves. *J. Fluid Mech.* **1980**, *98*, 509–525. [CrossRef]
18. Mirie, R.M.; Su, C.H. Collisions between two solitary waves. part 2. A numerical study. *J. Fluid Mech.* **1982**, *115*, 475–492. [CrossRef]
19. Dai, S.Q. Solitary waves at the interface of a two-layer fluid. *Appl. Math. Mech.* **1982**, *3*, 777–788.
20. Mirie, R.M.; Su, C.H. Internal solitary waves and their head-on collision. i. *J. Fluid Mech.* **1984**, *147*, 213–231. [CrossRef]
21. Mirie, R.M.; Su, C.H. Internal solitary waves and their head-on collision. ii. *Phys. Fluids (1958–1988)* **1986**, *29*, 31–37. [CrossRef]
22. Ozden, A.E.; Demiray, H. Re-visiting the head-on collision problem between two solitary waves in shallow water. *Int. J. Non-Linear Mech.* **2015**, *69*, 66–70. [CrossRef]
23. Marin, M.; Öchsner, A. The effect of a dipolar structure on the hölder stability in Green–Naghdi thermoelasticity. *Contin. Mech. Thermodynam.* **2017**, *29*, 1365–1374. [CrossRef]
24. Zhu, Y.; Dai, S.Q. On head-on collision between two gKdV solitary waves in a stratified fluid. *Acta Mech. Sin.* **1991**, *7*, 300–308.
25. Zhu, Y. Head-on collision between two mKdV solitary waves in a two-layer fluid system. *Appl. Math. Mech.* **1992**, *13*, 407–417.

26. Huang, G.; Lou, S.; Xu, Z. Head-on collision between two solitary waves in a Rayleigh-Bénard convecting fluid. *Phys. Rev. E* **1993**, *47*, R3830. [CrossRef]
27. Dai, H.H.; Dai, S.Q.; Huo, Y. Head-on collision between two solitary waves in a compressible Mooney-Rivlin elastic rod. *Wave Motion* **2000**, *32*, 93–111. [CrossRef]
28. Bhatti, M.M.; Lu, D.Q. Head-on collision between two hydroelastic solitary waves with Plotnikov-Toland's plate model. *Theor. Appl. Mech. Lett.* **2018**, *8*, 384–392. [CrossRef]
29. Bhatti, M.M.; Lu, D.Q. Head-on collision between two hydroelastic solitary waves in shallow water. *Qual. Theory Dynam. Syst.* **2018**, *17*, 103–122. [CrossRef]

© 2019 by the authors. Licensee MDPI, Basel, Switzerland. This article is an open access article distributed under the terms and conditions of the Creative Commons Attribution (CC BY) license (http://creativecommons.org/licenses/by/4.0/).

Article
MHD Nanofluids in a Permeable Channel with Porosity

Ilyas Khan [1] and Aisha M. Alqahtani [2,*]

[1] Faculty of Mathematics and Statistics, Ton Duc Thang University, Ho Chi Minh 72915, Vietnam; ilyaskhan@tdt.edu.vn
[2] Department of Mathematics, Princess Nourah bint Abdulrahman University, Riyadh 11564, Saudi Arabia
* Correspondence: alqahtani@pnu.edu.sa

Received: 19 January 2019; Accepted: 8 March 2019; Published: 14 March 2019

Abstract: This paper introduces a mathematical model of a convection flow of magnetohydrodynamic (MHD) nanofluid in a channel embedded in a porous medium. The flow along the walls, characterized by a non-uniform temperature, is under the effect of the uniform magnetic field acting transversely to the flow direction. The walls of the channel are permeable. The flow is due to convection combined with uniform suction/injection at the boundary. The model is formulated in terms of unsteady, one-dimensional partial differential equations (PDEs) with imposed physical conditions. The cluster effect of nanoparticles is demonstrated in the $C_2H_6O_2$, and H_2O base fluids. The perturbation technique is used to obtain a closed-form solution for the velocity and temperature distributions. Based on numerical experiments, it is concluded that both the velocity and temperature profiles are significantly affected by ϕ. Moreover, the magnetic parameter retards the nanofluid motion whereas porosity accelerates it. Each H_2O-based and $C_2H_6O_2$-based nanofluid in the suction case have a higher magnitude of velocity as compared to the injections case.

Keywords: Permeable walls; suction/injection; nanofluids; porous medium; mixed convection; magnetohydrodynamic (MHD)

1. Introduction

Heat transport in unsteady laminar flows has numerous real-world applications, particularly flows in a porous channel with permeable walls, which include medical devices, aerodynamic heating, chemical industry, electrostatic precipitation, petroleum industry, nuclear energy, and polymer technology. Based on this motivation, many researchers have considered the porous channel problem with suction and injection under different physical conditions. In earlier studies, Torda [1] studied the boundary layer flow with the suction/injection effect. Berman [2] derived an exact solution for the channel flow taking into consideration the uniform suction/injection at the boundary wall of the channel. The suction and injection and the combined effect of heat and mass transfer on a moving continuous flat surface were analyzed by Erickson et al. [3]. Alamri et al. [4] studied the Poiseuille flow of nanofluid in a channel under Stefan blowing and the second-order slip effect. Zeeshan et al. [5] reported analytical solutions for the Poiseuille flow of nanofluid in a porous wavy channel. Hassan et al. [6] investigated the flow of H_2O based nanofluid on a wavy surface. Ellahi et al. [7] studied the boundary layer Poiseuille plan flow of kerosene oil based nanofluid fluid with variable thermal conductivity. Ijaz et al. [8] presented a comprehensive study on the interaction of nanoparticles in the flow of nanofluid in a finite symmetric channel. Some recent important and interesting studies can be found in [9–12].

Magnetohydrodynamic (MHD) is referred to as the magnetic properties of the fluids under the influence of an electromagnetic force. MHD flows have numerous applications in MHD bearings and

MHD pumps. Many studies have been carried out on MHD flow in the literature. Abbas et al. [13] investigated the MHD flow of Maxwell fluid in a porous channel. The convective MHD flow of second-grade fluid was reported by Hayat and Abbas [14]. The effect of a transverse magnetic field on different flows in a semi-porous channel was presented by Sheikholeslami et al. [15]. Ravikumar et al. [16] studied three dimensional MHD due to the pressure gradient over the porous plate. Batti et al. [17] analyzed the heat transfer flow of nanofluid in a channel. They studied the effect of thermal radiation and the MHD effect by using Roseland's approximation, Ohm's law, and Maxwell equations. Ma et al. [18] study the MHD flow of nanofluid in a U-shaped enclosure using the Koo–Kleinstreuer–Li (KKL) correlation approximation for the effective thermal conductivity. Opreti [19] studied water-based silver nanofluid over a stretching sheet. They considered the effect of MHD, suction/injection, and heat generation/absorption in their study. Hosseinzadeh et al. [20] investigated the MHD squeezing flow of nonfluid in a channel. They presented analytical solutions by using similarity transformation and the perturbation technique. Narayana et al. [21] developed a mathematical model for the MHD stagnation point flow of Watler's-B fluid nanofluid.

Nano-sized particles of (Ag) nanoparticles inside H_2O-based fluids are commonly known as silver-based nanofluids. The viscosity of the nanofluids containing metallic nanoparticles has a much higher thermal conductivity than the nanofluids containing metallic oxide and non-metallic nanoparticles. Because of this, the interest of researchers in investigating nanofluids containing metallic nanoparticles has increased recently. The first exact solutions for different types of nanofluid were developed by Loganathan et al. [22]. Qasim et al. [23] reported numerical solutions for MHD ferrofluid in a stretching cylinder. Amsa et al. [24] investigated nanofluid flow near a vertical plate containing five different nanoparticles. The radiative heat transfer in the natural convection flow of oxide nanofluid was studied by Das and Jana [25]. Dhanai et al. [26]. Numerically studied the MHD mixed convection flow of nanofluid in a cylindrical coordinate system. The MHD rotational flow of nanofluid taking into consideration the effect of a porous medium, thermal radiation, and the chemical reaction was presented by Reddy et al. [27]. For some other interesting studies, readers are referred to [28–40].

Motivated by the above-discussed literature, the present study focused on the MHD channel flow of nanofluid in a porous medium with the suction and injection effect. The flow of electrically conducting nanofluid is considered under the influence of a transverse magnetic field. The analytical solutions for the proposed model are developed by using the perturbation method. The solutions are numerically computed, and the influence of various flow parameters is studied graphically.

2. Problem Description

Consider a porous channel of a width, d, filled with incompressible H_2O and $C_2H_6O_2$ based nanofluids with Ag nanoparticles. The channel walls are stationary with isothermal temperature conditions. The flow in the x-direction due to the temperature gradient is shown in Figure 1. Under the assumption of [11], the governing equations are as follows:

$$\rho_{nf}(\frac{\partial v}{\partial t} - v_w \frac{\partial v}{\partial y}) = -\frac{\partial p}{\partial x} + \mu_{nf}\frac{\partial^2 v}{\partial y^2} - \left(\sigma_{nf} B_0^2 + \frac{\mu_{nf}}{k_1}\right)u + (\rho\beta)_{nf} g(T - T_0), \qquad (1)$$

$$(\rho C_p)_{nf}(\frac{\partial T}{\partial t} - v_w \frac{\partial T}{\partial y}) = k_{nf}\frac{\partial^2 T}{\partial y^2} - \frac{\partial q}{\partial y}, \qquad (2)$$

together with the following physical conditions:

$$v(0,t) = 0, \; v(d,t) = 0, \qquad (3)$$

$$T(0,t) = T_0, \; T(d,t) = T_w, \qquad (4)$$

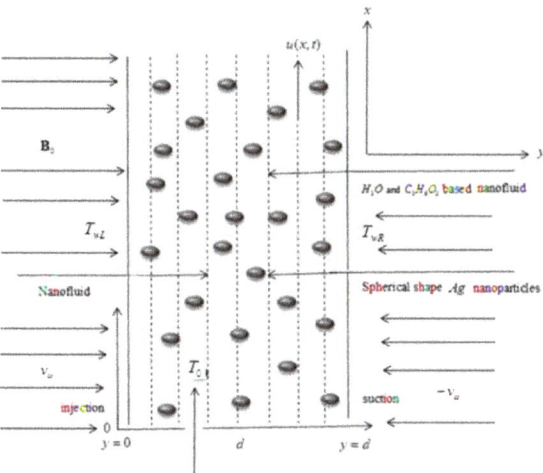

Figure 1. Physical configuration and coordinate system.

By using Xuan et al.'s [28] model, the effective thermal conductive, k_{nf}, and dynamic viscosity, μ_{nf}, of nanofluids are defined as:

$$k_{nf} = k_{static} + k_{Brownian},$$

$$k_{static} = k_f \left[\frac{(k_s + 2k_f) - 2\phi(k_f - k_s)}{(k_s + 2k_f) + \phi(k_f - k_s)} \right], k_{brownian} = \frac{\rho_s \phi c_{pf}}{2k_s} \sqrt{\frac{k_b T}{3\pi r_c \mu_f}}, \quad (5)$$

$$\mu_{nf} = \mu_{static} + \mu_{Brownian},$$

$$\mu_{static} = \frac{\mu_f}{(1-\phi)^{2.5}}, \mu_{Brownian} = \frac{\phi \rho_s (c_p)_s}{2k_s} \sqrt{\frac{k_b T}{3\pi r_c \mu_f}}, \quad (6)$$

where $k_b = 1.3807 \times 10^{-23} JK^{-1}$ and $300K > T > 325K$ are used, ϕ is the nanoparticles' volume fraction, and r_c is the radius of gyration for a number of particles. The static part in the effective thermal conductivity is derived from Maxwell's [29] model and the effective viscosity is derived from Brinkman's [30] model. Xuan et al.'s [28] model:

$$\begin{aligned}
\rho_{nf} &= (1-\phi)\rho_f + \phi\rho_s, (\rho\beta)_{nf} = (1-\phi)(\rho\beta)_f + \phi(\rho\beta)_s, \\
(\rho c_p)_{nf} &= (1-\phi)(\rho c_p)_f + \phi(\rho c_p)_s, \sigma_{nf} = \alpha_{nf}(\rho c_p)_{nf}, \\
\sigma_{nf} &= \sigma_f \left[1 + \frac{3(\sigma-1)\phi}{(\sigma+2)-(\sigma-1)\phi} \right], \sigma = \frac{\sigma_s}{\sigma_f},
\end{aligned} \quad (7)$$

where the numerical values of the thermo-physical of base fluid and nanoparticles are given in Table 1 [11,31]. The radiative heat flux is given by:

$$-\frac{\partial q}{\partial y} = 4\alpha^2 (T - T_0), \quad (8)$$

Substituting Equation (8) into Equation (2), gives:

$$(\rho c_p)_{nf} \left(\frac{\partial T}{\partial t} - v_w \frac{\partial T}{\partial y} \right) = k_{nf} \frac{\partial^2 T}{\partial y^2} + 4\alpha^2 (T - T_0), \quad (9)$$

The dimensionless variables:

$$x^* = \frac{x}{d}, \; y^* = \frac{y}{d}, \; u^* = \frac{u}{U_0}, \; t^* = \frac{tU_0}{d}, \; T^* = \frac{T-T_0}{T_w - T_0},$$
$$p^* = \frac{d}{\mu U_0}p, \; \omega^* = \frac{d\omega_1}{U_0}, \; v_0 = \frac{v_w}{U_0}, \quad (10)$$

Table 1. Thermo-physical properties of base fluid and nanoparticles.

Model	C_P (kg^{-1} K^{-1})	ρ (kg m^{-3})	k (Wm^{-1} K^{-1})	$\beta \times 10^{-5}$ (K^{-1})	(σ S/m)
Water (H_2O)	4179	997.1	0.613	21	5.5×10^{-6}
EG ($C_2H_6O_2$)	0.58	1.115	0.1490	6.5	1.07×10^{-6}
Alumina (Al_2O_3)	756	3970	40	0.85	1.07×10^{-6}
Silver (Ag)	235	10,500	429	1.89	6.30×10^{7}
Copper (Cu)	385	8933	401	1.67	59.6×10^{6}
Titanium Dioxide (TiO_2)	686.2	4250	8.9528	0.9	2.6×10^{6}

Are introduced into Equations (1) and (9), we get:

$$a_0\left(\frac{\partial u}{\partial t} - v_0\frac{\partial u}{\partial y}\right) = \lambda\varepsilon\exp(i\omega t) + \phi_2\frac{\partial^2 u}{\partial y^2} - m_0^2 u + a_1 T, \quad (11)$$

$$u(0,t) = 0; \; u(1,t) = 0; \; t > 0, \quad (12)$$

$$b_0\left(\frac{\partial T}{\partial t} - v_0\frac{\partial T}{\partial y}\right) = \frac{\partial^2 T}{\partial y^2} + b_1 T, \quad (13)$$

$$T(0,t) = 0; \; T(1,t) = 1; \; t > 0, \quad (14)$$

where:

$$a_0 = \phi_1 Re, \; \phi_1 = (1-\phi) + \phi\frac{\rho_s}{\rho_f}, \; Re = \frac{U_0 d}{v}, \; \phi_2 = \frac{1}{(1-\phi)^{2.5}}, \; m_0^2 = \phi_5 M^2,$$

$$\phi_5 = \left[1 + \frac{3(\sigma-1)\phi}{(\sigma+2)-(\sigma-1)\phi}\right], \; M^2 = \frac{\sigma_f B_0^2 d^2}{\mu_f}, \; a_1 = \phi_3 Gr, \; \phi_3 = (1-\phi)\rho_f + \phi\frac{(\rho\beta)_s}{\beta_f},$$

$$Gr = \frac{g\beta_f d^2(T_w-T_0)}{v_f U_0}, \; b_0^2 = \frac{Pe\phi_4}{\lambda_n}, \; Pe = \frac{U_0 d(\rho c_p)_f}{k_f},$$

$$\lambda_n = \frac{k_{nf}}{k_f} = \frac{(k_s + 2k_f) - 2\phi(k_f - k_s)}{(k_s - 2k_f) + \phi(k_f - k_s)}, \; \phi_4 = \left[(1-\phi) + \phi\frac{(\rho c_p)_s}{(\rho c_p)_f}\right], \; b_1^2 = \frac{N^2}{\lambda_n}, \; N^2 = \frac{4d^2\alpha_0^2}{k_f}.$$

The following general perturbed solutions are considered for Equations (11)—(14), the following type of solutions are assumed:

$$u(y,t) = [u_0(y) + \varepsilon\exp(i\omega t)u_1(y)], \quad (15)$$

$$T(y,t) = [T_0(y) + \varepsilon\exp(i\omega t)T_1(y)]. \quad (16)$$

Which lead to the following solutions:

$$\frac{d^2 u_0(y)}{dy^2} + \frac{a_0 v_0}{\phi_2}\frac{\partial u(y)}{\partial y} - \frac{m_0^2}{\phi_2}u_0(y) = -a_2 T_0, \quad (17)$$

$$u_0(0) = 0; \; u_0(1) = 0, \quad (18)$$

$$\frac{d^2 u_1(y)}{dy^2} + \frac{v_0}{\phi_2}\frac{\partial u_1(y)}{\partial y} - m_0^2 u_1(y) = -\frac{\lambda}{\phi_2}, \quad (19)$$

$$u_1(0) = 0; u_1(1) = 0, \tag{20}$$

$$\frac{d^2 T_0(y)}{dy^2} + b_0 v_0 \frac{\partial T_0(y)}{\partial y} + b_1^2 T_0(y) = 0, \tag{21}$$

$$T_0(0) = 0; T_0(1) = 1, \tag{22}$$

$$\frac{d^2 T_1(y)}{dy^2} + v_0 \frac{\partial T_1(y)}{\partial y} + (b_1 - b_0 i\omega) T_1(y) = 0, \tag{23}$$

$$T_1(0) = 0; T_1(1) = 0, \tag{24}$$

where:

$$m_1 = \sqrt{\frac{m_0^2}{\phi_2}}, a_2 = \frac{a_1}{\phi_2}, m_2 = \sqrt{\frac{m_0^2 + i\omega a_0}{\phi_2}}, m_3 = \sqrt{b_1 - i\omega b_0}.$$

The solutions of Equations (21) and (23) under the boundary conditions, (22) and (24), are obtained as:

$$T_0(y) = e^{-\alpha y} e^{\alpha} \frac{\sin(\beta y)}{\sin(\beta)}, \tag{25}$$

$$T_1(y) = 0, \tag{26}$$

where:

$$\alpha = \frac{b_0 v_0}{2}, \beta = \frac{1}{2}\sqrt{b_0 v_0 - 4b_1}.$$

Using Equations (25) and (26), Equation (16) becomes:

$$T(y,t) = T(y) = e^{-\alpha y} e^{\alpha} \frac{\sin(\beta y)}{\sin(\beta)}. \tag{27}$$

The solutions of Equations (17) and (19) after substituting Equation (25) under the boundary conditions, (18) and (20), are obtained as:

$$\begin{aligned} u_0(y) = & \; e^{-\alpha_2 y}(c_5 \sinh(\beta_2 y) + c_6 \cosh(\beta_2 y)) \\ & + a_1 e^{-\alpha y} e^{\alpha} \frac{[A \sin(\beta y) - B \cos(\beta y)]}{[A^2 + B^2]}, \end{aligned} \tag{28}$$

$$u_1(y) = e^{-\alpha_3 y}(c_7 \sinh(\beta_3 y) + c_8 \cosh(\beta_3 y)) + \frac{\lambda}{(m_2^2 \phi_2)}, \tag{29}$$

With:

$$\alpha_2 = \frac{a_0 v_0}{2\phi_2}, \beta_2 = \frac{1}{2}\sqrt{\frac{a_0^2 b_0^2}{\phi_2^2} + \frac{4m_0^2}{\phi_2}}, A = \alpha^2 - \beta^2 - \alpha\frac{a_0 v_0}{\phi_2} - \frac{m_0^2}{\phi_2},$$

$$B = -2\alpha\beta - \beta\frac{a_0 v_0}{(\phi_2)_3}, \alpha_3 = \frac{v_0}{2\phi_2}, \beta_3 = \frac{1}{2}\sqrt{\frac{v_0^2}{\phi_2^2} + 4m_0^2},$$

$$c_5 = \frac{1}{\sinh(\beta_2)}\left[\left(\frac{a_1 e^{\alpha} \beta}{[A^2 + B^2]}\right)\cosh(\beta_2) + \frac{e^{\alpha_2}[A\sin(\beta) - B\cos(\beta)]}{[A^2 + B^2]}\right] \tag{30}$$

$$c_6 = -\frac{a_1 e^{\alpha} \beta}{[A^2 + B^2]}, c_7 = \frac{\lambda}{(m_2^2 \phi_2)\sinh(\beta_3)}\cosh(\beta_3) - \frac{\lambda e^{\alpha_3}}{(m_2^2 \phi_2)}\frac{1}{\sinh(\beta_3)},$$

$$c_8 = -\frac{\lambda}{(m_2^2 \phi_2)}.$$

Finally, substituting Equations (28) to (30) into Equation (16), we get:

$$u(y,t) = e^{-\alpha_2 y}\left(\left(\frac{\sinh(\beta_2 y)}{\sinh(\beta_2)}\right)\left(\begin{array}{c}\left(\frac{a_1 e^{\alpha}\beta}{[A^2+B^2]}\right)\cosh(\beta_2) \\ +\frac{e^{\alpha_2}[A\sin(\beta)-B\cos(\beta)]}{[A^2+B^2]}\end{array}\right) - \left(\frac{a_1 e^{\alpha}\beta}{[A^2+B^2]}\right)\cosh(\beta_2 y)\right)$$
$$-a_1 e^{-\alpha y}e^{\alpha}\frac{[A\sin(\beta)-B\cos(\beta)]}{[A^2+B^2]}$$
$$+\exp(i\omega t)\left[e^{-\alpha_3 y}\left(\left(\frac{\frac{\lambda}{(m_2^2\phi_2)\sinh(\beta_3)}\cosh(\beta_3)}{\lambda e^{\alpha_3}} - \frac{1}{(m_2^2\phi_2)\sinh(\beta_3)}\right)\sinh(\beta_3 y)\right) - \left(\frac{\lambda}{(m_2^2\phi_2)}\right)\cosh(\beta_3 y)\right) + \frac{\lambda}{(m_2^2\phi_2)}\right]. \quad (31)$$

3. Nusselt Number

The dimensionless expression for the Nusselt number is given by:

$$Nu = \frac{\beta_1 e^{\alpha}}{\sin(\beta_1)}, \quad (32)$$

4. Skin-Friction

From Equation (31), the skin friction is calculated as:

$$\tau_t(t) = \frac{a_1 e^{\alpha}\beta_1\beta_2\cosh(\beta_2)}{[A^2+B^2]\sinh(\beta_2)} - \frac{\alpha_2 e^{\alpha_2}[A\sin(\beta_1)-B\cos(\beta_1)]}{[A^2+B^2]} + \frac{\alpha_2 a_1 e^{\alpha}\beta_1}{[A^2+B^2]}$$
$$-a_1 e^{\alpha}\frac{[\beta_1 A+\alpha B]}{[A^2+B^2]} + \exp(i\omega t)\left(\frac{\lambda\beta_3}{\left((m_2)_3^2(\phi_2)_3\right)\sinh(\beta_3)}\cosh(\beta_3) - \frac{1}{\left((m_2)_3^2(\phi_2)_3\right)\sinh(\beta_3)} - \frac{\alpha_3\lambda}{\left((m_2)_3^2(\phi_2)_3\right)}\right). \quad (33)$$

5. Results and Discussion

In this section, the graphs of the velocity and temperature for H_2O and $C_2H_6O_2$ based nanofluids containing Ag nanoparticles were plotted for different values of volume fraction, ϕ, and buoyancy parameter, Gr, permeability parameter, K, magnetic parameter, M, and radiation parameter, N, for both cases of suction and injection. The thermophysical properties of the base fluids and Ag nanoparticles are mentioned in Table 1. For this purpose, Figures 2–19 were plotted. Figures 2–5 were prepared to study the effects of the velocity for the cases of suction and injection of Ag in H_2O and $C_2H_6O_2$ based nanofluids, respectively. It was found that the velocity increases with increasing ϕ for both cases of suction and injection. However, no variation is observed in the velocity of Ag in $C_2H_6O_2$ based nanofluids in the case of injection. This behavior of velocity is found to be similar qualitatively to the results of Hajmohammadi et al. [32], however, they used Cu in water-based nanofluids.

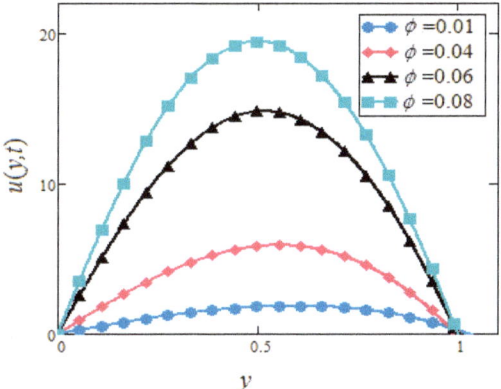

Figure 2. Velocity profiles for different values of ϕ of Ag in water based nanofluids when $Gr = 0.1$, $N = 0.1$, $r_c = 20$ nm, $Pe = 0.1$, $\lambda = 1$, $M = 1$, $K = 0.3$, $v_0 = 2$, $t = 5$, $\omega = 0.2$.

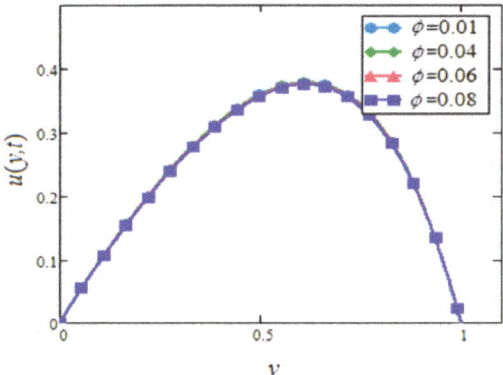

Figure 3. Velocity profiles for different values of ϕ of Ag in water based nanofluids when $Gr = 0.1$, $N = 0.1$, $r_c = 20$ nm, $Pe = 0.1$, $\lambda = 1$, $M = 1$, $K = 0.3$, $v_0 = -0.01$, $t = 5$, $\omega = 0.2$.

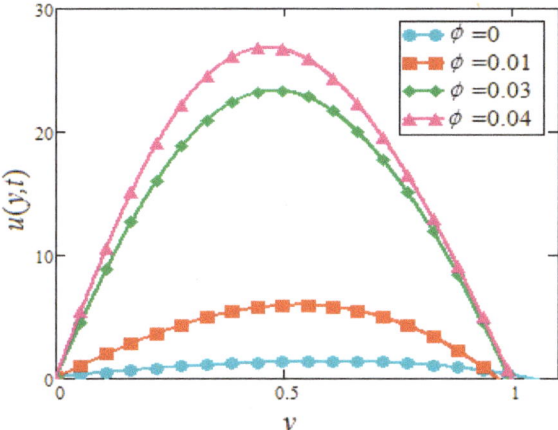

Figure 4. Velocity profiles for different values of ϕ of Ag in EG based nanofluids when $Gr = 0.1$, $N = 0.1$, $r_c = 20$ nm, $Pe = 0.1$, $\lambda = 1$, $M = 2$, $K = 3$, $v_0 = 4$, $t = 5$, $\omega = 0.2$.

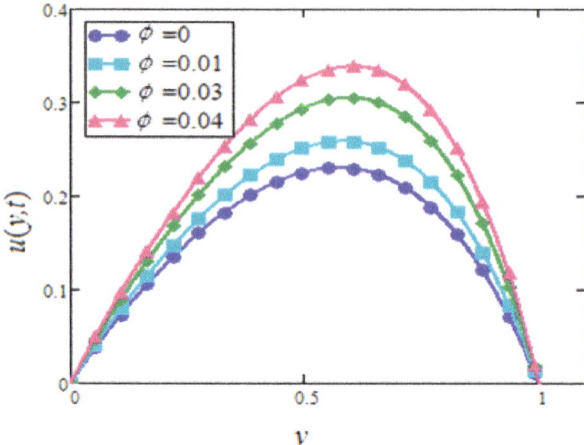

Figure 5. Velocity profiles for different values of ϕ of Ag in EG based nanofluids when $Gr = 0.1$, $N = 0.1$, $r_c = 20$ nm, $Pe = 0.1$, $\lambda = 1$, $M = 2$, $K = 3$, $t = 5$, $v_0 = -0.01$, $\omega = 0.2$.

Figures 6 and 7 are plotted for different values of Gr for both cases of suction and injection. It is noted from Figure 6 that the velocity of Ag in water-based nanofluids increases with the increase of Gr in the case of suction for Ag in water-based nanofluids while the velocity is decreased in the case of injection. The velocity in Figure 6, where $Gr = 0$, is not linear. However, the increasing values of Gr make it look like linear.

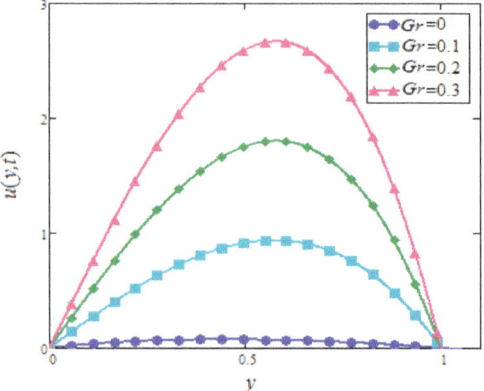

Figure 6. Velocity profiles for different values of Gr of Ag in water based nanofluids when $N = 0.1$, $Pe = 0.1$, $r_c = 20$ nm, $\phi = 0.04$, $\lambda = 1$, $M = 2$, $K = 3$, $v_0 = 10$, $t = 5$, $\omega = 0.2$.

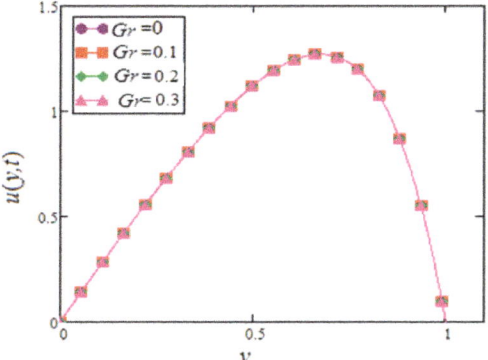

Figure 7. Velocity profiles for different values of Gr of Ag in water based nanofluids when $N = 0.1$, $Pe = 0.1$, $r_c = 20$ nm, $\phi = 0.04$, $Re = 0.1$, $\lambda = 1$, $M = 2$, $K = 3$, $v_0 = -1$, $t = 5$, $\omega = 0.2$.

Figures 8 and 9 were plotted to check the effect of K, the velocity of Ag in water-based nanofluids, for both cases of suction and injection. One can see from Figures 8 and 9 that the effect of suction, K, on the velocity of nanofluids is opposite to the case of injection.

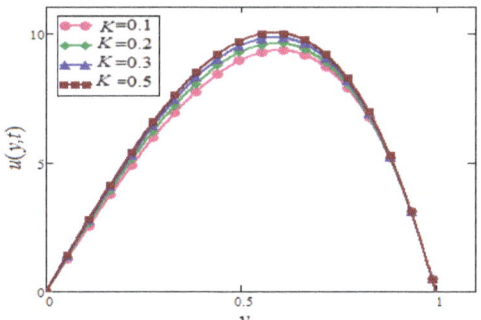

Figure 8. Velocity profiles for different values of K of Ag in water based nanofluids when $Gr = 0.1$, $N = 0.1$, $Pe = 0.1$, $r_c = 20$ nm, $\phi = 0.04$, $\lambda = 1$, $M = 2$, $v_0 = 6$, $t = 10$, $\omega = 0.2$.

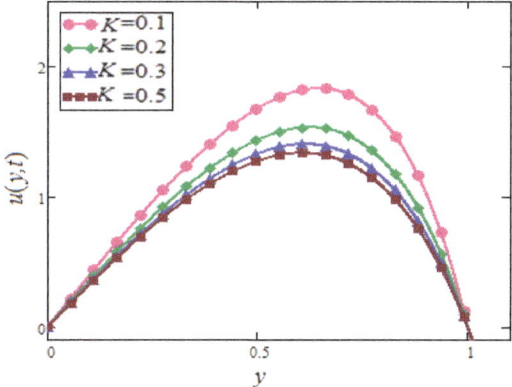

Figure 9. Velocity profiles for different values of K of Ag in water based nanofluids when $Gr = 0.1$, $N = 0.1$, $Pe = 0.1$, $r_c = 20$ nm, $\phi = 0.04$, $\lambda = 1$, $M = 2$, $v_0 = -0.01$, $t = 10$, $\omega = 0.2$.

The effect of the magnetic parameter, M, on the velocity profile is studied in Figures 10 and 11. For the case of suction, the $Ag - H_2O$ nanofluid's velocity profile decreases with increasing values of M. This effect is due to the Lorentz forces. Greater values of M correspond to stronger Lorentz forces, which reduces the nanofluid velocity. However, this trend reverses for the injection case.

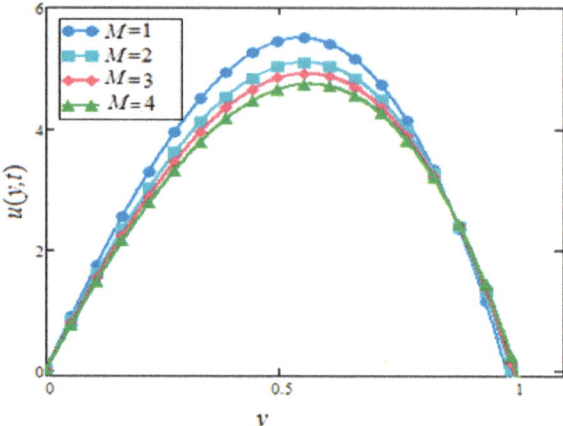

Figure 10. Velocity profiles for different values of M of Ag in water based nanofluids when $Gr = 0.1$, $N = 0.1$, $Pe = 0.1$, $r_c = 20$ nm, $\phi = 0.04$, $\lambda = 1$, $K = 0.3$ $v_0 = 5$, $t = 10$, $\omega = 0.2$.

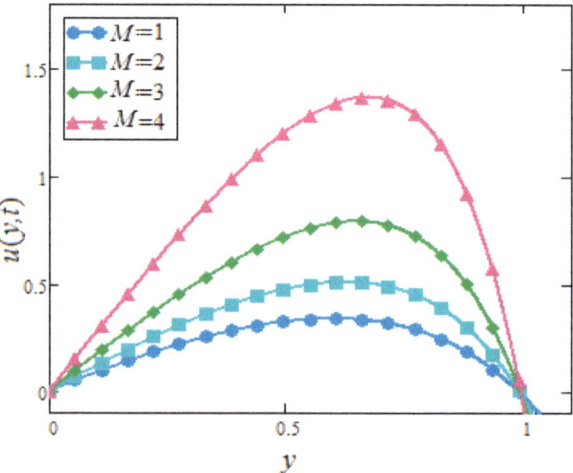

Figure 11. Velocity profiles for different values of M of Ag in water based nanofluids when $Gr = 0.1$, $N = 0.1$, $Pe = 0.1$, $r_c = 20$ nm, $\phi = 0.04$, $\lambda = 1$, $K = 0.3$ $v_0 = -0.01$ $t = 10$, $\omega = 0.2$.

Figures 12 and 13 shows that the velocity profiles of Ag in $Ag - H_2O$ nanofluids increase with the decrease of N in the case of suction. This physically means that an increase in N increases the conduction, which in turn decreases the viscosity of nanofluids. Decreasing the viscosity of nanofluids increases the velocity of nanofluids. However, the effect is the opposite due to the suction, whereas no variation is observed for injection. However, the velocity of zero radiation is greater than the velocity of nanofluids with radiation.

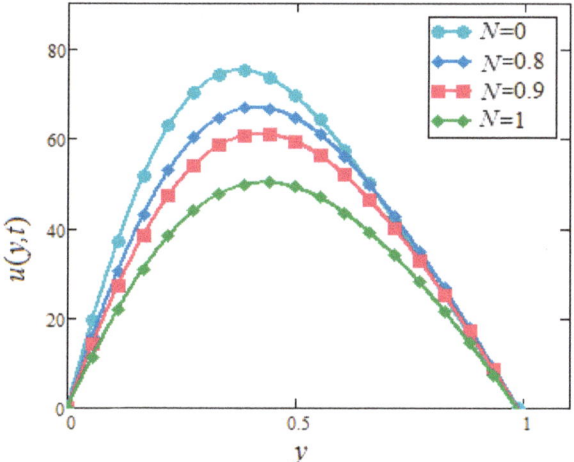

Figure 12. Velocity profiles for different values of N of Ag in water based nanofluids when $Gr = 0.1$, $Pe = 0.1$, $r_c = 20$ nm, $\phi = 0.04$, $\lambda = 1$, $M = 1$, $K = 1$, $v_0 = 7$, $t = 2$, $\omega = 0.2$.

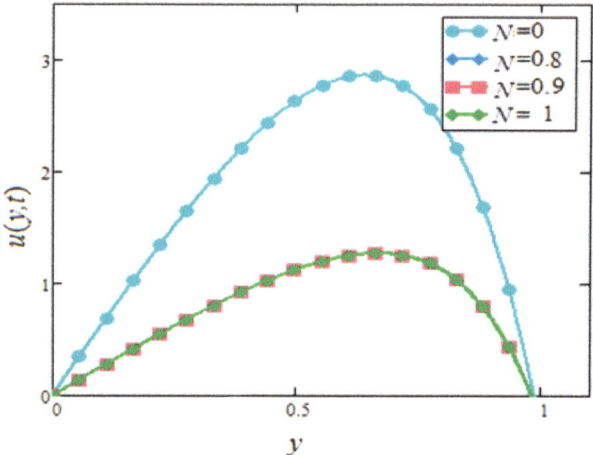

Figure 13. Velocity profiles for different values of N of Ag in water based nanofluids when $Gr = 0.1$, $Pe = 0.1$, $r_c = 20$ nm, $\phi = 0.04$, $\lambda = 1$, $M = 1$, $K = 1$, $v_0 = -1$, $t = 2$, $\omega = 0.2$.

The velocity profiles for different types of nanoparticles in water-based nanofluids are represented in Figure 14. It is clear that the velocity of Ag and Cu in water-based nanofluids is greater than TiO_2 and Al_2O_3 in water-based nanofluids. As mentioned in previous problems, different types of nanoparticles have different thermal conductivities and viscosities. It was concluded in our previous problems [31,32] that metallic nanoparticles, like Ag and Cu, had smaller velocities as compared to metallic oxide nanoparticles, like TiO_2 and Al_2O_3, due to high thermal conductivities and viscosities. However, the effect is the opposite to this problem because of the condition of permeable walls or suction. Due to these situations, different velocities have been observed.

The effect of ϕ in $Ag - H_2O$ nanofluids on the temperature profiles is shown in Figures 15 and 16 for the cases of suction and injection. It was found that the temperature of nanofluids increases with the increase of ϕ for the suction velocity whereas no significant variation is observed for the injection case.

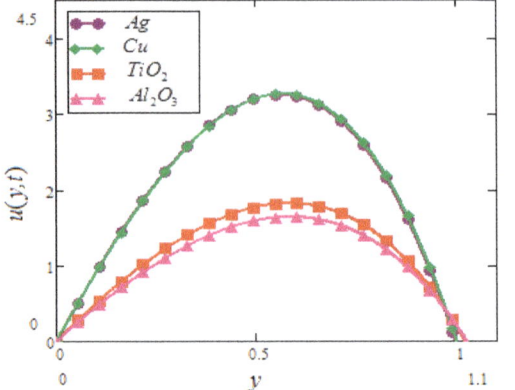

Figure 14. Velocity profiles for different types of nanoparticles in water based nanofluids when $Gr = 0.1, N = 0.1, Pe = 0.1, r_c = 20$ nm, $\lambda = 1, M = 2, K = 3, t = 5, \omega = 0.2$.

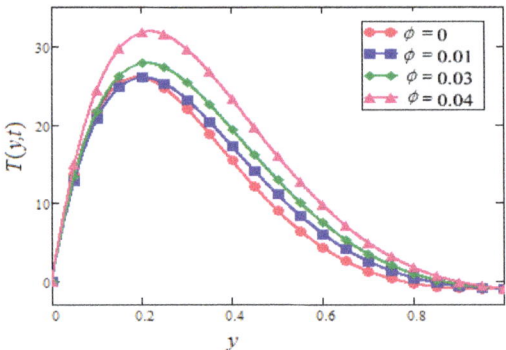

Figure 15. Temperature profiles for different values of ϕ of Ag in water based nanofluids when $r_c = 20$ nm, $N = 1, t = 1, v_0 = 10, \omega = 0.2$.

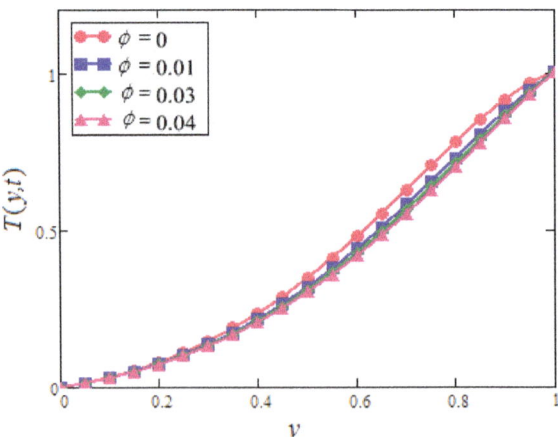

Figure 16. Temperature profiles for different values of ϕ of Ag in water based nanofluids when $r_c = 20$ nm, $N = 1, t = 1, v_0 = -1, \omega = 0.2$.

Figures 17 and 18 are sketched to show the effect of N on the temperature profiles of Ag in H_2O based nanofluids for both cases of suction and injection. The effects of different types of nanoparticles on the temperature of H_2O based nanofluids are plotted in Figure 19 for injection. It is observed that Cu in water-based nanofluid has the highest temperature followed by Ag, Al_2O_3, and TiO_2 in H_2O based nanofluids. This is due to the higher thermal conductivities of copper followed by Ag, Al_2O_3, and TiO_2 in water-based nanofluids. Due to these situations, different velocities were observed.

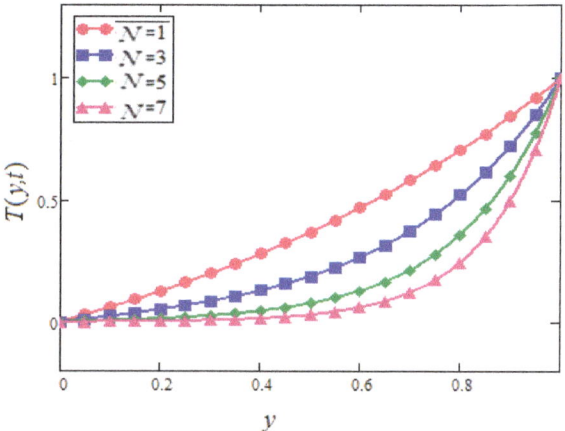

Figure 17. Temperature profiles for different values of N of Ag in water based nanofluids when $r_c = 20$ nm, $\phi = 0.04$, $t = 1$, $v_0 = -1$, $\omega = 0.2$.

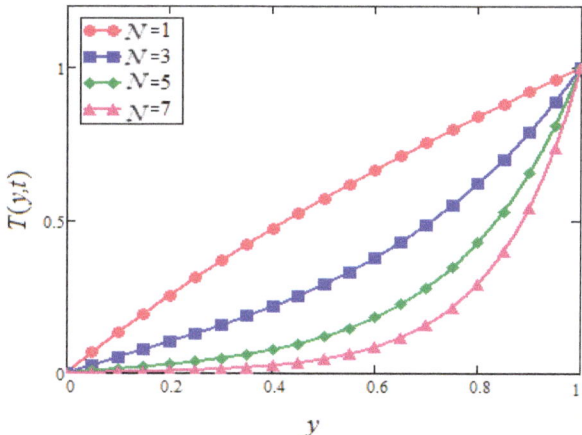

Figure 18. Temperature profiles for different values of N of Ag in water based nanofluids when $r_c = 20$ nm, $\phi = 0.04$, $t = 1$, $v_0 = 10$, $\omega = 0.2$.

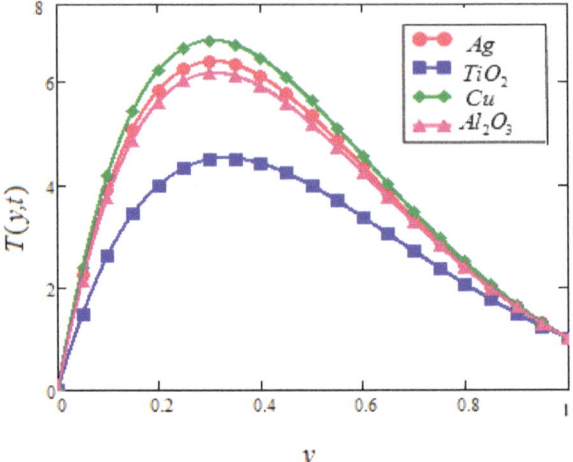

Figure 19. Temperature profiles for different types of nanoparticles in water based nanofluids when $r_c = 20$ nm, $N = 1$, $t = 1$, $v_0 = -1$, $\omega = 0.2$.

6. Conclusions

The channel flow of nanofluids in a porous medium with permeable walls was studied. The focal point of this research was to study the influence of permeable walls on momentum and heat transfer. The permeable parameter, which physically corresponds to suction and injection, was incorporated in both the momentum and energy equations. Expressions for the velocity and temperature were obtained. The effects of various parameters, such as thermal Grashof number, volume fraction, different types of nanoparticles, radiation, permeability, magnetic, suction, and injection, were studied in different plots. The concluding remarks are as follows:

1. It was found that the velocity of nanofluids increases with an increase of the volume fraction, radiation, and permeability parameter in the case of suction whereas an opposite behavior was noted in the case of injection.
2. The velocity of Ag nanofluids decreases with an increase of the magnetic parameter while the opposite behavior was noted in the case of injection.
3. The temperature of Ag nanofluids was found to decrease with an increase of ϕ for the extraction of fluid from the walls whereas a very small change was observed in the case of injection.
4. Finally, it was noticed that different types of nanoparticles have different effects on the velocity and temperature due to suction and injection.

Author Contributions: All the authors contributed equally to the conception of the idea, implementing and analyzing the experimental results, and writing the manuscript.

Funding: This research was funded by Deanship of Scientific Research at Princess Nourah bint Abdulrahman University. (Grant No# ٢٦٩/ص/٣٩).

Conflicts of Interest: The authors declare no conflict of interest.

Nomenclature

H_2O	Water
$C_2H_6O_2$	Ethelyn glycol
$v(y,t)$	Velocity component in the x-direction
$T(y,t)$	Temperature
$v_\omega > 0$	Suction
$v_\omega < 0$	Injection
ρ_{nf}	Density of nanofluid
M	Magnetic parameter
Pe	Peclet number
μ_{nf}	Dynamic viscosity of nanofluid
$(\rho\beta)_{nf}$	thermal expansion coefficient
g	Acceleration due to gravity
$(\rho c_p)_{nf}$	Heat capacitance of nanofluids
k_{nf}	The thermal conductivity of nanofluid
α	Mean radiation absorption coefficient
Re	Reynolds' number
Gr	Grashof number
N	Radiation parameter

References

1. Torda, T.P. Boundary Layer Control by Distributed Surface Suction or Injection. Bi-Parametric General Solution. *J. Math. Phys.* **1953**, *32*, 312–314. [CrossRef]
2. Berman, A.S. Laminar flow in channels with porous walls. *J. Appl. Phys.* **1953**, *24*, 1232–1235. [CrossRef]
3. Erickson, L.E.; Fan, L.T.; Fox, V.G. Heat and mass transfer on moving continuous flat plate with suction or injection. *Ind. Eng. Chem. Fundam.* **1966**, *5*, 19–25. [CrossRef]
4. Alamri, S.Z.; Ellahi, R.; Shehzad, N.; Zeeshan, A. Convective radiative plane Poiseuille flow of nanofluid through porous medium with slip: An application of Stefan blowing. *J. Mol. Liq.* **2019**, *273*, 292–304. [CrossRef]
5. Zeeshan, A.; Shehzad, N.; Ellahi, R.; Alamri, S.Z. Convective Poiseuille flow of Al 2 O 3-EG nanofluid in a porous wavy channel with thermal radiation. *Neural Comput. Appl.* **2018**, *30*, 3371–3382. [CrossRef]
6. Hassan, M.; Marinb, M.; Alsharifc, A.; Ellahide, R. Convective heat transfer flow of nanofluid in a porous medium over wavy surface. *Phys. Lett. A* **2018**, *382*, 2749–2753. [CrossRef]
7. Ellahi, R.; Zeeshan, A.; Shehzad, N.; Alamri, S.Z. Structural impact of kerosene-Al_2O_3 nanoliquid on MHD Poiseuille flow with variable thermal conductivity: Application of cooling process. *J. Mol. Liq.* **2018**, *264*, 607–615. [CrossRef]
8. Ijaz, N.; Zeeshan, A.; Bhatti, M.M.; Ellahi, R. Analytical study on liquid-solid particles interaction in the presence of heat and mass transfer through a wavy channel. *J. Mol. Liq.* **2018**, *250*, 80–87. [CrossRef]
9. Ali, F.; Aamina, A.; Khan, I.; Sheikh, N.A.; Saqib, M. Magnetohydrodynamic flow of brinkman-type engine oil based MoS 2-nanofluid in a rotating disk with Hall Effect. *Int. J. Heat Technol.* **2017**, *4*, 893–902.
10. Jan, S.A.A.; Ali, F.; Sheikh, N.A.; Khan, I.; Saqib, M.; Gohar, M. Engine oil based generalized brinkman-type nano-liquid with molybdenum disulphide nanoparticles of spherical shape: Atangana-Baleanu fractional model. *Numer. Methods Partial Differ. Equ.* **2018**, *34*, 1472–1488. [CrossRef]
11. Saqib, M.; Ali, F.; Khan, I.; Sheikh, N.A.; Khan, A. Entropy Generation in Different Types of Fractionalized Nanofluids. *Arab. J. Sci. Eng.* **2019**, *44*, 1–10. [CrossRef]
12. Saqib, M.; Ali, F.; Khan, I.; Sheikh, N.A.; Shafie, S.B. Convection in ethylene glycol-based molybdenum disulfide nanofluid. *J. Therm. Anal. Calorim.* **2019**, *135*, 523–532. [CrossRef]
13. Saqib, M.; Khan, I.; Shafie, S. Natural convection channel flow of CMC-based CNTs nanofluid. *Eur. Phys. J. Plus* **2018**, *133*, 549. [CrossRef]
14. Saqib, M.; Khan, I.; Shafie, S. Application of Atangana–Baleanu fractional derivative to MHD channel flow of CMC-based-CNT's nanofluid through a porous medium. *Chaos Solitons Fractals* **2018**, *116*, 79–85. [CrossRef]

15. Abbas, Z.; Sajid, M.; Hayat, T. MHD boundary-layer flow of an upper-convected Maxwell fluid in a porous channel. *Theor. Comput. Fluid Dyn.* **2006**, *20*, 229–238. [CrossRef]
16. Hayat, T.; Abbas, Z. Heat transfer analysis on the MHD flow of a second grade fluid in a channel with porous medium. *Chaos Solitons Fractals* **2008**, *38*, 556–567. [CrossRef]
17. Sheikholeslami, M.; Ashorynejad, H.R.; Domairry, D.; Hashim, I. Investigation of the laminar viscous flow in a semi-porous channel in the presence of uniform magnetic field using optimal homotopy asymptotic method. *Sains Malays.* **2012**, *41*, 1177–1229.
18. Ravikumar, V.; Raju, M.C.; Raju, G.S.S. MHD three dimensional Couette flow past a porous plate with heat transfer. *IOSR J. Math.* **2012**, *1*, 3–9. [CrossRef]
19. Bhatti, M.M.; Zeeshan, A.; Ellahi, R. Heat transfer with thermal radiation on MHD particle–fluid suspension induced by metachronal wave. *Pramana* **2017**, *89*, 48. [CrossRef]
20. Ma, Y.; Mohebbi, R.; Rashidi, M.M.; Yang, Z.; Sheremet, M.A. Numerical study of MHD nanofluid natural convection in a baffled U-shaped enclosure. *Int. J. Heat Mass Transf.* **2019**, *130*, 123–134. [CrossRef]
21. Upreti, H.; Pandey, A.K.; Kumar, M. MHD flow of Ag-water nanofluid over a flat porous plate with viscous-Ohmic dissipation, suction/injection and heat generation/absorption. *Alex. Eng. J.* **2018**, *57*, 1839–1847. [CrossRef]
22. Hosseinzadeh, K.; Alizadeh, M.; Ganji, D.D. Hydrothermal analysis on MHD squeezing nanofluid flow in parallel plates by analytical method. *Int. J. Mech. Mater. Eng.* **2018**, *13*, 4. [CrossRef]
23. Narayana, P.V.; Tarakaramu, N.; Makinde, O.D.; Venkateswarlu, B.; Sarojamma, G. MHD Stagnation Point Flow of Viscoelastic Nanofluid Past a Convectively Heated Stretching Surface. *Defect Diffus. Forum* **2018**, *387*, 106–120. [CrossRef]
24. Loganathan, P.; Chand, P.N.; Ganesan, P. Radiation effects on an unsteady natural convective flow of a nanofluid past an infinite vertical plate. *Nano* **2013**, *8*, 1350001. [CrossRef]
25. Qasim, M.; Khan, Z.H.; Khan, W.A.; Shah, I.A. MHD boundary layer slip flow and heat transfer of ferrofluid along a stretching cylinder with prescribed heat flux. *PLoS ONE* **2014**, *9*, e83930. [CrossRef] [PubMed]
26. Khalid, A.; Khan, I.; Shafie, S. Exact solutions for free convection flow of nanofluids with ramped wall temperature. *Eur. Phys. J. Plus* **2015**, *130*, 57. [CrossRef]
27. Das, S.; Jana, R.N. Natural convective magneto-nanofluid flow and radiative heat transfer past a moving vertical plate. *Alex. Eng. J.* **2015**, *54*, 55–64. [CrossRef]
28. Dhanai, R.; Rana, P.; Kumar, L. MHD mixed convection nanofluid flow and heat transfer over an inclined cylinder due to velocity and thermal slip effects: Buongiorno's model. *Powder Technol.* **2016**, *288*, 140–150. [CrossRef]
29. Reddy, J.V.R.; Sugunamma, V.; Sandeep, N.; Sulochana, C. Influence of chemical reaction, radiation and rotation on MHD nanofluid flow past a permeable flat plate in porous medium. *J. Niger. Math. Soc.* **2016**, *35*, 48–65. [CrossRef]
30. Xuan, Y.; Li, Q.; Hu, W. Aggregation structure and thermal conductivity of nanofluids. *AIChE J.* **2003**, *49*, 1038–1043. [CrossRef]
31. Maxwell, J.C.; Thompson, J.J. *A Treatise on Electricity and Magnetism*; Oxford University Press: Oxford, UK, 1904.
32. Brinkman, H.C. The viscosity of concentrated suspensions and solutions. *J. Chem. Phys.* **1952**, *20*, 571. [CrossRef]
33. Ali, F.; Saqib, M.; Khan, I.; Sheikh, N.A. Heat Transfer Analysis in Ethylene Glycol Based Molybdenum Disulfide Generalized Nanofluid via Atangana–Baleanu Fractional Derivative Approach. In *Fractional Derivatives with Mittag-Leffler Kernel*; Springer: Cham, Switzerland, 2019; pp. 217–233.
34. Hajmohammadi, M.R.; Maleki, H.; Lorenzini, G.; Nourazar, S.S. Effects of Cu and Ag nano-particles on flow and heat transfer from permeable surfaces. *Adv. Powder Technol.* **2015**, *26*, 193–199. [CrossRef]
35. Saqib, M.; Khan, I.; Shafie, S. New Direction of Atangana–Baleanu Fractional Derivative with Mittag-Leffler Kernel for Non-Newtonian Channel Flow. In *Fractional Derivatives with Mittag-Leffler Kernel*; Springer: Cham, Switzerland, 2019; pp. 253–268.
36. Ellahi, R.; Zeeshan, A.; Hussain, F.; Abbas, T. Study of shiny film coating on multi-fluid flows of a rotating disk suspended with nano-sized silver and gold particles: A comparative analysis. *Coatings* **2018**, *8*, 422. [CrossRef]

37. Ellahi, R.; Alamri, S.Z.; Basit, A.; Majeed, A. Effects of MHD and slip on heat transfer boundary layer flow over a moving plate based on specific entropy generation. *J. Taibah Univ. Sci.* **2018**, *12*, 476–482. [CrossRef]
38. Saqib, M.; Khan, I.; Shafie, S. Application of fractional differential equations to heat transfer in hybrid nanofluid: Modeling and solution via integral transforms. *Adv. Differ. Equ.* **2019**, *52*. [CrossRef]
39. Ellahi, R.; Tariq, M.H.; Hassan, M.; Vafai, K. On boundary layer magnetic flow of nano-Ferroliquid under the influence of low oscillating over stretchable rotating disk. *J. Mol. Liq.* **2017**, *229*, 339–345. [CrossRef]
40. Ellahi, R.; Raza, M.; Akbar, N.S. Study of peristaltic flow of nanofluid with entropy generation in a porous medium. *J. Porous Med.* **2017**, *20*, 461–478. [CrossRef]

© 2019 by the authors. Licensee MDPI, Basel, Switzerland. This article is an open access article distributed under the terms and conditions of the Creative Commons Attribution (CC BY) license (http://creativecommons.org/licenses/by/4.0/).

Article

Stability Analysis of Darcy-Forchheimer Flow of Casson Type Nanofluid Over an Exponential Sheet: Investigation of Critical Points

Liaquat Ali Lund [1,2], Zurni Omar [1], Ilyas Khan [3,*], Jawad Raza [4], Mohsen Bakouri [5] and I. Tlili [6]

1. School of Quantitative Sciences, Universiti Utara Malaysia, 06010 Sintok, Malaysia; balochliaqatali@gmail.com (L.A.L.); zurni@uum.edu.my (Z.O.)
2. KCAET Khairpur Mir's Sindh Agriculture University, Tandojam Sindh 70060, Pakistan
3. Faculty of Mathematics and Statistics, Ton Duc Thang University, Ho Chi Minh City 72915, Vietnam
4. Department of Mathematics and Statistics, Institute of Southern Punjab (ISP), Multan 32100, Pakistan; jawad_6890@yahoo.com
5. Department of Medical Equipment Technology, College of Applied Medical Science, Majmaah University, Al-Majmaah 11952, Saudi Arabia; m.bakouri@mu.edu.sa
6. Department of Mechanical and Industrial Engineering, College of Engineering, Majmaah University, Al-Majmaah 11952, Saudi Arabia.; l.tlili@mu.edu.sa
* Correspondence: ilyaskhan@tdt.edu.vn

Received: 31 December 2018; Accepted: 8 March 2019; Published: 20 March 2019

Abstract: In this paper, steady two-dimensional laminar incompressible magnetohydrodynamic flow over an exponentially shrinking sheet with the effects of slip conditions and viscous dissipation is examined. An extended Darcy Forchheimer model was considered to observe the porous medium embedded in a non-Newtonian-Casson-type nanofluid. The governing equations were converted into nonlinear ordinary differential equations using an exponential similarity transformation. The resultant equations for the boundary values problem (BVPs) were reduced to initial values problems (IVPs) and then shooting and Fourth Order Runge-Kutta method (RK-4th method) were applied to obtain numerical solutions. The results reveal that multiple solutions occur only for the high suction case. The results of the stability analysis showed that the first (second) solution is physically reliable (unreliable) and stable (unstable).

Keywords: dual solution; stability analysis; Darcy Forchheimer model; nanofluid; exponential sheet

---

1. Introduction

Many environmental and industrial systems, including geothermal energy system, catalytic reactors, fibrous insulation, heat exchanger designs, and geophysics, involve the convective flow of through porous surfaces. The classical Darcian model extended into the non-Darcian porous medium model includes tortuosity inertial drag, vorticity diffusion effects, as well as combinations of both effects [1]. The Darcy-Forchheimer (DF) model is the extension or modification of Darcian flow, which is used similarly to inertia effects. To determine the inertia effect, the velocity square term in the momentum equation must be added, and the resultant term is known as a Forchheimer's extension.

Flow over a porous surface is encountered in several applications, such as nuclear and gas waste storage, hydrocarbon recovery, hydrology, soil physics, transfer in living tissues, transfer in food products, soil mechanics, drying of the wood, and many others. Flow phenomena in the porous surfaces are complex given the interaction between the fluid and the packing particles, fluid and the column wall, and the particles and column wall. Muskat [2] called this interaction a Forchheimer factor. Some of the studies involving the Darcy-Forchheimer flow have been published [3–7]. Hayat et al. [8] considered Darcy-Forchheimer flow over a curve stretching surface and found that the porosity parameter produced high temperatures. Seth and Mandal [9] studied the Casson fluid in the presence of Darcy-Forchheimer, and observed the effect of the Casson and rotation parameter on the primary velocity. The behavior of primary velocity is reverse to the secondary velocity and close to that of the stretching surface. Ganesh et al. [10] considered a hydro-magnetic nanofluid under Darcy-Forchheimer flow on stretching and shrinking surfaces.

The investigation of the magnetic field impacts has many applications in the engineering, chemistry, and physics sectors. Industrial instruments, like bearings and pumps, boundary layer control, and magnetohydrodynamic generators are mostly affected due to interactions between a magnetic field and an electrically conducting fluid [10–12]. Viscous dissipation is usually an ignorable effect, but that contribution can be significant at very high fluid viscosity. The energy source usually changes temperature distributions, which affects the rates of heat transfer. Hsiao [13] investigated magnetohydrodynamic (MHD) flow and the effect of thermal radiation and viscous dissipation. Sheikholeslami et al. [14] examined MHD nanofluid flow as well as heat transfer in the presence of viscous dissipation.

Various researchers considered viscous dissipation effects in their studies [15–18]. Due to the importance of viscous dissipation, we also considered its effect in our considered model. At a microscopic level, in the boundary conditions, no-slip conditions are important when the flowing fluid layer adjacent to the solid boundary reaches the velocity of the solid boundary. Yet, no-slip conditions are not based on physical principles [19]. No-slip conditions occur in many macroscopic flows that have been proven experimentally. Nearly two centuries ago, Navier presented the general boundary conditions that cover the situation of slippery boundaries, where the velocity of the fluid is proportional to the shear stress on the surface [20]. In general, velocity slip occurs when the velocity of the fluid flow and the surface are different, indicating that different slip conditions exist when velocity, temperature, and concentration in the fluid and surface are different from each other. Hence, slip boundary conditions are of great importance due to their applications in the various fields of science and the technology, such as in microchannels or nanochannels as well as in applications where the surface is coated by a thick monolayer of hydrophobic octadecyl trichlorosilane, or when oil-moving plates are considered [19]. Wall slip occurs in working fluids with concentrated suspensions [21]. Non-Newtonian fluids, such as polymer melts, usually show wall slip. Many researchers studied different slip effects on fluid flow [22–27]. Motivated from the above-mentioned investigations, in this work, we focused on the velocity, thermal, and concentration slip effects on Casson fluid flow with Extended-Darcy-Forchheimer porous medium and viscous dissipation.

Unlike the flow over a stretching sheet, which received the attention of numerous researchers since it was first presented by Crane [28], the flow on the shrinking surface was viewed almost 50 years ago, in 1970, when Miklavčič and Wang [29] considered viscous flow on a shrinking sheet for the first time. Since flow is probably not going to occur on a shrinking surface, they added sufficient suction at a boundary to create vorticity in the boundary layer. Many researchers have considered a shrinking surface, including Naveed et al. [30], Jusoh et al. [31], Othman et al. [32], Khan and Hafeez [33], Naganthran et al. [34], and Qing et al. [35]. Rahman et al. [36,37] investigated Buongiorno's model on exponentially shrinking surfaces and found dual solutions. Some other interesting studies are given in [38–44]. In this study, we extend the work of Rahman et al. [36,37] to a permeable shrinking surface embedded in an Extended-Darcy-Forchheimer porous medium in the presence of viscous dissipation and velocity, and thermal and concentration slip effects over a shrinking surface, where the occurrence of dual solutions is possible. To produce a stable and physically reliable solution, we performed a stability analysis. Notably, shrinking sheet flow is basically a backward flow [38] that defines physical phenomena relatively differently from a stretching sheet.

2. Mathematical Description of the Problem

We considered the steady incompressible two-dimensional flow of a Casson electrically conductive nanofluid over an exponentially shrinking surface in Extended Darcy Forchheimer porous medium along with the effects of viscous dissipations and slip conditions, as shown in Figure 1. According to Nakamura and Sawada [39], the rheological equation of the state for isotropic and incompressible flow of a Casson fluid are:

$$\tau_{ij} = \begin{cases} \left(\mu_B + \left(\frac{P_y}{\sqrt{2\pi}}\right)\right)2e_{ij}, & \pi > \pi_c \\ \left(\mu_B + \left(\frac{P_y}{\sqrt{2\pi_c}}\right)\right)2e_{ij}, & \pi < \pi_c \end{cases} \quad (1)$$

where μ_B denotes the plastic dynamic viscosity, P_y denotes the yield stress of the fluid, π denotes the product of deformation rate component, where $\pi = e_{ij}. e_{ij}$ is the (i, j)th deformation rate component and π_c is a critical value of π, which is based on the non-Newtonian model. A Cartesian coordinate system is considered, where the x-axis is assumed along with a shrinking sheet and the y-axis is perpendicular to it. The uniform magnetic field of strength B_0 was applied normal to a shrinking sheet. The induced magnetic field was ignored due to the small value of the magnetic Reynolds number. According to these conditions, the governing equations for steady Casson nanofluid flow can be written as:

$$\frac{\partial u}{\partial x} + \frac{\partial v}{\partial y} = 0 \quad (2)$$

$$u\frac{\partial u}{\partial x} + v\frac{\partial u}{\partial y} = \vartheta\left(1 + \frac{1}{\beta}\right)\frac{\partial^2 u}{\partial y^2} - \frac{\vartheta\varphi}{K}u - \frac{b}{\sqrt{K}}u^2 - \frac{\sigma B^2 u}{\rho} \quad (3)$$

$$u\frac{\partial T}{\partial x} + v\frac{\partial T}{\partial y} = \alpha\frac{\partial^2 T}{\partial y^2} + \tau_1\left[D_B\frac{\partial C}{\partial y}\frac{\partial T}{\partial y} + \frac{D_T}{T_\infty}\left(\frac{\partial T}{\partial y}\right)^2\right] + \frac{\mu}{\rho c_p}\left(1 + \frac{1}{\beta}\right)\left(\frac{\partial u}{\partial y}\right)^2 \quad (4)$$

$$u\frac{\partial C}{\partial x} + v\frac{\partial C}{\partial y} = D_B\frac{\partial^2 C}{\partial y^2} + \frac{D_T}{T_\infty}\frac{\partial^2 T}{\partial y^2} \quad (5)$$

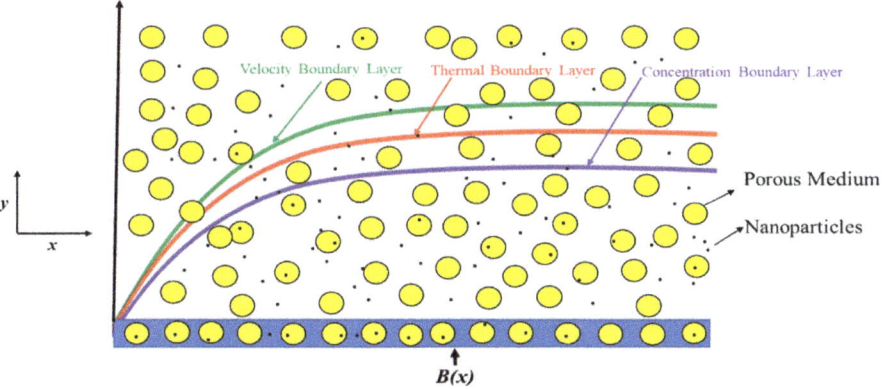

Figure 1. Physical model of flow.

The associated boundary conditions related to Equations (2)–(5) are:

$$\begin{cases} v = v_w, \ u = u_w + A\vartheta\left(1 + \frac{1}{\beta}\right)\frac{\partial u}{\partial y}, T = T_w(x) + D\frac{\partial T}{\partial y}, \ C = C_w(x) + N\frac{\partial C}{\partial y} \text{ at } y = 0 \\ u \to 0, \ T \to T_\infty, \ C \to C_\infty \text{ as } y \to \infty \end{cases} \quad (6)$$

where u and v represent velocity components in the x- and the y-directions, respectively; ρ, β, ϑ, σ, φ, b, K, T, and α are the density of the fluid, Casson fluid parameter, kinematic viscosity, the electrical conductivity of fluid, porosity, the local inertia coefficient, porosity, permeability of the porous medium, fluid temperature, and the thermal diffusivity of the Casson nanofluid, respectively; and $B = B_0 e^{\frac{x}{2l}}$ is the magnetic field by the constant magnetic strength B_0. $\tau_1 = \frac{(\rho c)_p}{(\rho c)_f}$ is the ratio between the effective heat capacity of the nanoparticle material and the capacity of the fluid; and D_T are the coefficients of Brownian diffusion and thermophoretic diffusion, respectively; $T_w = T_\infty + T_0 e^{\frac{x}{2l}}$ and $C_w = C_\infty + C_0 e^{\frac{x}{2l}}$ are the temperature and concentration of the wall, respectively; where T_∞ and C_∞ are the ambient temperature and concentration, respectively. $v_w = -\sqrt{\frac{\theta a}{2l}} e^{\frac{x}{2l}} S$, where S is the suction and blowing parameter, $u_w = -a\, e^{\frac{x}{l}}$ is the shrinking velocity of surface, $A = A_1 e^{\frac{-x}{2l}}$ is the velocity slip factor, $D = D_1 e^{\frac{-x}{2l}}$ is the thermal slip factor, and $N = N_1 e^{\frac{-x}{2l}}$ is the concentration slip factor.

To obtain the similarity solutions, the following similarity transformations are used:

$$\psi = \sqrt{2\theta l a} e^{\frac{x}{2l}} f(\eta), \ \theta(\eta) = \frac{(T - T_\infty)}{(T_w - T_\infty)}, \ \varnothing(\eta) = \frac{(C - C_\infty)}{(C_w - C_\infty)}, \ \eta = y\sqrt{\frac{a}{2\theta l}} e^{x/2l} \tag{7}$$

The stream function ψ is written as a velocity component as:

$$u = \frac{\partial \psi}{\partial y}, \ v = -\frac{\partial \psi}{\partial x} \tag{8}$$

The permeability of porous medium is taken as $K = 2K_0\, e^{\frac{-x}{l}}$. Note, the similarity transformation is mostly used to reduce the number of variables; the resultant equations are reduced to simple form. The similarity transformation is used to obtain the similarity solution. From a physical point of view, the meaning of similarity solutions is that "the velocity, temperature, and concentration profiles of the flow remain geometrically similar in each transversal section of the surface". By applying Equations (7) and (8) into Equations (2)–(6), the continuity equation is satisfied and momentum, energy, and the concentration equations can be expressed as:

$$f''' + \frac{\beta}{(1+\beta)} f f'' - \frac{\beta(2+F_S)}{(1+\beta)} f'^2 - \frac{\beta(K_1+M)}{(1+\beta)} f' = 0 \tag{9}$$

$$\frac{1}{Pr}\theta'' + f\theta' - f'\theta + Nb\varnothing'\theta' + Nt(\theta')^2 + Ec\left(1+\frac{1}{\beta}\right)(f'')^2 = 0 \tag{10}$$

$$\varnothing'' + Sc(f\varnothing' - f'\varnothing) + \frac{Nt}{Nb}\theta'' = 0 \tag{11}$$

along the boundary conditions

$$\begin{array}{c} f(0) = S,\ f'(0) = -1 + \delta\left(1 + \frac{1}{\beta}\right) f''(0), \\ \theta(0) = 1 + \delta_T \theta'(0),\ \varnothing(0) = 1 + \delta_C \varnothing'(0), \\ f'(\eta) \to 0,\ \theta(\eta) \to 0,\ \varnothing(\eta) \to 0\ \text{as}\ \eta \to \infty \end{array} \tag{12}$$

where M (range 0 to 0.5), K_1 (0–0.2), F_S (range 0.05–1.15), Pr, (0.7–2.5), N_b, (0.05–0.5), N_t, (0–0.5), Ec, (0–0.7) and Sc (0–1) denote the Hartmann number, permeability parameter, Forchheimer parameter, Prandtl number, Brownian motion parameter, thermophoresis parameter, Eckert number, and Schmidt number, respectively; δ (range 0–0.5), δ_T (0–0.5), and δ_C (0–0.5) are the velocity, thermal, and concentration slip parameters, respectively. The dimensionless quantities are defined as:

$$\left\{ \begin{array}{l} M = \frac{2l\sigma(B_0)^2}{\rho a},\ K_1 = \frac{l\theta}{aK_0},\ F_S = \frac{2lb}{\sqrt{K}},\ Pr = \frac{\theta}{\alpha},\ Sc = \frac{\theta}{D_B} \\ N_b = \frac{\tau_1 D_B(C_w-C_\infty)}{\theta},\ N_t = \frac{\tau_1 D_T(T_w-T_\infty)}{\theta T_\infty} \\ Ec = \frac{a^2}{C_p T_0},\ \delta = A_1\sqrt{\frac{\theta a}{2l}},\ \delta_T = D_1\sqrt{\frac{a}{2\theta l}},\ \delta_C = N_1\sqrt{\frac{a}{2\theta l}} \end{array} \right. \tag{13}$$

Physical quantities of interest include coefficient of skin friction, the local Nusselt number, and local Sherwood number, which are given by:

$$C_f = \frac{\left[\mu\left(1+\frac{1}{\beta}\right)\frac{\partial u}{\partial y}\right]_{y=0}}{\rho a^2}, \quad Nu = \frac{-x\left(\frac{\partial T}{\partial y}\right)_{y=0}}{(T_w - T_\infty)}, \quad S_h = \frac{-x\left(\frac{\partial C}{\partial y}\right)_{y=0}}{(C_w - C_\infty)} \quad (14)$$

$$C_f(Re_x)^{\frac{1}{2}} = \left(1+\frac{1}{\beta}\right)f''(0), \quad Nu(Re_x)^{-\frac{1}{2}} = -\theta'(0), \quad S_h(Re_x)^{-\frac{1}{2}} = -\varnothing'(0)$$

3. Linear Stability Analysis

Recently, many authors [39–41] investigated multiple solutions for different types of fluids under various fluid flow conditions. From an experimental point of view, it is worth investigating if the solutions are physically reliable. Therefore, linear stability is required to check the reliability of the solutions. To perform stability analysis, the governing boundary layer in Equations (3)–(5) were reduced to the following unsteady form, as suggested by Merkin [43] and Yasin et al. [38]:

$$\frac{\partial u}{\partial t} + u\frac{\partial u}{\partial x} + v\frac{\partial u}{\partial y} = \vartheta\left(1+\frac{1}{\beta}\right)\frac{\partial^2 u}{\partial y^2} - \frac{\vartheta \varphi}{K}u - \frac{b}{\sqrt{K}}u^2 - \frac{\sigma B^2 u}{\rho} \quad (15)$$

$$\frac{\partial T}{\partial t} + u\frac{\partial T}{\partial x} + v\frac{\partial T}{\partial y} = \alpha\frac{\partial^2 T}{\partial y^2} + \tau\left[D_B\frac{\partial C}{\partial y}\frac{\partial T}{\partial y} + \frac{D_T}{T_\infty}\left(\frac{\partial T}{\partial y}\right)^2\right] + \frac{\mu}{\rho c_p}\left(1+\frac{1}{\beta}\right)\left(\frac{\partial u}{\partial y}\right)^2 \quad (16)$$

$$\frac{\partial C}{\partial t} + u\frac{\partial C}{\partial x} + v\frac{\partial C}{\partial y} = D_B\frac{\partial^2 C}{\partial y^2} + \frac{D_T}{T_\infty}\frac{\partial^2 T}{\partial y^2} \quad (17)$$

where t denotes the time. A new similarity transformation is introduced:

$$\psi = \sqrt{2\vartheta l a}e^{\frac{x}{2l}}f(\eta,\tau), \quad \theta(\eta,\tau) = \frac{(T-T_\infty)}{(T_w-T_\infty)}$$
$$\varnothing(\eta,\tau) = \frac{(C-C_\infty)}{(C_w-C_\infty)}, \quad \eta = y\sqrt{\frac{a}{2\vartheta l}}e^{\frac{x}{2l}}, \quad \tau = \frac{a}{2l}e^{\frac{x}{l}}.t \quad (18)$$

Using Equation (18), Equations (15)–(17) can be written as:

$$\left(1+\frac{1}{\beta}\right)\frac{\partial^3 f(\eta,\tau)}{\partial \eta^3} + f(\eta,\tau)\frac{\partial^2 f(\eta,\tau)}{\partial \eta^2} - (2+F_S)\left(\frac{\partial f(\eta,\tau)}{\partial \eta}\right)^2 - (K_1+M)\frac{\partial f(\eta,\tau)}{\partial \eta} - \frac{\partial^2 f(\eta,\tau)}{\partial \tau \partial \eta} = 0 \quad (19)$$

$$\frac{\partial^2 \theta(\eta,\tau)}{\partial \eta^2} + Pr\left(f(\eta,\tau)\frac{\partial \theta(\eta,\tau)}{\partial \eta} - \frac{\partial f(\eta,\tau)}{\partial \eta}\theta(\eta,\tau) + Nb\frac{\partial \varnothing(\eta,\tau)}{\partial \eta}\frac{\partial \theta(\eta,\tau)}{\partial \eta}\right.$$
$$\left. + Nt\left(\frac{\partial \theta(\eta,\tau)}{\partial \eta}\right)^2 + Ec.\left(1+\frac{1}{\beta}\right)\left(\frac{\partial^2 f(\eta,\tau)}{\partial \eta^2}\right)^2 - \frac{\partial \theta(\eta,\tau)}{\partial \tau}\right) = 0 \quad (20)$$

$$\frac{\partial^2 \varnothing(\eta,\tau)}{\partial \eta^2} + Sc\left(f(\eta,\tau)\frac{\partial \varnothing(\eta,\tau)}{\partial \eta} - \frac{\partial f(\eta,\tau)}{\partial \eta}\varnothing(\eta,\tau)\right) + \frac{Nt}{Nb}\frac{\partial^2 \theta(\eta,\tau)}{\partial \eta^2} - \frac{\partial \varnothing(\eta,\tau)}{\partial \tau} = 0 \quad (21)$$

along with new boundary conditions:

$$f(0,\tau) = S, \quad \frac{\partial f(0,\tau)}{\partial \eta} = -1 + \delta\left(1+\frac{1}{\beta}\right)\frac{\partial^2 f(0,\tau)}{\partial \eta^2}, \quad \theta(0,\tau) = 1 + \delta_T\frac{\partial \theta(0,\tau)}{\partial \eta}, \quad \varnothing(0,\tau)$$
$$= 1 + \delta_C\frac{\partial \varnothing(0,\tau)}{\partial \eta} \quad (22)$$
$$\frac{\partial f(\eta,\tau)}{\partial \eta} \to 0, \quad \theta(\eta,\tau) \to 0, \quad \varnothing(\eta,\tau) \to 0 \quad \text{as } \eta \to \infty$$

To check the stability of the steady flow solutions, where $f(\eta) = f_0(\eta), \theta(\eta) = \theta_0(\eta)$ and satisfy the boundary value problem in Equations (9)–(12), we write:

$$f(\eta,\tau) = f_0(\eta) + e^{-\varepsilon\tau}F(\eta,\tau)$$
$$\theta(\eta,\tau) = \theta_0(\eta) + e^{-\varepsilon\tau}G(\eta,\tau) \quad (23)$$
$$\varnothing(\eta,\tau) = \varnothing_0(\eta) + e^{-\varepsilon\tau}H(\eta,\tau)$$

where $F(\eta)$, $G(\eta)$, and $H(\eta)$ are small relative values of $f_0(\eta)$, $\theta_0(\eta)$, and $\varnothing_0(\eta)$ respectively; and ε is the unknown eigenvalues. When we solve the eigenvalue problem in Equations (19)–(22), we have an infinite set of eigenvalues. From that set, we chose the smallest eigenvalue. If the smallest eigenvalue (ε) is negative, the flow is unstable and the disturbances grow, which is physically not possible. If the smallest eigenvalue is positive, it suggests that the solution is stable and physically reliable. Applying the relations in Equation (23) into Equations (19)–(22), the following equations are obtained:

$$\left(1 + \frac{1}{\beta}\right)F_0''' + f_0 F_0'' + F_0 f_0'' - 2(2 + F_S)f_0' F_0' - (K_1 + M)F_0' + \varepsilon F_0' = 0 \tag{24}$$

$$\frac{1}{Pr}G_0'' + f_0 G_0' + F_0 \theta_0' - f_0' G_0 - F_0' \theta_0 + Nb\varnothing_0' G_0' + NbH_0' \theta_0' + 2Nt\theta_0' G_0'$$
$$+ 2Ec.\left(1 + \frac{1}{\beta}\right)f_0'' F_0'' + \varepsilon G_0 = 0 \tag{25}$$

$$H_0''' + Sc\{(f_0\varnothing_0' + F_0 H_0') - (f_0' H_0 + F_0' \varnothing_0)\} + \frac{Nt}{Nb}G_0'' + Sc\varepsilon H_0 = 0 \tag{26}$$

subject to boundary condition:

$$F_0(0) = 0, \quad F_0'(0) = \delta\left(1 + \frac{1}{\beta}\right)F_0''(0), \quad G_0(0) = \delta_T G_0'(0), \quad H_0(0) = \delta_C H_0'(0),$$
$$F_0'(\eta) \to 0, \quad G_0(\eta) \to 0, \quad H_0(\eta) \to 0, \text{ as } \eta \to \infty \tag{27}$$

According to Haris et al. [41], to determine the stability of Equations (24)–(27), we need to relax one boundary condition on $F_0'(\eta)$, $G_0(\eta)$, and $H_0(\eta)$. We relaxed $F_0'(\eta) \to 0$ as $\eta \to \infty$ into $F_0''(0) = 1$ in this problem. We fixed the all parameters to: $\beta = 1.5$, $F_S = 0.2$, $K_1 = 0.1$, $Pr = 1$, $Nt = 0.15$, $Nb = 0.2$, $Ec = 0.1$, $Sc = 1$, $\delta = 0.1$, and $\delta_C = 0.1$, and varied the values of M and δ_T.

4. Result and Discussion

With the help of the shooting method, the transformed ordinary differential equations (BVPs) in Equations (9)–(11) along with the boundary conditions in Equation (12) were converted to initial value problems (IVPs). Equations of IVPs were solved via the Runge Kutta (RK) method. Another method, the three-stage Lobatto IIIa formula, was developed in *bvp4c* with the help of finite difference code. Later, stability analysis was conducted using the *bvp4c* solver function. According to Rehman et al. [36], "this collocation formula and the collocation polynomial provides a C^1 continuous solution that is fourth-order accurate uniformly in [a,b]. Mesh selection and error control are based on the residual of the continuous solution". The impacts of various physical parameters, such as Forchheimer parameter, thermal slip parameter, Casson parameter, magnetic parameter, permeability parameter, Prandtl number, Brownian motion, and thermophoresis parameter, on the flow and heat transfer characteristics were explored. Figure 1 shows the physical model of the problem.

Figure 2 illustrates the existence of multiple solutions for the variation of suction parameter S for three different values of the Forchheimer parameter F_S. For all three values of the Forchheimer parameter $F_S = 0.2, 0.7, 1.15$, there were critical points S_{ci}, $i = 1, 2, 3$, where multiple solutions exist. From a mathematical point of view, we know that the second solution cannot be produced experimentally, but the second solution is a part of the solution to the system of differential equations and therefore should be considered. Overall, we focused on the investigation of the occurrence of multiple solutions for the considered problem. From this profile, we concluded that there are only dual solutions if suction parameter S satisfies this relation $S \geq S_{ci}$, $i = 1, 2, 3$. For the case of the first solution, the skin friction coefficient decreases strictly monotonically as the Forchheimer parameter F_S increases. However, the opposite trend was observed for the second solution. Figure 3 depicts the occurrence of multiple solutions against the values of suction parameter S for two values of thermal slip parameter δ_T on heat transfer rate $-\theta'(0)$. Multiple solutions exist for the variation in the thermal slip parameter $\delta_T = 0.1, 0.5$ only when the suction parameter $S \geq S_{c1} = 2.19377$ and $S \geq S_{c2} = 2.19358$. The heat transfer coefficient declines gradually for the variation in the thermal slip parameter δ_T against the values of suction parameter S. The occurrence of multiple solutions for the values of mass slip parameter δ_C against suction parameter can be seen in the concentration profile in Figure 4. From this profile, the critical point where multiple solutions exist is the same for the two different values of mass slip parameter δ_C. The influence of Casson parameter β on velocity profile $f\prime(\eta)$ for the variation in different physical parameters is shown in Figure 5. From this profile, the velocity profile and its thickness of boundary layer increase with increasing values of the Casson parameter for the first solution. However, the momentum boundary layer decreases for $0 \leq \eta < 3$, due to the increase in a β plastic dynamic, and the viscosity increased, causing resistance to the fluid motion. The opposite behavior was observed for the second solution.

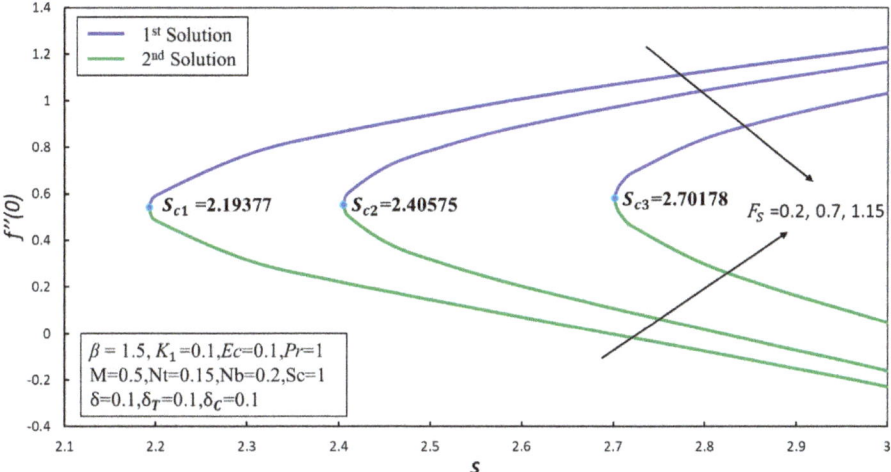

Figure 2. Skin-friction coefficient $f''(0)$ against suction for three values of the Forchheimer parameter F_S.

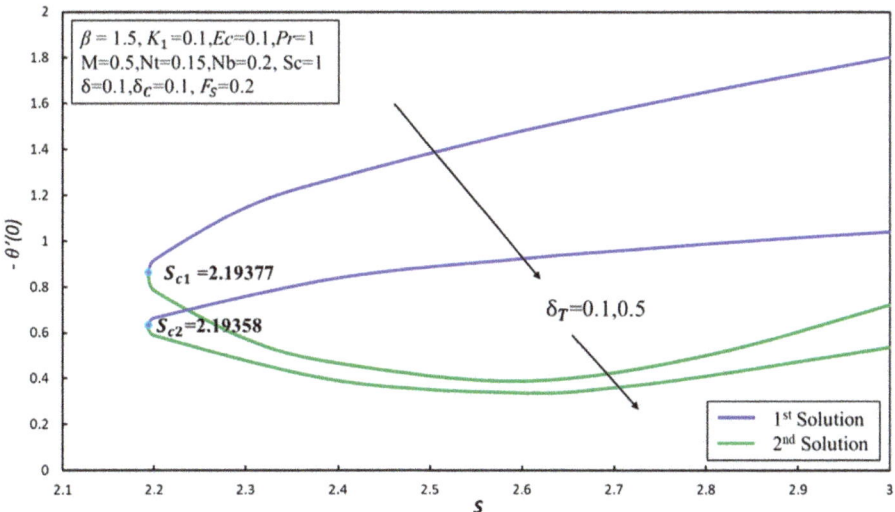

Figure 3. Heat transfer rate $-\theta\prime(0)$ against suction for two values of the thermal slip parameter δ_T.

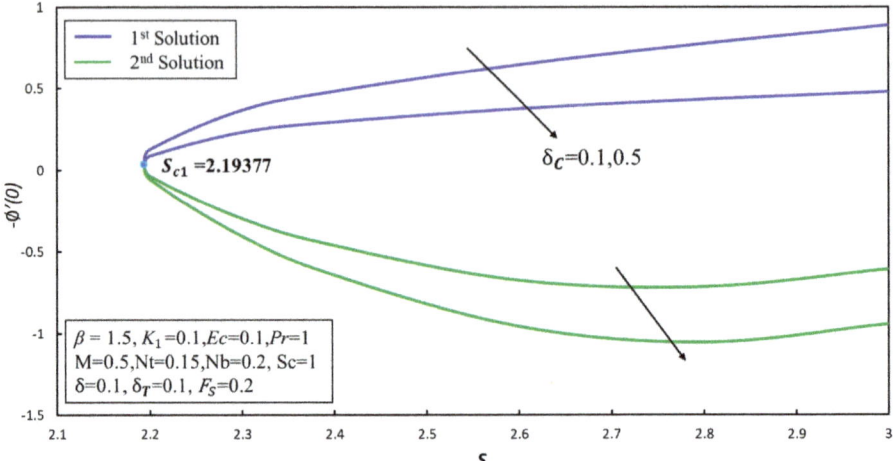

Figure 4. $-\varnothing\prime(0)$ against suction for two values of the mass slip parameter δ_C.

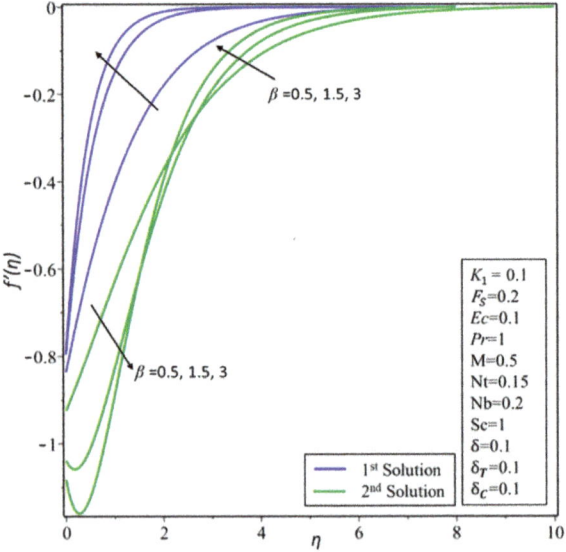

Figure 5. Variation of velocity $f\prime(\eta)$ with η for several values of the Casson parameter β.

The effect of the magnetic parameter M on velocity profile is shown in Figure 6. The thickness of the boundary layer and the velocity of the nanofluid flow are enhanced for the first solution and reduced for the second solution by increasing the strength of the magnetic parameter. Physically, the thickness of boundary layer decreases by increasing the values of the magnetic parameter for the second solution because the fluid particle motion diffuses quickly into the neighboring fluid layers as the values of M increase. Figures 7–9 depict the effect of permeability parameter K_1, velocity slip parameter, and Forchheimer parameter F_S on the velocity profile for fixed values of various physical parameters. These profiles show that the hydrodynamic boundary layer increases in the first solution and decreases in the second solution by increasing the strength of the porosity. However the opposite trend was observed by increasing the Forchheimer parameter F_S. With increasing the velocity slip parameter δ, the velocity profile for both cases (first and second solutions) increases gradually. Figure 10 depicts the effects of the Casson parameter on temperature profile. The temperature and boundary layer thickness of the nanofluid flow decrease for the first solution and increase for the second solution. We concluded that, according to the physical point of view, due to an increase in elasticity, stress parameter thickening of the thermal boundary layer occurred. The Prandtl number Pr effects on the temperature profile are exhibited in Figure 11. The temperature of the Casson nanofluid decreases with increasing Pr and the thermal boundary layer thickness decreases. The Prandtl number can be defined as "the ratio of momentum diffusivity to thermal diffusivity", which means a Casson nanofluid with a higher Prandtl number decreases thermal conductivity, which causes the reduction in the thermal boundary layer thickness.

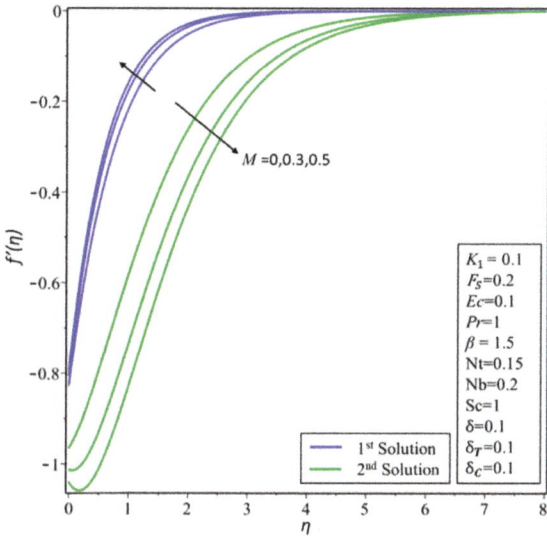

Figure 6. Variation of velocity $f\prime(\eta)$ with η for several values of magnetic parameter M.

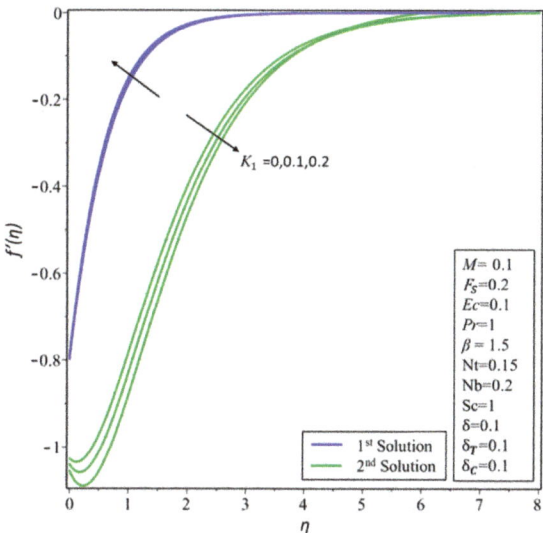

Figure 7. Variation of velocity $f\prime(\eta)$ with η for several values of porosity parameter K_1.

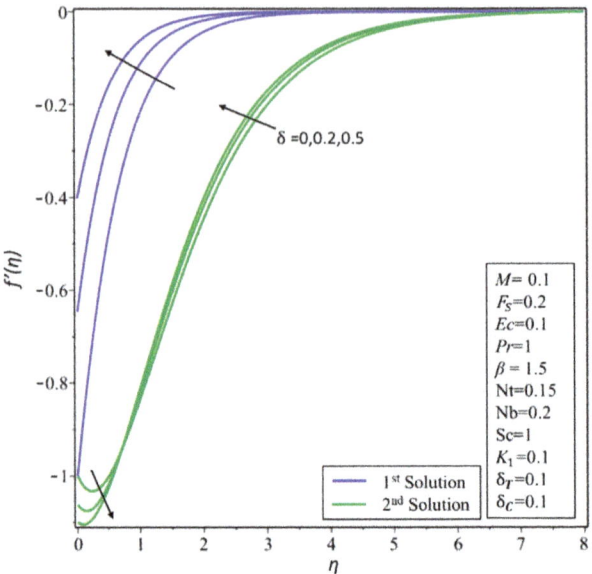

Figure 8. Variation of velocity $f\prime(\eta)$ with η for several values of velocity slip parameter δ.

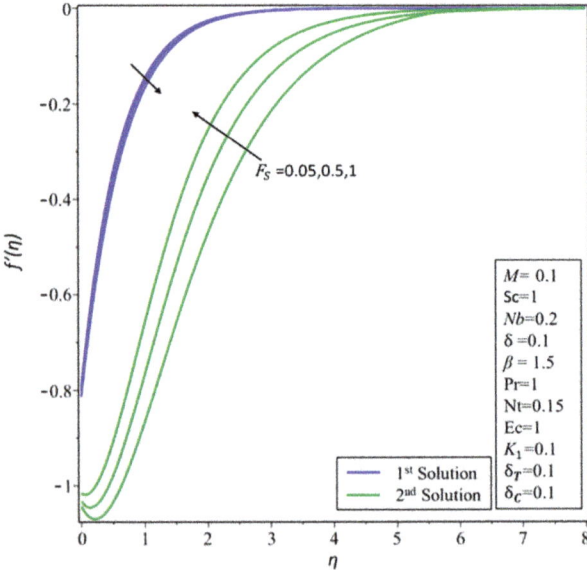

Figure 9. Variation of velocity $f\prime(\eta)$ with η for several values of the Forchheimmer parameter F_S.

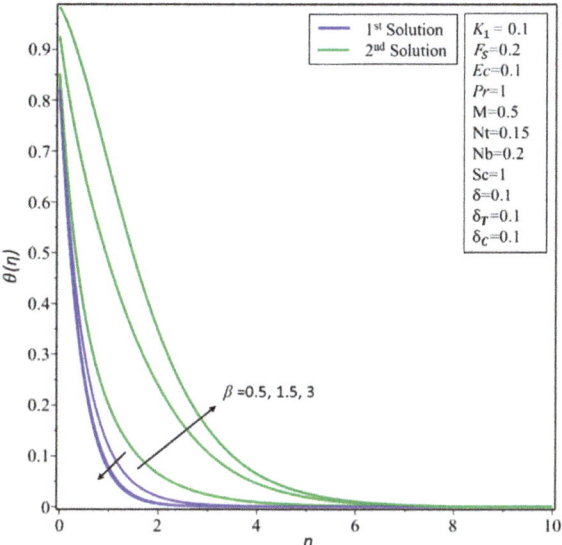

Figure 10. Variation of temperature $\theta(\eta)$ with η for several values of the Casson parameter β.

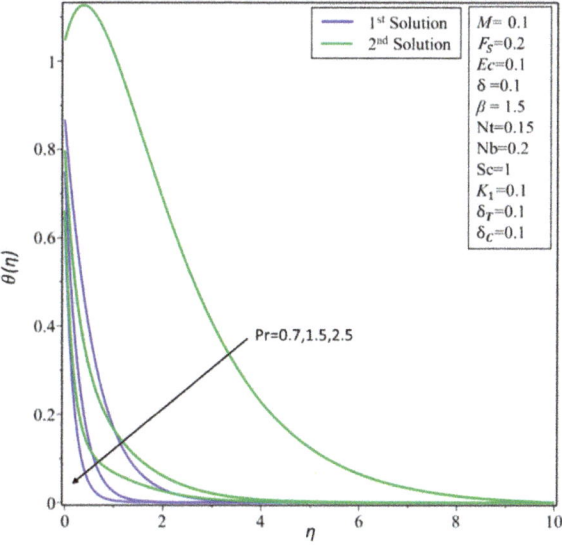

Figure 11. Variation of temperature $\theta(\eta)$ with η for several values of Prandtl number Pr.

Due to the increase in the thermophoresis parameter, the temperature profile and the thermal boundary layer of the nanofluid for the first and second solutions increase gradually (Figure 12). With the increase in thermophoresis parameter N_t thermophoresis force increased, which helped nanoparticles to travel from hot to cold areas. Subsequently, the temperature of the nanofluid increased. Figure 13 shows the effects of the Brownian motion parameter N_b on the temperature profile. This profile shows that temperature increased due to the increase in the Brownian motion parameter; therefore, thermal boundary layer thickness increased. Temperature profile increases as Eckert number increases; therefore, the thermal boundary layer increases gradually because an expansion in dissipation enhances the thermal conductivity of the flow, which upgrades the thermal boundary layers (Figure 14). The impact of thermophoresis parameter N_t on the nanoparticle volume fraction $\varphi(\eta)$ is depicted in Figure 15. The profile of the nanoparticle volume fraction increases with increasing values of the thermophoresis parameter N_t. Figure 16 presents the effects of the Brownian motion parameter N_b on the nanoparticle volume fraction. This profile shows that the nanoparticle volume boundary layer thickness decreases as N_b increases gradually. Figure 17 shows the effects of Schmidt number on concentration profile. Concentration profiles decrease as Sc increases. The comparison of the numerical results of our problem drawn from $bvp4c$ and the shooting method is outlined in Table 1; the results from both methods showed an excellent agreement. The smallest eigenvalue ε for some values of M and δ_T are shown in Table 2. These values show that all values of ε are positive for the first solution and negative for the second solution. Therefore, we concluded that the first solution is stable while the second solution is unstable.

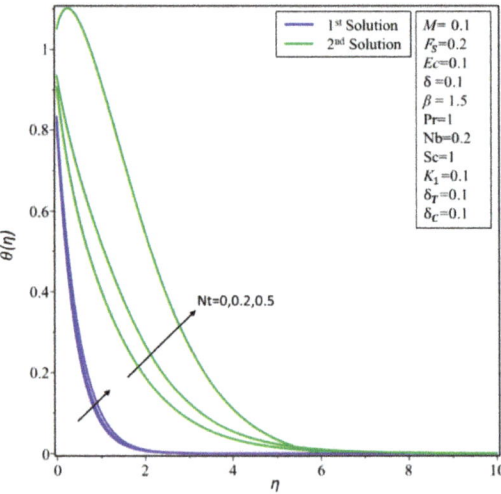

Figure 12. Variation of temperature $\theta(\eta)$ with η for several values of thermophoresis parameter N_t.

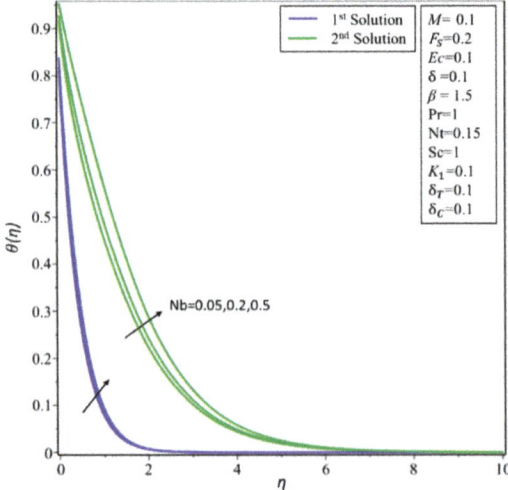

Figure 13. Variation of temperature $\theta(\eta)$ with η for several values of the Brownian motion parameter N_b.

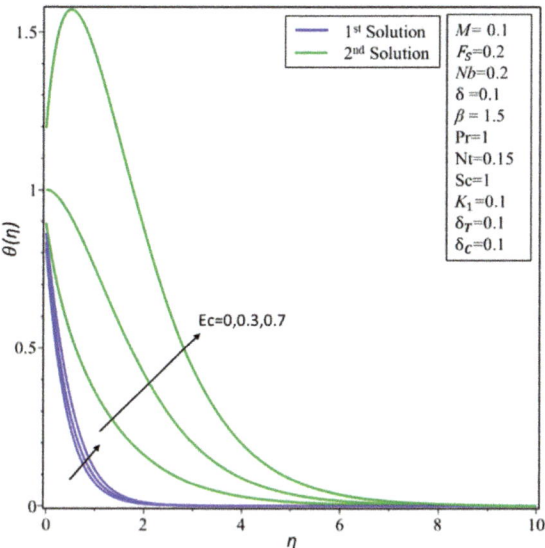

Figure 14. Variation of temperature $\theta(\eta)$ with η for several values of the Eckert number E_c.

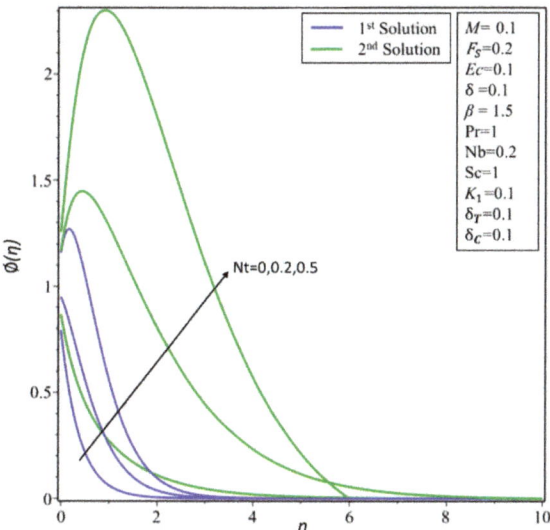

Figure 15. Variation of nanoparticle volume fraction $\varphi(\eta)$ with η for several values of thermophoresis parameter N_t.

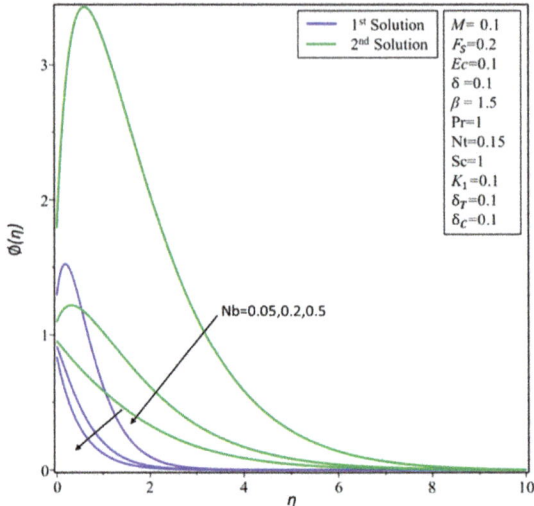

Figure 16. Variation of concentration $\theta(\eta)$ with η for several values of Brownian motion parameter N_b.

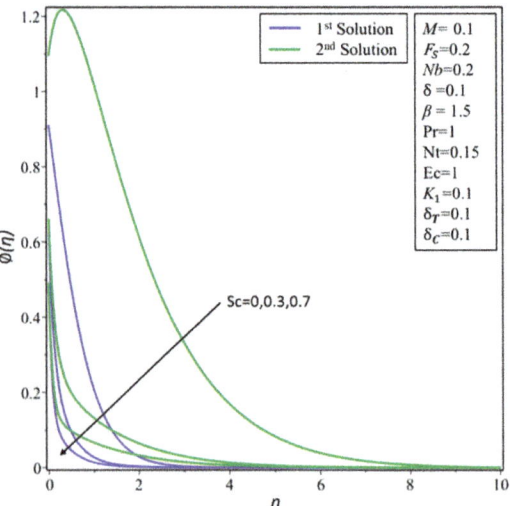

Figure 17. Variation of concentration $\theta(\eta)$ with η for several values of Schmitt number Sc.

Table 1. Comparison between *bvp4c* method and the shooting method for different values of K_1 and M when $\beta = 1.5$, $F_S = 0.2$, $Pr = 1$, $Nt = 0.15$, $Nb = 0.2$, $Ec = 0.1$, $Sc = 1$, $\delta = 0.1$, $\delta_T = 0.1$, and $\delta_C = 0.1$.

M	K_1	$f'(0)$	
		bvp4c Method	Shooting Method
0.5	0.1	1.636957	1.636908
	0.2	1.651884	1.651959
	0.3	1.666407	1.666527
0	0.1	1.552846	1.552775
1		1.707631	1.707671
1.5		1.769173	1.769238

Table 2. Smallest eigenvalues for different values of M and δ_T.

M	δ_T	ε_1	
		First Solution	Second Solution
0.5	0	0.87456	−1.04592
	0.1	0.73948	−1.00248
0.7	0	0.94310	−1.294601
	0.1	0.79092	−1.12253

5. Conclusions

Two-dimensional MHD flow of a Casson nanofluid over a shrinking surface in an Extended Darcy Forchheimer porous medium with the effects of viscous dissipation, velocity, thermal, and concentration slip were examined numerically in this study. The governing boundary layer equations were converted into ordinary differential equations before solving them using the shooting method with the Runge Kutta method. The numerical results showed the existence of dual solutions. To determine which solutions were stable and unstable, stability analysis was conducted. The values of the smallest eigenvalues indicated that only first solution was stable. We found ranges of dual solutions, solutions that depended on a suction parameter, and no solution. Notably, as the Forchheimmer parameter F_S increased, strong mass suction was required to obtain solutions. We found that the velocity profile is indirectly proportional to the velocity slip parameter in the first solution, and that the hydrodynamic boundary layer increases in the first solution and decreases in the second solution by increasing the strength of the porosity.

Author Contributions: Z.O. modelled the problem. L.A.L numerically computed results. I.K discussed the results physically. J.R. computed the tabulated results. M.B and I.T wrote the manuscript and proof read it.

Funding: No specific funding received for this work.

Acknowledgments: This work was supported in part by the Deanship of Scientific Research/ CAMS - Majmaah University under the Project Number 38/149. The authors would also like to thank Universiti Utara Malaysia (UUM) for the moral and financial support in conducting this research.

Conflicts of Interest: The authors declare no conflict of interest.

Nomenclature

u, v	velocity components	Ec	Eckert number
K	permeability of the porous medium	C_w	variable concentration at the sheet
b	local inertia coefficient	Re_x	local Reynolds number
T	Temperature	C_f	skin friction coefficient
T_0	a constant	Nu	local Nusselt number
T_w	variable temperature at the sheet	S	injunction/suction parameter
T_∞	ambient temperature	**Greek letters**	
C	Concentration	β	Casson parameter
C_0	a constant	ε_1	smallest eigen value
C_∞	ambient concentration	τ	Stability transformed variable
P_y	Fluid's yield stress	ε	unknown eigen value
$B(x)$	magnetic field	ψ	stream function
M	Hartmann number	δ	Velocity slip
Pr	Prandtl number	δ_T	Thermal slip
D_B	Brownian diffusion	δ_C	Concentration slip
D_T	thermophoretic diffusion	μ_B	Plastic dynamic viscosity
v_w	suction/injection velocity	φ	Porosity
S_h	local Sherwood number	η	transformed variable
N_b	Brownian motion parameter	α	thermal diffusivity
N_t	thermophoresis parameter	π	The product of the component of deformation rate with itself
Sc	Schmidt number		

References

1. Nield, D.A.; Bejan, A. *Convection in Porous Media*; Springer: New York, NY, USA, 2006; Volume 3.
2. Muskat, M. The Flow of Homogeneous Fluids through Porous Media. No. 532.5 M88. 1946. Available online: https://catalog.hathitrust.org/Record/009073808 (accessed on 30 December 2018).
3. Bakar, S.A.; Arifin, N.M.; Nazar, R.; Ali, F.M.; Pop, I. Forced convection boundary layer stagnation-point flow in Darcy-Forchheimer porous medium past a shrinking sheet. *Front. Heat Mass Transf.* **2016**, *7*, 38.
4. Hayat, T.; Muhammad, T.; Al-Mezal, S.; Liao, S.J. Darcy-Forchheimer flow with variable thermal conductivity and Cattaneo-Christov heat flux. *Int. J. Numer. Methods Heat Fluid Flow* **2016**, *26*, 2355–2369. [CrossRef]
5. Muhammad, T.; Alsaedi, A.; Shehzad, S.A.; Hayat, T. A revised model for Darcy-Forchheimer flow of Maxwell nanofluid subject to convective boundary condition. *Chin. J. Phys.* **2017**, *55*, 963–976. [CrossRef]
6. Hayat, T.; Saif, R.S.; Ellahi, R.; Muhammad, T.; Alsaedi, A. Simultaneous effects of melting heat and internal heat generation in stagnation point flow of Jeffrey fluid towards a nonlinear stretching surface with variable thickness. *Int. J. Therm. Sci.* **2018**, *132*, 344–354. [CrossRef]
7. Alshomrani, A.S.; Ullah, M.Z. Effects of homogeneous-heterogeneous reactions and convective condition in Darcy-Forchheimer flow of carbon nanotubes. *J. Heat Transf.* **2019**, *141*, 012405. [CrossRef]
8. Hayat, T.; Saif, R.S.; Ellahi, R.; Muhammad, T.; Ahmad, B. Numerical study for Darcy-Forchheimer flow due to a curved stretching surface with Cattaneo-Christov heat flux and homogeneous-heterogeneous reactions. *Results Phys.* **2017**, *7*, 2886–2892. [CrossRef]
9. Seth, G.S.; Mandal, P.K. Hydromagnetic rotating flow of Casson fluid in Darcy-Forchheimer porous medium. In *MATEC Web of Conferences*; EDP Sciences: Les Ulis, France, 2018; Volume 192, p. 02059.
10. Alarifi, I.M.; Abokhalil, A.G.; Osman, M.; Lund, L.A.; Ayed, M.B.; Belmabrouk, H.; Tlili, I. MHD Flow and Heat Transfer over Vertical Stretching Sheet with Heat Sink or Source Effect. *Symmetry* **2019**, *11*, 297. [CrossRef]
11. Ellahi, R.; Riaz, A. Analytical solutions for MHD flow in a third-grade fluid with variable viscosity. *Math. Comput. Model.* **2010**, *52*, 1783–1793. [CrossRef]
12. Ellahi, R.; Tariq, M.H.; Hassan, M.; Vafai, K. On boundary layer nano-ferroliquid flow under the influence of low oscillating stretchable rotating disk. *J. Mol. Liq.* **2017**, *229*, 339–345. [CrossRef]
13. Hsiao, K.L. Combined electrical MHD heat transfer thermal extrusion system using Maxwell fluid with radiative and viscous dissipation effects. *Appl. Therm. Eng.* **2017**, *112*, 1281–1288. [CrossRef]
14. Sheikholeslami, M.; Abelman, S.; Ganji, D.D. Numerical simulation of MHD nanofluid flow and heat transfer considering viscous dissipation. *Int. J. Heat Mass Transf.* **2014**, *79*, 212–222. [CrossRef]
15. Charrier, E.E.; Pogoda, K.; Wells, R.G.; Janmey, P.A. Control of cell morphology and differentiation by substrates with independently tunable elasticity and viscous dissipation. *Nat. Commun.* **2018**, *9*, 449. [CrossRef] [PubMed]
16. Kumar, R.; Kumar, R.; Shehzad, S.A.; Sheikholeslami, M. Rotating frame analysis of radiating and reacting ferro-nanofluid considering Joule heating and viscous dissipation. *Int. J. Heat Mass Transf.* **2018**, *120*, 540–551. [CrossRef]
17. Reddy, M.; Reddy, S.; Rama, G. Micropolar fluid flow over a nonlinear stretching convectively heated vertical surface in the presence of Cattaneo-Christov heat flux and viscous dissipation. *Front. Heat Mass Transf. (FHMT)* **2017**, *8*, 20.
18. Khan, M.I.; Hayat, T.; Khan, M.I.; Alsaedi, A. A modified homogeneous-heterogeneous reaction for MHD stagnation flow with viscous dissipation and Joule heating. *Int. J. Heat Mass Transf.* **2017**, *113*, 310–317. [CrossRef]
19. Saqib, M.; Ali, F.; Khan, I.; Sheikh, N.A. Heat and mass transfer phenomena in the flow of Casson fluid over an infinite oscillating plate in the presence of first-order chemical reaction and slip effect. *Neural Comput. Appl.* **2018**, *30*, 2159–2172. [CrossRef]
20. le Roux, C. Existence and uniqueness of the flow of second-grade fluids with slip boundary conditions. *Arch. Ration. Mech. Anal.* **1999**, *148*, 309–356. [CrossRef]
21. Soltani, F.; Yilmazer, Ü. Slip velocity and slip layer thickness in flow of concentrated suspensions. *J. Appl. Polym. Sci.* **1998**, *70*, 515–522. [CrossRef]
22. Khan, W.A.; Ismail, A.I.; Uddin, M.J. Melting and second order slip effect on convective flow of nanofluid past a radiating stretching/shrinking sheet. *Propuls. Power Res.* **2018**, *7*, 60–71.
23. Alamri, S.Z.; Ellahi, R.; Shehzad, N.; Zeeshan, A. Convective radiative plane Poiseuille flow of nanofluid through porous medium with slip: An application of Stefan blowing. *J. Mol. Liq.* **2019**, *273*, 292–304. [CrossRef]

24. Yunianto, B.; Tauviqirrahman, M.; Wicaksono, W.A. CFD analysis of partial slip effect on the performance of hydrodynamic lubricated journal bearings. In *MATEC Web of Conferences*; EDP Sciences: Les Ulis, France, 2018; Volume 204, p. 04014.
25. Ellahi, R.; Hussain, F. Simultaneous effects of MHD and partial slip on peristaltic flow of Jeffery fluid in a rectangular duct. *J. Magn. Magn. Mater.* **2015**, *393*, 284–292. [CrossRef]
26. Khan, M.I.; Hayat, T.; Waqas, M.; Khan, M.I.; Alsaedi, A. Entropy generation minimization (EGM) in nonlinear mixed convective flow of nanomaterial with Joule heating and slip condition. *J. Mol. Liq.* **2018**, *256*, 108–120. [CrossRef]
27. Ullah, I.; Shafie, S.; Khan, I. Effects of slip condition and Newtonian heating on MHD flow of Casson fluid over a nonlinearly stretching sheet saturated in a porous medium. *J. King Saud Univ. Sci.* **2017**, *29*, 250–259. [CrossRef]
28. Crane, L.J. Flow past a stretching plate. *Zeitschrift für angewandte Mathematik und Physik ZAMP* **1970**, *21*, 645–647. [CrossRef]
29. Miklavčič, M.; Wang, C. Viscous flow due to a shrinking sheet. *Q. Appl. Math.* **2006**, *64*, 283–290. [CrossRef]
30. Naveed, M.; Abbas, Z.; Sajid, M.; Hasnain, J. Dual Solutions in Hydromagnetic Viscous Fluid Flow Past a Shrinking Curved Surface. *Arab. J. Sci. Eng.* **2018**, *43*, 1189–1194. [CrossRef]
31. Jusoh, R.; Nazar, R.; Pop, I. Three-dimensional flow of a nanofluid over a permeable stretching/shrinking surface with velocity slip: A revised model. *Phys. Fluids* **2018**, *30*, 033604. [CrossRef]
32. Othman, N.A.; Yacob, N.A.; Bachok, N.; Ishak, A.; Pop, I. Mixed convection boundary-layer stagnation point flow past a vertical stretching/shrinking surface in a nanofluid. *Appl. Therm. Eng.* **2017**, *115*, 1412–1417. [CrossRef]
33. Khan, M.; Hafeez, A. A review on slip-flow and heat transfer performance of nanofluids from a permeable shrinking surface with thermal radiation: Dual solutions. *Chem. Eng. Sci.* **2017**, *173*, 1–11. [CrossRef]
34. Naganthran, K.; Nazar, R.; Pop, I. Stability analysis of impinging oblique stagnation-point flow over a permeable shrinking surface in a viscoelastic fluid. *Int. J. Mech. Sci.* **2017**, *131*, 663–671. [CrossRef]
35. Qing, J.; Bhatti, M.M.; Abbas, M.A.; Rashidi, M.M.; Ali, M.E.S. Entropy generation on MHD Casson nanofluid flow over a porous stretching/shrinking surface. *Entropy* **2016**, *18*, 123. [CrossRef]
36. Rahman, M.M.; Roşca, A.V.; Pop, I. Boundary layer flow of a nanofluid past a permeable exponentially shrinking/stretching surface with second order slip using Buongiorno's model. *Int. J. Heat Mass Transf.* **2014**, *77*, 1133–1143. [CrossRef]
37. Rahman, M.M.; Rosca, A.V.; Pop, I. Boundary layer flow of a nanofluid past a permeable exponentially shrinking surface with convective boundary condition using Buongiorno's model. *Int. J. Numer. Methods Heat Fluid Flow* **2015**, *25*, 299–319. [CrossRef]
38. Yasin, M.H.M.; Ishak, A.; Pop, I. Boundary layer flow and heat transfer past a permeable shrinking surface embedded in a porous medium with a second-order slip: A stability analysis. *Appl. Therm. Eng.* **2017**, *115*, 1407–1411. [CrossRef]
39. Raza, J.; Rohni, A.M.; Omar, Z. A note on some solutions of copper-water (cu-water) nanofluids in a channel with slowly expanding or contracting walls with heat transfer. *Math. Comput. Appl.* **2016**, *21*, 24. [CrossRef]
40. Raza, J.; Rohni, A.M.; Omar, Z.; Awais, M. Rheology of the Cu-H2O nanofluid in porous channel with heat transfer: Multiple solutions. *Phys. E: Low-Dimens. Syst. Nanostruct.* **2017**, *86*, 248–252. [CrossRef]
41. Raza, J.; Rohni, A.M.; Omar, Z. Rheology of micropolar fluid in a channel with changing walls: Investigation of multiple solutions. *J. Mol. Liq.* **2016**, *223*, 890–902. [CrossRef]
42. Nakamura, M.; Sawada, T. Numerical study on the flow of a non-Newtonian fluid through an axisymmetric stenosis. *J. Biomech. Eng.* **1988**, *110*, 137–143. [CrossRef]
43. Merkin, J.H. On dual solutions occurring in mixed convection in a porous medium. *J. Eng. Math.* **1986**, *20*, 171–179. [CrossRef]
44. Harris, S.D.; Ingham, D.B.; Pop, I. Mixed convection boundary-layer flow near the stagnation point on a vertical surface in a porous medium: Brinkman model with slip. *Transp. Porous Media* **2009**, *77*, 267–285. [CrossRef]

© 2019 by the authors. Licensee MDPI, Basel, Switzerland. This article is an open access article distributed under the terms and conditions of the Creative Commons Attribution (CC BY) license (http://creativecommons.org/licenses/by/4.0/).

Article

Effects MHD and Heat Generation on Mixed Convection Flow of Jeffrey Fluid in Microgravity Environment over an Inclined Stretching Sheet

Iskander Tlili

Department of Mechanical and Industrial Engineering, College of Engineering, Majmaah University, Al-Majmaah 11952, Saudi Arabia; l.tlili@mu.edu.sa; Tel.: +966-509107698

Received: 14 March 2019; Accepted: 22 March 2019; Published: 25 March 2019

Abstract: In this paper, Jeffrey fluid is studied in a microgravity environment. Unsteady two-dimensional incompressible and laminar g-Jitter mixed convective boundary layer flow over an inclined stretching sheet is examined. Heat generation and Magnetohydrodynamic MHD effects are also considered. The governing boundary layer equations together with boundary conditions are converted into a non-similar arrangement using appropriate similarity conversions. The transformed system of equations is resolved mathematically by employing an implicit finite difference pattern through quasi-linearization method. Numerical results of temperature, velocity, local heat transfer, and local skin friction coefficient are computed and plotted graphically. It is found that local skin friction and local heat transfer coefficients increased for increasing Deborah number when the magnitude of the gravity modulation is unity. Assessment with previously published results showed an excellent agreement.

Keywords: Jeffrey fluid; laminar g-Jitter flow; inclined stretching sheet; heat source/sink

---

1. Introduction

Nanoparticles composed of nanosized metals, oxides, and carbon nanotubes form fluid suspensions called nanofluids. Nanofluids are widely used in many practical applications including nano-electro-mechanical systems and in the industrial, manufacturing and medical sectors because compared to the conventional liquids, they are characterized by great thermal conductivities leading to an enhanced heat transfer rate. In this context, several researches were carried out in the last few decades to analyze theoretically and experimentally transport phenomena related to heat and nanofluid flow while considering diverse geometries, velocity, and temperature slip boundary conditions [1–6].

The gravity impact is abolished significantly in space, but the microgravity environment is able to reduce convective flows. g-Jitter which refers to the inertia impacts caused by the residual, oscillatory or transitory accelerations originating after squad waves, mechanical pulsations atmospheric slog and microgravity environments was investigated showing that microgravity is correlated to the frequency and magnitude of the periodical gravity modulation [7–11].

One of the major concerns of gravity is the study of the controversial impact of g-Jitter convective flow in various aspects. The impact of a gravitational field representative of g-Jitter was investigated [12–15]. It was demonstrated that g-Jitter has a great influence on the configuration of the three-dimensional boundary layer and the flow properties especially the skin friction and the heat rate. It was demonstrated that the skin friction coefficient is decreased when the magnetic field parameter rises [12,16]. Khoshnevis [17] studied the effect of residual and g-Jitter accelerations and reported that the diffusion procedure is more probably to be affected by the low-frequency involvement of g-Jitter. The effect of the gravity variations falls down against the rise of the forced flow velocity.

However, fuel injection velocity has the opposite effect when it augments [16]. Kumar et al. [18] studied the impact of gravity inflection in a couple stress liquid by using Ginzburg–Landau equations. It was underlined that to improve the rate of heat and mass transfer, the influence of the Prandtl number; concentration Rayleigh number; Lewis number and couple stress parameter on the Nusselt number and Sherwood number has to be increased. Mass transfer has the opposite behavior against Lewis quantity [15].

Magnetohydrodynamic convective boundary layer flow over an exponentially inclined permeable stretching surface was investigated by many researchers [9–11]. Reddy numerically studied [10] the scheme of joined boundary conditions for nonlinear ordinary differential equations. It was revealed that the momentum boundary layer thickness is reduced when the Casson fluid parameter augments. The thermal boundary layer thickness is improved when augmenting rates of the non-uniform heat source or sink parameters [12].

An extensive investigation has been conducted considering the different fluid physical conditions. Both the gravitational field and temperature gradient generate convective flows in porous and clear media. Greater temperature can be observed in the pure field as associated to the porous field in Couette, Poiseuille and widespread to Couette flows of an incompressible magneto hydrodynamic Jeffrey fluid among similar platters over homogeneous porous field employing slip boundary circumstances [18]. Convection flows of nanofluids in porous media have attracted great concern driven by material treating and solar energy gatherer uses. Bhadauria [19] focused on the impacts of flow and G-Jitter on chaotic convection in an anisotropic porous field. It was concluded that heat transfer is greater in the modulated system compared to the unmodulated system. Ghosh et al. [15] reported that sinusoidal g-Jitter leads to a flow streaming inside the porous cavity and time-dependent rolls privileged the encloser because of variances in thermal diffusivities among the solid matrix, wall, and fluid. The temperature profile has an opposite behavior for the Prandtl number [9,10,20] because of the influence of the slip parameter, non-Newtonian parameter, and Hartmann number [21] whereas it augments in parallel with the radiation parameter [4,11,12], Brinkman number, and Dufour and Soret numbers [14]. Compared to Maxwell and Oldroyd-B nanofluids, the Jeffrey nanofluid has superior heat transfer performance [22]. The heat transfer profile was studied in different conditions. Sandeep et al. analyzed numerically the momentum and heat transfer profile of Jeffrey, Maxwell, and Oldroyd-B nanofluids over a stretching surface to determine the impact of the transverse magnetic field, thermal radiation, non-uniform heat source/sink, and suction effects. Heat transfer rate was improved when the Biot number and suction parameter increase. The Jeffrey nanofluid has superior heat transfer performance than the Maxwell and Oldroyd-B nanofluids [23–30].

Some interesting studies concerning the flow of Jeffrey fluid have been conducted. Hayat et al. [8] considered a homogeneous–heterogeneous reaction in a nonlinear radiative flow of Jeffrey fluid between two stretchable rotating disks. The velocities augment in parallel with Deborah number. The thermal field and heat transfer rate are improved for the temperature ratio parameter. Khan et al. [14] analytically studied the Jeffrey liquid flow related to thermal-diffusion and diffusion-thermo characteristics using the homotopic method [31–34].

In the present work, we investigate the effects of Jeffrey fluid in a microgravity environment. Unsteady two-dimensional incompressible and laminar g-Jitter mixed convective boundary layer flow over an inclined stretching sheet is taken into account. The governing boundary layer equations together with boundary conditions are converted into a non-similar arrangement using appropriate similarity conversions. The transformed system of equations is solved numerically by using an implicit finite difference structure with quasi-linearization technique. Numerical results of velocity, temperature, local skin friction, and local heat transfer coefficient are computed and illustrated graphically.

2. Mathematical Formulation

The objective of this study is to investigate the unsteady incompressible flow of Jeffrey fluid past an inclined stretching sheet. In this problematic, the x-axis is ranging lengthways the surface with penchant angle γ to the perpendicular in the upward direction and y-axis is perpendicular to the surface. The plate is characterized with a linear speed $u_w(x)$ in x-direction. The temperature and flow of the platter varies linearly with the stretch x along the platter, anywhere $T_w(x) > T_\infty$ by means of $T_w(x)$ represent the temperature of the plate and T_∞ representing the uniform temperature of the ambient nanofluids.

The incessant stretching sheet is supposed to require the temperature and velocity in the arrangement of $u_w(x) = cx$ and $T_w(x) = T_\infty + ax$ where c and a are factors with $c > 0$. Mutually circumstances of reheating ($T_w(x) > T_\infty$) and cooling ($T_w(x) < T_\infty$) of the piece are considered, which settle to assisting flow for $a > 0$ and opposing flow for $a < 0$, correspondingly. The fluid conduct electricity in the control of a variable magnetic field $B(x) = \frac{B_0}{\sqrt{x}}$. And finally the heat generation effect is also considered.

In the typical boundary layer and Boussinesq calculations, the basic governing equations including the conservation of momentum, mass thermal energy equation of Jeffrey fluid can be written as [5,6,35],

$$\frac{\partial u}{\partial x} + \frac{\partial v}{\partial y} = 0 \tag{1}$$

$$\frac{\partial u}{\partial t} + u\frac{\partial u}{\partial x} + v\frac{\partial u}{\partial y} = \frac{v}{1+\lambda_1}\left[\frac{\partial^2 u}{\partial y^2}\right] + \frac{v\lambda_2}{1+\lambda_1}\left[\frac{\partial^3 u}{\partial t \partial y^2} + u\frac{\partial^3 u}{\partial x \partial y^2} - \frac{\partial u}{\partial x}\frac{\partial^2 u}{\partial y^2}\right]$$
$$+ \frac{v\lambda_2}{1+\lambda_1}\left[\frac{\partial u}{\partial y}\frac{\partial^2 u}{\partial x \partial y} + v\frac{\partial^3 u}{\partial y^3}\right] + g^*(t)\beta_T(T - T_\infty)\cos\alpha - \frac{\sigma B(x)^2}{\rho}u \tag{2}$$

$$\frac{\partial T}{\partial t} + u\frac{\partial T}{\partial x} + v\frac{\partial T}{\partial y} = \frac{k}{\rho c_p}\frac{\partial^2 T}{\partial y^2} + \frac{Q(x)}{\rho c_p}(T - T_\infty). \tag{3}$$

The suitable initial and boundary circumstances are prescribed as:

$$t = 0 : u = v = 0, T = T_\infty \text{ for any } x, y,$$
$$t > 0 : u_w(x) = ax, v = 0, T = T_w = T_\infty + bx \text{ at } y = 0, \tag{4}$$
$$u \to 0, \frac{\partial u}{\partial y} \to 0, T \to T_\infty \text{ as } y \to \infty,$$

where u and v are the speed modules lengthwise x and y axes, t represents time, and T characterizes the temperature of Jeffrey fluid. Meanwhile, ρ is the density, v is the kinematic viscosity, β_T is the volumetric coefficient of thermal expansion, C_p is the specific heat at constant pressure, k is the real thermal conductivity, λ_1 and λ_2 are two parameters related to the Jeffrey fluid which are, respectively, the fraction of relaxation to retardation times. It is worth mentioning that, for $\lambda_1 = \lambda_2 = 0$, and in the absence of MHD and heat generation terms, the problem reduces to the case of a regular viscous fluid, which is the same problem studied by Sharidan et al. (2006) [3]. The difficulty of the problematic is abridged by involving the subsequent similarity changes [3].

$$\tau = \omega t, \ \eta = \left(\frac{a}{v}\right)^{\frac{1}{2}} y, \ \psi = (av)^{\frac{1}{2}} x f(\tau, \eta), \ \theta(\tau, \eta) = \frac{(T - T_\infty)}{(T_w - T_\infty)}, \ g(\tau) = \frac{g(t)}{g_0}. \tag{5}$$

By employing the similarity transformations (5), Equation (1) is similarly fulfilled, and in addition, the following transformed governing equations are obtained:

185

$$\Omega\frac{\partial^2 f}{\partial\tau\partial\eta} + \left(\frac{\partial f}{\partial\eta}\right)^2 - f\frac{\partial^2 f}{\partial\eta^2} = \frac{1}{1+\lambda_1}\frac{\partial^3 f}{\partial\eta^3} + \frac{\beta}{1+\lambda_1}\Omega\left(\frac{\partial^4 f}{\partial\tau\partial\eta^3}\right) + \frac{\beta}{1+\lambda_1}\left(\left(\frac{\partial^2 f}{\partial\eta^2}\right)^2 - f\frac{\partial^4 f}{\partial\eta^4}\right)$$
$$+\lambda(1-\varepsilon\cos(\pi\tau))\cos\alpha\cdot\theta(\eta) - M\frac{\partial f}{\partial\eta},\tag{6}$$

$$\frac{1}{\Pr}\frac{\partial^2\theta}{\partial\eta^2} + f\frac{\partial\theta}{\partial\eta} - \theta\frac{\partial f}{\partial\eta} = \Omega\frac{\partial\theta}{\partial t} - Q_0\theta.\tag{7}$$

where

$$\Omega = \frac{\omega}{a},\ M = \frac{\sigma B_0^2}{a\rho},\ Q_0 = \frac{Q}{a\rho c_p}, \beta = a\lambda_2, \lambda = \frac{g_0\beta_T(T_w-T_\infty)\frac{x^3}{v^2}}{\left[u_w(x)\frac{x}{v}\right]^2} = \frac{Gr_x}{Re_x^2},$$

$$Gr_x = g_0\beta_T(T_w-T_\infty)\frac{x^3}{v^2},\ Re_x = u_w(x)\frac{x}{v},$$

$$\tau = 0: \tfrac{\partial f}{\partial\eta}(\tau,\eta) = 0,\ f(\tau,\eta) = 0,\ \theta(\tau,\eta) = 0 \text{ for any } x,y,$$

$$\eta > 0: \tfrac{\partial f}{\partial\eta}(\tau,\eta) = 1,\ f(\tau,\eta) = 0,\ \theta(\tau,\eta) = \tfrac{b}{a} \text{ at } \eta = 0,\tag{8}$$

$$\eta > 0: f(\tau,\eta) \to 0,\ \tfrac{\partial^2\theta}{\partial\eta^2}(\tau,\eta) \to 0,\ \theta(\tau,\eta) \to 0 \text{ as } \eta \to \infty,$$

where
$$\Pr = \frac{\mu c_p}{k},$$

The real amounts of main attention, such as the local Nusselt number, Nu_x and the skin friction coefficient, C_f, are defined as

$$C_f = \frac{\tau_w(x)}{\rho u_w^2} \text{ and } Nu_x = \frac{q_w(x)\,x}{k(T_w-T_\infty)}\tag{9}$$

where the $\tau_w(x)$ is the wall shear stress and $q_w(x)$ is the wall heat flux given by:

$$\tau_w = \frac{\mu}{1+\lambda_1}\left[\left(\frac{\partial u}{\partial y}\right) + \lambda_2\left(\frac{\partial^2 u}{\partial y\partial t} + u\frac{\partial^2 u}{\partial x\partial y} + v\frac{\partial^2 u}{\partial y^2}\right)\right]_{y=0} \text{ and } q_w(x) = -k\left(\frac{\partial T}{\partial y}\right)_{y=0}\tag{10}$$

The following skin friction coefficient and local Nusselt number are obtained as follows:

$$C_f Re_x^{1/2} = \frac{1}{(1+\lambda_1)}\left[\frac{\partial^2 f}{\partial\eta^2}(\tau,0) + \beta\left(\frac{\partial f}{\partial\eta}(\tau,0)\frac{\partial^2 f}{\partial\eta^2}(\tau,0) - f(\tau,0)\frac{\partial^3 f}{\partial\eta^3}(\tau,0)\right)\right],$$
$$\frac{Nu_x}{Re_x^{1/2}} = -\frac{\partial\theta}{\partial\eta}(\tau,0).\tag{11}$$

3. Numerical Method

The difficulty of the problematic is abridged by involving the subsequent similarity changes (Sharidan et al. (2006)) [3]. By employing the similarity transformations (5), Equations (1)–(3) with boundary conditions (4) transform to Equations (6)–(8). This model will be resolved numerically using Runge–Kutta–Fehlberg method of the seventh order (RKF7) joined with a shooting method. In the RKF7 method, several assessments are tolerable for each stage separately. For minor precision, this technique delivers the greatest effective outputs. A phase size $\Delta\eta = 0.001$ and a convergence condition of 10^{-6} were employed in the numerical calculations. The asymptotic boundary conditions given by Equation (11) were substituted using a value of 10 for the similarity variable η_{max} as follows:

$$f'(\eta,10) = \theta'(\eta,10) = 0$$

The choice of $\eta_{max} = 10$ guarantees that all mathematical solutions move toward the asymptotic values appropriately. The other details of this method can be found in [24]. To check the accuracy of the current technique, the found results are compared in special cases with the results obtained by Hayat et al. [5]. These comparisons are presented clearly in Table 1 in terms of the Nu_m at the heat source. A very good arrangement is established among the results.

Table 1. Assessment of the Nu_m.

Ra	Hayat et al. [5]	Current
1000	5.321	5.332
10,000	5.487	5.496
100,000	7.212	7.223
1,000,000	13.946	14.101

4. Results and Discussion

Unsteady two-dimensional incompressible and laminar g-Jitter mixed convective boundary layer flow over an inclined stretching sheet in the presence of heat source/sink, viscous dissipation and magnetic field for Jeffrey fluid has been explored numerically by means of a shooting scheme based Runge–Kutta–Fehlberg-integration algorithm.

Figure 1a,b present the variation of dimensionless velocity with pertinent parameters in which Figure 1a illustrates the effect of magnetic field and g-Jitter frequency period on the dimensionless velocity and Figure 1b shows the effect of thermal expansion coefficient and mixed convection parameter on the dimensionless velocity. As anticipated, the velocity profile is higher for dimensionless boundary layer thickness $\eta = 0$ then decreases significantly along the inclined stretching sheet to reach zero, this effect is mathematically noticeable in Equation (8) which fulfills the assigned boundary condition. Furthermore, it can be seen that the velocity profile decreases with both magnetic field and thermal expansion coefficient while the g-Jitter frequency period and mixed convection parameter have a negligible effect on velocity profile. Consequently, boundary layer thickness has a similar effect and behavior as the dimensionless velocity. It is important to note that in the absence of a magnetic field the velocity profile is maximum and declines with the application of magnetic field, this can be elucidated by the fact that a magnetic field generates Lorentz force which leads to opposing the flow and, therefore, reduces the velocity profile. Similarly, increasing a thermal expansion coefficient leads to generating more thermal buoyancy that affect the velocity profile but remains a small alteration compared to the effect of a magnetic field. Both of Figure 1a,b refer to the case of assisting flow ($\lambda > 0$) except that in Figure 1b the mixed convection parameter increases.

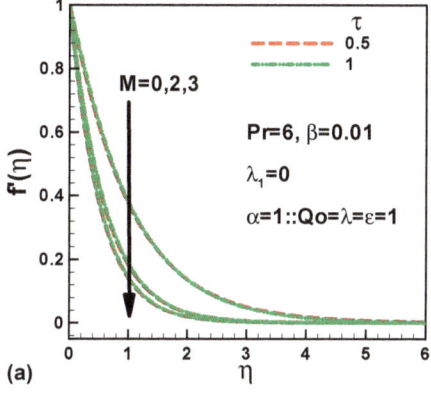
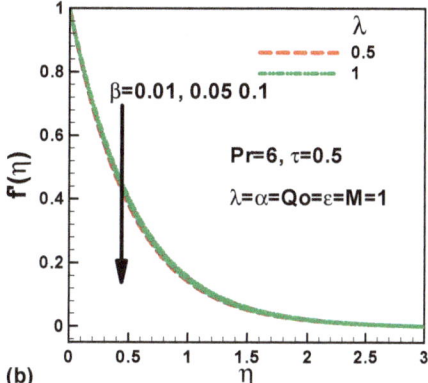

Figure 1. Variation of dimensionless velocity with (**a**) magnetic field and g-Jitter frequency period and with (**b**) thermal expansion coefficient and mixed convection parameter.

Figure 2a,b depict the variation of dimensionless temperature with pertinent parameters in which Figure 2a illustrates the effect of a magnetic field and g-Jitter frequency period on the dimensionless temperature and Figure 2b shows the effect of thermal expansion coefficient and mixed convection parameter on the dimensionless velocity. As expected, the temperature profile is higher for dimensionless boundary layer thickness $\eta = 0$ which then decreases significantly along the inclined stretching sheet to reach zero. This behavior is approved by the considered boundary condition and it is accurately harmonized with Equation (8). Moreover, tt can be seen that the dimensionless temperature variance of the velocity profile shown in Figure 1 augment slightly with both magnetic field and thermal expansion coefficient. Additionally, the thermal expansion coefficient and g-Jitter frequency have an insignificant effect on the temperature profile. Therefore, the thickness of the thermal boundary layer remains uniform in respect of the studied parameters. It should be pointed out that the overall effect of even very great magnetic fields, thermal expansion, and g-Jitter amplitudes on the temperature profile is very insignificant.

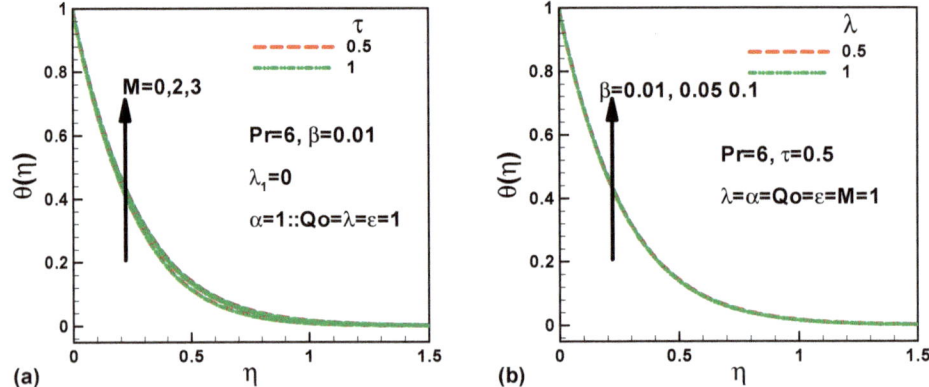

Figure 2. Variation of dimensionless temperature with (**a**) magnetic field and g-Jitter frequency period and with (**b**) thermal expansion coefficient and mixed convection parameter.

The effects of magnetic field and g-Jitter frequency on dimensionless skin friction are illustrated in Figure 3a,b. It is perceived that the dimensionless skin friction increases with both magnetic field and thermal expansion whereas it lessens with g-Jitter frequency. We realize that the velocity profile decreases with magnetic field; this effect will reduce the Reynold number. It can be interpreted from this fact that viscous force will be reduced compared to inertial force which, in turn, leads to augment dimensionless skin friction. Similarly, and for the same reason, it can be explained in Figure 1 that thermal expansion slightly reduces the velocity and consequently augments the dimensionless skin friction.

It is worthwhile to note that the skin friction coefficient Cf Rex-1/2 represents the velocity gradient at the surface; therefore, the velocity gradient at the surface will grow with thermal expansion and magnetic fields. It is very important and intriguing to note that the g-Jitter frequency period reduces skin friction. This effect can be physically elucidated by the fact that g-Jitter generates flow creating buoyancy forces due to the effect of vibration frequency distribution and density gradient which results to increase the acceleration of the fluid flow, this later owing to an augmented Reynolds number and consequently lessens skin friction.

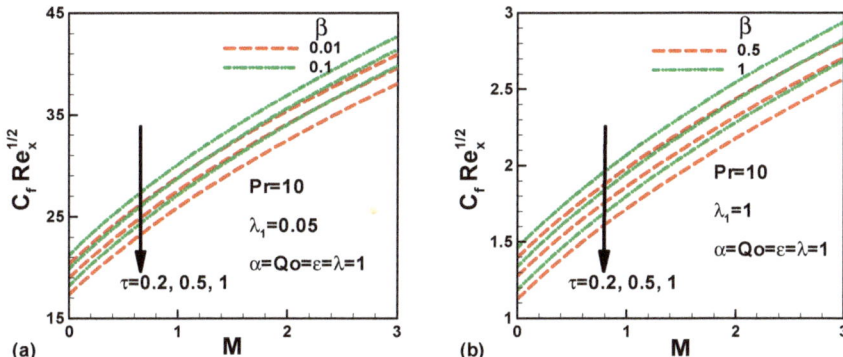

Figure 3. Variation of dimensionless skin friction with magnetic field and g-Jitter frequency period. (**a**) Assisting flow; (**b**) Opposing flow.

The effects of magnetic field, thermal expansion coefficient, and g-Jitter frequency period on the local Nusselt number are shown in Figure 4a,b. It is perceived that the local Nusselt number increases with both the thermal expansion coefficient and g-Jitter frequency period. Whereas, it decreases with magnetic field. It is clear that in the absence of a magnetic field, the Nusselt number is maximum and declines significantly until zero when the magnetic field augments, it can be interpreted on this fact that the increase of the magnetic field significantly enhances conduction heat transfer which, in turn, leads to lessening the local Nusselt number since it is well recognized that Nusselt numbers are the ratio of convection to conduction heat transfer. It is worthwhile to note that the local Nusselt number $Nu_x Re_x^{-1/2}$ represents the rate of heat transfer at the surface, consequently, heat transfer at the surface will decrease with magnetic field and increase with both the thermal expansion coefficient and g-Jitter frequency period. Applying a g-Jitter effect and increasing the thermal expansion coefficient, respectively, leads to acceleration of fluid flow and increases the thermal bouyancy effect, and both of them lead to increased convection heat transfer. Therefore, the local Nusselt number augments. Finally, it is realized that the Nusselt number at the inclined sheet surface with Jeffrey fluid and in a microgravity environment lessens with magnetic field and augments with the thermal expansion coefficient and g-Jitter frequency period. Thus, convection heat transfer will be enhanced by g-Jitter frequency and thermal expansion.

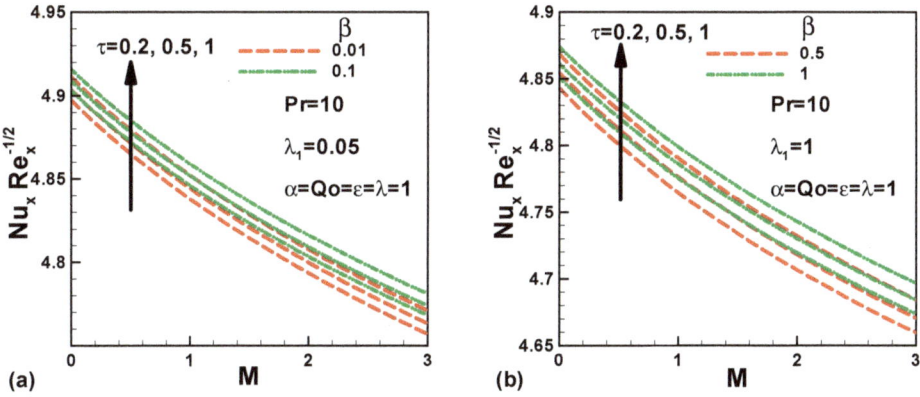

Figure 4. Variation of dimensionless heat transfer rate with several parameters in the presence of heat source for (**a**) assisting flow and (**b**) opposing flow.

5. Conclusions

Unsteady two-dimensional incompressible and laminar g-Jitter mixed convective boundary layer flow over an inclined stretching sheet in the presence of heat source/sink, viscous dissipation and magnetic field for Jeffrey fluid has been explored numerically by means of a shooting scheme based Runge–Kutta–Fehlberg-integration algorithm.

The results drawn from this study are mentioned as follows:

1. The velocity profile is higher for dimensionless boundary layer thickness $\eta = 0$ then decreases significantly along the inclined stretching sheet to reach zero. In the absence of a magnetic field, the velocity profile is maximum and declines with the application of a magnetic field
2. The temperature profile is higher for dimensionless boundary layer thickness $\eta = 0$ then decreases significantly along the inclined stretching sheet to reach zero. The thermal expansion coefficient and g-Jitter frequency have an insignificant effect on the temperature profile. Therefore, the thickness of the thermal boundary layer remains uniform in respect to the studied parameters.
3. The dimensionless skin friction increases with both magnetic field and thermal expansion whereas it lessens with g-Jitter frequency. The g-Jitter frequency period reduces skin friction; this effect can be physically elucidated by the fact that g-Jitter generates flow creating buoyancy forces due to the effect of the vibration frequency distribution and density gradients which results to increase the acceleration of the fluid flow.
4. The local Nusselt number increases with both the thermal expansion coefficient and g-Jitter frequency period. Whereas, it decreases with magnetic field. The Nusselt number at the inclined sheet surface with Jeffrey fluid and in microgravity environment lessens with magnetic field and augments with the thermal expansion coefficient and g-Jitter frequency period.

Funding: This research received no external funding.

Acknowledgments: Iskander Tlili thanks the Deanship of Scientific Research and the Deanship of Community Service at Majmaah University, Kingdom of Saudi Arabia, for supporting this work. Iskander Tlili thanks Dana Ibrahim Alrabiah, Renad Fahad Aba Hussein, Ritaj Al Anazi, Maryam Khaled Al Mizani and Wateen Satam Al-Mutairi.

Conflicts of Interest: The author declares no conflict of interest.

Nomenclature

Bi	Biot number
C	Nanoparticle volume fraction
C_p	Specific heat at constant pressure
C_f	Local skin-friction coefficient
Dn	Diffusivity of the microorganisms
D_B	Brownian diffusion coefficient
D_T	Thermophoretic diffusion coefficient of the microorganisms
f'	Dimensionless velocity
g	Gravitational acceleration
Gr_{x^*}	Local Grashof number
h_f	Convective heat transfer coefficient
k	Thermal conductivity
Le	Lewis number
Lb	Bioconvection Lewis number
Nb	Brownian motion number
Nr	Buoyancy ratio parameter
Nt	Thermophoresis number
Nu_{x^*}	Local Nusselt number
n	Density of motile microorganisms

Nn_x	Local density number
Pe	Bioconvection Péclet number
Pe_x	Local Péclet number
Pr	Prandtl number, ν/α
q_w	Wall heat flux
r	Local radius of the truncated cone
Ra_x	Modified Rayleigh number
Rb	Bioconvection Rayleigh number
Sh_x	Local Sherwood number
T	Temperature
u	Velocity component in the x-direction
U_r	Reference velocity
v	Velocity component in the y-direction
w_c	Maximum cell swimming speed
x	Streamwise coordinate
x_0	Distance of the leading edge of truncated cone measured from the origin
x^*	Distance measured from the leading edge of the truncated cone, $x-x_0$
y	Transverse coordinate
α	Thermal diffusivity
β	Coefficient of thermal expansion
γ	Average volume of a microorganism
σ	Motile parameter
η	Pseudo-similarity variable
θ	Dimensionless temperature
ϕ	Dimensionless nanoparticle volume fraction
ψ	Stream function
χ	Dimensionless density of motile microorganisms
ζ	Dimensionless distance
μ	Dynamic viscosity
ν	Kinematic viscosity
Ω	Half angle of the truncated cone
ρ_f	Density of the fluid
$\rho_{f\infty}$	Density of the base fluid
ρ_p	Density of the particles
$\rho_{m\infty}$	Density of the microorganism
$(\rho c)_f$	Heat capacity of the fluid
$(\rho c)_p$	Effective heat capacity of the nanoparticle material
ρ	Density
ψ	Stream function
w	Condition at the wall
∞	Condition at infinity

References

1. Rawi, N.A.; Kasim, A.R.M.; Isa, M.; Shafie, S. G-Jitter induced mixed convection flow of heat and mass transfer past an inclined stretching sheet. *J. Teknol.* **2014**, *71*. [CrossRef]
2. Shafie, S.; Amin, N.S.; Pop, I. G-Jitter free convection boundary layer flow of a micropolar fluid near a three-dimensional stagnation point of attachment. *Int. J. Fluid Mech. Res.* **2005**, *32*, 291–309. [CrossRef]
3. Sharidan, S.; Amin, N.; Pop, I. G-Jitter mixed convection adjacent to a vertical stretching sheet. *Microgravity-Sci. Technol.* **2006**, *18*, 5–14. [CrossRef]
4. Ramesh, K. Effects of viscous dissipation and Joule heating on the Couette and Poiseuille flows of a Jeffrey fluid with slip boundary conditions. *Propuls. Power Res.* **2018**, *7*, 329–341. [CrossRef]

5. Hayat, T.; Waqas, M.; Khan, M.I.; Alsaedi, A. Impacts of constructive and destructive chemical reactions in magnetohydrodynamic (MHD) flow of Jeffrey liquid due to nonlinear radially stretched surface. *J. Mol. Liq.* **2017**, *225*, 302–310. [CrossRef]
6. Selvi, R.K.; Muthuraj, R. MHD oscillatory flow of a Jeffrey fluid in a vertical porous channel with viscous dissipation. *Ain Shams Eng. J.* **2018**, *9*, 2503–2516. [CrossRef]
7. Dalir, N. Numerical study of entropy generation for forced convection flow and heat transfer of a Jeffrey fluid over a stretching sheet. *Alex. Eng. J.* **2014**, *53*, 769–778. [CrossRef]
8. Hayat, T.; Qayyum, S.; Imtiaz, M.; Alsaedi, A. Homogeneous-heterogeneous reactions in nonlinear radiative flow of Jeffrey fluid between two stretchable rotating disks. *Results Phys.* **2017**, *7*, 2557–2567. [CrossRef]
9. Khan, M.; Shahid, A.; Malik, M.Y.; Salahuddin, T. Thermal and concentration diffusion in Jeffery nanofluid flow over an inclined stretching sheet: A generalized Fourier's and Fick's perspective. *J. Mol. Liq.* **2018**, *251*, 7–14. [CrossRef]
10. Reddy, P.B.A. Magnetohydrodynamic flow of a Casson fluid over an exponentially inclined permeable stretching surface with thermal radiation and chemical reaction. *Ain Shams Eng. J.* **2016**, *7*, 593–602. [CrossRef]
11. Ramzan, M.; Bilal, M.; Chung, J.D. Effects of thermal and solutal stratification on jeffrey magneto-nanofluid along an inclined stretching cylinder with thermal radiation and heat generation/absorption. *Int. J. Mech. Sci.* **2017**, *131–132*, 317–324. [CrossRef]
12. Sandeep, N.; Sulochana, C. Momentum and heat transfer behaviour of Jeffrey, Maxwell and Oldroyd-B nanofluids past a stretching surface with non-uniform heat source/sink. *Ain Shams Eng. J.* **2018**, *9*, 517–524. [CrossRef]
13. Ahmad, K.; Ishak, A. Magnetohydrodynamic (MHD) Jeffrey fluid over a stretching vertical surface in a porous medium. *Propuls. Power Res.* **2017**, *6*, 269–276. [CrossRef]
14. Khan, M.I.; Waqas, M.; Hayat, T.; Alsaedi, A. Soret and Dufour effects in stretching flow of Jeffrey fluid subject to Newtonian heat and mass conditions. *Results Phys.* **2017**, *7*, 4183–4188. [CrossRef]
15. Ghosh, P.; Ghosh, M.K. Streaming flows in differentially heated square porous cavity under sinusoidal g-jitter. *Int. J. Therm. Sci.* **2009**, *48*, 514–520. [CrossRef]
16. Rouvreau, S.; Cordeiro, P.; Torero, J.L.; Joulain, P. Influence of g-jitter on a laminar boundary layer type diffusion flame. *Proc. Combust. Inst.* **2005**, *30*, 519–526. [CrossRef]
17. Khoshnevis, A.; Ahadi, A.; Saghir, M.Z. On the influence of g-jitter and prevailing residual accelerations onboard International Space Station on a thermodiffusion experiment. *Appl. Therm. Eng.* **2014**, *68*, 36–44. [CrossRef]
18. Kumar, A.; Gupta, V.K. Study of heat and mass transport in Couple-Stress liquid under G-jitter effect. *Ain Shams Eng. J.* **2018**, *9*, 973–984. [CrossRef]
19. Bhadauria, B.S.; Singh, A. Through flow and G-jitter effects on chaotic convection in an anisotropic porous medium. *Ain Shams Eng. J.* **2018**, *9*, 1999–2013. [CrossRef]
20. Shafie, S.; Amin, N.; Pop, I. G-Jitter free convection flow in the stagnation-point region of a three-dimensional body. *Mech. Res. Commun.* **2007**, *34*, 115–122. [CrossRef]
21. Asif, M.; Haq, S.U.; Islam, S.; Khan, I.; Tlili, I. Exact solution of non-Newtonian fluid motion between side walls. *Results Phys.* **2018**, *11*, 534–539. [CrossRef]
22. Agaie, B.G.; Khan, I.; Yacoob, Z.; Tlili, I. A novel technique of reduce order modelling without static correction for transient flow of non-isothermal hydrogen-natural gas mixture. *Results Phys.* **2018**, *10*, 532–540. [CrossRef]
23. Tlili, I.; Hamadneh, N.N.; Khan, W.A. Thermodynamic Analysis of MHD Heat and Mass Transfer of Nanofluids Past a Static Wedge with Navier Slip and Convective Boundary Conditions. *Arab. J. Sci. Eng.* **2018**, 1–13. [CrossRef]
24. Khalid, A.; Khan, I.; Khan, A.; Shafie, S.; Tlili, I. Case study of MHD blood flow in a porous medium with CNTS and thermal analysis. *Case Stud. Therm. Eng.* **2018**, *12*, 374–380. [CrossRef]
25. Khan, I.; Abro, K.A.; Mirbhar, M.N.; Tlili, I. Thermal analysis in Stokes' second problem of nanofluid: Applications in thermal engineering. *Case Stud. Therm. Eng.* **2018**, *12*, 271–275. [CrossRef]
26. Afridi, M.I.; Qasim, M.; Khan, I.; Tlili, I. Entropy generation in MHD mixed convection stagnation-point flow in the presence of joule and frictional heating. *Case Stud. Therm. Eng.* **2018**, *12*, 292–300. [CrossRef]

27. Khan, Z.A.; Haq, S.U.; Khan, T.S.; Khan, I.; Tlili, I. Unsteady MHD flow of a Brinkman type fluid between two side walls perpendicular to an infinite plate. *Results Phys.* **2018**, *9*, 1602–1608. [CrossRef]
28. Aman, S.; Khan, I.; Ismail, Z.; Salleh, M.Z.; Tlili, I. A new Caputo time fractional model for heat transfer enhancement of water based graphene nanofluid: An application to solar energy. *Results Phys.* **2018**, *9*, 1352–1362. [CrossRef]
29. Khan, Z.; Khan, I.; Ullah, M.; Tlili, I. Effect of thermal radiation and chemical reaction on non-Newtonian fluid through a vertically stretching porous plate with uniform suction. *Results Phys.* **2018**, *9*, 1086–1095. [CrossRef]
30. Imran, M.A.; Miraj, F.; Khan, I.; Tlili, I. MHD fractional Jeffrey's fluid flow in the presence of thermo diffusion, thermal radiation effects with first order chemical reaction and uniform heat flux. *Results Phys.* **2018**, *10*, 10–17. [CrossRef]
31. Rees, D.A.S.; Pop, I. The effect of g-jitter on vertical free convection boundary-layer flow in porous media. *Int. Commun. Heat Mass Transf.* **2000**, *27*, 415–424. [CrossRef]
32. Pandey, A.K.; Kumar, M. Natural convection and thermal radiation influence on nanofluid flow over a stretching cylinder in a porous medium with viscous dissipation. *Alex. Eng. J.* **2017**, *56*, 55–62. [CrossRef]
33. Ashorynejad, H.R.; Sheikholeslami, M.; Pop, I.; Ganji, D.D. Nanofluid flow and heat transfer due to a stretching cylinder in the presence of magnetic field. *Heat Mass Transf.* **2013**, *49*, 427–436. [CrossRef]
34. Buongiomo, J. Convective transport in nanofluids. *ASME Trans. J. Heat Transf.* **2006**, *128*, 240–250. [CrossRef]
35. Hayat, T.T.; Bibi, A.; Yasmin, H.H.; Alsaadi, F.E. Magnetic Field and Thermal Radiation Effects in Peristaltic Flow with Heat and Mass Convection. *ASME. J. Therm. Sci. Eng. Appl.* **2018**, *10*, 051018. [CrossRef]

© 2019 by the author. Licensee MDPI, Basel, Switzerland. This article is an open access article distributed under the terms and conditions of the Creative Commons Attribution (CC BY) license (http://creativecommons.org/licenses/by/4.0/).

Article

Influence of Cattaneo–Christov Heat Flux on MHD Jeffrey, Maxwell, and Oldroyd-B Nanofluids with Homogeneous-Heterogeneous Reaction

Anwar Saeed [1], Saeed Islam [1], Abdullah Dawar [2], Zahir Shah [1], Poom Kumam [3,4,5,*] and Waris Khan [6]

1. Department of Mathematics, Abdul Wali Khan University Mardan, Mardan 23200, Pakistan; anwarsaeed769@gmail.com (A.S.); saeedislam@awkum.edu.pk (S.I.); zahir1987@yahoo.com (Z.S.)
2. Department of Mathematics, Qurtuba University of Science and Information Technology, Peshawar 25000, Pakistan; abdullah.mathematician@gmail.com
3. KMUTTFixed Point Research Laboratory, Room SCL 802 Fixed Point Laboratory, Science Laboratory Building, Department of Mathematics, Faculty of Science, King Mongkut's University of Technology Thonburi (KMUTT), Bangkok 10140, Thailand
4. KMUTT-Fixed Point Theory and Applications Research Group, Theoretical and Computational Science Center (TaCS), Science Laboratory Building, Faculty of Science, King Mongkut's University of Technology Thonburi (KMUTT), Bangkok 10140, Thailand
5. Department of Medical Research, China Medical University Hospital, China Medical University, Taichung 40402, Taiwan
6. Department of Mathematics, Kohat University of Science and Technology, Kohat 26000, Pakistan; wariskhan758@yahoo.com
* Correspondence: poom.kum@kmutt.ac.th; Tel.: +66-2-4708-994

Received: 4 March 2019; Accepted: 19 March 2019; Published: 25 March 2019

Abstract: This research article deals with the determination of magnetohydrodynamic steady flow of three combile nanofluids (Jefferey, Maxwell, and Oldroyd-B) over a stretched surface. The surface is considered to be linear. The Cattaneo–Christov heat flux model was considered necessary to study the relaxation properties of the fluid flow. The influence of homogeneous-heterogeneous reactions (active for auto catalysts and reactants) has been taken in account. The modeled problem is solved analytically. The impressions of the magnetic field, Prandtl number, thermal relaxation time, Schmidt number, homogeneous–heterogeneous reactions strength are considered through graphs. The velocity field diminished with an increasing magnetic field. The temperature field diminished with an increasing Prandtl number and thermal relaxation time. The concentration field upsurged with the increasing Schmidt number which decreased with increasing homogeneous-heterogeneous reactions strength. Furthermore, the impact of these parameters on skin fraction, Nusselt number, and Sherwood number were also accessible through tables. A comparison between analytical and numerical methods has been presented both graphically and numerically.

Keywords: Magnetohydrodynamic (MHD); Jefferey, Maxwell and Oldroyd-B fluids; Cattaneo–Christov heat flux; homogeneous–heterogeneous reactions; analytical technique; Numerical technique

1. Introduction

A fluid composed of nanoparticles is called nanofluid. Nanoparticles of materials such as metallic oxides, carbide ceramics, nitride metals, ceramics, semiconductors, single, double or multi walled carbon nanotubes, alloyed, nanoparticles, etc. have been used for the preparation of nanofluids. Nanofluids have many characteristics in heat transfer, including microelectronics, local refrigerator, cooler, machining, and heat exchanger. The idea of nanofluid was introduced by Choi [1]. The Fourier's [2] recommended law of heat conduction normally works for heat transmission features from the time it was presented in the literature. By including the relaxation time parameter Cattaneo [3] it has improved this law and this term overwhelms the paradox of heat conduction. Christov [4] has named this theory Cattaneo-Christov heat flux theory, by further modifying the Cattaneo theory by exchanging the time derivative with Oldroyd-B upper convicted derivative. Mustafa [5] scrutinized the model [4] for heat transmission in a rotating Maxwell nanofluid flow. Chen [6] probed the influence of heat transfer and viscous dissipation of nanofluid flow over a stretching sheet. Sheikholeslami et al. [7–11] deliberated the three-dimensional magnetohydrodynamics (MHD) nanofluid flow in parallel rotating plates. Sheikholeslami [11–15] analytically and numerically deliberated the applications of nanofluids with different properties, behavior, and influences. Dawar et al. [16] examined the flow Williamson nanofluid over a stretching surface. Shah et al. [17] examined the micropolar nanofluid flow in rotating parallel plates with Hall current impact. Maleki et al. [18] scrutinized the non-Newtonian nanofluids flow and heat transfer over a porous surface. Nasiri et al. [19] deliberated the smoothed particle by a hydrodynamics approach for numerical simulation of nanofluid flows. Rashidi et al. [20] used the nanofluids in a circular tube heat exchanger and examined the entropy generation. Safaei et al. [21] studied numerically and experimentally the nanofluids convective heat transfer in closed conduits. Mahian et al. [22,23] presented the advances in the simulation and modeling of the flows of nanofluids.

Due to its relaxation properties, Jeffrey, Maxwell, and Oldroyd-B nanofluids have significant applications in the area of fluid mechanics. Ahmad et al. [24] scrutinized the flow of Jeffrey nanofluid with Magnetohydrodynamic impact. Ahmad and Ishak [25] deliberated the flow of Jeffrey nanofluid with MHD and transverse magnetic field impacts in a porous medium. Hayat et al. [26] probed the Oldroyd-B nanofluid flow with heat transfer and thermal radiation impacts. Raju et al. [27] deliberated the flow of Jeffrey nanofluid with a homogenous-heterogeneous reaction and non-linear thermal radiation impacts. The articles that are related to Jeffrey nanofluid can be found in [28–32]. Hayat et al. [33] inspected the MHD Maxwell nanofluid flow using suction/injection. Raju et al. [34] presented the heat and mass transmission in three-dimensional non-Newtonian nanofluid and Ferrofluid. Sandeep and Sulochana [35] investigated the mixed convection micropolar nanofluid flow over a stretching sheet. Raju et al. [36] deliberated the impacts of an inclined magnetic field, thermal radiation and cross diffusion on the two-dimensional flow. Nadeem et al. [37] presented the heat and mass transfer in Jeffrey nanofluid. Makinde et al. [38] deliberated the unsteady fluid flow with convective boundary conditions. Sheikholeslami [39] discussed the hydro-thermal behavior of nanofluids flow because of its external heated plates. Shah et al. [40] presented the Darcy-Forchheimer flow of radiative carbon nanotubes in a rotating frame. Chai et al. [41] presented a review of the heat transfer and hydrodynamic characteristics of nano/microencapsulated phase. Shah et al. [42] examined the electro-magneto micropoler Casson Ferrofluid over a stretching/shrinking sheet. Dawar et al. [43] analyzed the MHD CNTs Casson nanofluid in rotating channels. Khan et al. [44] deliberated the Williamson nanofluid flow over a linear stretching surface. Imtiaz et al. [45] examined the unsteady MHD flow due to a curved stretchable surface with homogeneous–heterogeneous reactions. Hayat et al. [46] deliberated the flow of nanofluids with homogeneous–heterogeneous reaction impacts over a non-linear stretched sheet with variable thickness. The recent study about nanofluid with application can be seen [47–50].

The present work is based on an analysis of MHD flow of three combine nanofluids (Maxwell, Oldroyd-B, and Jeffrey) over a linear stretching surface. The present model composed of Cattaneo–Christov heat flux. The impact of homogeneous-heterogeneous reactions were taken in this model. A boundary layer methodology was used in the mathematical expansion. The impact of dimensionless parameters on the fluid flow have been presented through graphs and tables.

2. Mathematical Modeling and Formulation

The incompressible electrically conducted three combined nanofluids (Jeffrey, Maxwell, and Oldroyd-B) were confined by a linear stretched surface. The fluid flow was taken in a two-dimensional steady state with stable surface temperature. x-axis was considered parallel to the surface, while y-axis was orthogonal to x-axis in the chosen coordinate system. The stretching velocity in x-axes direction was defined as $U_w(x) = \zeta x$. The conclusion of homogeneous-heterogeneous reactions on the fluid flows of two chemical species I and J were taken in account. In the y-axis direction a uniform magnetic field B_0 was acting. The heat transmission procedure was applied through Cattaneo–Christov heat flux theory.

In case of cubic autocatalysis, the Homogeneous reaction is [45,46]

$$I + 2J \rightarrow 3J, rate = k_c i j^2, \tag{1}$$

While on the catalyst surface, the heterogeneous reaction has been defined by

$$I \rightarrow J, rate = k_s i, \tag{2}$$

where k_c, k_s, I, J, i, j are the rate constants, chemical species, and concentrations of chemical species, respectively.

In the absence of viscous dissipation and thermal radiation, the boundary layer equations leading to the flow of viscoelastic fluids can be written as follows [47]:

$$u_x + v_y = 0, \tag{3}$$

$$uu_x + vu_y = -\lambda_1\left(u^2 u_{xx} + 2uv u_{xy} + v^2 u_{yy}\right) + \frac{v_f}{1+\lambda_2}\{u_{yy} + \lambda_3(uu_{xyy} + u_y u_{xy} + v u_{yyy} - u_x u_{yy})\} - \frac{\sigma B_0^2}{\rho_f}u, \tag{4}$$

$$\rho c_p(uT_x + vT_y) = -\nabla \cdot q, \tag{5}$$

$$ui_x + vi_y = D_I i_{yy} - k_c i j^2, \tag{6}$$

$$uj_x + vj_y = D_J j_{yy} + k_c i j^2. \tag{7}$$

Here u, v, μ, ρ_f, v_f are velocity components in their respective directions, dynamic viscosity, density, and kinematic viscosity respectively. $\lambda_1, \lambda_2, \lambda_3$ are the relaxation time, a proportion of the relaxation to retardation times, respectively. T, σ_f, B$_0$ indicated the temperature, electrical conductivity and the transverse magnetic field.

The problem is studied based on the following conditions:

i   Oldroyd-B nanofluid when $\lambda_1 \neq 0, \lambda_2 = 0$ and $\lambda_3 \neq 0$.
ii Maxwell nanofluid when $\lambda_1 \neq 0, \lambda_2 = 0$ and $\lambda_3 = 0$.
iii Jeffrey nanofluid when $\lambda_1 = 0, \lambda_2 \neq 0$ and $\lambda_3 \neq 0$.

The heat flux theory which was presented by Cattaneo–Christov:

$$q + \lambda_1\left\{\frac{\partial q}{\partial t} + V \cdot \nabla q - q \cdot \nabla V + (\nabla \cdot V)q + q_t\right\} = -k\nabla T, \tag{8}$$

where k, q represented thermal conductivity and heat flux. Classical Fourier's law was assumed by setting $\lambda_1 = 0$ in Equation (8). By assuming the condition $(\nabla \cdot V = 0)$ and steady flow with $(q_t = 0)$, Equation (8) became:

$$q + \lambda_1(V \cdot \nabla q - q \cdot \nabla V) = -k\nabla T. \tag{9}$$

The heat transfer equation proceeded as:

$$uT_x + vT_y + \lambda_1 \Phi_E = \alpha(T_{yy}), \quad (10)$$

where Φ_E is given as:

$$\Phi_E = uu_x T_x + vv_y T_y + uv_x T_y + vu_y T_x + 2uv T_{xy} + u^2 T_{xx} + v^2 T_{yy}. \quad (11)$$

The accompanying boundary conditions were:

$$\begin{aligned} u = U_w(x) = \zeta x, v = 0, T = T_w, D_I i_y = k_s i, D_J j_y = -k_s i \text{ at } y = 0, \\ u \to 0, \; T \to T_\infty, i \to i_0, j \to 0 \text{ at } y \to \infty, \end{aligned} \quad (12)$$

where $\alpha = \frac{k}{\rho c_p}$ indicated the thermal diffusivity, D_I and D_J indicated the diffusion coefficients, T_w denoted the temperature at the surface, T_∞ for the surrounding fluid temperature and ζ for non-negative stretching rate constant with T^{-1} as the dimension.

$$\begin{aligned} u = \zeta x F'(\eta), v = -(\zeta v)^{\frac{1}{2}} F(\eta), \eta = \left(\frac{\zeta}{v}\right)^{\frac{1}{2}} y, \\ G(\eta) = \frac{T-T_\infty}{T_w - T_\infty}, i = i_0 \phi(\eta), j = i_0 h(\eta). \end{aligned} \quad (13)$$

Apparently the equation of continuity is satisfied and Equations (4)–(13) become:

$$F''' + \kappa_2 \left(F''^2 - FF'''' \right) - M(1+\lambda_2) F' - (1+\lambda_2)\left\{ F'^2 - FF'' + \kappa_1 \left(F^2 F''' - 2FF'F''\right)\right\} = 0, \quad (14)$$

$$G'' + \Pr\left\{ FG' - \Omega\left(FF'G' + F^2 G''\right)\right\} = 0, \quad (15)$$

$$\phi'' + Sc\left(F\phi' - K\phi h^2\right) = 0, \quad (16)$$

$$\delta h'' + Sc\left(Fh' + K\phi h^2\right) = 0, \quad (17)$$

with boundary conditions:

$$\begin{aligned} F = 0, F' = 1, G = 1, \phi' = K_s \phi, \delta h' = -K_s \phi \text{ at } \omega = 0, \\ F' \to 0, G \to 0, \phi \to 1, h \to 0 \text{ at } \eta \to \infty, \end{aligned} \quad (18)$$

In the above equations, $M = \frac{\sigma_{nf} B_0^2}{\rho_f \zeta}$ indicated the magnetic field, $\kappa_1 = \lambda_1 \zeta$ and $\kappa_2 = \lambda_3 \zeta$ were the Debora numbers with respect to relaxation and retardation time, $\Pr = \frac{v_f}{\alpha}$ represented the Prandtl number, $\Omega = \zeta \lambda_1$ indicated the thermal relaxation time, $Sc = \frac{v_f}{D_I}$ is the Schmidt number, $K = \frac{k_c i_0^2}{U_w}$ indicating the homogeneous reaction strength, $K_s = \frac{k_s}{D_I i_0}\sqrt{\frac{\zeta}{v_f}}$ represented the heterogeneous reaction strength, and $\delta = \frac{D_J}{D_I}$ indicated the diffusion coefficient, When $D_I = D_J$ then $\delta = 1$ and as a result:

$$\phi(\eta) + h(\eta) = 1. \quad (19)$$

Now Equations (16) and (17) yield:

$$\phi'' + Sc\left\{F\phi' - K\phi(1-\phi)^2\right\} = 0. \quad (20)$$

The subjected boundary conditions are:

$$\phi'(0) = K_s \phi(0), \phi(\infty) \to 1. \quad (21)$$

Skin friction coefficient through the dimensionless scale is:

$$\mathrm{Re}_x^{\frac{1}{2}} Cf_x = \left(\frac{1+\kappa_1}{1+\kappa_2}\right) F''(0). \tag{22}$$

where Re_x is called the local Reynolds number.

The dimensionless form of Nu_x and Sh_x were found as:

$$\mathrm{Nu}_x = -G'(0), \quad \mathrm{Sh}_x = -\phi'(0). \tag{23}$$

3. Solution by Homtopy Analysis Method (HAM)

To evaluate the Equations (14), (15) and (20) with boundary conditions (18) and (21) using HAM with the following procedure.

The initial assumptions were picked as below:

$$\overline{F}_0(\eta) = 1 - e^{-\eta}, \ \overline{G}_0(\eta) = e^{-\eta}, \overline{\phi}_0(\eta) = e^{-\eta}. \tag{24}$$

The linear operators were taken as $L_{\overline{F}}, L_{\overline{G}}$ and $L_{\overline{\phi}}$:

$$L_{\overline{F}}(\overline{F}) = \overline{F}''' - \overline{F}', L_{\overline{G}}(\overline{G}) = \overline{G}'' - \overline{G}, L_{\overline{\phi}}(\overline{\phi}) = \overline{\phi}'' - \overline{\phi}, \tag{25}$$

With the following properties:

$$L_{\overline{F}}(r_1 + r_2 e^{-\eta} + r_3 e^{\eta}) = 0, \ L_{\overline{G}}(r_4 e^{-\eta} + r_5 e^{\eta}) = 0, L_{\overline{\phi}}(r_6 e^{-\eta} + r_7 e^{\eta}) = 0, \tag{26}$$

where $r_i (i = 1 - 7)$ were the constants:

The resulting non-linear operators $N_{\overline{F}}, N_{\overline{G}}$ and $N_{\overline{\phi}}$ were specified as:

$$N_{\overline{F}}[\overline{F}(\eta;\tau)] = \frac{\partial^3 \overline{F}(\eta;\tau)}{\partial \eta^3} + \kappa_2 \left\{ \left(\frac{\partial^2 \overline{F}(\eta;\tau)}{\partial \eta^2}\right)^2 - \frac{\partial \overline{F}(\eta;\tau)}{\partial \eta} \frac{\partial^4 \overline{F}(\eta;\tau)}{\partial \eta^4} \right\}$$
$$-M(1+\lambda_2) \left\{ \begin{array}{c} \left(\frac{\partial \overline{F}(\eta;\tau)}{\partial \eta}\right)^2 - \overline{F}(\eta;\tau) \frac{\partial^2 \overline{F}(\eta;\tau)}{\partial \eta^2} + \\ \kappa_1 \left(\overline{F}^2(\eta;\tau) \frac{\partial^3 \overline{F}(\eta;\tau)}{\partial \eta^3} - 2\overline{F}(\eta;\tau) \frac{\partial \overline{F}(\eta;\tau)}{\partial \eta} \frac{\partial^2 \overline{F}(\eta;\tau)}{\partial \eta^2}\right) \end{array} \right\}, \tag{27}$$

$$N_G[\overline{F}(\eta;\tau), \overline{G}(\eta;\tau)] = \frac{\partial^2 \overline{G}(\eta;\tau)}{\partial \eta^2} +$$
$$\mathrm{Pr} \left\{ \overline{F}(\eta;\tau) \frac{\partial \overline{G}(\eta;\tau)}{\partial \eta} - \Omega \left(\overline{F}(\eta;\tau) \frac{\partial \overline{F}(\eta;\tau)}{\partial \eta} \frac{\partial \overline{G}(\eta;\tau)}{\partial \eta} + \overline{F}^2(\eta;\tau) \frac{\partial^2 \overline{G}(\eta;\tau)}{\partial \eta^2} \right) \right\}, \tag{28}$$

$$N_{\overline{\phi}}[\overline{F}(\eta;\tau), \overline{\phi}(\eta;\tau)] = \frac{\partial^2 \overline{\phi}(\eta;\tau)}{\partial \eta^2} + Sc \left\{ \overline{F}(\eta;\tau) \frac{\partial \overline{\phi}(\eta;\tau)}{\partial \eta} - K\left(\overline{\phi}^3(\eta;\tau) - 2\overline{\phi}^2(\eta;\tau) + \overline{\phi}(\eta;\tau)\right) \right\}, \tag{29}$$

The zeroth-order problem for Equations (14), (15) and (20) were:

$$(1-\tau) L_{\overline{F}}[\overline{F}(\eta;\tau) - \overline{F}_0(\eta)] = \tau \hbar_{\overline{F}} N_{\overline{F}}[\overline{F}(\eta;\tau)], \tag{30}$$

$$(1-\tau) L_{\overline{G}}[\overline{G}(\eta;\tau) - \overline{G}_0(\eta)] = \tau \hbar_{\overline{G}} N_{\overline{G}}[\overline{F}(\eta;\tau), \overline{G}(\eta;\tau)], \tag{31}$$

$$(1-\tau) L_{\overline{\phi}}[\overline{\phi}(\eta;\tau) - \overline{\phi}_0(\eta)] = \tau \hbar_{\overline{\phi}} N_{\overline{\phi}}[\overline{F}(\eta;\tau), \overline{\phi}(\eta;\tau)]. \tag{32}$$

The related boundary conditions where:

$$\overline{F}(\eta;\tau)|_{\eta=0} = 0, \ \frac{\partial \overline{F}(\eta;\tau)}{\partial \eta}\bigg|_{\eta=0} = 1, \ \frac{\partial \overline{F}(\eta;\tau)}{\partial \eta}\bigg|_{\eta \to \infty} = 0,$$
$$\overline{G}(\eta;\tau)|_{\eta=0} = 1, \ \overline{G}(\eta;\tau)|_{\eta \to \infty} = 0, \tag{33}$$
$$\frac{\partial \overline{\phi}(\eta;\tau)}{\partial \eta}\bigg|_{\eta=0} = K_s \overline{\phi}(\eta;\tau)|_{\eta=0}, \ \overline{\phi}(\eta;\tau)|_{\eta \to \infty} = 1,$$

where $\tau \in [0,1]$ is the embedding parameter, $\hbar_{\overline{F}}, \hbar_{\overline{G}}, \hbar_{\overline{\phi}}$ that were used to control the solution convergence. When $\tau = 0$ and $\tau = 1$ we have:

$$\overline{F}(\eta;1) = \overline{F}(\eta), \; G(\eta;1) = \overline{G}(\eta) \text{ and } \overline{\phi}(\eta;1) = \overline{\phi}(\eta), \tag{34}$$

Expanding $\overline{F}(\eta;\tau)$, $G(\eta;\tau)$ and $\overline{\phi}(\eta;\tau)$ by Taylor's series:

$$\begin{aligned}\overline{F}(\eta;\tau) &= \overline{F}_0(\eta) + \sum_{q=1}^{\infty} \overline{F}_q(\eta)\tau^q, \\ \overline{G}(\eta;\tau) &= \overline{G}_0(\eta) + \sum_{q=1}^{\infty} \overline{G}_q(\eta)\tau^q, \\ \overline{\phi}(\eta;\tau) &= \overline{\phi}_0(\eta) + \sum_{q=1}^{\infty} \overline{\phi}_q(\eta)\tau^q.\end{aligned} \tag{35}$$

where:

$$\overline{F}_q(\eta) = \frac{1}{q!}\frac{\partial \overline{F}(\eta;\tau)}{\partial \eta}\bigg|_{\tau=0}, \; \overline{G}_q(\eta) = \frac{1}{q!}\frac{\partial \overline{G}(\eta;\tau)}{\partial \eta}\bigg|_{\tau=0} \text{ and } \overline{\phi}_q(\eta) = \frac{1}{q!}\frac{\partial \overline{\phi}(\eta;\tau)}{\partial \eta}\bigg|_{\tau=0}. \tag{36}$$

The $\hbar_{\overline{F}}, \hbar_{\overline{G}}$ and $\hbar_{\overline{\phi}}$ are taken in such a way that the series (35) converges at $\tau = 1$, we have:

$$\begin{aligned}\overline{F}(\eta) &= \overline{F}_0(\eta) + \sum_{q=1}^{\infty} \overline{F}_q(\eta), \\ \overline{G}(\eta) &= \overline{G}_0(\eta) + \sum_{q=1}^{\infty} \overline{G}_q(\eta), \\ \overline{\phi}(\eta) &= \overline{\phi}_0(\eta) + \sum_{q=1}^{\infty} \overline{\phi}_q(\eta).\end{aligned} \tag{37}$$

The following are satisfied by the qth-order problem.

$$\begin{aligned}L_{\overline{F}}[\overline{F}_q(\eta) - \chi_q \overline{F}_{q-1}(\eta)] &= \hbar_{\overline{F}} U_q^{\overline{F}}(\eta), \\ L_{\overline{G}}[\overline{G}_q(\eta) - \chi_q \overline{G}_{q-1}(\eta)] &= \hbar_{\overline{G}} U_q^{\overline{G}}(\eta), \\ L_{\overline{\phi}}[\overline{\phi}_q(\eta) - \chi_q \overline{\phi}_{q-1}(\eta)] &= \hbar_{\overline{\phi}} U_q^{\overline{\phi}}(\eta).\end{aligned} \tag{38}$$

Which have the following boundary conditions:

$$\begin{aligned}\overline{F}_q(0) = \overline{F}'_q(0) = \overline{F}'_q(\infty) &= 0, \\ \overline{G}_q(0) = \overline{G}_q(\infty) &= 0, \\ \overline{\phi}'_q(0) - K_s \overline{\phi}_q(0) = \overline{\phi}_q(\infty) &= 0.\end{aligned} \tag{39}$$

Here

$$U_q^{\overline{F}}(\eta) = \overline{F}'''_{q-1} + \kappa_2 \left(\sum_{k=0}^{q-1} \overline{F}_{q-1-k}\overline{F}''_k - \sum_{k=0}^{q-1} \overline{F}_{q-1-k}\overline{F}^{iv}_k \right) - M(1+\lambda_2)\overline{F}'_{q-1} -$$
$$(1+\lambda_2)\left\{ \sum_{k=0}^{q-1} \overline{F}_{q-1-k}\overline{F}' - \sum_{k=0}^{q-1} \overline{F}_{q-1-k}\overline{F}''_k + \kappa_1 \left(\sum_{k=0}^{q-1} \overline{F}_{q-1-k}\sum_{j=0}^{k} \overline{F}_{k-j}\overline{F}'''_j - 2\sum_{k=0}^{q-1} \overline{F}_{q-1-k}\sum_{j=0}^{k} \overline{F}'_{k-j}\overline{F}''_j \right) \right\}, \tag{40}$$

$$U_q^{\overline{G}}(\eta) = \overline{G}''_{q-1} + \Pr\left\{ \sum_{k=0}^{q-1} \overline{F}_{q-1-k}\overline{G}'_k - \Omega\left(\sum_{k=0}^{q-1} \overline{F}_{q-1-k}\sum_{j=0}^{k} \overline{F}'_{k-j}\overline{G}'_j + \sum_{k=0}^{q-1} \overline{F}_{q-1-k}\sum_{j=0}^{k} \overline{F}_{k-j}\overline{G}''_j \right) \right\}, \tag{41}$$

$$U_q^{\overline{\phi}}(\eta) = \overline{\phi}''_{q-1} - KSc\left(\sum_{k=0}^{q-1} \overline{\phi}_{q-1-k}\sum_{j=0}^{k} \overline{\phi}_{k-j}\overline{\phi}_j - 2\sum_{k=0}^{q-1} \overline{\phi}_{q-1-k}\overline{\phi}_k + \overline{\phi}_{q-1} \right), \tag{42}$$

where:

$$\chi_q = \begin{cases} 0, & \text{if } \tau \leq 1 \\ 1, & \text{if } \tau > 1 \end{cases} \quad (43)$$

4. HAM Solution Convergences

In this segment we graphically discussed the superior effect of the concerned parameters. The convergence of Equation (36) was subjected entirely through the auxiliary constraints $\hbar_F, \hbar_G, \hbar_\phi$. This is a collection in a way that it controls and converges the series solutions. The optional division of $\hbar$, was plotted through $\hbar$-curves $F''(0), G'(0), \phi'(0)$ for the 2nd ordered approximated solution of HAM. The operational region of $\hbar$ is $-2.2 < \hbar_F < 0.2, -2.1 < \hbar_G < -0.1, -2.4 < \hbar_\phi < 0.1$. The convergence of HAM through the $\hbar$-curve on velocity profile $F''(0)$, temperature profile $G'(0)$ and concentration profile $\phi'(0)$ is presented in Figure 1.

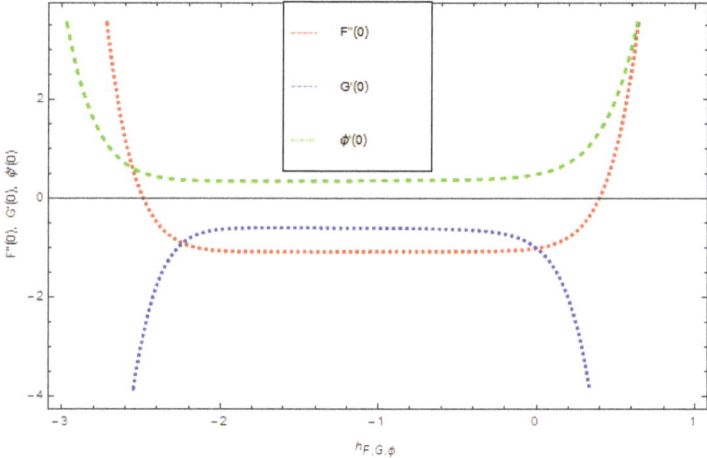

Figure 1. The combined $\hbar$-curves for $F''(0), G'(0)$ and $\phi'(0)$.

5. Results and Discussion

In this segment the impact of emerging parameters on velocity function $F'(\eta)$, temperature function $G(\eta)$ and concentration function $\phi(\eta)$ within the defined domain have been discussed. The impact of M on $F'(\eta)$ is deliberated in Figure 2. The Lorentz force theory deliberated that the magnetic field grows at a reversed force to the fluids flow. This force reduced the momentum boundary layer while it improved the thickness of the boundary layer. Therefore, with the escalating magnetic field M the velocity profile $F'(\eta)$ declined. From here we concluded that Jeffrey nanofluid was greatly subjected by the magnetic field compared to the other two. In Figures 3 and 4 the impact of Pr and Ω on $G(\eta)$ were presented respectively. In Figure 3 we perceived that $G(\eta)$ diminished with the rise in Pr. Physically the thickness of the boundary layer increased with the reduction in thermal diffusion. In addition, it can also be seen from the figure that Pr is more effective on Jeffrey and Maxwell nanofluids compared to the Oldroyd-B nanofluid. In Figure 4 the effect of thermal relaxation parameter Ω on $G(\eta)$ has been described. From here we saw that $G(\eta)$ reduced with the escalation in Ω. This was attributable to the fact that as we escalate Ω, the material particles need more time for heat transmission to its nearest particles. In addition, it can be stated that this material shows a non-conducting behavior with higher values of Ω which results in a reduction in $G(\eta)$. The impact of Sc, K and K_s on $\phi(\eta)$ are schemed in Figures 5–7 respectively. In Figure 5 the effect of Sc on $\phi(\eta)$ has been described. Schmidt number is the ratio of momentum diffusivity to mass diffusivity. Physically, the Schmidt number is related to hydrodynamic layer's thickness and boundary layer. The escalating

Sc intensifies the momentum of the boundary layer flow which results in an increase in concentration profile. It is clear from the figure that $\phi(\eta)$ upsurges with the rise in Sc. In Figure 6 the impact of K on $\phi(\eta)$ has been described. From here we concluded that larger K results in a reduction in $\phi(\eta)$. This may have been caused by the fact that the reaction rates controlled the diffusion coefficients. To a certain extent similar results are displayed in Figure 7. In Figure 7 the impact of K_s on $\phi(\eta)$ has been described. From this figure we have concluded that the growing values of K_s showed a drop in behavior in $\phi(\eta)$. This results from an agreement with the general physical behavior of the homogeneous reaction K and the heterogeneous reaction K_s. In Figures 8 and 9 the impact of M on Cf_x and Nu_x for Jeffrey, Maxwell and Oldroyd-B nanofluids have been described. It is clear from the figures that the growing values of M were decreasing for both Cf_x and Nu_x. The magnetic field was applied perpendicular to the flow of fluids and had an inverse variation with the skin friction of the fluid flow. This is the reason why the increasing magnetic field reduced the skin friction of the fluids flow. Similarly, the behavior of the heat transfer rate was due to the growing magnetic force on the fluids flow phenomena, with the fluid particles requiring more time to transfer the heat to the nearest particle. This was because the heat transfer rate reduced with the escalating magnetic field. The impact of Pr and Ω on Nu_x for the nanofluids flow has been described in Figures 10 and 11. From here we have concluded that the escalation in Pr increased the heat transfer rate while the increased Ω reduced the heat transfer rate for the nanofluids flow. Figure 12 shows the Total Residual error for the three types of nanofluid flow.

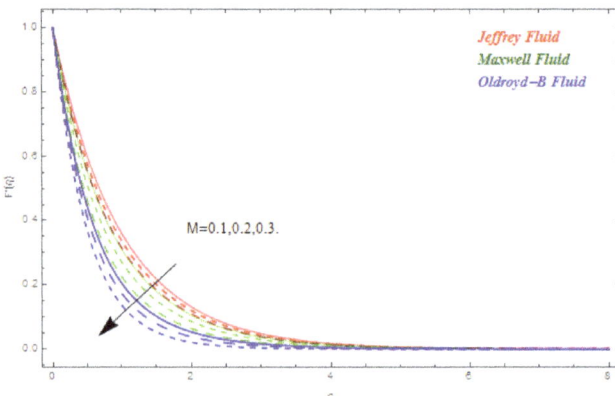

Figure 2. Impact of M on $F'(\eta)$, when $\Omega = 0.5, Sc = 0.6, K = 0.8, Pr = 0.7, K_s = 0.9$.

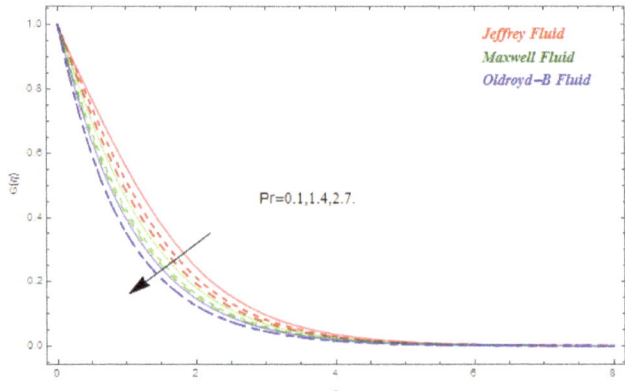

Figure 3. Impact of Pr on $G(\eta)$, when $\Omega = 0.5, Sc = 0.6, K = 0.8, M = 0.1, K_s = 0.9$.

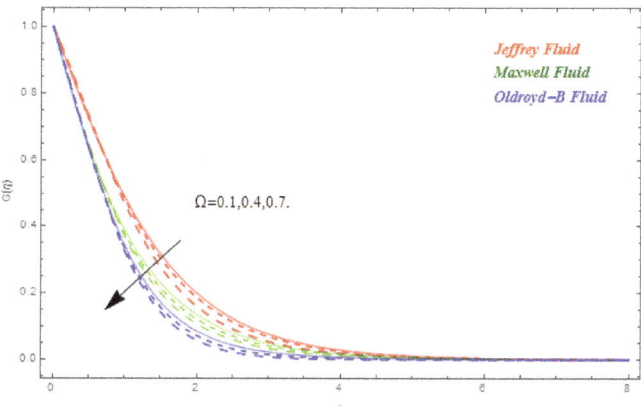

Figure 4. Impact of Ω on $G(\eta)$, when $Sc = 0.6, Pr = 0.7, K = 0.8, M = 0.1, K_s = 0.9$.

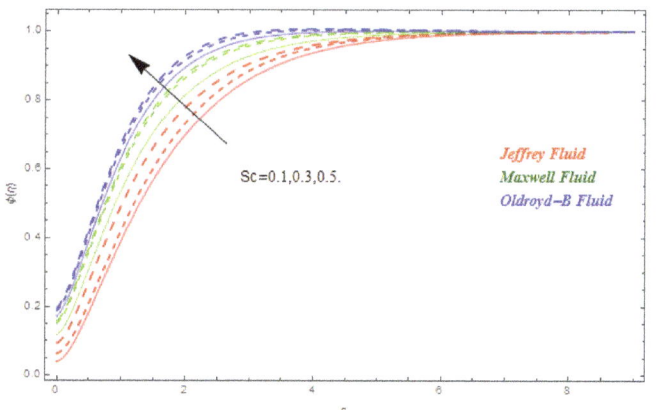

Figure 5. Impact of Sc on $\phi(\eta)$, when $\Omega = 0.1, Pr = 0.7, K = 0.8, M = 0.1, K_s = 0.9$.

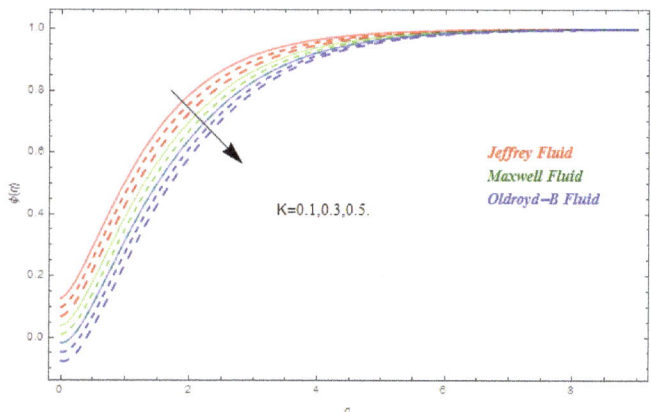

Figure 6. Impact of K on $\phi(\eta)$, when $\Omega = 0.1, Pr = 0.7, Sc = 0.6, M = 0.1, K_s = 0.9$.

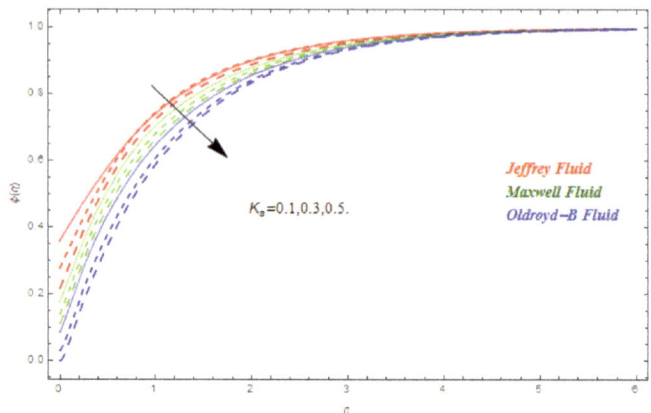

Figure 7. Impact of K_s on $\phi(\eta)$, when $\Pr = 0.7, \Omega = 0.1, Sc = 0.6, M = 0.1, K = 0.8$.

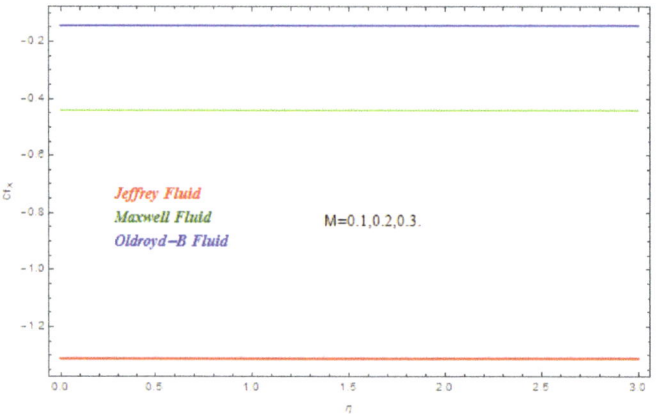

Figure 8. Impact of M on Cf_x.

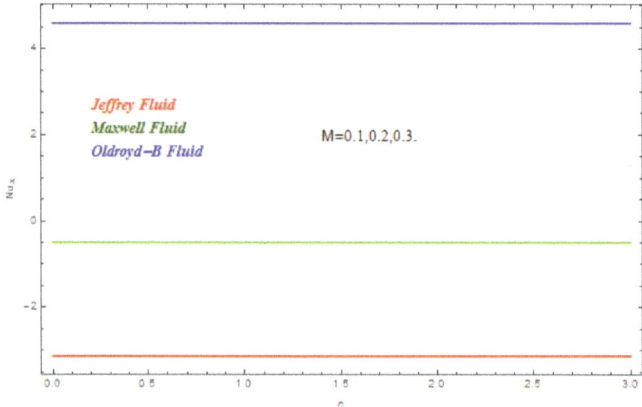

Figure 9. Impact of M on Nu_x.

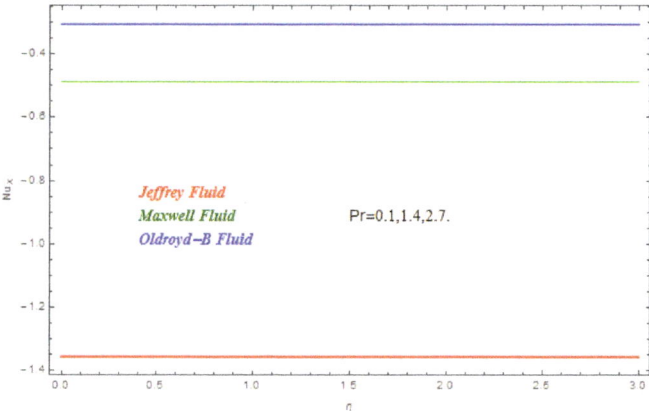

Figure 10. Impact of Pr on Nu_x.

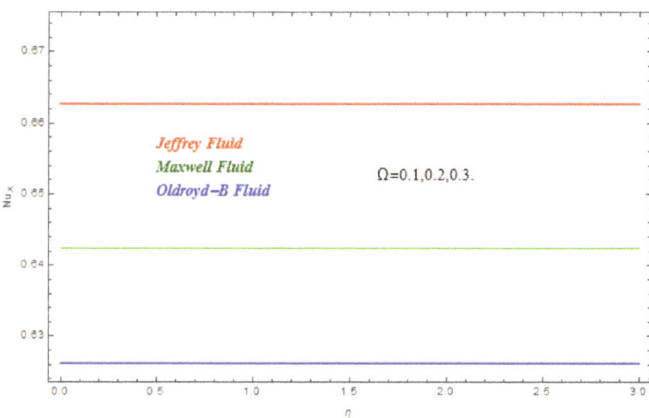

Figure 11. Impact of Ω on Nu_x.

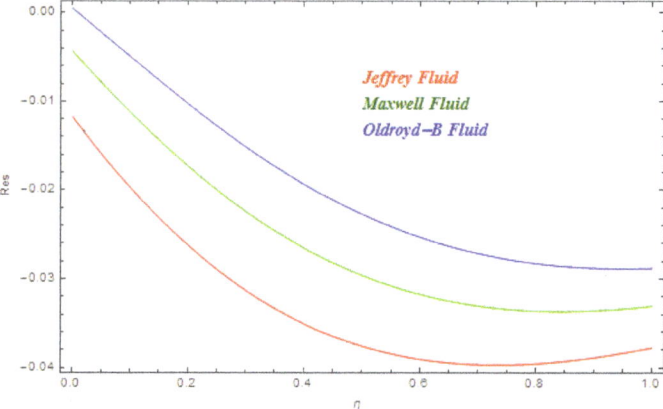

Figure 12. Total residual error for Jeffery, Maxwell and Oldroyd-B nanofluids.

Tables Discussion

In this section we have demonstrated the effect of emerging dimensionless parameters on the presented model of nanofluids. Table 1 displayed the conclusions associated with emerging parameters of skin fraction coefficients. This shows the impression of magenatic field parameter M on skin fraction coefficients. The magenatic field parameter shows a reduction in the skin fraction coefficient. Table 2 demonstrated the conclusion of incipient parameters on local Nusselt numbers. The heat transfer rate decreases with the rise in thermal relaxation parameter Ω while escalates with the increase in Prandtl number Pr. Table 2 shows that the thermal relaxation parameter has more effect on Jefferey nanofluids in comparison to Maxwell and Oldroyd-B nanofluids. Table 3 demonstrated the conclusion of an emerging parameter on the Sherwood number. The Sherwood number reduces with its rise, which upsurges with the escalation of the strength of homogeneous reaction K and the strength of heterogeneous reaction K_s.

Table 1. Distinction in $-Cf_x$ for different M.

M	Ref. [47]	Present Results for Jeffrey Nanofluid	Ref. [47]	Present Results for Maxwell Nanofluid	Ref. [47]	Present Results for Oldroyd-B Nanofluid
1.0	1.210458	0.210462	1.504151	1.504153	1.071019	1.071022
2.0	1.431584	1.431587	1.804788	1.804791	1.248081	1.248084

Table 2. Distinction in Nu_x for different Ω and Pr.

Ω	Pr	Ref. [47]	Jeffrey Nanofluid	Ref. [47]	Maxwell Nanofluid	Ref. [47]	Oldroyd-B Nanofluid
1.0		———	0.610394	———	0.595298	———	0.610846
1.2		———	0.607503	———	0.593311	———	0.607993
	6.0	0.418081	0.513786	0.421167	0.511247	0.426476	0.5154367
	7.0	0.439695	0.626865	0.441919	0.548966	0.447670	0.5477974

Table 3. Distinction in Sh_x for different Sc, K and K_s.

Sc	K	K_s	Jeffrey	Maxwell	Oldroyd-B
1.2			−0.096477	−0.095593	−0.096771
1.5			−0.096782	−0.095890	−0.097081
	1.5		−0.058030	−0.047238	−0.049135
	1.7		−0.056699	−0.037230	−0.039262
		0.5	−0.018933	−0.037233	0.012399
		0.8	−0.160028	0.046603	−0.125205

6. Comparison of Analytical Solutions and Numerical Solutions

An analytical solution means an exact solution. To study the behavior of systems, an analytical solution can be used with varying properties. Regrettably there are many practical systems that lead to an analytical solution, and analytical solutions are often of limited use. This is why we use a numerical approach to generate answers that are closer to practical results. These solutions which cannot be used as complete mathematical expressions are numerical solutions. In the natural worldthere are almost no problems that are exactly solvable, which makes the problem more difficult than all the exactly solvable problems. There are three or four of them in nature that have already been solved, unfortunately even numerical methods cannot always give an exact solution. Numerical techniques can handle any completed physical geometries which are often impossible to solve analytically. In this article both analytical and numerical approaches are tested to solve the modeled problem. A comparison of HAM and ND-Solve technique for $F'(\eta)$, $G(\eta)$ and $\phi(\eta)$ are deliberated in Figures 13–15 and Tables 4–6, respectively.

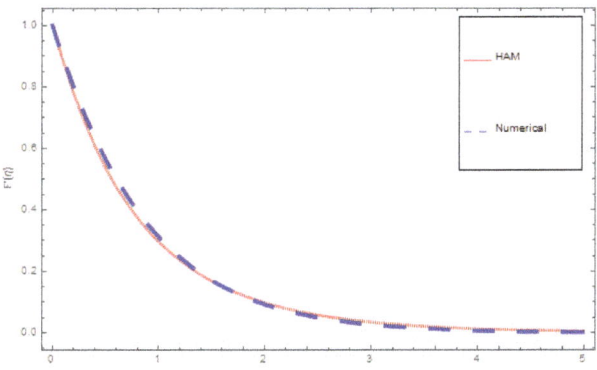

Figure 13. HAM versus numerical comparison for $F'(\eta)$.

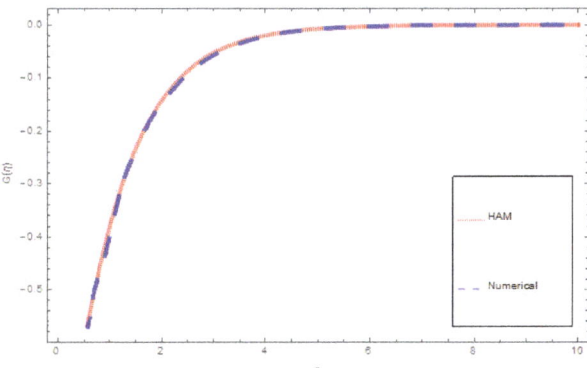

Figure 14. HAM versus numerical comparison for $G(\eta)$.

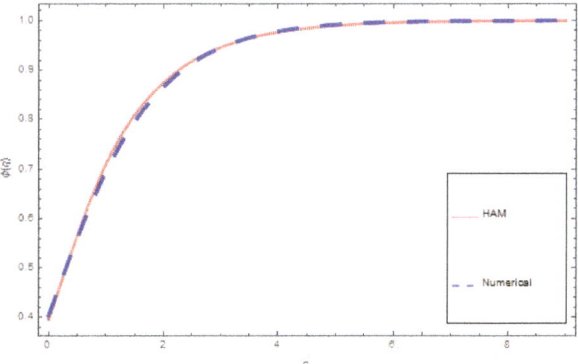

Figure 15. HAM versus numerical comparison for $\phi(\eta)$.

Table 4. Symmetry of HAM versus numerical solutions for $F'(\eta)$, when $Sc = Pr = K_2 = 1.0$, $\kappa = \kappa_1 = \kappa_2 = \lambda_2 = M = K = 0.1$.

η	HAM $F'(\eta)$	Numerical $F'(\eta)$	Absolute Error AE
0.0	3.33067×10^{-16}	0.000000	3.33067×10^{-16}
0.1	0.0950433	0.093334	0.002876
0.2	0.180827	0.173972	0.009647
0.3	0.258260	0.242791	0.016488
0.4	0.328165	0.300578	0.020276
0.5	0.391281	0.348033	0.019089
0.6	0.448277	0.385777	0.011828
0.7	0.499753	0.414357	0.085396
0.8	0.546251	0.434262	0.111989
0.9	0.588258	0.445923	0.142335
1.0	0.626213	0.449730	0.176483

Table 5. Symmetry of HAM versus numerical solutions for $G(\eta)$, when $Sc = Pr = K_s = 1.0$, $\kappa = \kappa_1 = \kappa_2 = \lambda_3 = M = K = 0.1$.

η	HAM $G(\eta)$	Numerical $G(\eta)$	Absolute Error AE
0.0	1.000000	1.00000	0.000000
0.1	0.915165	0.887477	0.002876
0.2	0.835725	0.775917	0.009647
0.3	0.761845	0.666195	0.016488
0.4	0.693506	0.559077	0.020276
0.5	0.630563	0.45522	0.019089
0.6	0.572791	0.355165	0.011828
0.7	0.519912	0.259342	0.250570
0.8	0.471618	0.168073	0.303545
0.9	0.427594	0.0815811	0.346013
1.0	0.387518	5.60459×10^{-9}	0.387517

Table 6. Symmetry of HAM versus numerical solutions for $\phi(\eta)$, when $Sc = Pr = K_s = 1.0$, $\kappa = \kappa_1 = \kappa_2 = \lambda_2 = M = K = 0.1$.

η	HAM $\phi(\eta)$	Numerical $\phi(\eta)$	Absolute Error AE
0.0	0.396762	0.404080	0.007318
0.1	0.429884	0.436425	0.006841
0.2	0.464667	0.468606	0.003939
0.3	0.499830	0.500349	0.000519
0.4	0.534490	0.531418	0.003072
0.5	0.568056	0.561618	0.006438
0.6	0.600146	0.590789	0.009357
0.7	0.630533	0.618810	0.011723
0.8	0.659102	0.645589	0.013513
0.9	0.685811	0.671065	0.014746
1.0	0.710675	0.695202	0.015473

7. Conclusions

In this article the MHD flow of three combined nanofluids (Jefferey, Maxwell, and Oldroyd-B) over a linear stretched surface have been scrutinized. The problem was solved analytically by HAM.

The convergence of HAM has been presented through graphical presentations. The concluding remarks are as follows:

- ➢ The upsurges in magnetic field diminishes the velocity field.
- ➢ The upsurges in Prandtl number and thermal relaxation parameters diminish the temperature field.
- ➢ The upsurges in Schmidt number upsurges the concentration field.
- ➢ The larger homogeneous reaction and heterogeneous reaction strengths falloff from the concentration field.

Author Contributions: A.S. and Z.S. modeled the problem and wrote the manuscript. S.I., A.D. and P.K. thoroughly checked the mathematical modeling and English corrections. W.K. and A.S. solved the problem using Mathematica software. Z.S., S.I. and P.K. contributed to the results and discussions. All authors finalized the manuscript after its internal evaluation.

Conflicts of Interest: The authors declare no conflict of interest.

Nomenclature

B_0	Magnetic field strength (NmA^{-1})
Cf_x	Skin friction coefficient
D_I, D_J	Diffusion coefficients
F	Velocity profile
G	Temperature profile
I, J	Chemical species
i, j	Concentration
K	Strength of homogenous reaction
K_s	Strength of heterogeneous reaction
k	Thermal conductivity ($Wm^{-1}K^{-1}$)
M	Magnetic parameter
Nu_x	Nusselt number
Pr	Prandtl number
q	Heat flux (Wm^{-2})
Re_x	Local Reynolds number
Sc	Schmidt number
Sh_x	Sherwood number
T	Fluid temperature (K)
T_w	Surface temperature (K)
T_∞	Temperature at infinity (K)
u, v	Velocity components (ms^{-1})
x, y	Coordinates
α	Thermal diffusivity (m^2s^{-1})
η	Similarity variable
μ	Dynamic viscosity (mPa)
v_f	Kinematic viscosity (mPa)
ρ_f	Density (Kgm^{-3})
λ_1	Relaxation time
λ_2	Relaxation to retardation time
λ_3	Retardation time
ζ	Stretching rate
κ	Deborah number
Ω	Thermal relaxation parameter
σ	Electrical conductivity (Sm^{-1})
ϕ	Dimensional concentration profile

References

1. Choi, S.U.S. Enhancing thermal conductivity of fluids with nanoparticles. *Int. Mech. Eng. Congr. Expo.* **1995**, *66*, 99–105.
2. Fourier, J.B.J. *TheÂorie Analytique De La Chaleur*; Chez Firmin Didot, père et fils: Paris, France, 1822.
3. Cattaneo, C. Sulla conduzione delcalore. *Atti Semin Mat Fis Univ Modena Reggio Emilia* **1948**, *3*, 83–101.
4. Christov, C.I. On frame indifferent formulation of the Maxwell-Cattaneo model of finite-speed heat conduction. *Mech. Res. Commun.* **2009**, *36*, 481–486. [CrossRef]
5. Mustafa, M. Cattaneo-Christov heat flux model for rotating flow and heat transfer of upper-convected Maxwell fluid. *AIP Adv.* **2015**, *5*, 047109. [CrossRef]
6. Chen, C.H. Effect of viscous dissipation on heat transfer in a non-Newtonian liquid film over an unsteady stretching sheet. *J. Non-Newton. Fluid Mech.* **2005**, *135*, 128–135. [CrossRef]
7. Sheikholeslami, M.; Shah, Z.; Shafi, A.; Khan, I.; Itili, I. Uniform magnetic force impact on water based nanofluid thermal behavior in a porous enclosure with ellipse shaped obstacle. *Sci. Rep.* **2019**, *9*, 1196. [CrossRef] [PubMed]
8. Sheikholeslami, M.; Shah, Z.; Tassaddiq, A.; Shafee, A.; Khan, I. Application of Electric Field for Augmentation of Ferrofluid Heat Transfer in an Enclosure Including Double Moving Walls. *IEEE Access* **2019**, *7*, 21048–21056. [CrossRef]
9. Sheikholeslami, M.; Haq, R.L.; Shafee, A.; Zhixiong, L. Heat transfer behavior of Nanoparticle enhanced PCM solidification through an enclosure with V shaped fins. *Int. J. Heat Mass Transf.* **2019**, *130*, 1322–1342. [CrossRef]
10. Sheikholeslami, M.; Gerdroodbary, M.B.; Moradi, R.; Shafee, A.; Zhixiong, L. Application of Neural Network for estimation of heat transfer treatment of Al_2O_3-H_2O nanofluid through a channel. *Comput. Methods Appl. Mech. Eng.* **2019**, *344*, 1–12. [CrossRef]
11. Sheikholeslami, M. Numerical study of heat transfer enhancement in a pipe filled with porous media by axisymmetric TLB model based on GPU. *Int. J. Heat Mass. Trans.* **2014**, *70*, 1040–1049.
12. Shah, Z.; Islam, S.; Gul, T.; Bonyah, E.; Altaf Khan, M. The Elcerical MHD And Hall Current Impact on Micropolar Nanofluid Flow Between Rotating Parallel Plates. *Results Phys.* **2018**. [CrossRef]
13. Shah, Z.; Bonyah, E.; Islam, S.; Gul, T. Impact of thermal radiation on electrical mhd rotating flow of carbon nanotubes over a stretching sheet. *AIP Adv.* **2019**, *9*, 015115. [CrossRef]
14. Shah, Z.; Tassaddiq, A.; Islam, S.; Alklaibi, A.; Khan, I. Cattaneo–Christov Heat Flux Model for Three-Dimensional Rotating Flow of SWCNT and MWCNT Nanofluid with Darcy–Forchheimer Porous Medium Induced by a Linearly Stretchable Surface. *Symmetry* **2019**, *11*, 331. [CrossRef]
15. Shah, Z.; Bonyah, E.; Islam, S.; Khan, W.; Ishaq, M. Radiative MHD thin film flow of Williamson fluid over an unsteady permeable stretching. *Heliyon* **2018**, *4*, e00825. [CrossRef] [PubMed]
16. Dawar, A.; Shah, Z.; Khan, W.; Islam, S.; Idrees, M. An optimal analysis for Darcy-Forchheimer 3-D Williamson Nanofluid Flow over a stretching surface with convective conditions. *Adv. Mech. Eng.* **2019**, *11*, 1–15. [CrossRef]
17. Shah, Z.; Islam, S.; Ayaz, H.; Khan, S. Radiative Heat and Mass Transfer Analysis of Micropolar Nanofluid Flow of Casson Fluid between Two Rotating Parallel Plates with Effects of Hall Current. *ASME J. Heat Transf.* **2019**. [CrossRef]
18. Maleki, H.; Safaei, M.R.; Alrashed, A.A.A.A.; Kasaeian, A. Flow and heat transfer in non-Newtonian nanofluids over porous surfaces. *J. Anal. Calorim.* **2019**, *135*, 1655. [CrossRef]
19. Nasiri, H.; Abdollahzadeh Jamalabadi, M.Y.; Sadeghi, R.; Safaei, M.R.; Nguyen, T.K.; Shadloo, M.S. A smoothed particle hydrodynamics approach for numerical simulation of nano-fluid flows. *J. Anal. Calorim.* **2019**, *135*, 1733. [CrossRef]
20. Rashidi, M.M.; Nasiri, M.; Shadloo, M.S.; Yang, Z. Entropy Generation in a Circular Tube Heat Exchanger Using Nanofluids: Effects of Different Modeling Approaches. *Heat Transf. Eng.* **2017**, *38*, 853–866. [CrossRef]
21. Safaei, M.R.; Shadloo, M.S.; Goodarzi, M.S.; Hadjadj, A.; Goshayeshi, H.R.; Afrand, M.; Kazi, S.N. A survey on experimental and numerical studies of convection heat transfer of nanofluids inside closed conduits. *Adv. Mech. Eng.* **2016**. [CrossRef]

22. Mahian, O.; Kolsi, L.; Amani, M.; Estelle, P.; Ahmadi, G.; Kleinstreuer, C.; Marshall, J.S.; Siavashi, M.; Taylor, R.A.; Niazmand, H.; et al. Recent advances in modeling and simulation of nanofluid flows-Part I: Fundamentals and theory. *Phys. Rep.* **2019**, *790*, 1–48. [CrossRef]
23. Mahian, O.; Kolsi, L.; Amani, M.; Estelle, P.; Ahmadi, G.; Kleinstreuer, C.; Marshall, J.S.; Taylor, R.A.; Abu-Nada, E.; Rashidi, S.; et al. Recent advances in modeling and simulation of nanofluid flows-Part II: Applications. *Phys. Rep.* **2019**, *791*, 1–59. [CrossRef]
24. Kartini, A.; Zahir, H.; Anuar, I. Mixed convection Jeffrey fluid flow over an exponentially stretching sheet with magnetohydrodynamic effect. *AIP Adv.* **2016**, *6*, 035024.
25. Kartini, A.; Anuar, I. Magnetohydrodynamic Jeffrey fluid over a stretching vertical surface in a porous medium. *Propuls. Power Res.* **2017**, *6*, 269–276.
26. Hayat, T.; Hussain, T.; Shehzad, S.A.; Alsaedi, A. Flow of Oldroyd-B fluid with nano particles and thermal radiation. *Appl. Math. Mech. Engl. Ed.* **2015**, *36*, 69–80. [CrossRef]
27. Raju, C.S.K.; Sandeep, N.; Gnaneswar, R.M. Effect of nonlinear thermal radiation on 3D Jeffrey fluid flow in the presence of homogeneous–heterogeneous reactions. *Int. J. Eng. Res. Afr.* **2016**, *21*, 52–68. [CrossRef]
28. Raju, C.S.K.; Sandeep, N. Heat and mass transfer in MHD non-Newtonian bio-convection flow over a rotating cone/plate withcross diffusion. *J. Mol. Liq.* **2016**, *215*, 115–126. [CrossRef]
29. Alla, A.M.A.; Dahab, S.M.A. Magnetic field and rotation effects on peristaltic transport on Jeffrey fluid in an asymmetric channel. *J. Mag. Magn. Mater.* **2015**, *374*, 680–689. [CrossRef]
30. Khan, A.A.; Ellahi, R.; Usman, M. Effects of variable viscosity on the flow of non-Newtonian fluid through a porous medium in an inclined channel with slip conditions. *J. Porous Media* **2013**, *16*, 59–67. [CrossRef]
31. Ellahi, R.; Bhatti, M.M.; Riaz, A.; Sheikholeslami, M. Effects of magnetohydrodynamics on peristaltic flow of Jeffrey fluid in a rectangular duct through a porous medium. *J. Porous Media* **2014**, *17*, 143–147. [CrossRef]
32. Hayat, T.; Ali, N. A mathematical description of peristaltic hydromagnetic flow in a tube. *Appl. Math. Comput.* **2007**, *188*, 1491–1502. [CrossRef]
33. Hayat, T.; Abbas, Z.; Sajid, M. Series solution for the upper convected Maxwell fluid over a porous stretching plate. *Phys. Lett. A* **2006**, *358*, 396–403. [CrossRef]
34. Raju, C.S.K.; Sandeep, N. Heat and mass transfer in 3D non-Newtonian nano and Ferro fluids over a bidirectional stretching surface. *Int. J. Eng. Res. Afr.* **2016**, *21*, 33–51. [CrossRef]
35. Sandeep, N.; Sulochana, C. Dual solutions for unsteady mixed convection flow of MHD micropolar fluid over a stretching/ shrinking sheet with non-uniform heat source/sink. *Eng. Sci. Technol. Int. J.* **2015**, *18*, 738–745. [CrossRef]
36. Raju, C.S.K.; Sandeep, N.; Sulochana, C.; Sugunamma, V.; Jayachandra, B.M. Radiation, inclined magnetic field and cross diffusion effects on flow over a stretching surface. *J. Niger. Math. Soc.* **2015**, *34*, 169–180. [CrossRef]
37. Nadeem, S.; Akbar, N.S. Influence of haet and mass transfer on a peristaltic motion of a Jeffrey-six constant fluid in an annulus. *Heat Mass Transf.* **2010**, *46*, 485–493. [CrossRef]
38. Makinde, O.D.; Chinyoka, T.; Rundora, L. Unsteady flow of a reactive variable viscosity non-Newtonian fluid through a porous saturated medium with asymmetric convective boundary conditions. *Comput. Math. Appl.* **2011**, *62*, 3343–3352. [CrossRef]
39. Sheikholeslami, M. KKL correlation for simulation of nanofluid flow and heat transfer in a permeable channel. *Phys. Lett. A* **2014**, *378*, 3331–3339. [CrossRef]
40. Shah, Z.; Dawar, A.; Islam, S.; Khan, I.; Ching, D.L.C. Darcy-Forchheimer Flow of Radiative Carbon Nanotubes with Microstructure and Inertial Characteristics in the Rotating Frame. *Case Stud. Therm. Eng.* **2018**, *12*, 823–832. [CrossRef]
41. Chai, L.R.; Shaukat, L.; Wang, H.; Wang, S. A review on heat transfer and hydrodynamic characteristics of nano/microencapsulated phase change slurry (N/MPCS) in mini/microchannel heat sinks. *Appl. Therm. Eng.* **2018**. [CrossRef]
42. Shah, Z.; Dawar, A.; Islam, S.; Khan, I.; Ching, D.L.C.; Khan, A.Z. Cattaneo-Christov model for Electrical Magnetite Micropoler Casson Ferrofluid over a stretching/shrinking sheet using effective thermal conductivity model. *Case Stud. Therm. Eng.* **2018**. [CrossRef]
43. Dawar, A.; Shah, Z.; Islam, S.; Idress, M.; Khan, W. Magnetohydrodynamic CNTs Casson Nanofluid and Radiative heat transfer in a Rotating Channels. *J. Phys. Res. Appl.* **2018**, *1*, 017–032.

44. Khan, A.S.; Nie, Y.; Shah, Z.; Dawar, A.; Khan, W.; Islam, S. Three-Dimensional Nanofluid Flow with Heat and Mass Transfer Analysis over a Linear Stretching Surface with Convective Boundary Conditions. *Appl. Sci.* **2018**, *8*, 2244. [CrossRef]
45. Imtiaz, M.; Hayat, T.; Alsaedi, A.; Hobiny, A. Homogeneous-heterogeneous reactions in MHD flow due to an unsteady curved stretching surface. *J. Mol. Liq.* **2016**, *221*, 245–253. [CrossRef]
46. Hayat, T.; Hussain, Z.; Muhammad, T.; Alsaedi, A. Effects of homogeneous and heterogeneous reactions in flow of nanofluids over a nonlinear stretching surface with variable surface thickness. *J. Mol. Liq.* **2016**, *221*, 1121–1127. [CrossRef]
47. Nasir, S.; Shah, Z.; Islam, S.; Khan, W.; Khan, S.N. Radiative flow of magneto hydrodynamics single-walled carbon nanotube over a convectively heated stretchable rotating disk with velocity slip effect. *Adv. Mech. Eng.* **2019**, *11*, 1–11.
48. Nasir, S.; Shah, Z.; Islam, S.; Khan, W.; Bonyah, E.; Ayaz, M.; Khan, A. Three dimensional Darcy-Forchheimer radiated flow of single and multiwall carbon nanotubes over a rotating stretchable disk with convective heat generation and absorption. *AIP Adv.* **2019**, *9*, 035031. [CrossRef]
49. Hammed, K.; Haneef, M.; Shah, Z.; Islam, I.; Khan, W.; Asif, S.M. The Combined Magneto hydrodynamic and electric field effect on an unsteady Maxwell nanofluid Flow over a Stretching Surface under the Influence of Variable Heat and Thermal Radiation. *Appl. Sci.* **2018**, *8*, 160. [CrossRef]
50. Ullah, A.; Alzahrani, E.O.; Shah, Z.; Ayaz, M.; Islam, S. Nanofluids Thin Film Flow of Reiner-Philippoff Fluid over an Unstable Stretching Surface with Brownian Motion and Thermophoresis Effects. *Coatings* **2019**, *9*, 21. [CrossRef]

© 2019 by the authors. Licensee MDPI, Basel, Switzerland. This article is an open access article distributed under the terms and conditions of the Creative Commons Attribution (CC BY) license (http://creativecommons.org/licenses/by/4.0/).

Article

Unsteady Flow of Fractional Fluid between Two Parallel Walls with Arbitrary Wall Shear Stress Using Caputo–Fabrizio Derivative

Muhammad Asif [1], Sami Ul Haq [2], Saeed Islam [1], Tawfeeq Abdullah Alkanhal [3], Zar Ali Khan [4], Ilyas Khan [5],* and Kottakkaran Sooppy Nisar [6]

1. Department of Mathematics, Abdul Wali Khan University, Mardan 23200, Pakistan; asif_best1986@yahoo.com (M.A.); saeed.sns@gmail.com (S.I.)
2. Department of Mathematics, Islamia College University, Peshawar 25000, Pakistan; samiulhaqmaths@yahoo.com
3. Department of Mechatronics and System Engineering, College of Engineering, Majmaah University, Majmaah 11952, Saudi Arabia; t.alkanhal@mu.edu.sa
4. Department of Mathematics, University of Peshawar, Peshawar 25000, Pakistan; zaralikhangmk@gmail.com
5. Faculty of Mathematics and Statistics, Ton Duc Thang University, Ho Chi Minh City 72915, Vietnam
6. Department of Mathematics, College of Arts and Science, Prince Sattam bin Abdulaziz University, Wadi Al-Dawaser 11991, Saudi Arabia; n.sooppy@psau.edu.sa
* Correspondence: ilyaskhan@tdt.edu.vn

Received: 3 January 2019; Accepted: 17 March 2019; Published: 1 April 2019

Abstract: In this article, unidirectional flows of fractional viscous fluids in a rectangular channel are studied. The flow is generated by the shear stress given on the bottom plate of the channel. The authors have developed a generalized model on the basis of constitutive equations described by the time-fractional Caputo–Fabrizio derivative. Many authors have published different results by applying the time-fractional derivative to the local part of acceleration in the momentum equation. This approach of the fractional models does not have sufficient physical background. By using fractional generalized constitutive equations, we have developed a proper model to investigate exact analytical solutions corresponding to the channel flow of a generalized viscous fluid. The exact solutions for velocity field and shear stress are obtained by using Laplace transform and Fourier integral transformation, for three different cases namely (i) constant shear, (ii) ramped type shear and (iii) oscillating shear. The results are plotted and discussed.

Keywords: viscous fluid; Caputo–Fabrizio time-fractional derivative; Laplace and Fourier transformations; side walls; oscillating shear stress

1. Introduction

The branch of mathematics that studies derivatives and integrals is called calculus, i.e., discussing integer order derivatives and integrals. When the order of derivatives changes from integer order to real (non-integer) order a new branch of calculus comes into being, called fractional calculus. Fractional order derivatives occur in many physical problems for example, frequency-dependent damping behavior of objects, velocity of infinite thin plate in a viscous fluid, creeping and relaxation functions of viscoelastic materials, and the control of dynamical systems as mentioned in [1–4]. Fractional calculus provides more generalized derivatives, and therefore it has more applications as compared with the classical or integer order derivatives. Fractional differential equations also explain the phenomena in electrochemistry, acoustics, electromagnetics, viscoelasticity, and material science [5–10]. For the last twenty years, a lot of work has been done on fractional calculus. Some authors [11–13] have used the formal definitions of fractional calculus like Riemann–Liouville and Caputo operators. These definitions provide a strong basis for the modern approach of Caputo and Fabrizio who have presented the definition without singular kernel [14].

The Caputo–Fabrizio differential operator is used by many authors to obtain exact solutions concerning real life problems [15–17]. All the benefits of Reimann–Liouville and Caputo definitions are also included in Caputo–Fabrizio, which is a worthy point of this definition. The Caputo–Fabrizio definition has been used by different authors in the medical sciences for example, the cancer treatment model and the flow of blood through veins under the effect of magnetic field [18,19]. Shah et al. [20] investigated the exact solutions over an isothermal vertical plate of free convectional flow of viscous fluids by using a definition of the Caputo–Fabrizio time-fractional derivative. Free convectional time-fractional flow with Newtonian heating near a vertical plate including mass diffusion has been investigated by Vieru et al. [21].

The effect of side walls over the velocity of a non-Newtonian fluid while the motion is produced due to the oscillation of the lower plate has been investigated by Fetecau et al. [22]. In addition, Haq et al. [23] have analyzed the exact solution of viscous fluid over an infinite plate using the Caputo–Fabrizio fractional order derivatives. Most of the authors have discussed different fluids using the fractional order differential operator defined by Caputo, Caputo–Fabrizio etc. and published many interesting results by applying the fractional order definition only to the local part of acceleration. Henry et al. [24] and Hristov [25,26] have suggested a generalized Fourier law for the thermal heat flux. It is clear from their discussion that the fractional differential operator has been employed in the constitutive relation of energy equation, rather than directly using it in the governing equation. This approach is appealing to mathematical and physical aspects of fluid mechanics. Hameid et al. [27] applied the definition of the fractional order derivative to the convective part of a constitutive equation and explained their model in a very interesting way. Vieru et al. [28] have followed the discussion presented by Hameid et al. [27] by applying the fractional derivative definition in a constitutive equation.

Keeping in mind all the above discussions, we present this article exploring the effect of side walls on the motion of an incompressible fluid using generalized fractional constitutive equations and the Caputo–Fabrizio derivative through a rectangular channel.

2. Problem Formulation

Consider an incompressible fluid, which is viscous in nature, present over an infinite plate between two parallel side walls that are at right angles to the horizontal plate as shown in Figure 1. Initially both the plate and fluid are at rest for $t = 0$, after time $t > 0$, and the flow is generated by the shear stress given by $\tau_0\, f(t)$ which engender the velocity as

$$\mathbf{V} = v(y,z,t)i, \qquad (1)$$

where i stands for unit vector.

In the absence of the body forces and the pressure gradient in the flow direction, the linear momentum equation in the x-direction is:

$$\rho\frac{\partial u_x}{\partial t} + u_x\frac{\partial u_x}{\partial x} + u_y\frac{\partial u_x}{\partial y} + u_z\frac{\partial u_x}{\partial z} = \frac{\partial \tau_{xy}}{\partial y} + \frac{\partial \tau_{xz}}{\partial z} \qquad (2)$$

Therefore, in our case, the velocity field is the advection terms in Equation (2) that are zero with initial and boundary conditions as follows:

$$\begin{aligned}
& v(y,z,0) = 0,\ \text{for } y > 0 \text{ and } z \in [0,l]\,, \\
& v(y,0,t) = v(y,l,t) = 0\ \text{for } y,t > 0\,, \\
& \tau_{xy}(0,z,t) = \mu\ {}^{CF}D_t^\alpha \left.\frac{\partial v(y,z,t)}{\partial y}\right|_{y=0} = \tau_0\, f(t)\,,
\end{aligned} \qquad (3)$$

The above relation (Equation (3)), shows the shear stress which is non-trivial, where μ is dynamic viscosity, $v = \frac{\mu}{\rho}$, where τ_0 shows the constant parameter, and it is assumed that $f(\cdot)$ is a dimensionless, piecewise, continuous function such that $f(0) = 0$, $v(y,z,t)$ and $\frac{\partial v(y,z,t)}{\partial y} \to 0$ as $y \to \infty$, $z \in [0,l]$ for $t > 0$.

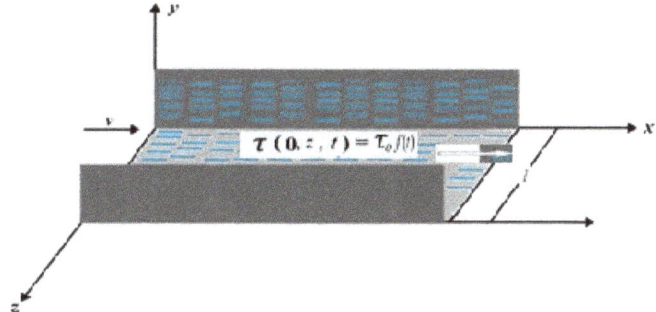

Figure 1. Geometry of the problem.

We have the Caputo–Fabrizio derivative operator of order α given by Zafar et al. [29]

$$^{CF}D_t^\alpha[h(t)] = \frac{1}{1-\alpha}\int_0^t h'(s)\exp\left[-\frac{\alpha(t-s)}{1-\alpha}\right]ds \text{ for } 0 \leq \alpha < 1. \quad (4)$$

In the present paper, we consider the generalized constitutive equations with the Caputo–Fabrizio time-fractional derivative, namely:

$$\tau_{xy} = \mu\, ^{CF}D_t^\alpha\left(\frac{\partial v}{\partial y}\right) \text{ and } \tau_{xz} = \mu\, ^{CF}D_t^\alpha\left(\frac{\partial v}{\partial z}\right) \text{ for } \alpha \in [0,1). \quad (5)$$

It is known that any constitutive equation must satisfy the principle of material objectivity, and therefore it must be frame-invariant with respect to Euclidean transformations. Yang et al. ([30], Equation (3.1)) have formulated a constitutive equation with fractional derivatives for generalized upper-convected Maxwell fluids on the basis of the convected coordinate system. They have proven that the proposed constitutive equation is frame-indifferent and have studied some particular cases of the proposed equation.

By applying the Laplace transform, the constitutive Equation (5) for the shear stress τ_{xy} can be written in the following equivalent form:

$$\tau_{xy} = \frac{1}{1-\alpha}\frac{\partial v}{\partial y} - \frac{\alpha}{(1-\alpha)^2}\int_0^t \exp\left(\frac{-\alpha(t-\tau)}{1-\alpha}\right)\frac{\partial v(y,\tau)}{\partial y}d\tau, \; 0 < \alpha < 1. \quad (6)$$

Equation (6) is equivalent to the equation studied by Yang et al. ([30], Equation (4.5)), therefore, the proposed constitutive equations given by Equation (5) satisfy the principle of material objectivity.

3. Problem Solution

Using Equation (5) in Equation (2), applying Laplace transform to the obtained form and simplifying the result we get:

$$v(y,z,q) = \frac{v}{(1-\alpha)q+\alpha}\left[\frac{\partial^2 v(y,z,q)}{\partial y^2} + \frac{\partial^2 v(y,z,q)}{\partial z^2}\right], \quad (7)$$

where $v(y,z,q) = \int_0^\infty v(y,z,t)\exp(-qt)dt$ is the Laplace transform with respect to t.

Applying the Fourier cosine transform with respect to variable y namely $v(\xi,z,q) = \int_0^\infty v(y,z,q)\cos(y\xi)dy$ and finite Fourier sine transform with respect to variable z $v(\xi,n,q) = \int_0^l v(\xi,z,q)\sin\left(\frac{n\pi z}{l}\right)dz$, $n = 1,2,\ldots$, we obtain:

$$v_{sc}(\xi,q) = \frac{v}{(1-\alpha)q + \alpha + v\left(\xi^2 + \lambda_n^2\right)}\frac{\tau_0}{\mu}\sqrt{\frac{2}{\pi}}f(q)\frac{(-1)^n - 1}{\lambda_n}, \tag{8}$$

where $\lambda_n = \frac{n\pi}{l}$ and subscript "sc" represents finite Fourier sine and infinite cosine transforms.

For simplification, Equation (8) can be written as:

$$v_{sc}(\xi,q) = \frac{\tau_0}{\mu}\frac{1-(-1)^n}{\lambda_n}\sqrt{\frac{2}{\pi}}\frac{f(q)}{(\xi^2+\lambda_n^2)} - \frac{\tau_0}{\mu}\frac{1-(-1)^n}{\lambda_n}\sqrt{\frac{2}{\pi}}f(q) \times \left[\frac{(1-\alpha)q+\alpha}{(\xi^2+\lambda_n^2)[(1-\alpha)q+\alpha+v(\xi^2+\lambda_n^2)]}\right]. \tag{9}$$

Applying the inverse Laplace transformation, we get:

$$v_{sc}(\xi,t) = \frac{\tau_0}{\mu}\sqrt{\frac{2}{\pi}}\frac{1-(-1)^n}{\lambda_n}\frac{f(t)}{(\xi^2+\lambda_n^2)} + \frac{1}{(1-\alpha)(\xi^2+\lambda_n^2)}e^{-A(\xi)t}$$
$$\times \left[\frac{\tau_0}{\mu}\sqrt{\frac{2}{\pi}}\frac{1-(-1)^n}{\lambda_n}(1-\alpha)f^\bullet(t) + \frac{\tau_0}{\mu}\sqrt{\frac{2}{\pi}}\frac{1-(-1)^n}{\lambda_n}\alpha f(t)\right], \tag{10}$$

where $A(\xi) = \frac{v(\xi^2+\lambda_n^2)+\alpha}{1-\alpha}$.

Applying the inverse Fourier transformation, we find:

$$v(y,z,t) = \frac{2}{\pi}\frac{2}{l}\frac{\tau_0}{\mu}\sum_{n=1}^\infty \frac{(-1)^n-1}{\lambda_n}\sin(\lambda_n z)\left[f(t)\int_0^\infty \frac{\cos(y\xi)}{\xi^2+\lambda_n^2}d\xi + \int_0^\infty \frac{\cos(y\xi)}{(\xi^2+\lambda_n^2)}e^{-A(\xi)t}d\xi\right], \tag{11}$$

where $m = 2n - 1$ and $l = 2h$. Changing the origin by using $z = z^* + h$,

$$v(y,z,t) = \frac{2\tau_0}{\mu\pi h}\sum_{n=1}^\infty \frac{(-1)^{n+1}?\cos(\gamma_m z^*)}{\gamma_m}\left[f(t)\int_0^\infty \frac{\cos(y\xi)}{\xi^2+\gamma_m^2}d\xi + \int_0^\infty \frac{\cos(y\xi)}{(\xi^2+\gamma_m^2)}e^{-A(\xi)t}d\xi\right], \tag{12}$$

ignoring the * notation, keeping in view the following result:

$$\int_0^\infty \frac{\cos(ax)}{b^2+x^2}dx = \frac{\pi}{2b}e^{-ab}, \text{ for } a > 0 \text{ } Re(b) > 0,$$

and putting in Equation (11) we get:

$$v(y,z,t) = \frac{2}{h}\sum_{n=1}^\infty (-1)^{n+1}\frac{\tau_0}{\mu}\frac{\cos\gamma_m z}{\gamma_m}f(t)\frac{e^{-y\gamma_m}}{\gamma_m} - \frac{4}{\pi h}\sum_{n=1}^\infty (-1)^{n+1}\frac{\tau_0}{\mu}\frac{\cos\gamma_m z}{\gamma_m}$$
$$\times \left[\int_0^t\int_0^\infty \frac{\cos(y\xi)}{\xi^2+\gamma_m^2}f^\bullet(t-\tau)e^{-A(\xi)\tau}d\tau d\xi + \frac{\alpha}{1-\alpha}\int_0^t\int_0^\infty \frac{\cos(y\xi)}{\xi^2+\gamma_m^2}f^\bullet(t-\tau)e^{-A(\xi)\tau}d\tau d\xi\right]. \tag{13}$$

where $\gamma_m = (2n-1)\frac{\pi}{2h}$

4. Graphical Illustration and Discussions

After finding the general solution for the velocity of the fluid, we discuss three different cases which are very useful in engineering. The obtained results are presented graphically for the three cases. Figures 2–5 show different profiles of velocity by taking case I (constant shear) into consideration. Figures 6–9 show the behavior of fluid velocity for case II (ramped type shear), and Figures 10–12 show the same discussion for Case III (oscillating shear).

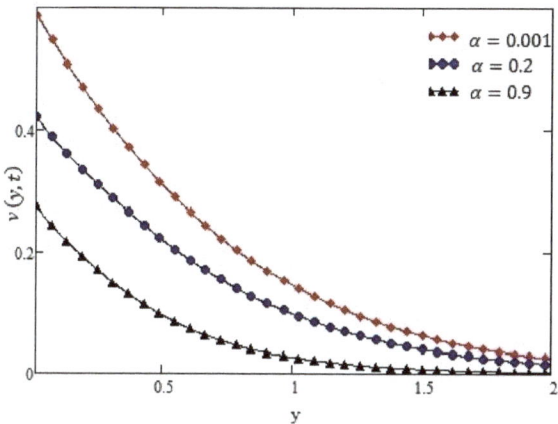

Figure 2. Profiles of velocity given by Equation (14) of constant shear for different values of α, for $t = 20$ s, $h = 0.6$ m, $\tau_0 = -1$ N/m^2, $\nu = 0.1$ m^2/s, and $\mu = 1.4$ kg·m/s.

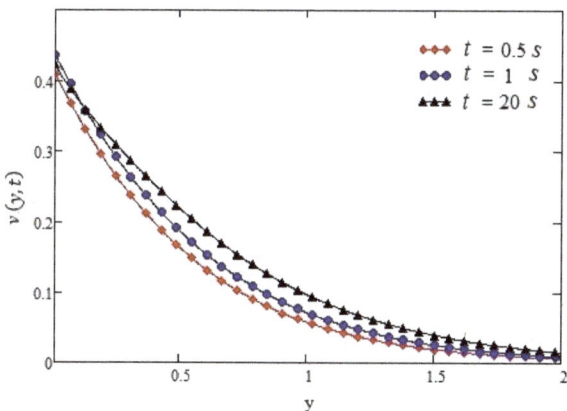

Figure 3. Profiles of velocity given by Equation (14) of constant shear for different values of t, for $\alpha = 0.2$, $h = 0.6$ m, $\tau_0 = -1$ N/m^2, $\nu = 0.1$ m^2/s, and $\mu = 1.4$ kg·m/s.

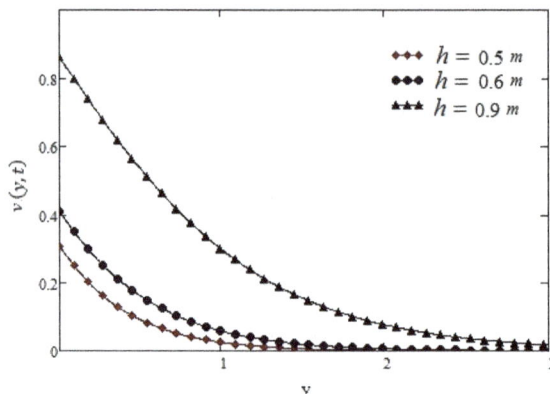

Figure 4. Profiles of velocity given by Equation (14) of constant shear for different values of h, for $t = 0.5$ s, $\alpha = 0.2$, $\tau_0 = -1$ N/m^2, $\nu = 0.1$ m^2/s, and $\mu = 1.4$ kg·m/s.

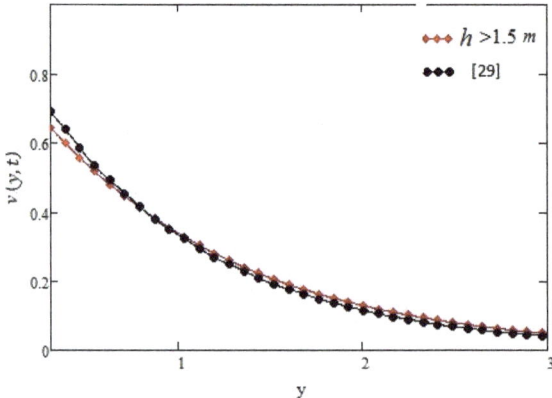

Figure 5. The value of h required for overlapping the velocity profiles of equation (14) and infinite plate [29] in constant shear for $t = 0.5$ s, $\alpha = 0.2$, $\tau_0 = -1$ N/m^2, $\nu = 0.1$ m^2/s, $\mu = 1.4$ kg·m/s, and $h > 1.5$ m.

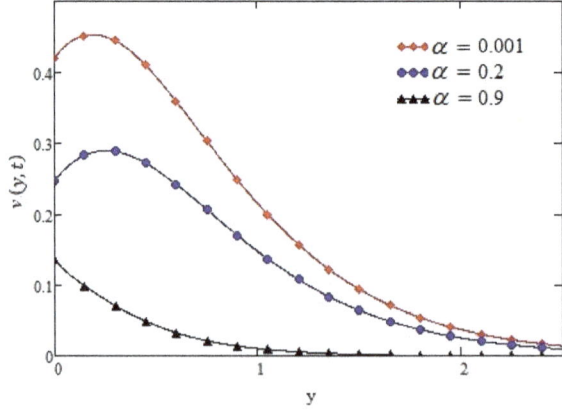

Figure 6. Profiles of velocity given by Equation (15) of ramped type shear velocity for different values of α, for $t = 0.5$ s, $h = 0.6$ m, $\tau_0 = -1$ N/m^2, $\nu = 0.1$ m^2/s, and $\mu = 1.4$ kg·m/s.

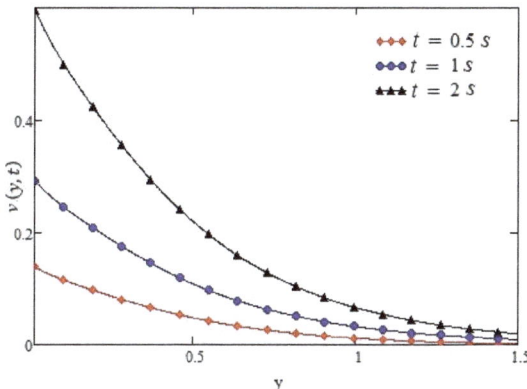

Figure 7. Profiles of velocity given by Equation (15) of ramped type shear for different values of t, for $\alpha = 0.9$, $h = 0.6$ m, $\tau_0 = -1$ N/m^2, $\nu = 0.1$ m^2/s, and $\mu = 1.4$ kg·m/s.

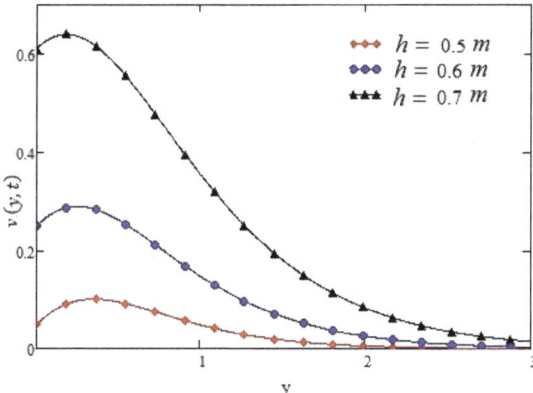

Figure 8. Profiles of velocity given by Equation (15) of ramped type shear for different values of h, for $t = 0.5$ s, $\alpha = 0.2$, $\tau_0 = -1$ N/m^2, $\nu = 0.1$ m^2/s, and $\mu = 1.4$ kg·m/s.

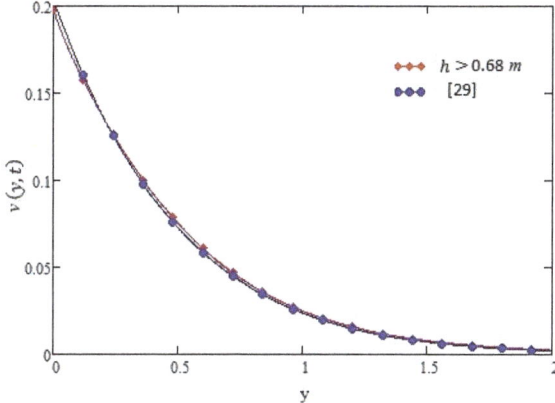

Figure 9. The value of h required for overlapping the velocity profiles of Equation (15) and infinite plate [29] in ramped type shear for $t = 0.4$ s, $\alpha = 0.9$, $\nu = 0.1$ m^2/s, $\tau_0 = -1.5$ N/m^2, $\mu = 1.4$ kg·m/s and $h \geq 0.68$.

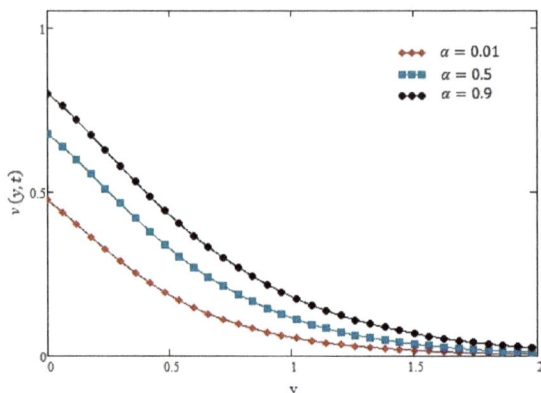

Figure 10. Profiles of velocity given by Equation (16) of oscillating shear for different values of α, for $t = 1$ s, $h = 0.6$ m, $\nu = 0.05$ m^2/s, $\tau_0 = -1$ N/m^2, $\mu = 1.4$ kg·m/s, and $\omega = 1$.

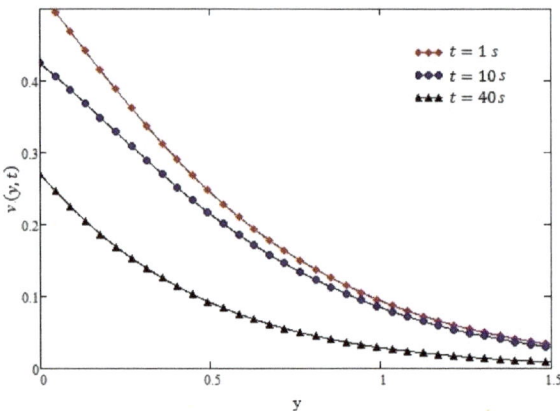

Figure 11. Profiles of velocity given by Equation (16) of oscillating shear for different values of t, for $\alpha = 0.5$, $h = 0.6$ m, $\tau_0 = -1$ N/m^2, $\nu = 0.1$ m^2/s, $\mu = 1.4$ kg·m/s, and $\omega = 1$.

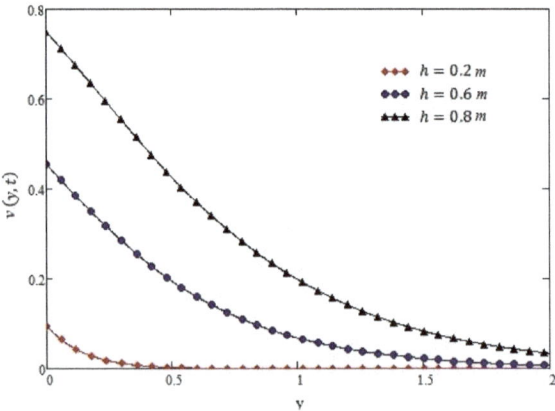

Figure 12. Profiles of velocity given by Equation (16) of oscillating shear for different values of h, for $t = 1$ s, $\alpha = 0.5$, $\tau_0 = -1$ N/m^2, $\nu = 0.1$ m^2/s, $\mu = 1.4$ kg·m/s, and $\omega = 1$.

4.1. Case I (Constant Shear)

Taking $f(t) = H(t)$, where $H(t)$ is Heaviside unit step function we find the velocity profile as:

$$v(y,z,t) = \frac{2}{h}\sum_{n=1}^{\infty}(-1)^{n+1}\frac{\tau_0}{\mu}\frac{\cos\gamma_m z}{\gamma_m}\left[\begin{array}{c}e^{-y\gamma_m}+(1-\alpha)(t-\alpha)\\ \times e^{-yA}\frac{B_n\cos(yB_n)+A_n\sin(yB_n)}{c_n\left(\sqrt{b_n^4+c_n^2}\right)}\end{array}\right]$$

$$-\frac{4}{\pi h}\sum_{n=1}^{\infty}(-1)^{n+1}\frac{\tau_0}{\mu}\frac{\cos\gamma_m z}{\gamma_m}\int_0^{\infty}\left(\frac{e^{-A(\xi)t}-1}{(A(\xi))^2}-\frac{\alpha e^{-A(\xi)t}}{(1-\alpha)A(\xi)}\right)\frac{\cos(y\xi)}{\xi^2+\gamma_m^2}d\xi \quad (14)$$

Taking the following identity into account:

$$\int_0^{\infty}\frac{\cos(y\xi)}{\left(\xi^2-b_n^2\right)^2+c_n^2}=\frac{\pi}{2c_n}e^{-yA}\frac{B_n\cos(yB_n)+A_n\sin(yB_n)}{\sqrt{b_n^4+c_n^2}},$$

where $2A_n^2 = \sqrt{b_n^4+c_n^2}+b_n^2$ and $2B_n^2 = \sqrt{b_n^4+c_n^2}-b_n^2$ in which: $b_n = \frac{v\gamma_m^2+\frac{\alpha}{2}}{v}$ and $c_n = \frac{v\gamma_m^4+\alpha\gamma_m^2}{v}-b_n^2$.

Figure 2 shows that as the value of α ($0 < \alpha < 1$) is increasing, the fluid velocity is decreasing. Figure 3 shows the velocity profiles for different times, which implies that as time passages the fluid velocity increases in constant case. Figure 4 shows the profiles of velocity for different values of h, if we increase the distance between side walls the fluid velocity will increase while keeping the other parameters constant. Figure 5 shows the curve of Equation (14) is overlapping as that of [29] constant case if we increase the distance between the side walls to $h \geq 1.5$, which gives our result for constant case more validity.

4.2. Case II (Ramped Type Shear)

Taking $f(t) = tH(t)$, we find the velocity profile as:

$$v(y,z,t) = \frac{2}{d}\sum_{n=1}^{\infty}(-1)^{n+1}\frac{\tau_0}{\mu}\frac{\cos\gamma_m z}{\gamma_m}\frac{e^{-y\gamma_m}}{\gamma_m}+\frac{2}{h}\sum_{n=1}^{\infty}(-1)^{n+1}\frac{\tau_0}{\mu}\frac{\cos\gamma_m z}{\gamma_m}\frac{e^{-yA}}{c_n}$$

$$\times\frac{B_n\cos(yB_n)+A_n\sin(yB_n)}{\left(\sqrt{b_n^4+c_n^2}\right)}\left[t(2-\alpha)-t^2(1-\alpha)\right]-\frac{4}{\pi h}\sum_{n=1}^{\infty}(-1)^{n+1} \quad (15)$$

$$\times\frac{\tau_0}{\mu}\frac{\cos\gamma_m z}{\gamma_m}\int_0^{\infty}\left(2+\frac{\alpha}{(1-\alpha)}\right)\frac{e^{-A(\xi)t}}{A(\xi)}+\frac{1}{(A(\xi))^2}\frac{\cos(y\xi)}{\xi^2+\gamma_m^2}d\xi.$$

Figure 6 shows that by increasing the value of the differential parameter the velocity of the fluid decreases. Figures 7 and 8 show similar behavior of t and h, as observed in Figures 3 and 4, whereas in Figure 9, we calculated the value of h in order to validate our result in case of ramped type shear.

4.3. Case III (Oscillating Shear Stress)

Taking $f(t) = \sin(\omega t)H(t)$, in Equation (13) we get the velocity profile as:

$$v(y,z,t) = \sum_{n=1}^{\infty}(-1)^{n+1}\frac{\tau_0}{\mu}\frac{\cos\gamma_m z}{\gamma_m}\left[\frac{2}{h}\sin(\omega t)\frac{e^{-y\gamma_m}}{\gamma_m}-\frac{4}{\pi h}\int_0^{\infty}\left\{\frac{1}{(A(\xi))^2+\omega^2}\right.\right.$$

$$\times\left(-A(\xi)e^{-A(\xi)t}+A(\xi)\cos(\omega t)+\omega\sin(\omega t)+\frac{\alpha}{1-\alpha}\right) \quad (16)$$

$$\left.\left.\times\left(\omega e^{-A(\xi)t}+\omega\cos(\omega t)+A(\xi)\sin(\omega t)\right)\right\}\frac{\cos(y\xi)}{\xi^2+\gamma_m^2}d\xi\right].$$

Figures 10 and 11 show opposite behavior as compared to those of constant and ramped type shear for α and time t parameters, and Figure 12 shows the same behavior as the previous two cases for parameter h, i.e., distance between side walls.

5. Conclusions

This article is presented to obtain the exact solution in general form for the velocity of the fluid present over an infinite plate between two side walls using the definition of the Caputo–Fabrizio of fractional order differential operator, while the motion is produced due to shear stress as $\tau_0 f(t)$. The results are obtained by using Laplace and Fourier transformations for three different cases:

1. Constant shear;
2. Ramped type shear;
3. Oscillating shear.

After the above discussion we have concluded the following results.

Firstly, keeping the condition on shear the response of the fractional fluid velocity is very quick, as compared with that of ordinary fluid velocity for all the cases. Secondly, in fractional fluid a small change in time parameter shows a clear difference for the profiles in all the cases listed above. Thirdly, for the above cases there are some values of h at which the motion of the fluid is unaffected by side walls.

Author Contributions: Cconceptualization, M.A. and S.U.H.; methodology, T.A. and S.I.; software, Z.A.K.; validation, I.K., and K.S.N.; formal analysis, M.A.; investigation, S.U.H.; resources, I.K.; writing—original draft preparation, T.A.A.; and writing—review and editing.

Funding: This research received no external funding.

Acknowledgments: The authors acknowledge the anonymous referees for their useful suggestions.

Conflicts of Interest: The authors declare no conflict of interest.

References

1. Podlubny, I. *Fractional Differential Equations: An Introduction to Fractional Derivatives, Fractional Differential Equations, to Methods of Their Solution and Some of Their Applications*; Elsevier: Amsterdam, The Netherlands, 1998; Volume 198.
2. Torvik, P.J.; Bagley, R.L. On the appearance of the fractional derivative in the behavior of real materials. *J. Appl. Mech.* **1984**, *51*, 294–298. [CrossRef]
3. Caputo, M. *Elasticità e dissipazione*; Zanichelli: Bologna, Italy, 1969.
4. Suarez, L.; Shokooh, A. An eigenvector expansion method for the solution of motion containing fractional derivatives. *J. Appl. Mech.* **1997**, *64*, 629–635. [CrossRef]
5. Michalski, M.W. *Derivatives of Noninteger Order and Their Applications*; Institute of Mathematics, Polish Academy of Sciences Warsaw: Warszawa, Poland, 1993.
6. Gloeckle, W.G.; Nonnenmacher, T.F. Fractional integral operators and fox functions in the theory of viscoelasticity. *Macromolecules* **1991**, *24*, 6426–6434. [CrossRef]
7. Ray, S.S.; Bera, R. An approximate solution of a nonlinear fractional differential equation by adomian decomposition method. *Appl. Math. Comput.* **2005**, *167*, 561–571.
8. Babenko, Y. Non integer differential equation. In Proceedings of the 3rd International Conference on Intelligence in Networks, Bordeaux, France, 11–13 October 1994.
9. Gaul, L.; Klein, P.; Kemple, S. Damping description involving fractional operators. *Mech. Syst. Signal Process.* **1991**, *5*, 81–88. [CrossRef]
10. Ochmann, M.; Makarov, S. Representation of the absorption of nonlinear waves by fractional derivatives. *J. Acoust. Soc. Am.* **1993**, *94*, 3392–3399. [CrossRef]
11. Kumar, S. A new fractional modeling arising in engineering sciences and its analytical approximate solution. *Alex. Eng. J.* **2013**, *52*, 813–819. [CrossRef]
12. Mainardi, F. *Fractional Calculus and Waves in Linear Viscoelasticity: An Introduction to Mathematical Models*; World Scientific: Singapore, 2010.
13. Podlubny, I.; Chechkin, A.; Skovranek, T.; Chen, Y.; Jara, B.M.V. Matrix approach to discrete fractional calculus ii: Partial fractional differential equations. *J. Comput. Phys.* **2009**, *228*, 3137–3153. [CrossRef]

14. Caputo, M.; Fabrizio, M. A new definition of fractional derivative without singular kernel. *Progr. Fract. Differ. Appl.* **2015**, *1*, 1–13.
15. Atangana, A. On the new fractional derivative and application to nonlinear fishers reaction–diffusion equation. *Appl. Math. Comput.* **2016**, *273*, 948–956.
16. Caputo, M.; Fabrizio, M. Applications of new time and spatial fractional derivatives with exponential kernels. *Progr. Fract. Differ. Appl.* **2016**, *2*, 1–11. [CrossRef]
17. Fetecau, C.; Vieru, D.; Fetecau, C.; Akhter, S. General solutions for magnetohydrodynamic natural convection flow with radiative heat transfer and slip condition over a moving plate. *Z. Naturforsch. A* **2013**, *68*, 659–667. [CrossRef]
18. Dokuyucu, M.A.; Celik, E.; Bulut, H.; Baskonus, H.M. Cancer treatment model with the caputo-fabrizio fractional derivative. *Eur. Phys. J. Plus* **2018**, *133*, 92. [CrossRef]
19. Riaz, M.; Zafar, A. Exact solutions for the blood flow through a circular tube under the influence of a magnetic field using fractional caputo-fabrizio derivatives. *Math. Model. Nat. Phenom.* **2018**, *13*, 8. [CrossRef]
20. Shah, N.A.; Imran, M.; Miraj, F. Exact solutions of time fractional free convection flows of viscous fluid over an isothermal vertical plate with caputo and caputo-fabrizio derivatives. *J. Prime Res. Math.* **2017**, *13*, 56–74.
21. Vieru, D.; Fetecau, C.; Fetecau, C. Time-fractional free convection flow near a vertical plate with newtonian heating and mass diffusion. *Therm. Sci.* **2015**, *19* (Suppl. 1), 85–98. [CrossRef]
22. Fetecau, C.; Vieru, D.; Fetecau, C. Effect of side walls on the motion of a viscous fluid induced by an infinite plate that applies an oscillating shear stress to the fluid. *Cent. Eur. J. Phys.* **2011**, *9*, 816–824. [CrossRef]
23. Haq, S.U.; Khan, M.A.; Shah, N.A. Analysis of magneto hydrodynamic flow of a fractional viscous fluid through a porous medium. *Chin. J. Phys.* **2018**, *56*, 261–269. [CrossRef]
24. Henry, B.I.; Langlands, T.A.; Straka, P. An introduction to fractional diffusion. In *Complex Physical, Biophysical and Econophysical Systems*; World Scientific: Singapore, 2010; pp. 37–89.
25. Hristov, J. Transient heat diffusion with a non-singular fading memory: From the cattaneo constitutive equation with jeffreys kernel to the caputofabrizio time-fractional derivative. *Therm. Sci.* **2016**, *20*, 757–762. [CrossRef]
26. Hristov, J. Derivatives with non-singular kernels from the caputo–fabrizio definition and beyond: Appraising analysis with emphasis on diffusion models. *Front. Fract. Calc.* **2017**, *1*, 270–342.
27. El-Lateif, A.M.A.; Abdel-Hameid, A.M. Comment on "solutions with special functions for time fractional free convection flow of brinkman-type fluid" by F. Ali et al. *Eur. Phys. J. Plus* **2017**, *132*, 407. [CrossRef]
28. Ahmed, N.; Shah, N.A.; Vieru, D. Natural convection with damped thermal flux in a vertical circular cylinder. *Chin. J. Phys.* **2018**, *56*, 630–644. [CrossRef]
29. Zafar, A.; Fetecau, C. Flow over an infinite plate of a viscous fluid with noninteger order derivative without singular kernel. *Alex. Eng. J.* **2016**, *55*, 2789–2796. [CrossRef]
30. Yang, P.; Lam, Y.C.; Zhu, K.Q. Constitutive equation with fractional derivatives for the generalized UCM model. *J. Non-Newton. Fluid Mech.* **2010**, *165*, 88–97. [CrossRef]

© 2019 by the authors. Licensee MDPI, Basel, Switzerland. This article is an open access article distributed under the terms and conditions of the Creative Commons Attribution (CC BY) license (http://creativecommons.org/licenses/by/4.0/).

Article

Modeling and Optimization of Gaseous Thermal Slip Flow in Rectangular Microducts Using a Particle Swarm Optimization Algorithm

Nawaf N. Hamadneh [1], Waqar A. Khan [2], Ilyas Khan [3],* and Ali S. Alsagri [4]

1. Department of Basic Sciences, College of Science and Theoretical Studies, Saudi Electronic University, Riyadh 11673, Saudi Arabia; nhamadneh@seu.edu.sa
2. Department of Mechanical Engineering, College of Engineering, Prince Mohammad Bin Fahd University, Al Khobar 31952, Saudi Arabia; wkhan@pmu.edu.sa
3. Faculty of Mathematics and Statistics, Ton Duc Thang University, Ho Chi Minh City 72915, Vietnam
4. Mechanical Engineering Department, Qassim University, Buraydah 51431, Saudi Arabia; a.alsagri@qu.edu.sa
* Correspondence: ilyaskhan@tdtu.edu.vn

Received: 17 January 2019; Accepted: 2 April 2019; Published: 4 April 2019

Abstract: In this study, pressure-driven flow in the slip regime is investigated in rectangular microducts. In this regime, the Knudsen number lies between 0.001 and 0.1. The duct aspect ratio is taken as $0 \leq \varepsilon \leq 1$. Rarefaction effects are introduced through the boundary conditions. The dimensionless governing equations are solved numerically using MAPLE and MATLAB is used for artificial neural network modeling. Using a MAPLE numerical solution, the shear stress and heat transfer rate are obtained. The numerical solution can be validated for the special cases when there is no slip (continuum flow), $\varepsilon = 0$ (parallel plates) and $\varepsilon = 1$ (square microducts). An artificial neural network is used to develop separate models for the shear stress and heat transfer rate. Both physical quantities are optimized using a particle swarm optimization algorithm. Using these results, the optimum values of both physical quantities are obtained in the slip regime. It is shown that the optimal values ensue for the square microducts at the beginning of the slip regime.

Keywords: forced convection; microducts; Knudsen number; Nusselt number; artificial neural networks; particle swarm optimization

---

1. Introduction

1.1. Rarefied Gas Flows

Flows in microducts are found in microelectromechanical systems, nanotechnology applications, therapeutic and superhydrophobic microchannels, low-pressure environments, biochemical applications and cryogenics. The rarefaction effect can be found in microchannels and can be expressed in terms of the Knudsen number. In this case, the deviations from continuum behavior are smaller. Therefore, the Navier–Stokes equations can be employed with slip boundary conditions. The difference between a fully developed flow in a rectangular duct and in a microchannel is that a rectangular microduct needs 2D analysis, whereas a microchannel entail a one-dimensional analysis.

Generally, the Navier–Stokes equations can be solved with slip boundary conditions at the microduct walls [1,2]. Previous studies [1,2] found that the solutions were validated by the experimental data in several microchannel flows. Otherwise, significant departures were observed between the numerical and experimental results without applying the slip conditions. Ameel et al. [3] confirmed the statement of Reference [1,2]. Zade et al. [4] considered variable physical properties and investigated thermal features of developing flows in rectangular microchannels. They employed a collocated finite volume method and studied several channel dimensions for different values of slip parameters. They found that variable properties show substantial changes in flow behavior. Ghodoossi and Eğrican [5] determined the heat transfer rate using an integral method. They predicted thermal features for isothermal boundary conditions. They noticed that, for any aspect ratio, the rarefaction had obvious effects on the heat transfer.

The effects of rarefaction on the Nusselt numbers were investigated by Hettiarachchi et al. [6] for thermally developing flows. They noticed that these Nusselt numbers increased with the slip velocity, and decline with the thermal jump boundary condition. The slip flow forced convection in rectangular microchannels was also considered by Yu and Ameel [7]. It was found that the velocity slip and aspect ratio enhance heat transfer. For rarefied gas flows, Renksizbulut et al. [8] observed exceptionally large reductions in the friction and heat transfer coefficients. They found that these coefficients are insensible to rarefaction effects in corner-dominated flows. Tamayol and Hooman [9] analyzed microchannels of polygonal, rectangular, and rhombic cross-sections and showed that the Poiseuille numbers decrease with increasing aspect ratios and Knudsen numbers for rectangular channels. Hooman [10] investigated forced convection in microducts using a superposition approach for both thermal boundary conditions and demonstrated that their results were valid for several cross-sections. Sadeghi and Saidi [11] considered the rarefaction effects in two different microgeometries and demonstrated that for both geometries, the Brinkman number reduced the heat transfer rate. In fact, the Brinkman number measures the viscous heating effects with reference to the heat conduction. They also developed a correlation for the Nusselt number.

Yovanovich and Khan [12] developed several slip flow models for microchannels of different cross-sections. The slip flow in circular and other cross-sections was investigated by a number of authors, including References [13–16]. They developed different models for envisaging the friction factor under both developing and fully developed conditions. Duan and Muzychka [13,14] proposed a simple model to predict the Poiseuille numbers in these microchannels. Duan and Muzychka [15] considered slip flow in the entrance of circular and parallel plate microchannels and noticed that the Poisseuille number depends on the Knudsen number. Yovanovich and Khan [16] modeled the Poiseuille flow in long channels of different cross-sections. Ebert and Sparrow [17] analyzed abstemiously rarefied gas flows in different ducts and noticed that the velocity flattened and the friction lessened due to the slip parameter.

Baghani et al. [18] treated rarefaction effects in microducts of different cross-sections and obtained Nusselt numbers under isothermal boundary condition. They tabulated Nusselt numbers for the entire slip flow range of the Knudsen number. Wang [19] considered the slip flow under an isothermal boundary condition in different ducts. It was found that both hydrodynamic and thermal boundary conditions have substantial effects on both the Poiseuille and Nusselt numbers. Also, both parameters showed an opposite trend with an increasing velocity slip. Beskok and Karniadakis [20] considered internal rarefied gas flows and presented mathematical models for different regimes. To account for heat transfer from the surface, they suggested a new boundary condition and demonstrated that it is applicable in all flow regimes. Yovanovich and Khan [21] proposed compact models for isothermal, incompressible slip flows in different microchannels. They used the principle of superposition in the analysis and introduced new flow parameters. They compared their results with previous numerical results for different cross-sections and found them in a very good agreement. Klášterka et al. [22] obtained analytical and numerical solutions for the flow variables and demonstrated the effect of slip parameters on the friction and heat transfer coefficients.

The above literature survey reveals that there are only a few studies related to rectangular microducts. However, there has been no study where the numerical models are developed using an artificial neural network (ANN) and are optimized for the duct geometry. This fact inspires us to model our numerical results using ANN and to acquire optimized values of friction factors and Nusselt numbers using a particle swarm optimization algorithm.

1.2. Artificial Neural Networks

Artificial neural networks (ANNs) are mathematical models, which are inspired by biological nervous systems [23,24]. The multilayer perceptron neural network (MLPNN) is one of the widely known feedforward neural networks. MLPNNs have three types of layers, which are the input layer, the hidden layers, and the output layer. Figure 1 shows the structure of MLPNN [25,26].

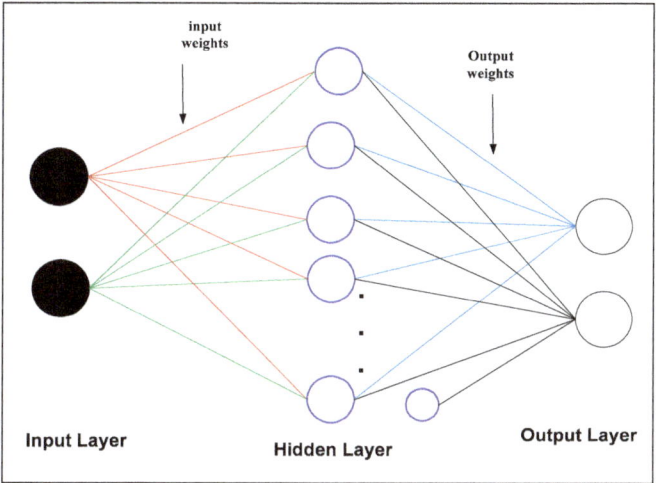

Figure 1. Structure of a multilayer perceptron neural network (MLPNN).

The activation function is defined as an output of a node or a neuron and can be written as:

$$Y = \frac{1}{1 + e^{-x}} \qquad (1)$$

where x is an input value from the input layer. Supervised learning of ANNs is useful for improving the performance of ANN models. Several algorithms are used for the training to find the best parameters of the neural networks [27,28]. The optimization algorithms are used to find the best values for inputs and output weights that are parameters of MLPNNs. In this study, the particle swarm optimization (PSO) algorithm is used to train the neural networks, because it is one of the most effective meta-heuristic algorithms used for optimization problems. Equation (2) is the sum of squared error (SSE) function, which was used as a fitness function for testing the performance of the ANNs [25].

$$SEE = \sum (Actual\ output - target\ output)^2 \qquad (2)$$

where the *target output* values are determined by Equation (3), while the *Actual output* values are determined by the neural network (NN) output values.

$$Output(y_k) = \sum_{j=1}^{m} \sum_{i=1}^{n} w_{jk} \sum_{k=1}^{s} w'_{ki} \frac{1}{1 + e^{-x_j}} \qquad (3)$$

where n, m, k, w_{jk}, and w'_{ki} represent the number of neurons in the hidden layer, number of input data, number of neurons in input layer, number of neurons in output layer, input weight—which is between the input neurons and hidden neurons—and output weight, which is between the hidden neurons and output neurons, respectively.

1.3. Particle Swarm Optimization Algorithm

The principle of the particle swarm optimization (PSO) algorithm came from the ideas of the social behavior of bird flocks, as well as from the shoaling behavior of fish [27–29]. All solution members of the algorithm are called particles. The particle flies by updating its velocity vectors v and position x, by using Equations (4) and (5), respectively [30].

$$v_{ik}(t+1) = v_{ik}(t) + c_1 * r_{1k}(t) * (y_{ik}(t) - x_{ik}(t)) + c_2 * r_{2k}(t) * (\hat{y}_k(t) - x_{ik}(t)) \tag{4}$$

$$x_i(t+1) = x_i(t) + v_i(t+1) \tag{5}$$

where $x_i(t)$, $v_i(t)$, $y_i(t)$, and $\hat{y}_k(t)$ represent the position at time t, the velocity at time t, the best personal position (p^{best}) at time t, and the global best position (g^{best}) of population at time t respectively, whereas c_1 and c_2 are the learning factors of (p^{best}) in interval between 0 and 2 and r_1 and r_2 are the random numbers in the interval 0 to 1.

It is important to note that the other particles follow the performance of the best particle [31–34]. In addition, each particle keeps the best position that it has achieved so far. The flowchart of Figure 2 shows the steps of the present PSO algorithm [27,32,35].

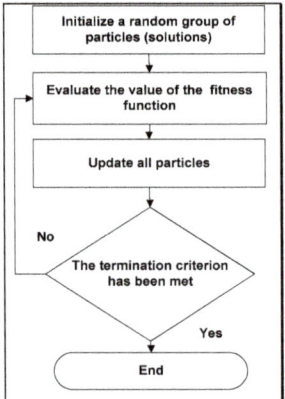

Figure 2. Structure of the particle swarm optimization (PSO) algorithm.

Thus, the particles should move towards the best directions, and utilize useful information from other particles along with the best particles [30,35].

In this study, the algorithm was used to find the optimal values of the Nusselt number (Nu) and Poiseuille number (Po) corresponding to the aspect ratio and the Knudsen number in the slip flow regime. The number of the iterations was set as 50, with initial populations 60, and $c_1 = c_2 = 0.05$ [30].

To the best of our knowledge, most of the existing analytical solutions for slip flows are limited to fluid flow in rectangular ducts. There is a lack of literature related to heat transfer in microducts. Furthermore, there is no study related to ANN modeling and optimization of pressure drop and heat transfer in microducts. The aim of this paper was to cover these aspects.

2. Mathematical Model

2.1. Hydrodynamic Analysis

Let us consider a fully developed flow in a rectangular duct, as shown in Figure 3. The semi-major and -minor axes are taken as a and b along the x- and y-axes, respectively. The aspect ratio is $\varepsilon = b/a$.

The gas is assumed to flow along the z-axis. The momentum equation for the flow in a microduct can be written as:

$$\frac{\partial^2 u}{\partial x^2} + \frac{\partial^2 u}{\partial y^2} = -\frac{1}{\mu}\frac{dP}{dz} \qquad (6)$$

which is a 2D equation for the axial velocity u, uniform gas viscosity μ and constant pressure gradient dP/dz. Due to double symmetry, only the OABCO is selected for the simulation. The velocity slip and symmetry boundary conditions for this problem can be written as:

$$u = -\lambda \frac{2-\sigma_v}{\sigma_v} \frac{\partial u}{\partial y}\bigg|_{y=b}, \quad u = -\lambda \frac{2-\sigma_v}{\sigma_v} \frac{\partial u}{\partial x}\bigg|_{x=a} \qquad (7)$$

and:

$$\frac{\partial u}{\partial y}\bigg|_{y=0} = 0; \quad \frac{\partial u}{\partial x}\bigg|_{x=0} = 0 \qquad (8)$$

where σ_v is the tangential–momentum accommodation coefficient, which lies between 0.9 and 1, and λ is the mean free path. If A is the area of cross-section and P is the perimeter of the duct, then the hydraulic diameter, D_h can be written as:

$$D_h = 4A/P = 4b/(1+\varepsilon) \qquad (9)$$

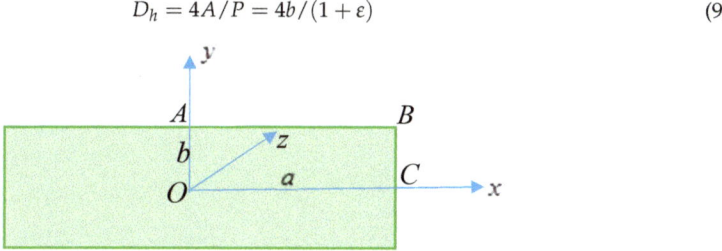

Figure 3. Schematic diagram of rectangular microduct.

At this stage, it is convenient to define the Knudsen number (the measure of rarefaction effects) as follows:

$$Kn = \lambda/D_h \Rightarrow \lambda = 4bKn/1+\varepsilon \qquad (10)$$

Following Ebert and Sparrow [17], Equations (6)–(8) can be non-dimensionalized as follows. Let:

$$\xi = x/a, \eta = y/b \qquad (11)$$

The dimensionless momentum equation can be written as:

$$\varepsilon^2 \frac{\partial^2 u}{\partial \xi^2} + \frac{\partial^2 u}{\partial \eta^2} = -1 \qquad (12)$$

where the axial velocity u is normalized with $u_o = \left(\frac{b^2}{\mu}\right)\left|\frac{dP}{dz}\right|$. Using Equation (11), dimensionless boundary conditions can be written as:

$$\begin{array}{l} u = -\frac{4Kn}{1+\varepsilon}\frac{2-\sigma_v}{\sigma_v}\frac{\partial u}{\partial \eta}\bigg|_{\eta=1}; 0 \leq \xi < 1 \\ u = -\frac{4\varepsilon Kn}{1+\varepsilon}\frac{2-\sigma_v}{\sigma_v}\frac{\partial u}{\partial \xi}\bigg|_{\xi=1}; 0 \leq \eta < 1 \\ \frac{\partial u}{\partial \eta}\bigg|_{\eta=0} = 0; 0 \leq \xi < 1; \frac{\partial u}{\partial x}\bigg|_{\xi=0} = 0; 0 \leq \eta < 1 \end{array} \qquad (13)$$

The mean velocity in the duct is defined as:

$$\bar{u} = u_0 \int_0^1 \int_0^1 u(\xi, \eta) \cdot d\xi \cdot d\eta \qquad (14)$$

The Poiseuille number Po is defined as:

$$Po = c_f \cdot Re_{D_h} \qquad (15)$$

where $c_f = \bar{\tau}_w / 0.5\rho\bar{u}^2$ is the coefficient of friction and $Re_{D_h} = \bar{u}D_h/\nu$ is the Reynolds number, based on mean velocity. For a fully developed flow, the force balance on a control volume of width dz gives:

$$-A \cdot dp = \bar{\tau}_w P dz \Rightarrow \bar{\tau}_w = (dp/dz)(D_h/4) \qquad (16)$$

Thus, the dimensionless shear stress can be expressed in terms of the Poiseuille number Po:

$$Po = -\frac{32}{(1+\varepsilon)^2} \cdot \frac{1}{(\bar{u}/u_0)} \qquad (17)$$

where $\bar{u}/u_0$ is the dimensionless average axial velocity in the microduct and can be determined from Equation (14).

2.2. Thermal Analysis

For fully thermally developed and steady flow in the rectangular microducts, the energy equation is given by:

$$\frac{k_f}{\rho c_p}\left(\frac{\partial^2 T}{\partial x^2} + \frac{\partial^2 T}{\partial y^2}\right) = u\frac{\partial T}{\partial z} \qquad (18)$$

with temperature jump and symmetry boundary conditions:

$$\begin{aligned} T - T_w &= -\frac{\lambda}{Pr}\frac{2-\sigma_t}{\sigma_t}\frac{2\gamma}{1+\gamma}\frac{\partial T}{\partial y}\bigg|_{y=b} \\ T - T_w &= -\frac{\lambda}{Pr}\frac{2-\sigma_t}{\sigma_t}\frac{2\gamma}{1+\gamma}\frac{\partial T}{\partial x}\bigg|_{x=a} \end{aligned} \qquad (19)$$

and:

$$\frac{\partial T}{\partial y}\bigg|_{y=0} = 0 \quad \frac{\partial T}{\partial x}\bigg|_{x=0} = 0 \qquad (20)$$

where Pr is the Prandtl number for the selected gas, σ_t is the energy accommodation coefficient, and γ is the ratio of specific heats of gas.

Using Equation (11), the energy balance in the flow direction can be written as:

$$\frac{k_f}{\rho c_p}\left(\varepsilon^2\frac{\partial^2 T}{\partial \xi^2} + \frac{\partial^2 T}{\partial \eta^2}\right) = u\frac{b^2 Q}{4ab\rho c_p \bar{u}} \qquad (21)$$

or:

$$\varepsilon^2\frac{\partial^2 T}{\partial \xi^2} + \frac{\partial^2 T}{\partial \eta^2} = \frac{u}{\bar{u}}\frac{\varepsilon}{4}\frac{Q}{k_f} \qquad (22)$$

where Q is the total energy per unit length of the duct and is assumed to be constant. Using $\theta = (T - Tw)/(Q/k_f)$, Equation (22) can be written as:

$$\varepsilon^2\frac{\partial^2 \theta}{\partial \xi^2} + \frac{\partial^2 \theta}{\partial \eta^2} = \frac{u}{\bar{u}}\frac{\varepsilon}{4} \qquad (23)$$

with transformed boundary conditions [17]:

$$\left. \begin{array}{l} \theta = -\frac{4}{1+\varepsilon} \frac{Kn}{Pr} \frac{2-\sigma_t}{\sigma_t} \frac{2\gamma}{1+\gamma} \frac{\partial \theta}{\partial \eta} \Big|_{\eta=1}; 0 \leq \xi < 1 \\ \theta = -\frac{4\varepsilon}{1+\varepsilon} \frac{Kn}{Pr} \frac{2-\sigma_t}{\sigma_t} \frac{2\gamma}{1+\gamma} \frac{\partial \theta}{\partial \xi} \Big|_{\xi=1}; 0 \leq \eta < 1 \\ \frac{\partial \theta}{\partial \eta} \Big|_{\eta=0} = 0; 0 \leq \xi < 1; \frac{\partial \theta}{\partial \xi} \Big|_{\xi=0} = 0; 0 \leq \eta < 1 \end{array} \right\} \quad (24)$$

For a constant temperature jump condition, the Nusselt number can be defined as:

$$Nu_T = \frac{\partial T/\partial y|_{\text{wall}}}{T - T_{\text{wall}}} \quad (25)$$

2.3. Solution Procedure

The solution procedure comprises three steps. In the first step, the numerical values of the Nu and Po numbers are determined, corresponding to different values of ε and Kn by solving the governing Equations (12) and (23) by a finite difference method using MAPLE, and numerical data was obtained. In the second step, the back-error-propagation algorithm is employed to train a multi-layer perceptron neural network (MLPNN) using the neural network toolbox in MATLAB. There are two inputs, namely, the aspect ratio ε and the Knudsen number Kn. The corresponding outputs are the predicted values of the Nu and Po numbers, where each output is represented by a separate model. In each model, different input, hidden and output neurons are used. Finally, the results are optimized using PSO algorithm in order to find the optimal values of Nu and Po without gaining to the numerical values. Accordingly, the dimensionless partial differential equations will not use again in any stage of our work. As a result, the PSO algorithm uses neural network models to obtain the optimum values of Nu and Po.

3. Results and Discussion

The numerical values of Poiseuille numbers for forced convection in microducts were compared with the available data for different values of Kn in Tables 1–3. The comparison showed excellent agreement with the available data. The predicted values of the Nusselt numbers are also provided in Table 4 for different values of ε and Kn. To build the NN models, 70% of data was used as the training data, 15% as the testing data, and 15% as the validation data. In NN MATLAB tools, the back-error-propagation algorithm (an optimization algorithm) was used to determine the best models that have the optimal errors. These models were based on the mean squared error function of MSE = 0.00020768 and MSE = 0.00022112 for the two models, respectively, as shown in Figures 4 and 5. The figures show that the best validation performance, in terms of the mean squared error function, were found after 187 and 452 iterations, respectively.

Table 1. Comparison of the Poiseuille numbers for forced convection when $Kn = 0.1$.

ε	Morini et al. [36]	Sadeghi et al. [37]	Present Results
0.2	9.46	9.464	9.519
0.4	8.65	8.654	8.721
0.6	8.25	8.248	8.331
0.8	8.08	8.076	8.178
1	8.04	8.033	8.159

Table 2. Comparison of numerical and exact values of Po when $Kn = 0.001$.

ε	Numerical	Exact [38]	% Difference
0.001	24.031	23.7	1.3774
0.01	23.743	23.41	1.4025
0.02	23.432	22.24	5.0871
0.1	21.259	20.95	1.4535
0.2	19.182	18.89	1.5223
0.3	17.641	17.36	1.5929
0.4	16.514	16.24	1.6592
0.5	15.712	15.43	1.7948
0.6	15.164	14.87	1.9388
0.7	14.811	14.5	2.0998
0.8	14.608	14.28	2.2453
0.9	14.519	14.17	2.4037
1	14.514	14.14	2.5768

Table 3. Comparison of numerical and exact values of Po when $Kn = 0.1$.

ε	Numerical	Exact [38]	% Difference
0.001	10.944	10.9	0.402
0.01	10.862	10.82	0.3867
0.02	10.773	10.47	2.8126
0.1	10.139	10.09	0.4833
0.2	9.519	9.46	0.6198
0.3	9.056	9	0.6184
0.4	8.721	8.66	0.6995
0.5	8.487	8.41	0.9073
0.6	8.331	8.25	0.9723
0.7	8.233	8.14	1.1296
0.8	8.178	8.08	1.1983
0.9	8.157	8.04	1.4344
1	8.159	8.04	1.4585

Table 4. Numerical values of Nu for different values of Kn.

ε	$Kn = 0.001$	$Kn = 0.05$	$Kn = 0.1$
0.001	6.1658	1.4213	0.7164
0.25	5.1047	1.7625	1.0819
0.35	4.6786	1.8996	1.2287
0.5	4.0394	2.1051	1.4488
0.65	3.8048	2.1830	1.5590
0.75	3.6485	2.2348	1.6325
0.85	3.6156	2.2745	1.6816
1	3.5662	2.3341	1.7552

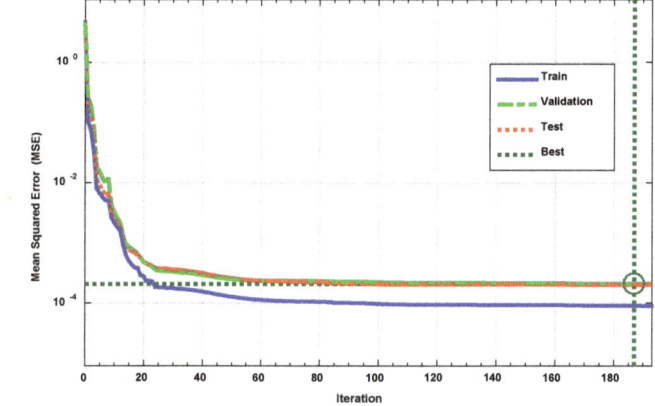

Figure 4. Best Validation Performance is 0.00020768 at iteration 187 of using MATLAB to generate the NN model of Nu.

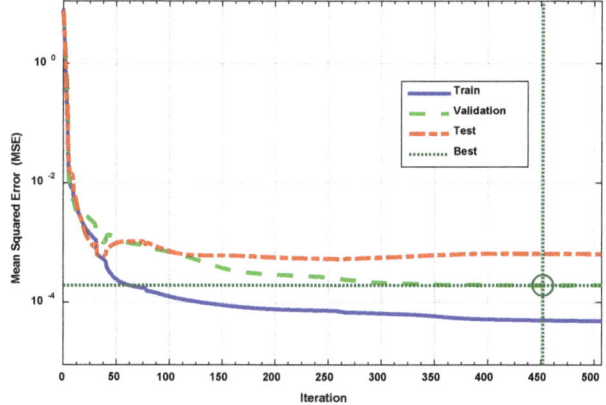

Figure 5. Best Validation Performance is 0.00022112 at iteration 452 using MATLAB to generate the neural network (NN) model of *Po*.

The correlation coefficients of the models are shown in Figures 6 and 7. As expected, the correlation coefficient of the models was greater than 0.95. In addition, Figures 6 and 7 show a significant convergence between the real values and the corresponding values that resulted from models. Accordingly, there was enough support to believe that the models were suitable to find the estimated values of the *Nu* and *Po* numbers. In addition, the coefficient of determination of the models was more than 0.99, which meant that both models could represent whole target data.

For optimization, the PSO algorithm was used to find the optimal values of *Po* and *Nu*, based on NN models. The optimal values of *Po* and *Nu* were achieved when $\varepsilon = 1$, and $Kn = 0.001$. The behavior of *Po* and *Nu* values in the PSO algorithm is shown in Figures 8 and 9. The optimal values of the curves were determined when the curves became stable. Accordingly, the optimal values for *Nu* and *Po* were found in iteration number 4 and 10, respectively. Note that, the data set values were converted into z-score values for use in the neural network toolbox in MATLAB.

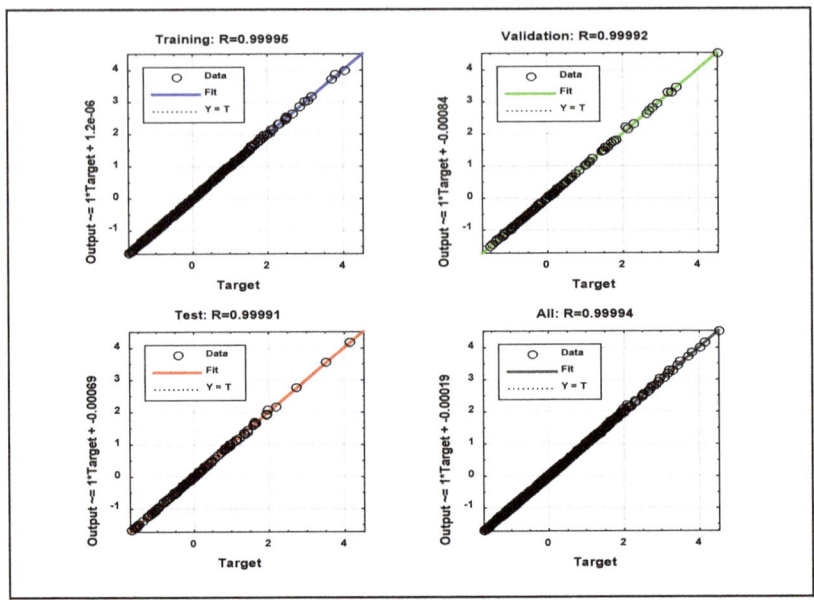

Figure 6. Correlation coefficient and the regression of the NN model values of Nu and the desired values.

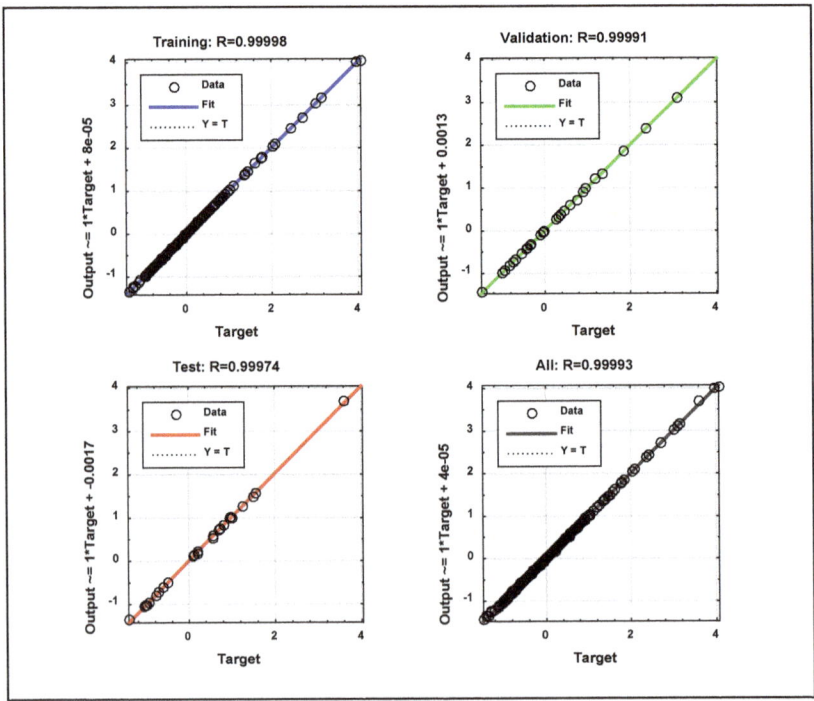

Figure 7. Correlation coefficient and the regression of the NN model values of Po and the desired values.

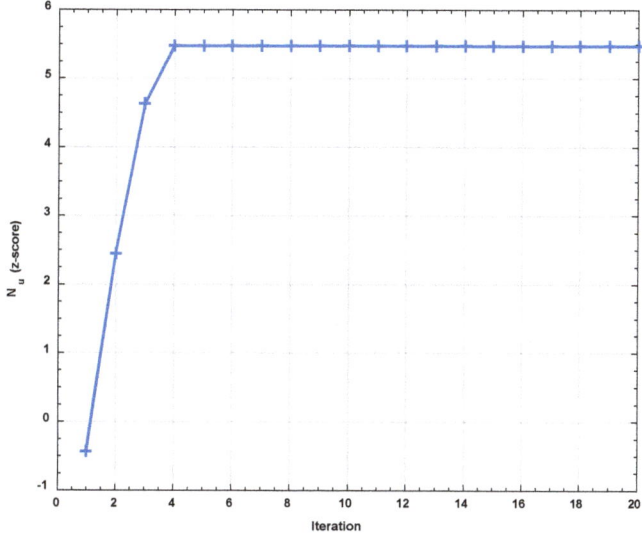

Figure 8. Best performance of the PSO algorithm in terms of Nu.

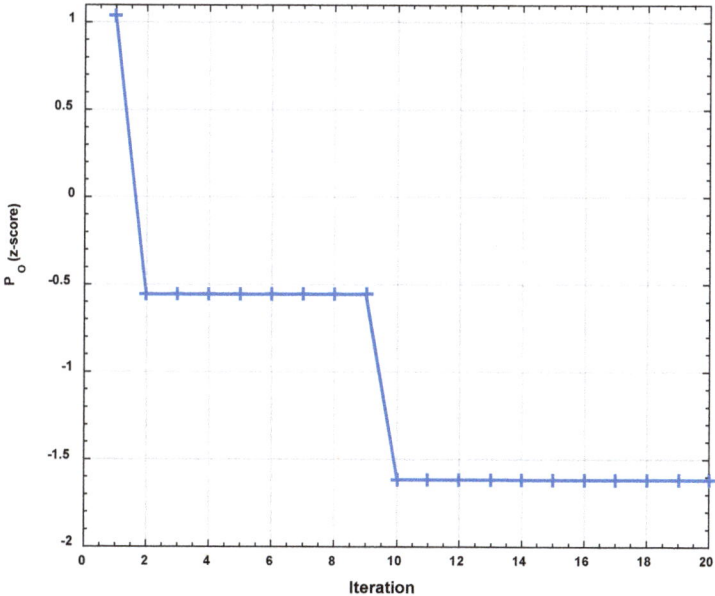

Figure 9. Best performance of the PSO algorithm in terms of Po.

4. Conclusions

In this study, the slip flow and heat transfer were investigated in rectangular microducts. The dimensionless governing partial differential equations with slip boundary conditions were solved numerically using MATLAB. The numerical data were used successfully for building ANN models. The outputs of the models were the corresponding Poiseuille Po and Nusselt Nu numbers to the input values ε and Kn. The main outcomes of the study were:

1. The small tolerance values for the mean squared error function values and a very close correlation coefficient of 1 in the NN models proved that the NN models were suitable to use for predicting the Poiseuille Po and Nusselt Nu numbers.
2. Without going back to the numerical data, the proposed models were used with the PSO algorithm to find the optimal values of Po and Nu.
3. The optimal values of Po and Nu are found when $\varepsilon = 1$ and $Kn = 0.001$.
4. The correlation coefficient of the models was greater than 0.95.
5. The optimal values of the curves were determined when the curves became stable.

Author Contributions: Optimization work: N.N.H.; Results and discussion: N.N.H. and I.K.; Introduction and literature review: A.S.A. and I.K.; Abstract, Conclusios and final revision: W.A.K. and A.S.A.

Funding: This research received no external funding.

Acknowledgments: The first author is grateful to the Deanship of Scientific Research at Saudi Electronic University for providing great assistance in completing this research.

Conflicts of Interest: The authors do not have any conflict of interest.

References

1. Liu, J.; Tai, Y.-C.; Pong, C.-M.H.-C. MEMS for pressure distribution studies of gaseous flows in microchannels. In Proceedings of the IEEE Micro Electro Mechanical Systems, Amsterdam, The Netherlands, 29 January–2 February 1995; p. 209.
2. Arkilic, E.B. Gaseous Flow in Micro-channels, Application of Microfabrication to Fluid Mechanics. *AsmeFed* **1994**, *197*, 57–66.
3. Ameel, T.A.; Wang, X.; Barron, R.F.; Warrington, R.O. Laminar forced convection in a circular tube with constant heat flux and slip flow. *Microscale Thermophys. Eng.* **1997**, *1*, 303–320.
4. Zade, A.Q.; Renksizbulut, M.; Friedman, J. Heat transfer characteristics of developing gaseous slip-flow in rectangular microchannels with variable physical properties. *Int. J. Heat Fluid Flow* **2011**, *32*, 117–127. [CrossRef]
5. Ghodoossi, L. Prediction of heat transfer characteristics in rectangular microchannels for slip flow regime and H1 boundary condition. *Int. J. Therm. Sci.* **2005**, *44*, 513–520. [CrossRef]
6. Hettiarachchi, H.M.; Golubovic, M.; Worek, W.M.; Minkowycz, W. Three-dimensional laminar slip-flow and heat transfer in a rectangular microchannel with constant wall temperature. *Int. J. Heat Mass Transf.* **2008**, *51*, 5088–5096. [CrossRef]
7. Yu, S.; Ameel, T.A. Slip-flow heat transfer in rectangular microchannels. *Int. J. Heat Mass Transf.* **2001**, *44*, 4225–4234. [CrossRef]
8. Renksizbulut, M.; Niazmand, H.; Tercan, G. Slip-flow and heat transfer in rectangular microchannels with constant wall temperature. *Int. J. Therm. Sci.* **2006**, *45*, 870–881. [CrossRef]
9. Tamayol, A.; Hooman, K. Slip-flow in microchannels of non-circular cross sections. *J. Fluids Eng.* **2011**, *133*, 091202. [CrossRef]
10. Hooman, K. A superposition approach to study slip-flow forced convection in straight microchannels of uniform but arbitrary cross-section. *Int. J. Heat Mass Transf.* **2008**, *51*, 3753–3762. [CrossRef]
11. Sadeghi, A.; Saidi, M.H. Viscous dissipation and rarefaction effects on laminar forced convection in microchannels. *J. Heat Transf.* **2010**, *132*, 072401. [CrossRef]
12. Yovanovich, M.; Khan, W.A. Compact Slip Flow Models for Gas Flows in Rectangular, Trapezoidal and Hexagonal Microchannels. In Proceedings of the ASME 2015 International Technical Conference and Exhibition on Packaging and Integration of Electronic and Photonic Microsystems, San Francisco, CA, USA, 6–9 July 2015.
13. Duan, Z.; Muzychka, Y. Slip flow in elliptic microchannels. *Int. J. Therm. Sci.* **2007**, *46*, 1104–1111. [CrossRef]
14. Duan, Z.; Muzychka, Y. Slip flow in non-circular microchannels. *Microfluid. Nanofluid.* **2007**, *3*, 473–484. [CrossRef]
15. Duan, Z.; Muzychka, Y. Slip flow in the hydrodynamic entrance region of circular and noncircular microchannels. *J. Fluids Eng.* **2010**, *132*, 011201. [CrossRef]

16. Yovanovich, M.M.; Khan, W.A. Friction and Heat Transfer in Liquid and Gas Flows in Micro-and Nanochannels. *Adv. Heat Transf.* **2015**, *47*, 203–307.
17. Ebert, W.; Sparrow, E.M. Slip flow in rectangular and annular ducts. *J. Basic Eng.* **1965**, *87*, 1018–1024. [CrossRef]
18. Baghani, M.; Sadeghi, A.; Baghani, M. Gaseous slip flow forced convection in microducts of arbitrary but constant cross section. *Nanoscale Microscale Thermophys. Eng.* **2014**, *18*, 354–372. [CrossRef]
19. Wang, C. Benchmark solutions for slip flow and H1 heat transfer in rectangular and equilateral triangular ducts. *J. Heat Transf.* **2013**, *135*, 021703. [CrossRef]
20. Beskok, A.; Karniadakis, G.E. Report: A model for flows in channels, pipes, and ducts at micro and nano scales. *Microscale Thermophys. Eng.* **1999**, *3*, 43–77.
21. Yovanovich, M.; Khan, W. Similarities of rarefied gas flows in elliptical and rectangular microducts. *Int. J. Heat Mass Transf.* **2016**, *93*, 629–636. [CrossRef]
22. Klášterka, H.; Vimmr, J.; Hajžman, M. Contribution to the gas flow and heat transfer modelling in microchannels. *Appl. Comput. Mech.* **2009**, *12*, 63–74.
23. Hamadneh, N.; Tilahun, S.; Sathasivam, S.; Choon, O.H. Prey-predator algorithm as a new optimization technique using in radial basis function neural networks. *Res. J. Appl. Sci.* **2013**, *8*, 383–387.
24. Rakhshandehroo, G.R.; Vaghefi, M.; Aghbolaghi, M.A. Forecasting groundwater level in Shiraz plain using artificial neural networks. *Arab. J. Sci. Eng.* **2012**, *37*, 1871–1883. [CrossRef]
25. Yeung, D.S.; Li, J.-C.; Ng, W.W.; Chan, P.P. MLPNN training via a multiobjective optimization of training error and stochastic sensitivity. *IEEE Trans. Neural Netw. Learn. Syst.* **2016**, *27*, 978–992. [CrossRef] [PubMed]
26. Mefoued, S. Assistance of knee movements using an actuated orthosis through subject's intention based on MLPNN approximators. In Proceedings of the 2013 International Joint Conference on Neural Networks, Dallas, TX, USA, 4–9 August 2013; pp. 1–6.
27. Shi, Y.; Eberhart, R.C. Empirical study of particle swarm optimization. In Proceedings of the 1999 Congress on Evolutionary Computation, Washington, DC, USA, 6–9 July 1999; pp. 1945–1950.
28. Trelea, I.C. The particle swarm optimization algorithm: Convergence analysis and parameter selection. *Inf. Process. Lett.* **2003**, *85*, 317–325. [CrossRef]
29. Liu, X. Radial basis function neural network based on PSO with mutation operation to solve function approximation problem. In Proceedings of the International Conference in Swarm Intelligence, Brussels, Belgium, 8–10 September 2010; pp. 92–99.
30. Eberhart, R.; Kennedy, J. A new optimizer using particle swarm theory. In Proceedings of the Sixth International Symposium on Micro Machine and Human Science, Nagoya, Japan, 4–6 October 1995; pp. 39–43.
31. Shi, Y. Particle swarm optimization: Developments, applications and resources. In Proceedings of the 2001 Congress on Evolutionary Computation, Seoul, Korea, 27–30 May 2001; pp. 81–86.
32. Marinke, R.; Araujo, E.; Coelho, L.S.; Matiko, I. Particle swarm optimization (PSO) applied to fuzzy modeling in a thermal-vacuum system. In Proceedings of the Fifth International Conference on Hybrid Intelligent Systems, Rio de Janeiro, Brazil, 6–9 November 2005; p. 6.
33. Nenortaite, J.; Butleris, R. Application of particle swarm optimization algorithm to decision making model incorporating cluster analysis. In Proceedings of the 2008 Conference on Human System Interactions, Krakow, Poland, 25–27 May 2008; pp. 88–93.
34. Hamadneh, N.; Khan, W.A.; Sathasivam, S.; Ong, H.C. Design optimization of pin fin geometry using particle swarm optimization algorithm. *PLoS ONE* **2013**, *8*, e66080. [CrossRef] [PubMed]
35. Bai, Q. Analysis of particle swarm optimization algorithm. *Comput. Inf. Sci.* **2010**, *3*, 180. [CrossRef]
36. Morini, G.L.; Spiga, M.; Tartarini, P. The rarefaction effect on the friction factor of gas flow in microchannels. *Superlattices Microstruct.* **2004**, *35*, 587–599. [CrossRef]
37. Sadeghi, M.; Sadeghi, A.; Saidi, M.H. Gaseous slip flow mixed convection in vertical microducts of constant but arbitrary geometry. *J. Thermophys. Heat Transf.* **2014**, *28*, 771–784. [CrossRef]
38. Shah, R.; London, A. *Laminar Flow Forced Convection in Ducts, Advances in Heat Transfer*; Academic Press: New York, NY, USA, 1978.

 © 2019 by the authors. Licensee MDPI, Basel, Switzerland. This article is an open access article distributed under the terms and conditions of the Creative Commons Attribution (CC BY) license (http://creativecommons.org/licenses/by/4.0/).

Article

MHD Slip Flow of Casson Fluid along a Nonlinear Permeable Stretching Cylinder Saturated in a Porous Medium with Chemical Reaction, Viscous Dissipation, and Heat Generation/Absorption

Imran Ullah [1], Tawfeeq Abdullah Alkanhal [2], Sharidan Shafie [3], Kottakkaran Sooppy Nisar [4], Ilyas Khan [5,*] and Oluwole Daniel Makinde [6]

1. College of Civil Engineering, National University of Sciences and Technology Islamabad, Islamabad 44000, Pakistan; ullahimran14@gmail.com
2. Department of Mechatronics and System Engineering, College of Engineering, Majmaah University, Majmaah 11952, Saudi Arabia; t.alkanhal@mu.edu.sa
3. Department of Mathematical Sciences, Faculty of Science, Universiti Teknologi Malaysia, UTM Johor Bahru 81310, Johor, Malaysia; sharidan@utm.my
4. Department of Mathematics, College of Arts and Science at Wadi Al-Dawaser, Prince Sattam bin Abdulaziz University, Al Kharj 11991, Saudi Arabia; n.sooppy@psau.edu.sa
5. Faculty of Mathematics and Statistics, Ton Duc Thang University, Ho Chi Minh City 72915, Vietnam
6. Faculty of Military Science, Stellenbosch University, Private Bag X2, Saldanha 7395, South Africa; makinded@gmail.com
* Correspondence: ilyaskhan@tdt.edu.vn

Received: 13 February 2019; Accepted: 8 March 2019; Published: 12 April 2019

Abstract: The aim of the present analysis is to provide local similarity solutions of Casson fluid over a non-isothermal cylinder subject to suction/blowing. The cylinder is placed inside a porous medium and stretched in a nonlinear way. Further, the impact of chemical reaction, viscous dissipation, and heat generation/absorption on flow fields is also investigated. Similarity transformations are employed to convert the nonlinear governing equations to nonlinear ordinary differential equations, and then solved via the Keller box method. Findings demonstrate that the magnitude of the friction factor and mass transfer rate are suppressed with increment in Casson parameter, whereas heat transfer rate is found to be intensified. Increase in the curvature parameter enhanced the flow field distributions. The magnitude of wall shear stress is noticed to be higher with an increase in porosity and suction/blowing parameters.

Keywords: Casson fluid; chemical reaction; cylinder; heat generation; magnetohydrodynamic (MHD); slip

1. Introduction

Boundary layer flow on linear or nonlinear stretching surfaces has a wide range of engineering and industrial applications, and has been used in many manufacturing processes, such as extrusion of plastic sheets, glass fiber production, crystal growing, hot rolling, wire drawing, metal and polymer extrusion, and metal spinning. The viscous flow past a stretching surface was first developed by Crane [1]. Later on, this pioneering work was extended by Gupta and Gupta [2] and Chen and Char [3], and the suction/blowing effects on heat transfer flow over a stretching surface were investigated. Gorla and Sidawi [4] analyzed three-dimensional free convection flow over permeable stretching surfaces. Motivated by this, the two-dimensional heat transfer flow of viscous fluid due to a nonlinear stretching sheet was investigated by Vajravelu [5]. The similarity solutions for viscous flow over a nonlinear stretching sheet was obtained by Vajravelu and Cannon [6]. On the other hand, Bachok and Ishak [7] studied the prescribed surface heat flux characteristics on boundary layer flow generated by a stretching cylinder. Hayat et al. [8] analyzed the heat and mass transfer features on two-dimensional flow due to a stretching cylinder placed through a porous media in the presence of convective boundary conditions. The heat transfer analysis in ferromagnetic viscoelastic fluid flow over a stretching sheet was discussed by Majeed et al. [9].

The study of magnetohydrodynamic (MHD) boundary layer flow towards stretching surface has gained considerable attention due to its important practical and engineering applications, such as MHD power generators, cooling or drying of papers, geothermal energy extraction, solar power technology, cooling of nuclear reactors, and boundary layer flow control in aerodynamics. Vyas and Ranjan [10] investigated two-dimensional flow over a nonlinear stretching sheet in the presence of thermal radiation and viscous dissipation. They predicted that stronger radiation boosts the fluid temperature field. The effect of magnetic field on incompressible viscous flow generated due to stretching cylinder was analyzed by Mukhopadhyay [11], and it was observed that a larger curvature parameter allowed more fluid to flow. Fathizadeh et al. [12] studied the MHD effect on viscous fluid due to a sheet stretched in a nonlinear way. Akbar et al. [13] developed laminar boundary layer flow induced by a stretching surface in the presence of a magnetic field. They noticed that the intensity of the magnetic field offered resistance to the fluid flow, because of which, skin friction was enhanced. In another study, Ellahi [14] demonstrated the effects of magnetic field on non-Newtonian nanofluid through a pipe.

The momentum slip at a stretching surface plays an important role in the manufacturing processes of several products, including emulsion, foams, suspensions, and polymer solutions. In recent years, researchers have avoided no-slip conditions and take velocity slip at the wall. The reason is that it has been proven through experiments that momentum slip at the boundary can enhance the heat transfer. Fang et al. [15] obtained the exact solution for two-dimensional slip flow due to stretching surface. The slip effects on stagnation point flow past a stretching sheet were numerically analyzed by Bhattacharyya et al. [16]. The slip effect on viscous flow generated due to a nonlinear stretching surface in the presence of first order chemical reaction and magnetic field was developed by Yazdi et al. [17]. They concluded that velocity slip at the wall reduced the friction factor. Hayat et al. [18] investigated the impact of hydrodynamic slip on incompressible viscous flow over a porous stretching surface under the influence of a magnetic field and thermal radiation. They predicted that suction and slip parameters have the same effect on fluid velocity. Seini and Makinde [19] analyzed the hydromagnetic boundary layer flow of a viscous fluid under the influence of velocity slip at the wall. They noticed that wall shear stress enhanced with the growth of the magnetic parameter. Motivated by this, Rahman et al. [20] discussed the slip mechanisms in boundary layer flow of Jeffery nanofluid through an artery, and the solutions were achieved by the homotopy perturbation method.

In the recent years, the analysis of non-Newtonian fluid past stretching surfaces has gained the attention of investigators due to its wide range practical applications in several industries, for instance, food processes, ground water pollution, crude oil extraction, production of plastic materials, cooling of nuclear reactors, manufacturing of electronic chips, etc. Due to the complex nature of these fluids, different models have been proposed. Among other non-Newtonian model, the Casson fluid model is one of them. The Casson fluid model was originally developed by Casson [21] for the preparation of printing inks and silicon suspensions. Casson fluid has important applications in polymer industries and biomechanics [22]. The Casson fluid model is also suggested as the best rheological model for blood and chocolate [23,24]. For this reason, many authors have considered Casson fluid for different geometries. Shawky [25] analyzed the heat and mass transfer mechanisms in MHD flow of Casson fluid over a linear stretching sheet saturated in a porous medium. Mukhopadhyay [26] and Medikare et al. [27] investigated heat transfer effects on Casson fluid over a nonlinear stretching sheet in the absence and presence of viscous dissipation, respectively. Mythili and Sivaraj [28] considered the geometry of cone and flat plate and studied the impact of chemical reaction on Casson fluid flow with thermal radiation. The impact of magnetic field and heat generation/absorption on heat transfer flow of Casson fluid through a porous medium was presented by Ullah et al. [29]. Imtiaz et al. [30] developed the mixed convection flow of Casson fluid due to a linear stretching cylinder filled with nanofluid with convective boundary conditions.

The above discussion and its engineering applications is the source of motivation to investigate the electrically conductive flow of Casson fluid due to a porous cylinder being stretched in a nonlinear way. It is also clear from the published articles that the mixed convection slip flow of Casson fluid for the geometry of a nonlinear stretching cylinder saturated in a porous medium in the presence of thermal radiation, viscous dissipation, joule heating, and heat generation/absorption has not yet been analyzed. It is worth mentioning that the current problem can be reduced to the flow over a flat plate ($n = 0$ and $\gamma = 0$), linear stretching sheet ($n = 1$ and $\gamma = 0$), nonlinear stretching sheet ($\gamma = 0$), and linear stretching cylinder ($n = 1$). Local similarity transformations are applied to transform the governing equations. The obtained system of equations are then computed numerically using the Keller box method [31] via MATLAB. The variations of flow fields for various pertinent parameters are discussed and displayed graphically. Comparison of the friction factor is made with previous literature results and close agreement is noted. The accuracy achieved has developed our confidence that the present MATLAB code is correct and numerical results are accurate.

2. Mathematical Formulation

Consider a steady, two-dimensional, incompressible mixed convection slip flow of Casson fluid generated due to a nonlinear stretching cylinder in a porous medium in the presence of chemical reaction, slip, and convective boundary conditions. The cylinder is stretched with the velocity of $u_w(x) = cx^n$, where c, n ($n = 1$ represents linear stretching and $n \neq 1$ corresponds to nonlinear stretching) are constants. The x-axis is taken along the axis of the cylinder and the r-axis is measured in the radial direction (see Figure 1). It is worth mentioning here that the momentum boundary layer develops when there is fluid flow over a surface; a thermal boundary layer must develop if the bulk temperature differs from the surface temperature and a concentration boundary layer develops above the surfaces of species in the flow regime.

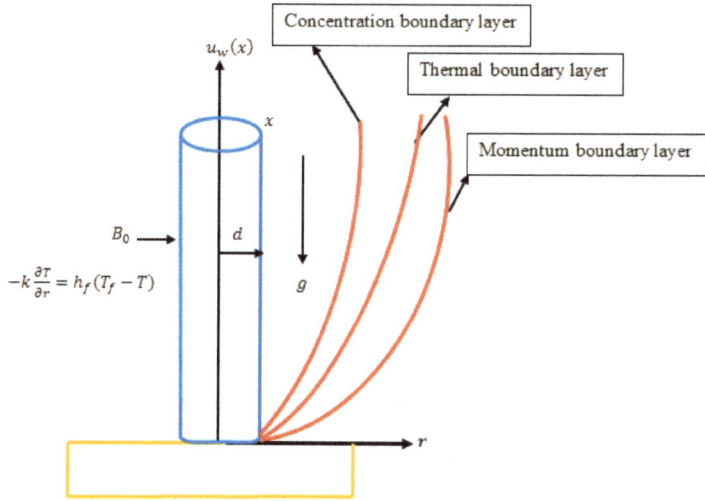

Figure 1. Coordinate system and physical model.

A transverse magnetic field $B(x) = B_0 x^{(n-1)/2}$ is applied in the radial direction with constant B_0. Further, it is also assumed that surface of cylinder is heated by temperature $T_f(x) = T_\infty + Ax^{2n-1}$, in which A is a reference temperature. Concentration is $C_s(x) = C_\infty + B^* x^{2n-1}$, where B^* is the reference concentration. The temperature and concentration at free stream are T_∞ and C_∞, respectively.

The rheological equation of state for an isotropic and incompressible flow of a Casson fluid is

$$\tau_{ij} = \begin{cases} 2(\mu_B + p_y/\sqrt{2\pi_1})e_{ij}, & \pi_1 > \pi_c, \\ 2(\mu_B + p_y/\sqrt{2\pi_c})e_{ij}, & \pi_1 < \pi_c, \end{cases}$$

Here, $\pi_1 = e_{ij} e_{ij}$ and e_{ij} is the $(i,j) - th$ component of the deformation rate, π_1 is the product of the component of deformation rate with itself, π_c is a critical value of this product based on the non-Newtonian model, μ_B is the plastic dynamic viscosity of the non-Newtonian fluid, and p_y is the yield stress of the fluid.

Under the above assumption, the governing equations for Casson fluid along with the continuity equation are given as

$$\frac{\partial(ru)}{\partial x} + \frac{\partial(rv)}{\partial r} = 0 \tag{1}$$

$$u\frac{\partial u}{\partial x} + v\frac{\partial u}{\partial r} = \nu\left(1 + \frac{1}{\beta}\right)\frac{1}{r}\frac{\partial}{\partial r}\left(r\frac{\partial u}{\partial r}\right) - \left(\frac{\sigma B^2(x)}{\rho}\right) + \left(1 + \frac{1}{\beta}\right)\frac{\nu\phi}{k_1}u + g\beta_T(T - T_\infty) + g\beta_C(C - C_\infty) \tag{2}$$

$$u\frac{\partial T}{\partial x} + v\frac{\partial T}{\partial r} = \alpha\left(1 + \frac{4}{3}R_d\right)\frac{1}{r}\frac{\partial}{\partial r}\left(r\frac{\partial T}{\partial r}\right) - \frac{\nu}{c_p}\left(1 + \frac{1}{\beta}\right)\left(\frac{\partial u}{\partial r}\right)^2 + \frac{\sigma B^2(x)}{\rho c_p}u^2 + \frac{Q}{\rho c_p}(T - T_\infty) \tag{3}$$

$$u\frac{\partial C}{\partial x} + v\frac{\partial C}{\partial r} = D\frac{1}{r}\frac{\partial}{\partial r}\left(r\frac{\partial C}{\partial r}\right) - k_c(C - C_\infty) \tag{4}$$

In the above expressions u and v denote the velocity components in x and r direction, respectively, ν is kinematic viscosity, σ is the electrically conductivity, β is the Casson parameter, ρ is the fluid density, ϕ is the porosity, $k_1(x) = k_0/x^{(n-1)}$ is the variable permeability of porous medium, g is the gravitational force due to acceleration, β_T is the volumetric coefficient of thermal expansion, β_C the coefficient of concentration expansion, $\alpha = \dfrac{k}{\rho c_p}$ is the thermal diffusivity of the Casson fluid, k is the thermal conductivity of fluid, c_p is the heat capacity of the fluid, $R_d = \dfrac{4\sigma^* T_\infty^3}{kk_1^*}$ is the radiation parameter, $Q(x) = Q_0 x^{n-1}$ is heat generation/absorption coefficient, D is the coefficient of mass diffusivity, $k_c(x) = ak_2 x^{n-1}$ is the variable rate of chemical reaction, k_2 is a constant reaction rate and a is the reference length along the flow.

The corresponding boundary conditions are written as follows

$$u = u_w(x) + N_1 \nu\left(1 + \frac{1}{\beta}\right)\frac{\partial u}{\partial r}, \; k\frac{\partial T}{\partial r} = -h_f(T_f - T), \; D\frac{\partial C}{\partial r} = -h_s(C_s - C) \text{ at } r = d \tag{5}$$

$$u \to 0, \; T \to T_\infty, \; C \to C_\infty \text{ as } r \to \infty. \tag{6}$$

Here $N_1(x) = N_0 x^{-(\frac{n-1}{2})}$ represents velocity slip with constant N_0, $h_f(x) = h_0 x^{\frac{n-1}{2}}$ and $h_s(x) = h_1 x^{\frac{n-1}{2}}$ represents the convective heat and mass transfer with h_0, h_1 being constants,

Now introduce the stream function ψ, a similar variable η and the following similarity transformations;

$$\psi = \sqrt{\frac{2\nu c}{(n+1)}} x^{\frac{n+1}{2}} df(\eta), \; \eta = \frac{r^2 - d^2}{2d}\sqrt{\frac{(n+1)c}{2\nu}} x^{\frac{n-1}{2}}, \; \theta(\eta) = \frac{T - T_\infty}{T_f - T_\infty}, \; \varphi(\eta) = \frac{C - C_\infty}{C_s - C_\infty} \tag{7}$$

Equation (1) is identically satisfied by the introduction of the equation

$$u = \frac{1}{r}\frac{\partial \psi}{\partial r}, \quad v = -\frac{1}{r}\frac{\partial \psi}{\partial x} \tag{8}$$

The system of Equations (2)–(4) will take the form

$$\left(1+\tfrac{1}{\beta}\right)\left[\left(1+2\sqrt{\tfrac{2}{n+1}}\gamma\eta\right)f''' + 2\sqrt{\tfrac{2}{n+1}}\gamma f''\right] + ff'' - \tfrac{2n}{n+1}f'^2 - \tfrac{2}{(n+1)}\left(M + \left(1+\tfrac{1}{\beta}\right)K\right)f' + \tfrac{2}{(n+1)}(Gr\theta + Gm\varphi) = 0 \tag{9}$$

$$\left(1+\tfrac{4}{3}R_d\right)\left[\left(1+2\sqrt{\tfrac{2}{n+1}}\gamma\eta\right)\theta'' + 2\sqrt{\tfrac{2}{n+1}}\gamma\theta'\right] + Prf\theta' - 2\left(\tfrac{2n-1}{n+1}\right)Prf'\theta + MEcf'^2\left(1+\tfrac{1}{\beta}\right)\left(1+2\sqrt{\tfrac{2}{n+1}}\gamma\eta\right)PrEcf''^2 + \left(\tfrac{2}{n+1}\right)\varepsilon\theta = 0 \tag{10}$$

$$\tfrac{1}{Sc}\left[\left(1+2\sqrt{\tfrac{2}{n+1}}\gamma\eta\right)\varphi'' + 2\sqrt{\tfrac{2}{n+1}}\gamma\varphi'\right] + f\varphi' - 2\left(\tfrac{2n-1}{n+1}\right)f'\varphi - \tfrac{2}{(n+1)}R\varphi = 0 \tag{11}$$

The associated boundary conditions in Equations (5) and (6) are transformed as

$$\left.\begin{array}{c} f(0) = \sqrt{\tfrac{2}{n+1}}S,\ f'(0) = 1 + \delta\sqrt{\tfrac{n+1}{2}}\left(1+\tfrac{1}{\beta}\right)f''(0),\ \theta'(0) = -\left(\sqrt{\tfrac{2}{n+1}}\right)Bi_1[1-\theta(0)] \\ \varphi'(0) = -\left(\sqrt{\tfrac{2}{n+1}}\right)Bi_2[1-\varphi(0)] \end{array}\right\}, \tag{12}$$

$$f'(\infty) = 0,\ \theta(\infty) = 0,\ \varphi(\infty) = 0. \tag{13}$$

In the above expressions, γ, M, K, Gr, Gm, S ($S > 0$ corresponds to suction and $S < 0$ indicates blowing), δ, Pr, Ec, ε ($\varepsilon > 0$ is for heat generation and $\varepsilon < 0$ denotes heat absorption), Sc, Bi_1, Bi_2, and R ($R > 0$ corresponds to destructive chemical reaction and $R = 0$ represents no chemical reaction) are the curvature parameter, magnetic parameter, porosity parameter, thermal Grashof number, mass Grashof number, suction/blowing parameter, slip parameter, Prandtl number, Eckert number, heat generation/absorption parameter, Schmidt number, Biot numbers and chemical reaction parameter, and are defined as

$$\gamma = \sqrt{\frac{\nu x^{1-n}}{cd^2}},\ M = \frac{\sigma B_0^2}{\rho c},\ K = \frac{\nu\phi}{k_0 c},\ Gr = \frac{g\beta_T A}{c^2},\ Gm = \frac{g\beta_C B^*}{c^2},$$

$$S = -V_0\sqrt{\frac{1}{c\nu}},\ \delta = N_0\sqrt{c\nu},\ Pr = \frac{\nu}{\alpha},\ Ec = \frac{u_w^2}{c_p(T_f - T_\infty)},\ \varepsilon = \frac{Q_0}{\rho c_p c},$$

$$Sc = \frac{\nu}{D},\ Bi_1 = \frac{h_0}{k}\left[\frac{\nu}{c}\right]^{1/2},\ Bi_2 = \frac{h_1}{D}\left[\frac{\nu}{c}\right]^{1/2},\ R = \frac{ak_2}{c}$$

The wall skin friction, wall heat flux, and wall mass flux, respectively, are defined by

$$\tau_w = \mu_B\left(1+\tfrac{1}{\beta}\right)\left[\frac{\partial u}{\partial r}\right]_{r=d},\ q_w = -\left(\left(\alpha + \frac{16\sigma^* T_\infty^3}{3\rho c_p k_1^*}\right)\frac{\partial T}{\partial r}\right)_{r=d}\ \text{and}\ q_s = -D\left(\frac{\partial C}{\partial y}\right)_{r=d}$$

The dimensionless skin friction coefficient $Cf_x = \frac{\tau_w}{\rho u_w^2}$, the local Nusselt number $Nu_x = \frac{xq_w}{\alpha(T_f - T_\infty)}$ and local Sherwood number $Sh_x = \frac{xq_s}{D_B(C_w - C_\infty)}$ on the surface along x—direction, local Nusselt number Nu_x and Sherwood number Sh_x are given by

$$(Re_x)^{1/2}Cf_x = \sqrt{\tfrac{n+1}{2}}\left(1+\tfrac{1}{\beta}\right)f''(0),\ (Re_x)^{-1/2}Nu_x = -\sqrt{\tfrac{n+1}{2}}\left(1+\tfrac{4}{3}R_d\right)\theta'(0),$$
$$(Re_x)^{-1/2}Sh_x = -\left(\sqrt{\tfrac{n+1}{2}}\right)\varphi'(0)$$

where $Re_x = \frac{cx^{n+1}}{\nu}$ is the local Reynold number.

3. Results and Discussion

The system of Equations (9)–(11) are solved numerically by using the Keller-box method [31] and numerical computations are carried out for different values of physical parameters including curvature parameter γ, Casson fluid parameter β, nonlinear stretching cylinder parameter n, magnetic parameter M, porosity parameter K, Grashof number Gr, mass Grashof number Gm, Prandtl number Pr, radiation parameter R_d, Eckert number Ec, heat generation/absorption parameter ε, Schmidt number Sc, chemical reaction parameter R, slip parameter δ, and Biot numbers Bi_1, Bi_2. In order to validate the algorithm developed in MATLAB software for the present method, the numerical results for skin friction coefficient are compared with the results of Akbar et al. [13], Fathizadeh et al. [12], Fang et al. [15], and Imtiaz et al. [30], and presented in Table 1. Comparison revealed a close agreement with them.

Table 1. Comparison of skin friction coefficient $f''(0)$ for different values of M with $\beta \to \infty$, $Bi_1 \to \infty$, $Bi_2 \to \infty$, $n = 1$ and $\gamma = M = K = Gr = Gm = S = \delta = R_d = Ec = \varepsilon = R = 0$.

M	$(1+\frac{1}{\beta})f''(0)$				
	Akbar et al. [13]	Fathizadeh et al. [12]	Fang et al. [15]	Imtiaz et al. [30]	Present Results
0	−1	−1	−1	−1	−1
1	−1.4142	−1.4142	−1.4142	−1.4142	−1.4142
5	−2.4495	−2.4494	−2.4494	−2.4494	−2.4495
10	−3.3166	-	-	-	−3.3166

Figures 2–10 are depicted to see the physical behavior of γ, β, n, M, K, Gr, Gm, δ, and S on velocity profile. Figure 2 exhibits the variation of γ on fluid velocity for $n = 1$ (linear stretching) and $n \neq 1$ (nonlinear stretching). It is noticed that fluid velocity is higher for increasing values of γ. Since the increase in γ leads to reduction in the radius of curvature, it also reduces cylinder area. Thus, the cylinder experiences less resistance from the fluid particles and fluid velocity is enhanced. It can also be seen that the momentum boundary layer is thicker with increased γ when $n \neq 1$. The influence of β on velocity profile for different values of S is depicted in Figure 3. In all cases, the fluid velocity is a decreasing function of β. The reason is that the fluid becomes more viscous with the growth of β. Therefore, more resistance is offered which reduces the momentum boundary layer thickness. Figure 4 elucidates the effect of n on velocity profile for $M = 0$ and $M \neq 0$. It is evident that increasing values of n enhance the fluid velocity. Also, this enhancement is more pronounced when $M \neq 0$. The momentum boundary layer is thicker when $n \neq 1$.

The variation of M for $K = 0$ and $K \neq 0$ on the velocity profile is presented in Figure 5. As expected, the strength of the magnetic field lowers the fluid flow. It is an agreement with the fact that increase in M produces Lorentz force that provides resistance to the flow, and apparently thins the momentum boundary layer across the boundary. It can also be seen that the fluid velocity is more influenced with M when $K = 0$. A similar kind of variation is observed on velocity profile for different values of K, as displayed in Figure 6. Since the porosity of porous medium provides resistance to the flow, fluid motion slows down and produces larger friction between the fluid particles and the cylinder surface. The impact of Gr for $M = 0$ and $M \neq 0$ on velocity profile is depicted in Figure 7. The convection inside the fluid rises as the temperature difference $\left(T_f - T_\infty\right)$ enhances due to the growth of Gr. In addition, increase in Gr leads to stronger buoyancy force, in which case, the momentum boundary layer becomes thicker. The same kind of physical explanation can be given for the effect of Gm on velocity profile (see Figure 8). The variation of S on velocity profile for both $n = 1$ and $n \neq 1$ is portrayed in Figure 9. Clearly, the fluid velocity declines when $S > 0$, whereas a reverse trend is noted when $S < 0$. Physically, stronger blowing forces the hot fluid away from the surface, in which case the viscosity reduces and the fluid gets accelerated. On the other hand, wall suction ($S > 0$) exerts a drag force at the surface and hence thinning of the momentum boundary layer. Figure 10 demonstrates the effect of δ on velocity profile for $K = 0$ and $K \neq 0$. It can be easily seen that fluid velocity falls with increase in δ. Since the resistance between the cylinder surface and the fluid particles rises with increase in δ, the momentum boundary layer become thinner.

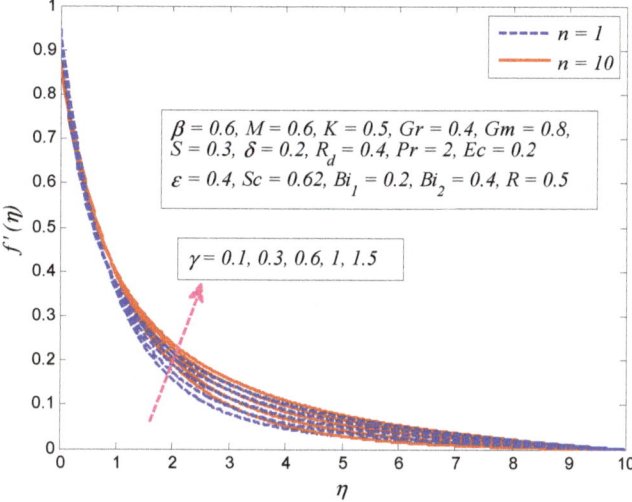

Figure 2. Effect of curvature parameter γ on velocity for linear ($n = 1$) and nonlinear $n = 10$ stretching parameter.

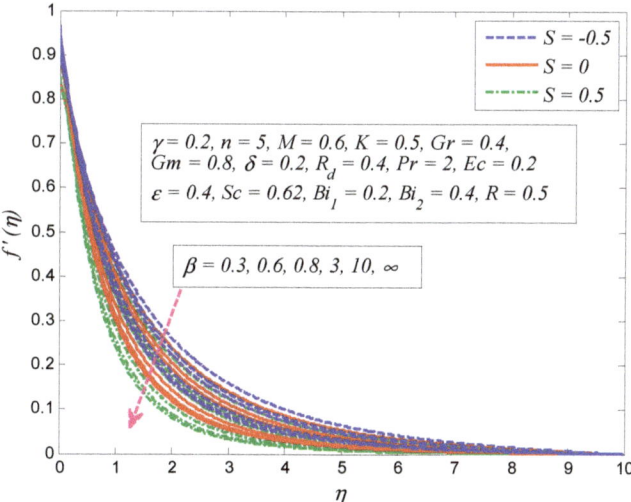

Figure 3. Effect of Casson fluid parameter β on velocity profile for different values of suction/blowing parameter S.

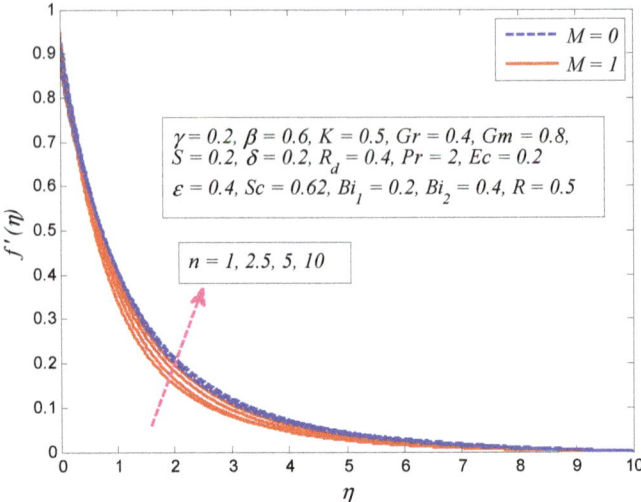

Figure 4. Effect of nonlinear stretching parameter n on velocity profile in the presence and absence of magnetic parameter M.

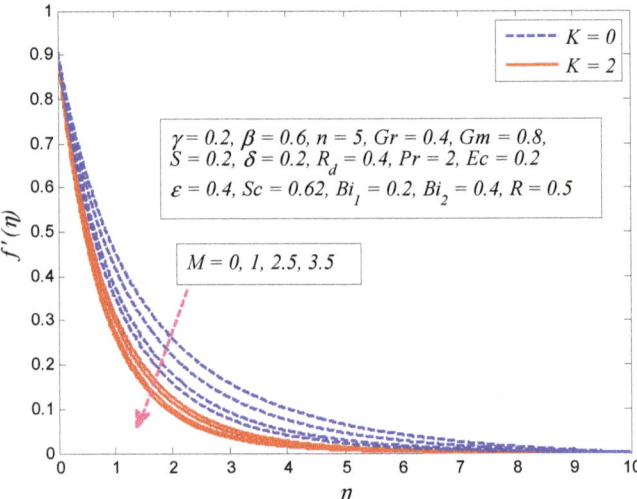

Figure 5. Effect of magnetic parameter M on velocity profile in the presence and absence of porosity parameter K.

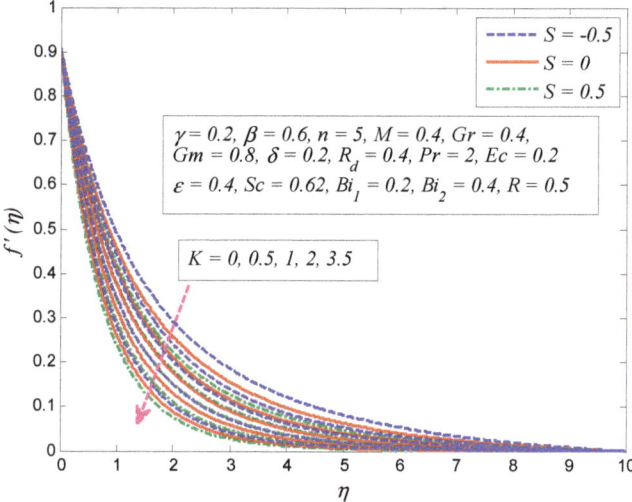

Figure 6. Effect of porosity parameter K on velocity profile for different values of suction/blowing parameter S.

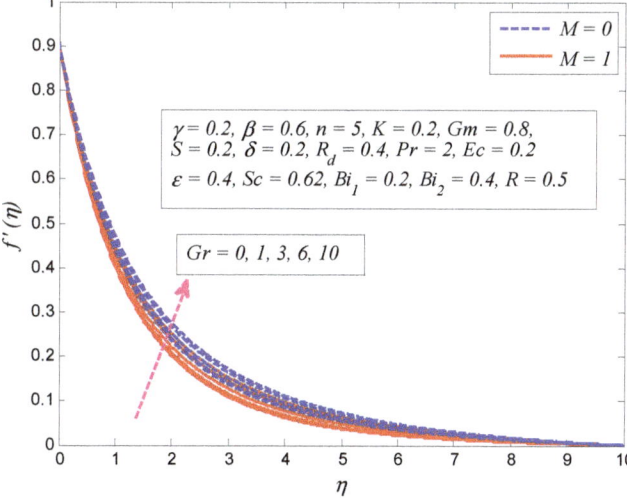

Figure 7. Effect of Grashof number Gr on velocity profile in the presence and absence of magnetic parameter M.

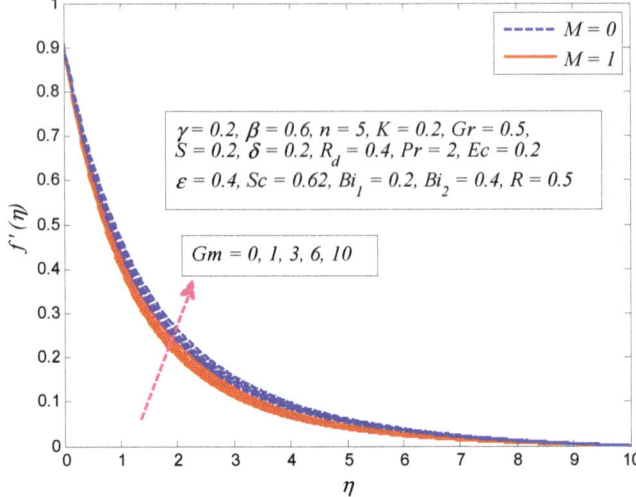

Figure 8. Effect of mass Grashof number Gm on velocity profile in the presence and absence of magnetic parameter M.

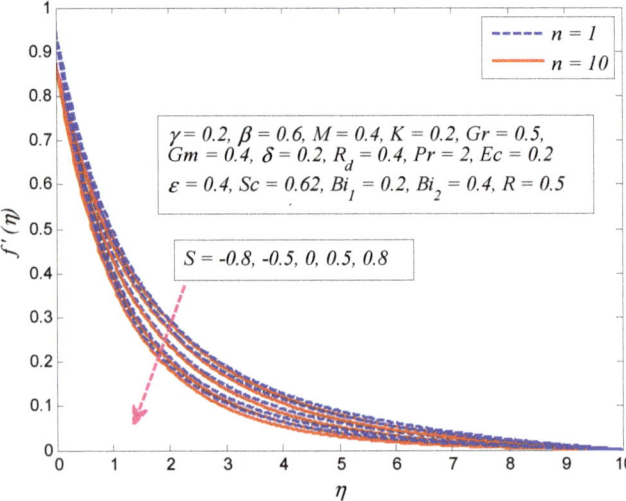

Figure 9. Effect of suction/blowing parameter S on velocity profile for nonlinear stretching parameter n.

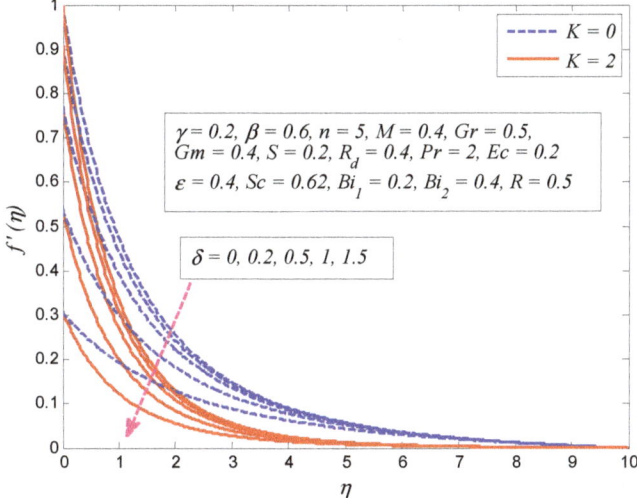

Figure 10. Effect of slip parameter δ on velocity profile in the presence and absence of porosity parameter K.

Figures 11–21 are plotted to get insight on the variation of γ, β, n, M, K, S, Pr, R_d, Ec, ε, and Bi_1 on the temperature profile. Figure 11 illustrates the variation of γ on dimensionless temperature profile for $\beta \to \infty$ (Newtonian fluid) and $\beta = 0.6$ (Casson fluid). It is noticed that temperature rises with increment in γ. A thermal boundary layer thickness is also noted. Figure 12 displays the influence of β on temperature profile for various values of S. It is noticeable that fluid temperature declines with the increase in β for all the three cases of S. The reason is that increase in β implies a reduction in yield stress, and consequently the thickness of the thermal boundary layer reduces. The effect of n on temperature profile for $M = 0$ and $M \neq 0$ is examined in Figure 13. It is clear from this figure that temperature is a decreasing function of n. It is also noticed that the fluid temperature thermal boundary layer is thicker for a linear stretching cylinder ($n = 1$) as compared to nonlinear stretching of the cylinder ($n \neq 1$). Figure 14 shows the variation of M on temperature profile for different values of S. It is noticeable that stronger magnetic field rises the fluid temperature in the vicinity of stretching cylinder. Because increasing M enhances the Lorentz force, this force makes the thermal boundary layer thicker. The same kind of behavior is noticed for the effect of K on dimensionless temperature profile for $\delta = 0$ and $\delta \neq 0$, as presented in Figure 15.

Figure 16 reveals the influence of S on temperature profile for $n = 1$ (linear stretching) and $n \neq 1$ (nonlinear stretching). Clearly, fluid temperature falls when $S > 0$, whereas it rises when $S < 0$. Since the wall suction offers resistance to fluid flow, the thermal boundary layer becomes thinner, and the opposite occurs when $S < 0$. The variation of Pr on dimensionless temperature profile for $Ec = 0$ and $Ec = 0.2$ is depicted in Figure 17. The Prandtl number is defined as the ratio of momentum diffusivity to thermal diffusivity. As expected, fluid temperature drops with the growth of Pr. It is a well-known fact that higher thermal conductivities are associated with lower Prandtl fluids, therefore heat diffuses quickly from the surface as compared to higher Prandtl fluids. Thus, Pr can be utilized to control the rate of cooling in conducting flows. Figure 18 exhibits the effect of R_d on the temperature profile for different values of S. It is noticeable that the strength of R_d boosts the temperature. The larger surface heat flux corresponds to larger values of R_d, causing the fluid to be warmer.

Figure 19 illustrates the influence of Ec on the temperature profile for $K = 0$ and $K \neq 0$. It is noted that the temperature is higher for higher values of Ec. Physically this is true, because viscous dissipation generates heat energy due to friction between fluid particles and thereby thickens the thermal boundary layer structure. It is also observed from this figure that in the presence of porous medium, the strength of Ec effectively enhances the fluid temperature. The influence of ε on temperature profile for $M = 0$ and $M \neq 0$ is displayed in Figure 20. It is clear from this graph that the temperature is enhanced when $\varepsilon > 0$ (heat generation), whereas the opposite trend is observed when $\varepsilon < 0$ (heat absorption). Internal heat generation causes the heat energy to be enhanced. Consequently, the heat transfer rate rises and thickens the thermal boundary layer. Besides, the heat absorption causes a reverse effect, i.e. the heat transfer rate and the thermal boundary layer thickness are reduced. Figure 21 reveals the variation of Bi_1 on the dimensionless temperature profile for $K = 0$ and $K \neq 0$. The Biot number is the ratio of the internal thermal resistance of a solid to the boundary layer thermal resistance. It is noticed that fluid temperature is higher for larger values of Bi_1. The reason is that increment in Bi_1 keeps the convection heat transfer higher and the cylinder thermal resistance lower. It is worth mentioning here that when $Bi_1 < 0.1$, the internal resistance to heat transfer is negligible, representing that the value of k is much larger than h_0, and the internal thermal resistance is noticeably lower than the surface resistance. On the other hand, when $Bi_1 \to \infty$ the higher Biot number intends that the external resistance to heat transfer reduces, indicating that the surface and the surroundings temperature difference is minor and a noteworthy contribution of temperature to the center comes from the surface of the stretching cylinder.

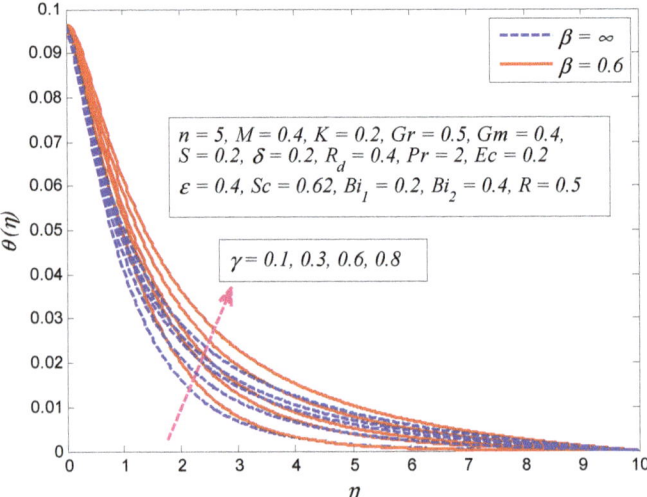

Figure 11. Effect of curvature parameter γ on temperature profile for Newtonian fluid $\beta = \infty$ and Casson fluid $\beta = 0.6$.

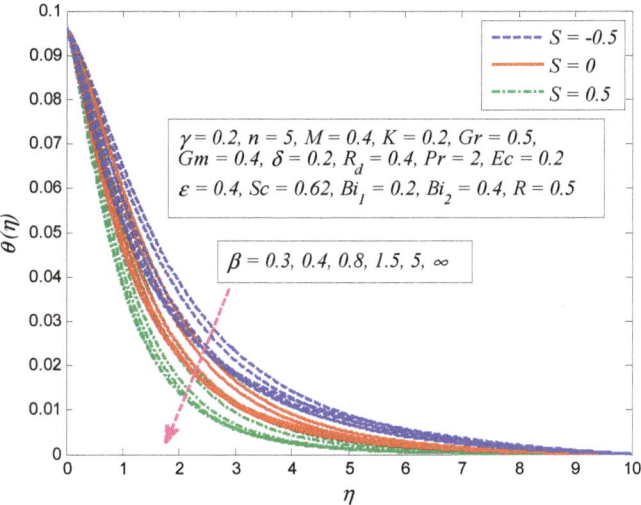

Figure 12. Effect of Casson fluid parameter β on temperature profile for different values of suction/blowing parameter S.

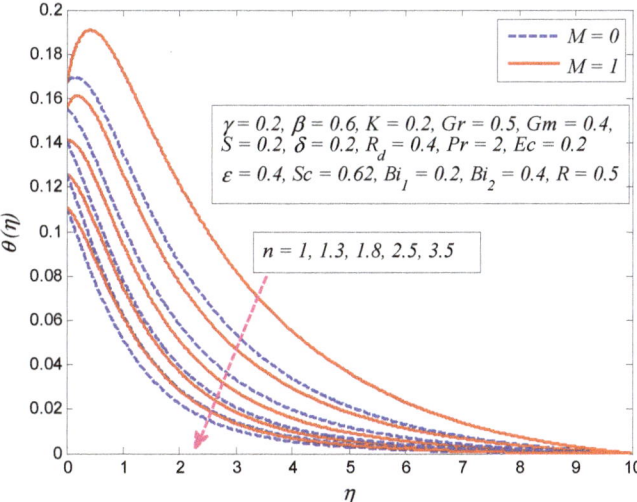

Figure 13. Effect of nonlinear stretching parameter n on temperature profile in the presence and absence of magnetic parameter M.

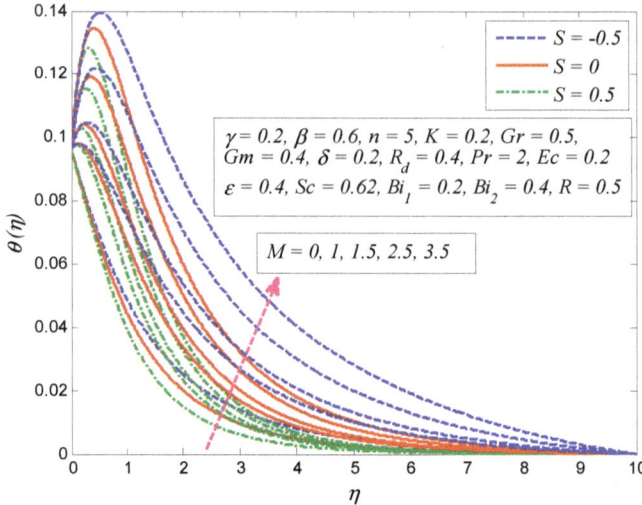

Figure 14. Effect of magnetic parameter M on temperature profile for different values of suction/blowing parameter S.

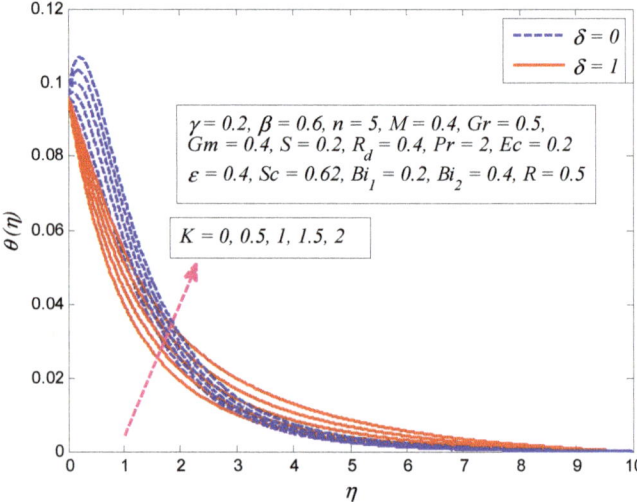

Figure 15. Effect of porosity parameter K on temperature profile in the presence and absence of slip parameter δ.

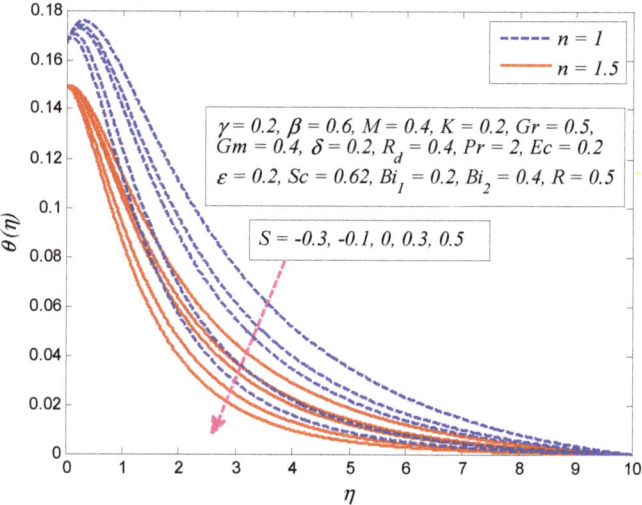

Figure 16. Effect of suction/blowing parameter S on temperature profile for different values of nonlinear stretching parameter n.

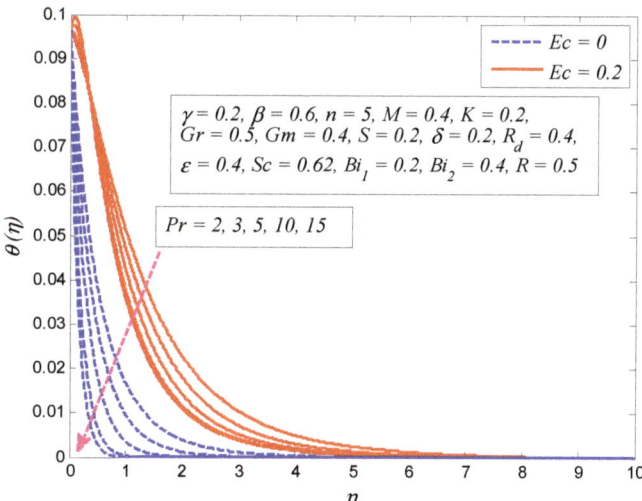

Figure 17. Effect of Prandtl number Pr on temperature profile in the presence and absence of Eckert number Ec.

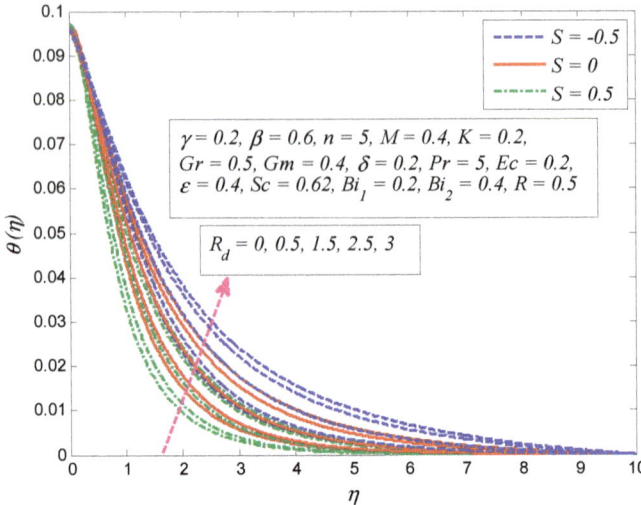

Figure 18. Effect of radiation parameter R_d on temperature profile for different values of suction/blowing parameter S.

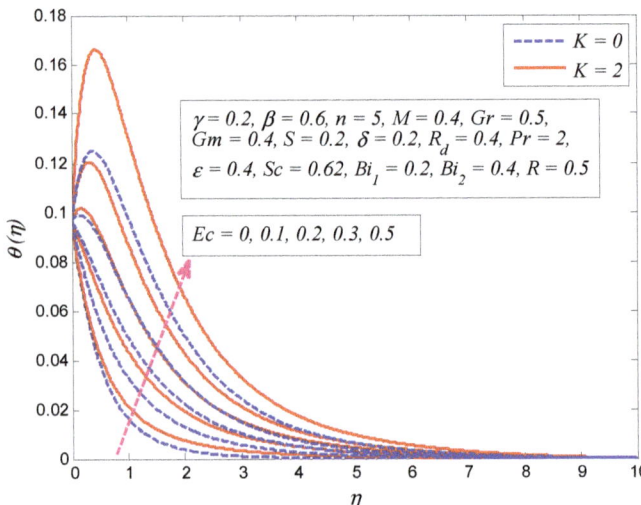

Figure 19. Effect of Eckert number Ec on temperature profile in the presence and absence of porosity parameter K.

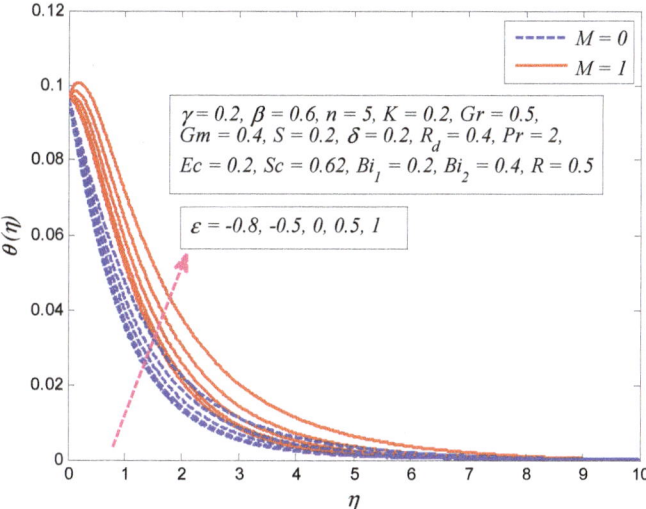

Figure 20. Effect of heat generation/absorption parameter ε on temperature profile in the presence and absence of magnetic parameter M.

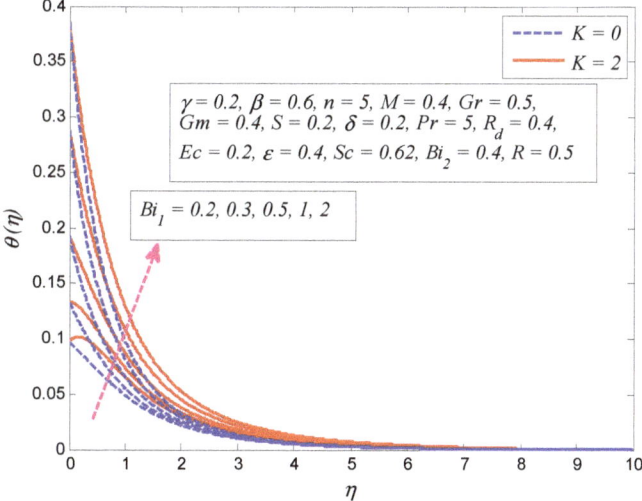

Figure 21. Effect of Biot number Bi_1 on temperature profile in the presence and absence of porosity parameter K.

Figures 22–31 display the variation of γ, β, n, M, K, δ, S, Sc, R, and Bi_1 on concentration profile, respectively. Figure 22 elucidates the effect of γ on concentration profile for $\beta \to \infty$ (Newtonian fluid) and $\beta = 0.6$ (Casson fluid). It is found that increasing values of γ enhances the fluid concentration and associated boundary layer thickness. Figure 23 demonstrates the influence of β on concentration profile for $M = 0$ and $M \neq 0$. It is noted that fluid concentration is higher as β grows. The viscosity of the fluid increases with increasing β, in which case the concentration rises and the concentration boundary layer becomes thicker. The opposite behavior is noticed for the effect of n on concentration profile for various values of S (see Figure 24). It is also observed that thickness of concentration boundary layer shortens for large n. Figure 25 determines the variation of M on the dimensionless concentration profile for $K = 0$ and $K \neq 0$. It is seen that fluid concentration is higher for higher values of M. As mentioned earlier for velocity and temperature profiles, fluid motion reduces due to magnetic field and results in an enhancement in thermal and concentration boundary layer thicknesses. A similar trend is observed for the effect of K and δ on the concentration profile, as plotted in Figures 26 and 27, respectively. The growth of both parameters offers resistance to the fluid particles and the concentration boundary layer becomes thicker. Figure 28 shows that fluid concentration reduces when $S > 0$, while it is enhanced when $S < 0$. Indeed, when mass suction occurs, some of the fluid is sucked through the wall which thins the boundary layer; on the contrary, blowing thickens the concentration boundary layer structure.

Figure 29 examines the variation of Sc ($Sc = 0.30, 0.62, 0.78, 0.94, 2.57$ corresponds to hydrogen, helium, water vapor, hydrogen sulphide, and propyl Benzene) on the dimensionless concentration profile when $\beta \to \infty$ (Newtonian fluid) and $\beta = 0.6$ (Casson fluid). For both fluids, an increase in Sc reduces the fluid concentration. Since higher values of Sc lead to higher mass transfer rate, the thickness of the concentration boundary layer declines. The effect of R on the concentration distribution for different values of S is depicted in Figure 30. It is clear that fluid concentration drops with the growth of R. Physically this makes sense, because the decomposition rate of reactant species enhances in the destructive chemical reaction ($R > 0$). Consequently, the mass transfer rate grows and thickens the concentration boundary layer. Figure 31 exhibits the variation of Bi_2 on concentration distribution for $M = 0$ and $M \neq 0$. It is noticeable that fluid concentration rises with increasing Bi_2. As increase in Biot number enhances the temperature field, the concentration field excites, making the solutal boundary layer thicker.

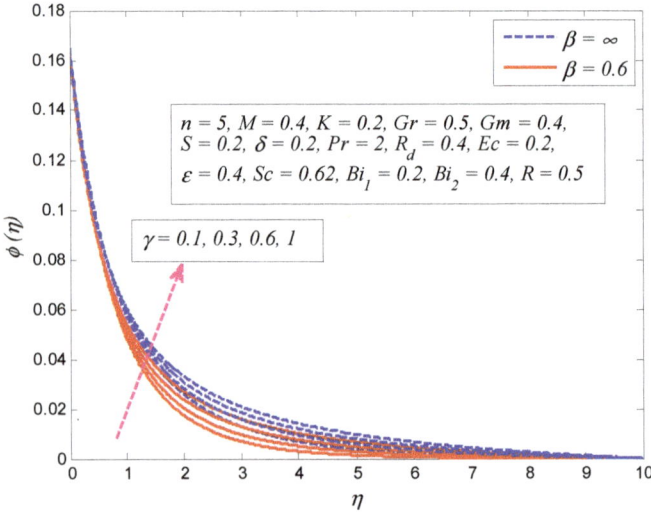

Figure 22. Effect of curvature parameter γ on concentration profile for Newtonian fluid $\beta = \infty$ and Casson fluid $\beta = 0.6$.

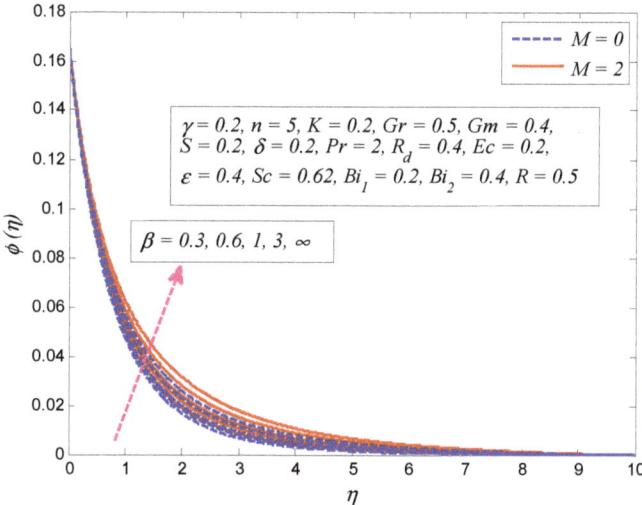

Figure 23. Effect of Casson parameter β on concentration profile in the presence and absence of magnetic parameter M.

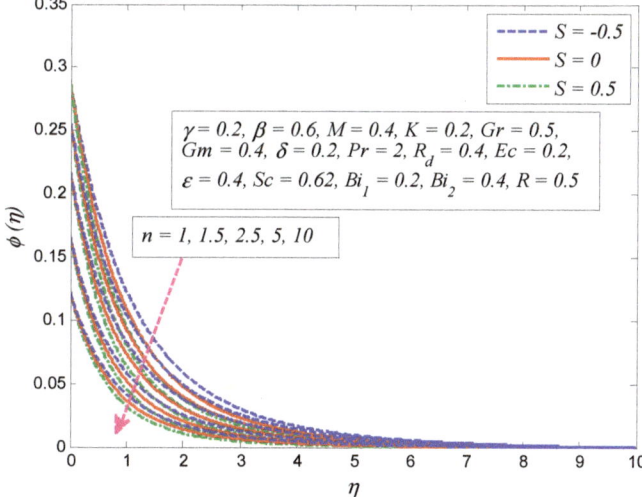

Figure 24. Effect of nonlinear stretching parameter n on concentration profile for different values of suction/blowing parameter S.

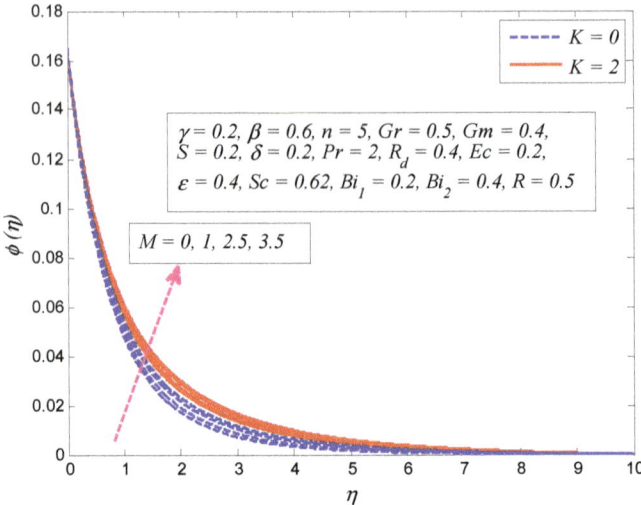

Figure 25. Effect of magnetic parameter M on concentration profile in the presence and absence of porosity parameter K.

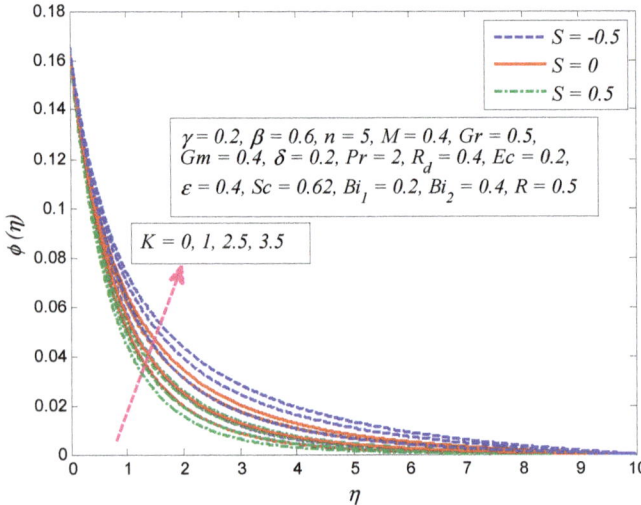

Figure 26. Effect of porosity parameter K on concentration profile for different values of suction/blowing parameter S.

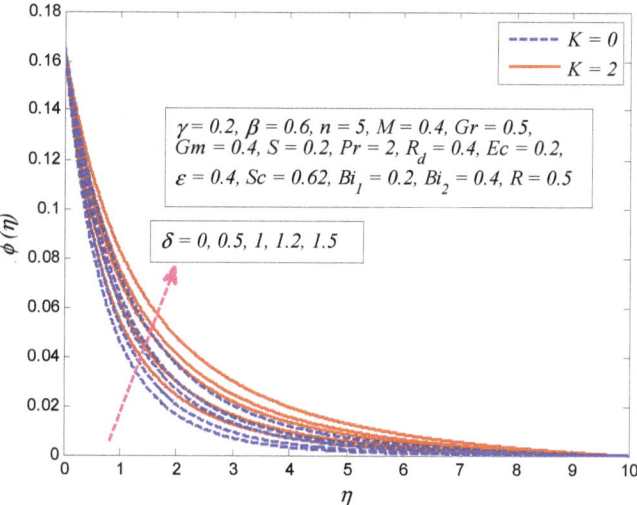

Figure 27. Effect of slip parameter δ on concentration profile in the presence and absence of porosity parameter K.

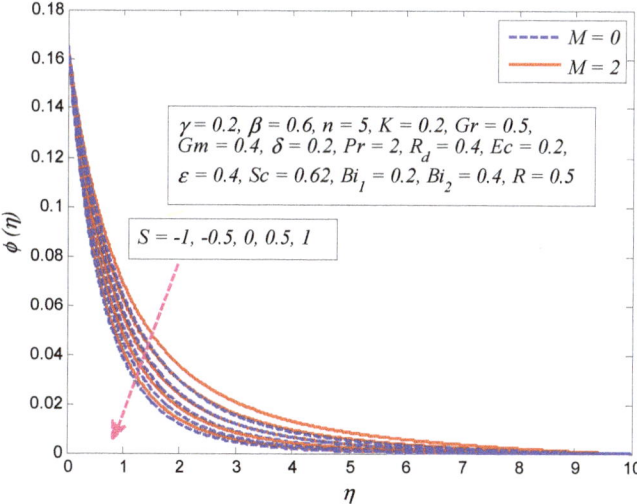

Figure 28. Effect of suction/blowing parameter S on concentration profile in the presence and absence of magnetic parameter M.

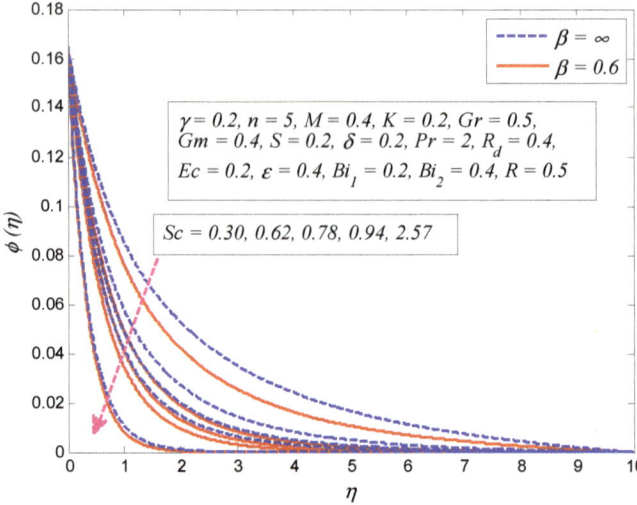

Figure 29. Effect of Schmidt number Sc on concentration profile for Newtonian fluid $\beta = \infty$ and Casson fluid $\beta = 0.6$.

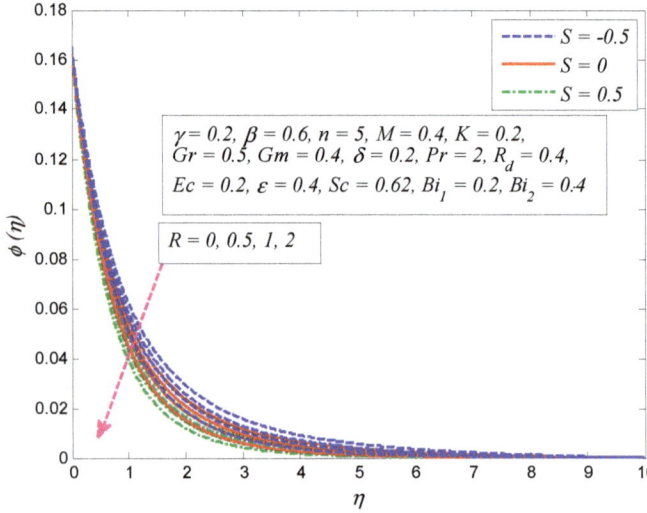

Figure 30. Effect of chemical reaction parameter R on concentration profile for different values of suction/blowing parameter S.

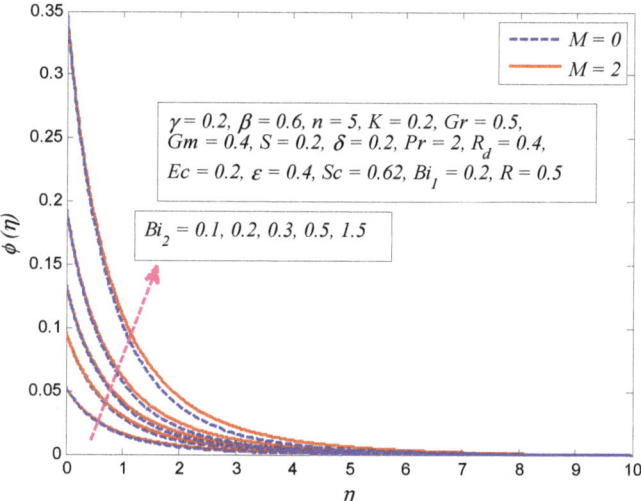

Figure 31. Effect of Biot number Bi_1 on concentration profile in the presence and absence of magnetic parameter M.

Figures 32–35 depict the effect of the skin friction coefficient, Nusselt number, and Sherwood number for different values of γ, β, M, K, n, S, Ec, and R, respectively. Figure 32 reveals the variation of wall shear stress for various values of γ, β, and M. It is noted that the absolute values of wall shear stress increase as γ and M increase, whereas the opposite is observed for the effect of β. It is also noticeable that the values of friction factor are negative, which shows that the stretching cylinder experiences a drag force from the fluid particles. Moreover, the effect of γ on wall shear stress is more pronounced for Casson fluid. The effect of K, n, and S on dimensionless skin friction coefficient is examined in Figure 33. This figure shows that friction factor absolute values decline as K, n, and S increase. Figure 34 portrays the variation of Nusselt number for various values of γ, β, and Ec. It is shown that heat transfer rate drops as γ and Ec increase, whereas they increase for larger values of β. However, the heat transfer rate is more influenced for Casson fluid. It is also noted that heat transfer rate is negative for higher values of Ec. These negative values show that heat is transferred from the working fluid to the stretching surface. Finally, the effect of Sherwood number for various γ, β, and R is illustrated in Figure 35. It is found that the mass transfer rate is an increasing function of γ and R and a decreasing function of β.

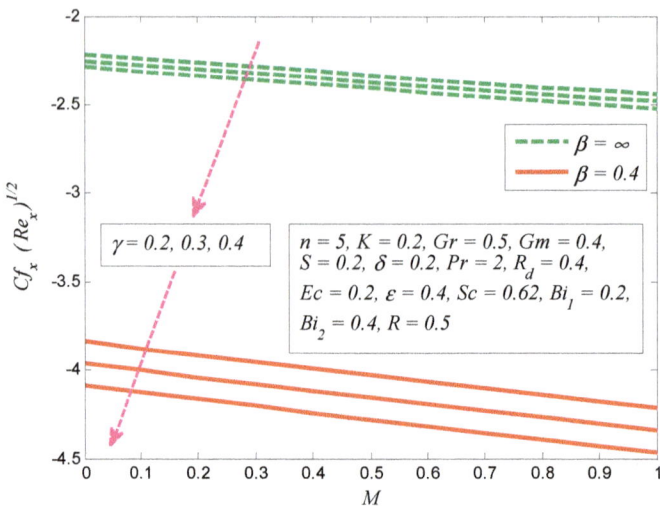

Figure 32. Variation of skin friction coefficient for various values of Casson fluid parameter β, curvature parameter γ, and magnetic parameter M.

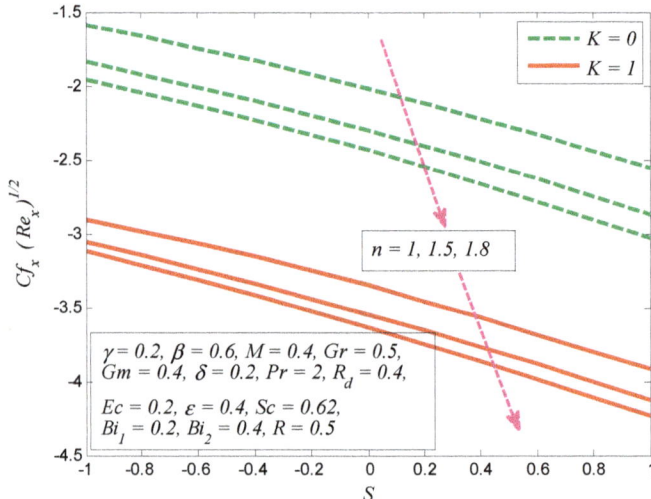

Figure 33. Variation of skin friction coefficient for various values of nonlinear stretching parameter n, porosity parameter K, and suction/blowing parameter S.

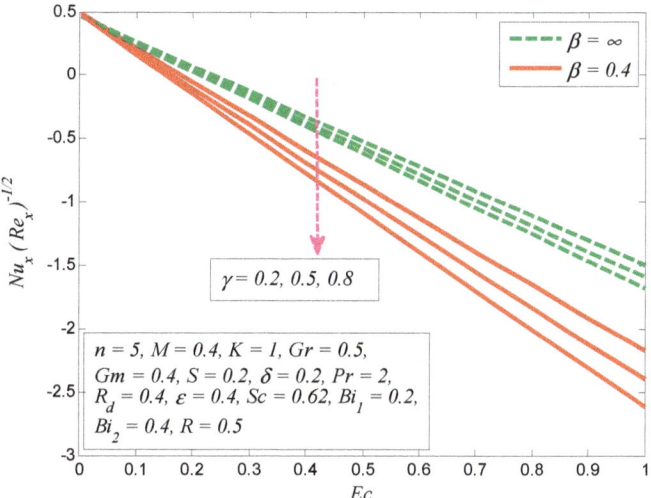

Figure 34. Variation of Nusselt number for various values of Casson parameter β, curvature parameter γ, and Eckert number Ec.

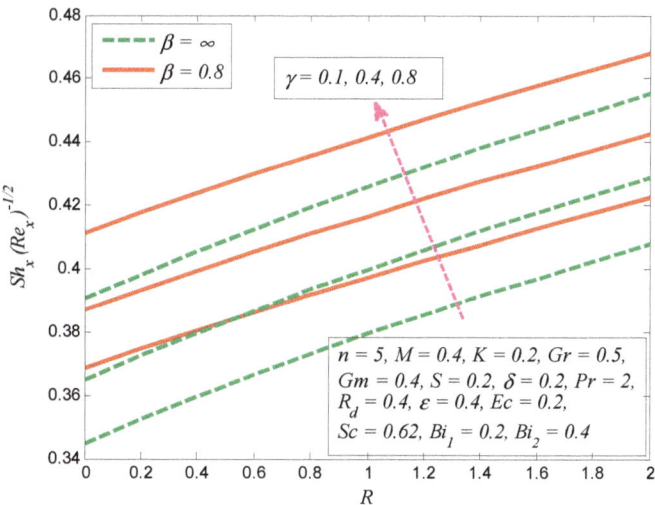

Figure 35. Variation of Sherwood number for various values of Casson fluid parameter β, curvature parameter γ, and chemical reaction parameter R.

4. Conclusions

In the present study, the influence of chemical reaction on MHD slip flow of Casson fluid due to nonlinear cylinder stretching was investigated numerically. Similarity solutions for velocity, temperature, and concentration distributions are achieved via the Keller box method. The numerical results of wall shear stress and heat transfer rate are also compared as a limiting case. The effect of physical parameters, namely, unsteadiness parameter γ, Casson parameter β, nonlinear stretching parameter n, magnetic parameter M, porosity parameter K, thermal Grashof number Gr, mass Grashof number Gm, Prandtl number Pr, radiation parameter R_d, Eckert number Ec, heat generation/absorption parameter ε, Schmidt number Sc, chemical reaction parameter R, suction/blowing parameter S, slip parameter δ, and Biot numbers Bi_1, Bi_2 are discussed and displayed graphically. Some interesting observations from the present analysis are as follow:

1. The fluid velocity, temperature, and concentration are found to increase with γ.
2. The magnitude of wall shear stress and mass transfer rate increase with the growth of β, whereas the heat transfer rate is enhanced.
3. The effect of M on fluid velocity is more pronounced when $K = 0$ (nonporous medium).
4. The temperature field is more influenced with increasing Ec when $K \neq 0$.
5. The velocity, temperature, and concentration distributions decrease when $S > 0$, while the reverse trend is seen when $S < 0$.
6. The concentration boundary layer is observed to be thinner during destructive chemical reaction.

Author Contributions: Conceptualization, I.U. and I.K.; methodology, I.U. and I.K.; software, I.U.; validation, I.U., I.K. and S.S.; formal analysis, I.U., O.D.M., T.A.A., S.S. and K.S.N.; investigation, I.U., T.A.A., K.S.N. and O.D.M; resources, I.U., I.K., and S.S; data curation, I.U.; writing—original draft preparation, I.U. and O.D.M.; writing—review and editing, I.U.; visualization, T.A.A., K.S.N., I.U. and I.K.; supervision, I.U. and S.S.

Funding: This research received no external funding.

Acknowledgments: The authors would like to thank the reviewers and editor for their constructive and insightful comments in relation to this work.

Conflicts of Interest: The authors declare no conflict of interest.

References

1. Crane, L.J. Flow past a stretching plate. *Z. Angew. Math. Phys.* **1970**, *21*, 645–647. [CrossRef]
2. Gupta, P.S.; Gupta, A.S. Heat and mass transfer on a stretching sheet with suction or blowing. *Can. J. Chem. Eng.* **1977**, *55*, 744–746. [CrossRef]
3. Chen, C.K.; Char, M.I. Heat transfer of a continuous, stretching surface with suction or blowing. *J. Math. Anal. Appl.* **1988**, *135*, 568–580. [CrossRef]
4. Gorla, I.; Sidawi, R.S.R. Free Convection on a Vertical Stretching Surface with Suction and Blowing. *Appl. Sci. Res.* **1994**, *52*, 247–257. [CrossRef]
5. Vajravelu, K. Viscous flow over a nonlinearly stretching sheet. *Appl. Math. Comput.* **2001**, *124*, 281–288. [CrossRef]
6. Vajravelu, K.; Cannon, J.R. Fluid flow over a nonlinearly stretching sheet. *Appl. Math. Comput.* **2006**, *181*, 609–618. [CrossRef]
7. Bachok, N.; Ishak, A. Flow and heat transfer over a stretching cylinder with prescribed surface heat flux. *Malays. J. Math. Sci.* **2010**, *4*, 159–169.
8. Hayat, T.; Saeed, Y.; Asad, S.; Alsaedi, A. Convective heat and mass transfer in flow by an inclined stretching cylinder. *J. Mol. Liq.* **2016**, *220*, 573–580. [CrossRef]
9. Majeed, A.; Zeeshan, A.; Alamri, S.Z.; Ellahi, R. Heat transfer analysis in ferromagnetic viscoelastic fluid flow over a stretching sheet with suction. *Neur. Comp. Appl.* **2018**, *30*, 1947–1955. [CrossRef]
10. Vyas, A.; Ranjan, P. Dissipative mhd boundary layer flow in a porous medium over a sheet stretching nonlinearly in the presence of radiation. *Appl. Math. Sci.* **2010**, *4*, 3133–3142.
11. Mukhopadhyay, S. MHD boundary layer flow along a stretching cylinder. *Ain Shams Eng. J.* **2013**, *4*, 317–324. [CrossRef]
12. Fathizadeh, M.; Madani, M.; Khan, Y.; Faraz, N.; Yildirim, A.; Tutkun, S. An effective modification of the homotopy perturbation method for MHD viscous flow over a stretching sheet. *J. King Saud Univ. Sci.* **2013**, *25*, 107–113. [CrossRef]
13. Akbar, N.S.; Ebaid, A.; Khan, Z.H. Numerical analysis of magnetic field effects on Eyring-Powell fluid flow towards a stretching sheet. *J. Magn. Magn. Mater.* **2015**, *382*, 355–358. [CrossRef]
14. Ellahi, R. The effects of MHD and temperature dependent viscosity on the flow of non-Newtonian nanofluid in a pipe: Analytical solutions. *Appl. Math. Model.* **2013**, *37*, 1451–1467. [CrossRef]
15. Fang, T.; Zhang, J.; Yao, S. Slip MHD viscous flow over a stretching sheet—An exact solution. *Commun. Nonlinear Sci. Numer. Simul.* **2009**, *14*, 3731–3737. [CrossRef]

16. Bhattacharyya, K.; Mukhopadhyay, S.; Layek, G.C. Slip effects on an unsteady boundary layer stagnation-point flow and heat transfer towards a stretching sheet. *Chin. Phys. Lett.* **2011**, *28*, 094702. [CrossRef]
17. Yazdi, M.H.; Abdullah, S.; Hashim, I.; Sopian, K. Slip MHD liquid flow and heat transfer over non-linear permeable stretching surface with chemical reaction. *Int. J. Heat Mass Transf.* **2011**, *54*, 3214–3225. [CrossRef]
18. Hayat, T.; Qasim, M.; Mesloub, S. MHD flow and heat transfer over permeable stretching sheet with slip conditions. *Int. J. Numer. Methods Fluids* **2011**, *66*, 963–975. [CrossRef]
19. Seini, I.Y.; Makinde, O.D. Boundary layer flow near stagnation-points on a vertical surface with slip in the presence of transverse magnetic field. *Int. J. Numer. Methods Heat Fluid Flow* **2014**, *24*, 643–653. [CrossRef]
20. Rahman, S.U.; Ellahi, R.; Nadeem, S.; Zia, Q.Z. Simultaneous effects of nanoparticles and slip on Jeffrey fluid through tapered artery with mild stenosis. *J. Mol. Liq.* **2016**, *218*, 484–493. [CrossRef]
21. Casson, N. *A Flow Equation for Pigment-Oil Suspensions of the Printing Ink Type*; Mill, C.C., Ed.; Rheol. Disperse Syst. Pergamon Press: Oxford, UK, 1959; pp. 84–104.
22. Nandy, S.K. Analytical Solution of MHD Stagnation-Point Flow and Heat Transfer of Casson Fluid over a Stretching Sheet with. *Thermodynamics* **2013**. [CrossRef]
23. Singh, S. Clinical significance of aspirin on blood flow through stenotic blood vessels. *J. Biomim. Biomater. Tissue Eng.* **2011**, *10*, 17–24.
24. Mukhopadhyay, S.; Bhattacharyya, K.; Hayat, T. Exact solutions for the flow of Casson fluid over a stretching surface with transpiration and heat transfer effects. *Chin. Phys. B* **2013**, *22*, 114701. [CrossRef]
25. Shawky, H.M. Magnetohydrodynamic Casson fluid flow with heat and mass transfer through a porous medium over a stretching sheet. *J. Porous Media* **2012**, *15*, 393–401. [CrossRef]
26. Mukhopadhyay, S. Casson fluid flow and heat transfer over a nonlinearly stretching surface. *Chin. Phys. B* **2013**, *22*, 074701. [CrossRef]
27. Medikare, M.; Joga, S.; Chidem, K.K. MHD Stagnation Point Flow of a Casson Fluid over a Nonlinearly Stretching Sheet with Viscous Dissipation. *Am. J. Comput. Math.* **2016**, 37–48. [CrossRef]
28. Mythili, D.; Sivaraj, R. Influence of higher order chemical reaction and non-uniform heat source/sink on Casson fluid flow over a vertical cone and flat plate. *J. Mol. Liq.* **2016**, *216*, 466–475. [CrossRef]
29. Ullah, I.; Khan, I.; Shafie, S. Hydromagnetic Falkner-Skan flow of Casson fluid past a moving wedge with heat transfer. *Alex. Eng. J.* **2016**, *55*, 2139–2148. [CrossRef]
30. Imtiaz, M.; Hayat, T.; Alsaedi, A. Mixed convection flow of Casson nanofluid over a stretching cylinder with convective boundary conditions. *Adv. Powder Technol.* **2016**, *27*, 2245–2256. [CrossRef]
31. Cebeci, T.; Bradshaw, P. *Physical and Computational Aspects of Convective Heat Transfer*, 1st ed.; Springer: New York, NY, USA, 1988.

© 2019 by the authors. Licensee MDPI, Basel, Switzerland. This article is an open access article distributed under the terms and conditions of the Creative Commons Attribution (CC BY) license (http://creativecommons.org/licenses/by/4.0/).

Article

MHD Boundary Layer Flow of Carreau Fluid over a Convectively Heated Bidirectional Sheet with Non-Fourier Heat Flux and Variable Thermal Conductivity

Dianchen Lu [1], Mutaz Mohammad [2], Muhammad Ramzan [3,4,*], Muhammad Bilal [5], Fares Howari [6] and Muhammad Suleman [1,7]

1. Department of Mathematics, Faculty of Science, Jiangsu University, Zhenjiang 212013, China; dclu@ujs.edu.cn (D.L.); suleman@ujs.edu.cn (M.S.)
2. Department of Mathematics & Statistics, College of Natural and Health Sciences, Zayed University, 144543 Abu Dhabi, UAE; Mutaz.Mohammad@zu.ac.ae
3. Department of Computer Science, Bahria University, Islamabad Campus, Islamabad 44000, Pakistan
4. Department of Mechanical Engineering, Sejong University, Seoul 143-747, Korea
5. Department of Mathematics, University of Lahore, Chenab Campus, Gujrat 50700, Pakistan; me.bilal.786@outlook.com
6. College of Natural and Health Sciences, Zayed University, 144543 Abu Dhabi, UAE; Fares.Howari@zu.ac.ae
7. Department of Mathematics, COMSATS University, Islamabad 45550, Pakistan
* Correspondence: mramzan@bahria.edu.pk; Tel.: +92-3005122700

Received: 1 April 2019; Accepted: 23 April 2019; Published: 2 May 2019

Abstract: In the present exploration, instead of the more customary parabolic Fourier law, we have adopted the hyperbolic Cattaneo–Christov (C–C) heat flux model to jump over the major hurdle of "parabolic energy equation". The more realistic three-dimensional Carreau fluid flow analysis is conducted in attendance of temperature-dependent thermal conductivity. The other salient impacts affecting the considered model are the homogeneous-heterogeneous (h-h) reactions and magnetohydrodynamic (MHD). The boundary conditions supporting the problem are convective heat and of h-h reactions. The considered boundary layer problem is addressed via similarity transformations to obtain the system of coupled differential equations. The numerical solutions are attained by undertaking the MATLAB built-in function bvp4c. To comprehend the consequences of assorted parameters on involved distributions, different graphs are plotted and are accompanied by requisite discussions in the light of their physical significance. To substantiate the presented results, a comparison to the already conducted problem is also given. It is envisaged that there is a close correlation between the two results. This shows that dependable results are being submitted. It is noticed that h-h reactions depict an opposite behavior versus concentration profile. Moreover, the temperature of the fluid augments for higher values of thermal conductivity parameters.

Keywords: Carreau fluid; Cattaneo–Christov heat flux model; convective heat boundary condition; temperature dependent thermal conductivity; homogeneous-heterogeneous reactions

1. Introduction

Non-Newtonian fluids have gained substantial attention of researchers and scientists owing to their widespread applications. A number of examples like apple sauce, chyme, emulsions, mud, soaps, shampoos and blood at low shear stress may be quoted as non-Newtonian fluids. Existing literature does not facilitate us to identify a single relation that exhibits numerous physiognomies of non-Newtonian fluids. This is why a variety of mathematical models have been suggested, as deemed appropriate, to the requirement. The viscosity of the fluid plays a vital role in the chemical engineering industry. In case of generalized Newtonian fluids, viscosity is dependent on shear stress. In some fluids, a change up to two to three orders in magnitude may not make a visible effect in some fluids, but its impact can't be ignored particularly in polymer industry and lubrication processes. Bird et al. [1] presented the idea of generalized Newtonian fluids with the idea that the viscosity fluctuates with the shear rate. Fluid flows over a solid surface have been frequently studied and the reports revealed that surface forces become significant on a micro level and lead to the enhanced fluid viscosity due to fluid layering [2–5]. The major shortcoming of the Power-law model is that it does not properly address the viscosity in case of very low or high shear rates. To overcome this hurdle, the Carreau fluid model is introduced [6]. Contrary to the Power law model, the viscosity remains finite as the shear rate vanishes. This is why the constitutive relation for the Carreau fluid model is more appropriate in case of free surface flows. Owing to such important characteristics, the Carreau fluid model has attracted researchers for many years. Chhabra and Uhlherr [7] deliberated the Carreau fluid flow over the spheres and this concept was extended by Bush and Phan-Thein [8]. The squeezing Carreau fluid flow past sphere is examined by Uddin et al. [9]. Tshehla [10] deliberated the flow of the Carreau fluid over an inclined plane. The nonlinear radiation impact on the 3D Carreau fluid flow was deliberated by Khan et al. [11]. Khan et al. [12] obtained the solution of the Carreau nanofluid flow analytically with entropy generation. Similar explorations discussing Carreau fluid flow may be found at [13–15] and many therein.

Flows under the influence of magnetohydrodynamics (MHD) have a wide range of applications including thermal insulators, blood flow measurements, petroleum and polymer technologies, nuclear reactors and MHD generators. Taking into account all such applications, many researchers have examined the flows stimulated by magnetohydrodynamics. Waqas et al. [16] conversed micropolar fluid's flow with the effect of convective boundary condition and magnetohydrodynamics. He also considered effects of viscous dissipation and mixed convection. Ramzan et al. [17] deliberated the series solution of micropolar fluid flow in attendance of MHD, partial slip and convective boundary condition over a porous stretching sheet. Besthapu et al. [18] calculated the numerical solution of double stratification nanofluid flow with MHD and viscous dissipation using the finite element method past an exponentially stretching sheet. Khan and Azam [19] explored the flow of Carreau nanofluid under the influence of magnetohydrodynamic using a numerical technique named bvp4c. Turkyilmazoglu [20] examined the exact solution of micropolar fluid flow with the mixed convection and magnetohydrodynamic past a permeable heated/cooled deformable plate. Hayat et al. [21] premeditated the flow of Oldroyd-B nanofluid in attendance of MHD and heat generation/absorption using Optimal Homotopy analysis method HAM. Some recent attempts highlighting effects of magnetohydrodynamic may be found at [22–25].

The importance of heat transfer is fundamental in many engineering processes like nuclear reactors, fuel cells, and microelectronics. Thermal conductivity is considered to be constant in all such procedures. Nevertheless, the requirement of variable characteristics is fundamental. A variation from $0°F$ to $400°F$ [26] in temperature is observed in such cases. Fourier law of heat conduction [27] has been a customary gauge for years in heat transfer applications. However, a major drawback of parabolic energy equation experiences a disruption in the beginning which prevails throughout the entire process, which forces the researchers to look for some modification to Fourier's law. Cattaneo [28] proposed an improved Fourier's law by instituting a thermal relaxation term. Later, Oldroyd's upper-convected derivatives [29] are considered as an alternative to the thermal relaxation time in Cattaneo's model. Recent attempts in various scenarios with an emphasis on C–C flux model may encompass a study by Ramzan et al. [30], highlighting effects of the 2D third grade-fluid flow accompanying Cattaneo–Christov heat flux and magnetohydrodynamics. Flow analysis is done in the presence of h-h reactions and convective boundary condition. Hayat et al. [31] found an analytical solution of Jeffrey fluid flow past a stretched cylinder with the effect of C–C heat flux and thermal stratification. Sui et al. [32] studied upper-convected Maxwell nanofluid flow with C–C heat flux and slip boundary condition past a linearly stretched sheet. Liu et al. [33] discussed a fractional C–C flux model numerically where the fractional derivative is represented by a weight coefficient.

Many chemical reactions necessitate the presence of h-h reactions. Fewer of these reactions act at a slow pace, whereas some absolutely not except in the attendance of the catalyst. These reactions are involved in many scenarios like fibrous insulations, production of polymers and ceramics, and air and water pollution. The heterogeneous reactions cover the complete phase evenly and is found in solid phase. Nevertheless, the homogeneous reactions are covered by catalysis and combustion. The homogeneous catalyst occurs in liquid and gaseous states but heterogeneous catalyst exists in solid form. The latest research discussing the h-h reactions effects may comprise the study by Kumar et al. [34] who examined the irreversibility process with h-h reactions of the flow of carbon nanotubes based nanofluid past a bi-directional stretched surface. The flow with h-h reactions of Blasius nanofluid is pondered by Xu [35]. Sithole et al. [36] used a Bivariate spectral local linearization method to investigate the effects of h-h reactions on the flow of time dependent micropolar nanofluid past a stretched surface. The numerical simulations are conducted for h-h reactions and nonlinear thermal radiation past a 3D crossfluid flow with MHD by Khan et al. [37]. In a gravity driven nanofluid film flow, the effects of h-h reactions with mixed convection are deliberated by Rasees et al. [38]. Ramzan et al. [39,40] highlighted the time dependent nanofluid squeezing flow with carbon nanotubes under the influence of h-h reactions and C–C heat flux, and in Micropolar nanofluid flow with thermal radiation past a nonlinear stretched surface and many therein [40–45].

In all the aforementioned literature surveys, it is observed that either the effect of only C–C heat flux or h-h reactions have been discussed in various geometries. Even if the simultaneous effects of C–C heat flux and h-h reactions have been discussed, it is in the two-dimensional case. However, much less literature is available featuring effects of both C–C and h-h reactions in 3D models. The present study discusses the 3D Carreau fluid model in attendance of temperature-dependent thermal conductivity, C–C heat flux and h-h reactions accompanied by the impact of convective heat with h-h boundary conditions. A MATLAB built-in bvp4c routine is betrothed to obtain series solutions. Graphs are drawn depicting effects of pertinent parameters on involved distributions. Validation of presented results in the limiting case is also an additional feature of this exploration.

2. Mathematical Formulation

Let us presume a 3D flow of Carreau fluid in x- and y-directions with respective velocities $u = u_w(x) = cx$ and $v = v_w(y) = dy$ occupying the region $z = 0$ under the influence of C–C heat flux and variable thermal conductivity past a bidirectional stretching surface as shown in Figure 1. Flow analysis is performed subject to h-h reactions with magnetohydrodynamics. Temperature at the surface T_w is considered to be more than the temperature away from the surface T_∞. A magnetic field with strength B_o is introduced along the z-axis. Electric and Hall effects are ignored. Small Reynolds number's assumption needs to omit an induced magnetic field. For two chemical species A and B, analysis is performed in the presence of h-h reactions. For homogeneous reaction, the cubic autocatalysis is epitomized by the following expression [46]:

$$A + 2B \rightarrow 3B, rate = k_c ab^2. \tag{1}$$

However, on the catalyst surface, the first order isothermal reaction is given by:

$$A \rightarrow B, rate = k_s a. \tag{2}$$

For both the h-h reaction processes, it is assumed that temperature is constant. Governing equations that abide by the above mentioned assumptions are given below:

$$u_x + v_y + w_z = 0, \tag{3}$$

$$uu_x + vu_y + wu_z = vu_{zz}\left[\beta^* + (1-\beta^*)\left\{1+\Gamma^2(u_z)^2\right\}^{\frac{n-1}{2}}\right] - \frac{\sigma B_0^2}{\rho}u$$
$$+v(n-1)(1-\beta^*)\Gamma^2(u_{zz})(u_z)^2\left\{1+\Gamma^2(u_z)^2\right\}^{\frac{n-3}{2}}, \quad (4)$$

$$uv_x + vv_y + wv_z = vv_{zz}\left[\beta^* + (1-\beta^*)\left\{1+\Gamma^2(v_z)^2\right\}^{\frac{n-1}{2}}\right] - \frac{\sigma B_0^2}{\rho}v$$
$$+v(n-1)(1-\beta^*)\Gamma^2(v_{zz})(v_z)^2\left\{1+\Gamma^2(v_z)^2\right\}^{\frac{n-3}{2}}, \quad (5)$$

$$\rho C_P \mathbf{V}.\nabla T = -\nabla.\mathbf{q}, \quad (6)$$

$$ua_x + va_y + wa_z = D_A a_{zz} - k_c ab^2, \quad (7)$$

$$ub_x + vb_y + wb_z = D_B b_{zz} + k_c ab^2, \quad (8)$$

with **q** being the heat flux satisfying the relation

$$\mathbf{q} + K_1\left(\mathbf{q}_t + \mathbf{V}.\nabla\mathbf{q} - \mathbf{q}.\nabla\mathbf{V} + (\nabla.\mathbf{V})\mathbf{q}\right) = -\nabla(\alpha T). \quad (9)$$

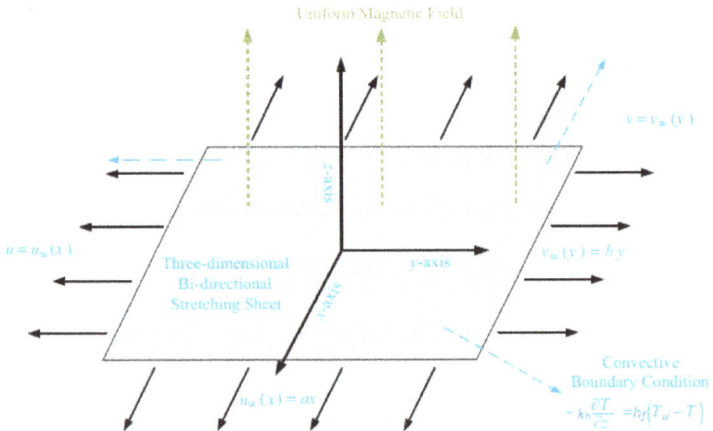

Figure 1. Geometry of the problem.

Using the fluid's incompressibility condition and Christov [29], Equations (6) and (9) take the following form after omission of **q**:

$$uT_x + vT_y + +wT_z = \frac{1}{\rho C_P}(\alpha T_z)_z,$$

$$-K_1\begin{pmatrix} u^2 T_{xx} + v^2 T_{yy} + w^2 T_{zz} + 2uv T_{xy} \\ +2vw T_{yz} + 2uw T_{xz} + (uu_x + vu_y + wu_z)T_x + \\ (uv_x + vv_y + wv_z)T_y + (uw_x + vw_y + ww_z)T_z \end{pmatrix}. \quad (10)$$

The supporting boundary conditions to the given model are

$$u = u_w(x) = cx, \ v = v_w(y) = dy, \ w = 0,$$

$$-k_h T_z = h_f(T_w - T), \ D_A a_z = k_s a, \ D_B b_z = -k_s a, \ at \ z = 0,$$

$$u \to 0, \ v \to 0, \ a \to a_o, \ b \to 0, T \to T_\infty as z \to \infty, \quad (11)$$

considering temperature dependent thermal conductivity $\epsilon = \frac{k_w - k_\infty}{k_\infty}$ as defined in [47].

Taking into account the transformations

$$u = cxf'(\eta), \ v = cyg'(\eta), \ w = -\sqrt{cv}\left(f(\eta) + g(\eta)\right), \ \alpha = \alpha_\infty(1 + \epsilon\theta),$$
$$\theta(\eta) = \frac{T - T_\infty}{T_w - T_\infty}, \ \eta = \sqrt{\frac{c}{v}}z, \ a = a_0\phi(\eta), \ b = a_0h(\eta). \quad (12)$$

The requirement of Equation (3) is met inevitably, whereas Equations (4), (5), (7), (8), (10) and (11) take the following form:

$$\left[\beta^* + (1 - \beta^*)\left\{1 + We_1^2(f'')^2\right\}^{\frac{n-3}{2}}\left\{1 + nWe_1^2(f'')^2\right\}\right]f''' + (f + g)f'' - f'^2 - M^2f' = 0, \quad (13)$$

$$\left[\beta^* + (1 - \beta^*)\left\{1 + We_2^2(g'')^2\right\}^{\frac{n-3}{2}}\left\{1 + nWe_2^2(g'')^2\right\}\right]g''' + (f + g)g'' - g'^2 - M^2g' = 0, \quad (14)$$

$$(1 + \epsilon\theta)\theta'' + \epsilon\theta'^2 + Pr(f + g)\theta' - Pr\lambda_1\left((f + g)^2\theta'' + (f + g)(f' + g')\theta'\right) = 0, \quad (15)$$

$$\phi'' + Sc(f + g)\phi' - Sc\gamma_1\phi h^2 = 0, \quad (16)$$

$$\zeta h'' + Sc(f + g)h' + Sc\gamma_1\phi h^2 = 0, \quad (17)$$

$$f(0) = 0, f'(0) = 1, \ g(0) = 0, g'(0) = \lambda,$$
$$\phi'(0) = \gamma_2\phi(0), \theta'(0) = -\delta(1 - \theta(0)),$$
$$\zeta h'(0) = -\gamma_2\phi(0), \ f'(\infty) \to 0, \ f''(\infty) \to 0, \ g'(\infty) \to 0,$$
$$g''(\infty) \to 0, \ \theta(\infty) \to 0, \ \phi(\infty) \to 1. \quad (18)$$

Different parameters used in the above equations are defined as follows:

$$\gamma_1 = \frac{k_c a_0^2}{c}, \ \gamma_2 = \frac{k}{D_A a_0}\sqrt{\frac{v}{c}}, \ Sc = \frac{v}{D_A}, \ Pr = \frac{\mu C_p}{k}, \ \lambda = \frac{d}{c},$$
$$We_1 = \sqrt{\frac{c\Gamma^2 U_w^2}{v}}, We_2 = \sqrt{\frac{c\Gamma^2 V_w^2}{v}}, \ \lambda_1 = K_1c, \ \zeta = \frac{D_B}{D_A}, \ \delta = \frac{h_f}{k}\sqrt{\frac{v}{c}}, M^2 = \frac{\sigma B_0^2}{c\rho}. \quad (19)$$

The expectation as in most applications that coefficients of chemical species A and B are of equivalent magnitude lead us to make a supplementary presumption that diffusion coefficients D_A and D_B are equivalent i.e., $\zeta = 1$, [46]. Thus, we get:

$$\phi(\eta) + h(\eta) = 1. \quad (20)$$

Now, Equations (16) and (17) take the following form:

$$\phi'' + Sc(f + g)\phi' - Sc\gamma_1\phi(1 - \phi)^2 = 0, \quad (21)$$

with boundary conditions

$$\phi'(0) = \gamma_2\phi(0), \ \phi(\infty) = 1. \quad (22)$$

The skin friction coefficient in dimensional form is

$$C_{fx} = \frac{\tau_{xz}}{\rho U_w^2(x)}, C_{fy} = \frac{\tau_{yz}}{\rho u_w^2(y)}. \quad (23)$$

Dimensionless forms of Skin friction coefficient is

$$C_{fx}Re_x^{1/2} = f''(0)\left[\beta^* + (1 - \beta^*)\left\{1 + We_1^2(f''(0))^2\right\}^{\frac{n-3}{2}}\right], \quad (24)$$

$$C_{fy}Re_x^{1/2} = g''(0)\left[\beta^* + (1 - \beta^*)\left\{1 + We_2^2(g''(0))^2\right\}^{\frac{n-3}{2}}\right]. \quad (25)$$

3. Numerical Solutions

The software MATLAB with built-in bvp4c function is engaged to solve the system of differential equations. It requires converting the differential equation with higher order to the first order along with their respective boundary conditions. We have considered f as y_1, g as y_4, θ as y_7 and ϕ as y_9 during the conversion as

$$
\begin{aligned}
y_1' &= y_2, \quad y_2' = y_3, \\
y_3' &= \frac{y_2^2 + My_2 - (y_1 + y_4)y_3}{\beta^* + (1 - \beta^*)\left(1 + We_1^2 y_3^2\right)^{\frac{n-3}{2}}\left(1 + nWe_1^2 y_3^2\right)}, \\
y_4' &= y_5, \quad y_5' = y_6, \\
y_6' &= \frac{y_5^2 + My_5 - (y_1 + y_4)y_6}{\beta^* + (1 - \beta^*)\left(1 + We_2^2 y_6^2\right)^{\frac{n-3}{2}}\left(1 + nWe_2^2 y_6^2\right)}, \\
y_7' &= y_8, \\
y_8' &= \frac{\Pr K_2 (y_1 + y_4)(y_2 + y_5)y_8 - \Pr(y_1 + y_4)y_8}{1 + \epsilon y_7 - \Pr K_2 (y_1 + y_4)^2}, \\
y_9' &= y_{10}, \\
y_{10}' &= Sc\gamma_1 y_9 (1 - y_9)^2 - Sc(y_1 + y_4)y_{10},
\end{aligned}
$$

accompanying the conditions

$$
\begin{aligned}
y_1(0) &= 0, y_2(0) = 1, y_4(0) = 0, y_5(0) = \lambda, y_2(\infty) = 0, y_5(\infty) = 0, \\
y_8(0) &= -\delta(1 - y_7(0)), y_7(\infty) = 0, y_{10}(0) = \gamma_2 y_9(0), y_9(\infty) = 1.
\end{aligned}
$$

This MATLAB built-in routine is verified by drawing Table 1, in which the results are compared with the previously published article in a limiting case. Previously, Khan et al. [11] have used the same bvp4c technique to tackle the 3D Carreau fluid model. In Table 1, the Skin friction coefficients for varied values of λ is calculated. It is found that all obtained values are in total alignment to [11].

Table 1. Comparison of $-f''(0)$ varied estimates of λ when $n = 3, We_1 = We_2 = 0$.

λ	Khan et al. [11]	Present (bvp4c)
0.1	1.020264	1.020264
0.2	1.039497	1.039497
0.3	1.057956	1.057956
0.4	1.075788	1.075788
0.5	1.093095	1.093095
0.6	1.109946	1.109946
0.7	1.126397	1.126397
0.8	1.142488	1.142488
0.9	1.158253	1.158253
1.0	1.173720	1.173720

In Table 2, a comparison is tabulated for various magnetic parameters and stretching ratio parameter values against the skin friction coefficient along vertical and horizontal directions. It is noted that the skin friction along the x-direction is gradually increasing for the mounting values of λ.

Table 2. Comparison of $-f''(0)$ and $-g''(0)$ for various values of M and λ.

M	$\lambda = 0$		$\lambda = 0.5$		$\lambda = 0.5$		$\lambda = 1.0$	
	$-f''(0)$		$-f''(0)$		$-g''(0)$		$-g''(0)$	
	[48]	Present	[48]	Present	[48]	Present	[48]	Present
0.0	1.0042	1.0045	1.0932	1.0930	0.4653	0.4652	1.1748	1.1742
10	3.3165	3.3149	3.3420	3.3137	1.6459	1.6440	3.3667	3.3654
100	10.0498	10.0427	10.0582	10.0531	5.0208	5.0201	10.0663	10.0654

4. Results and Discussion

This segment is dedicated to highlight the impacts of prominent parameters on all involved profiles. In all the figures, the solid lines show the effect of shear thickening ($n > 1$) fluid while the dashed lines show the shear thinning ($n < 1$) fluid properties. Figures 2 and 3 are illustrated to distinguish the impact of local Weissenberg numbers We_1 and We_2 on the velocity components $f'(\eta)$ and $g'(\eta)$ used for shear thickening and shear thinning fluids respectively. From these figures, it is noted that, for the augmented estimates of We_1, the velocity declines in the case of shear thickening phenomena, while, for the shear thinning phenomena, the velocity increases. Physically, We_1 denotes the proportion between the relaxation time of fluid and increment of viscosity growth of the liquid. For the shear thinning case, the fluid viscosity decreases; consequently, the velocity of the fluid increases. Moreover, for the shear thickening phenomenon, the thickness of the boundary layer escalates for higher values of We_1. In Figure 3, we observed the contradictory behavior for the velocity component $g'(\eta)$. Consequently, it is also found that shear thickening fluid increases the values of We_2, which results in increasing the velocity of fluid and thickness of its related boundary layer. In Figure 4, the effect of viscosity ratio parameter β^* on the velocity is profile $f'(\eta)$ is discussed for the case of shear thinning and shear thickening and keeping all other parameters fixed. An inverse relation is observed, in the case of the shear thinning fluid and for shear thickening fluid velocity of the fluid augmented with escalating values of the viscosity ratio parameter. Moreover, it has been observed that the corresponding boundary layer thickness is less in the case of shear thinning fluid as compared to the shear thickening fluid. Figures 5 and 6 depict declines in velocity profile against the mounting values of magnetic field strength M. A decline in velocity profile is being observed because of the fact that larger values of M enhance the Lorentz force, which increases the resistance for the fluid motion. This decrease in the thickness of the boundary layer is more vigorous for the shear thinning of fluids. Figures 7 and 8 exhibit the effect of stretching ratio parameter λ on $f'(\eta)$ and $g'(\eta)$ velocity profiles, respectively. In Figure 7, the mounting values of stretching ratio parameter resist the fluid flow along the x-axis and this decline is more prominent in shear thinning fluid. Figure 8 shows the opposite trend for the large values of λ on the velocity profile, as the velocity increases for both shear thinning and thickening of fluids. Stretching ratio parameter is the ratio of velocity components along the y-axis to the x-axis. An increase in λ implies the increment in the y-component of the velocity. The effect of Prandtl number Pr on temperature field is shown in Figure 9. Temperature profile decreases for higher values of Pr. The Prandtl number represents the fraction of momentum diffusivity to thermal diffusivity. Thus, an increase in Pr deteriorates the thermal conductivity; ultimately, it decreases the temperature distribution. The effect of thermal conductivity parameter ϵ on the temperature field is being displayed in Figure 10. It is observed that an increase in values of ϵ boosts the temperature distribution. It is an accepted truth that liquids with larger thermal conductivity possess higher temperature. The impact of Schmidt number Sc on concentration profile is being displayed in Figure 11. The augmented values of Sc number boosts the concentration profile and thickness of boundary layer for both the thinning and shear thickening fluids, respectively. The Schmidt number represents the ratio of the molecular diffusion to the viscous diffusion, and the viscous diffusion decreases upon increasing the Sc, which enhances the mass transfer in fluid flow. In Figure 12, the thermal relaxation time parameter λ_1 is portrayed against the temperature profile. It is witnessed that, for the increasing values of λ_1, the temperature profile and thickness of the thermal boundary layer decrease. Figure 13 portrays the effect of Biot number δ on temperature profile. It is noticed that larger values of Biot number escalate the temperature field. A direct relation of heat transfer coefficient with Biot number implies an increase in temperature profile for increasing values of δ. The strength of homogeneous and heterogeneous reactions γ_1 and γ_2 against concentration profile is shown in Figures 14 and 15 as reactants expend in homogeneous reactions. Thus, a reduction in concentration profile is seen for mounting values of γ_1. This fact is shown in Figure 14. An opposite behavior for concentration distribution is observed in Figure 15. Escalating values of heterogeneous reactions decrease diffusion and thereby decrement in concentration is perceived for less diffused particles.

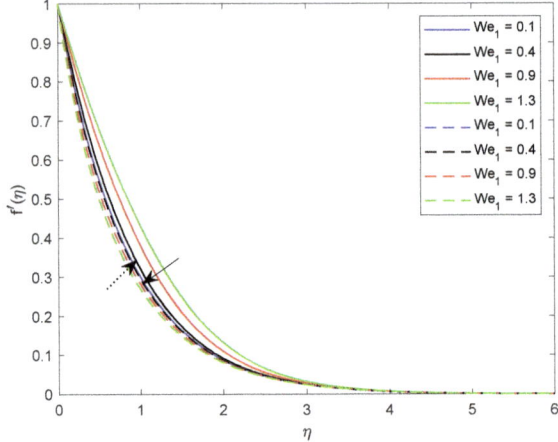

Figure 2. Illustration of We_1 versus $f'(\eta)$.

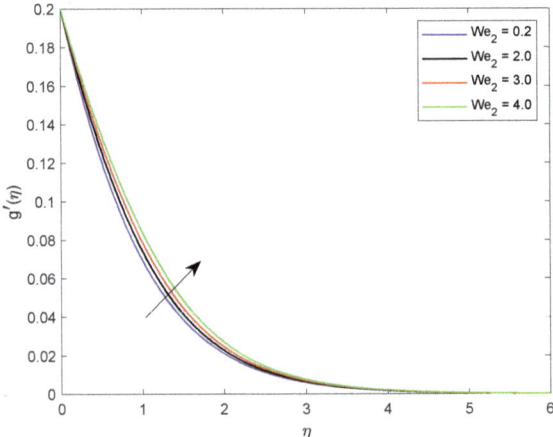

Figure 3. Illustration of We_2 versus $g'(\eta)$.

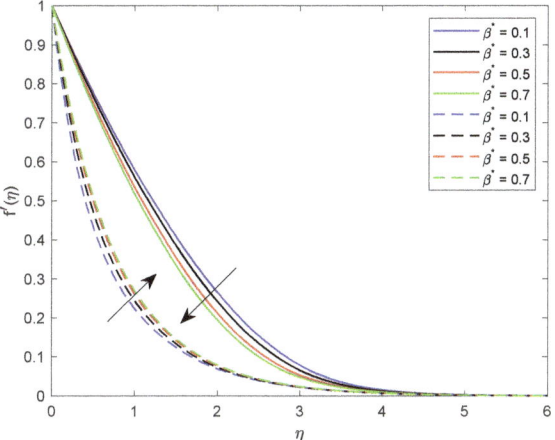

Figure 4. Illustration of β^* versus $f'(\eta)$.

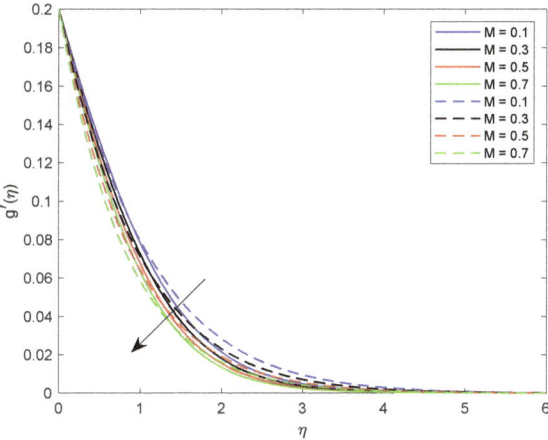

Figure 5. Illustration of M versus $g'(\eta)$.

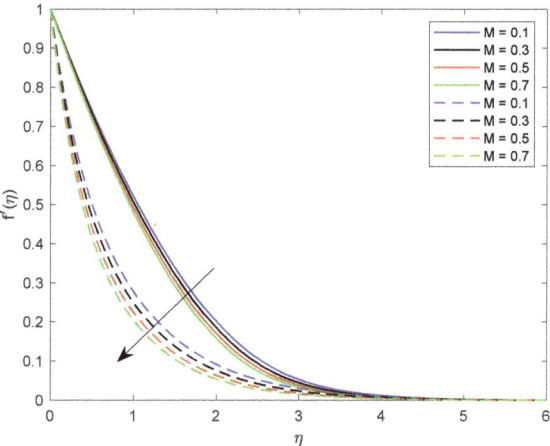

Figure 6. Illustration of M versus $f'(\eta)$.

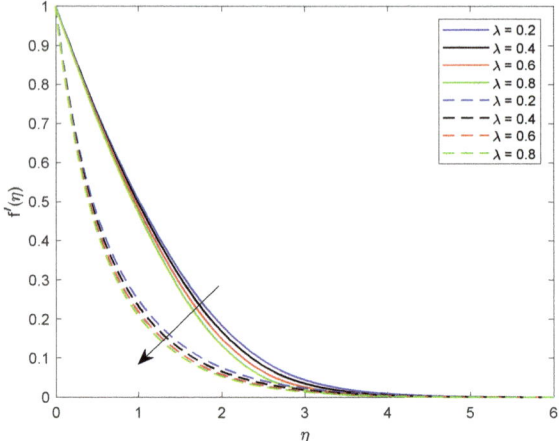

Figure 7. Illustration of λ versus $f'(\eta)$.

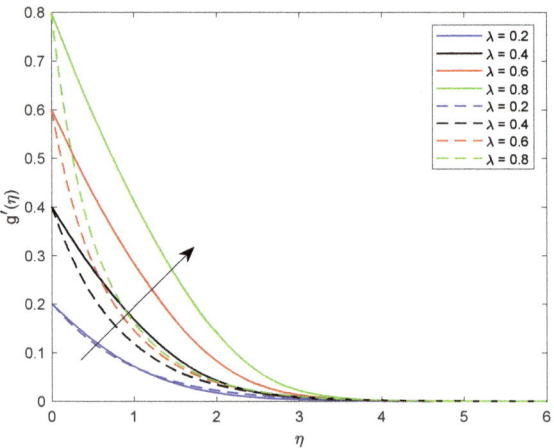

Figure 8. Influence of λ on $g'(\eta)$.

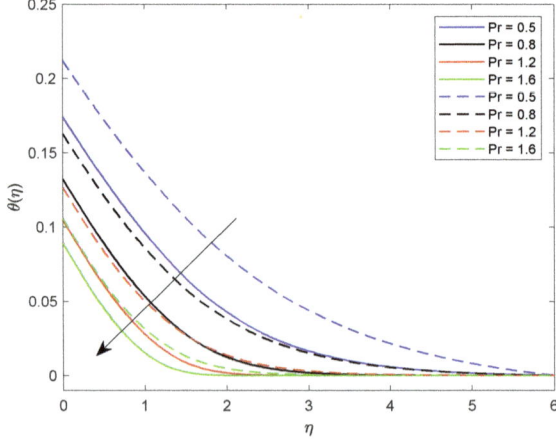

Figure 9. Illustration of Pr versus $\theta(\eta)$.

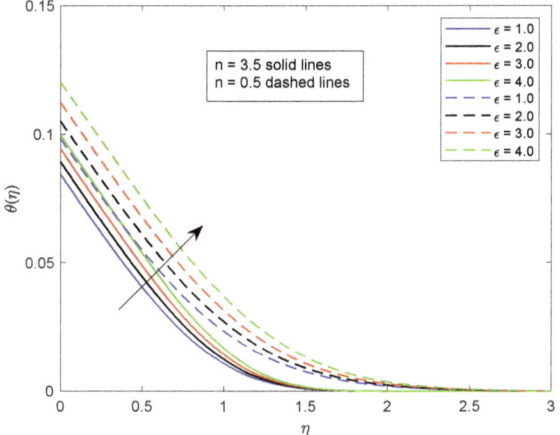

Figure 10. Illustration of ϵ versus $\theta(\eta)$.

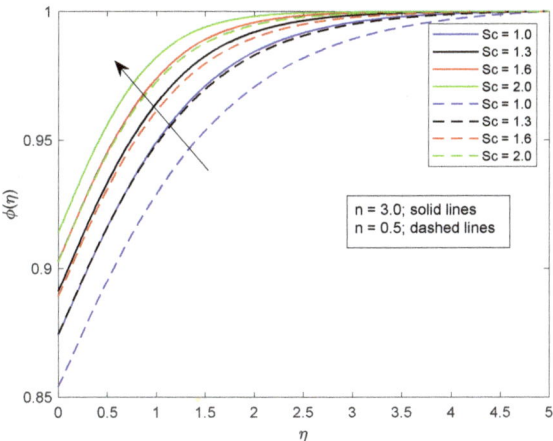

Figure 11. Illustration of Sc versus $\phi(\eta)$.

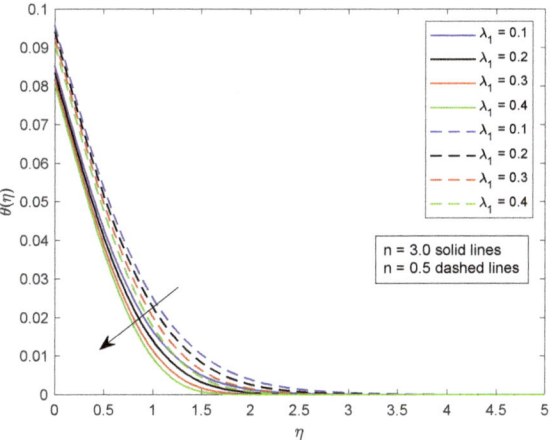

Figure 12. Illustration of λ_1 versus $\theta(\eta)$.

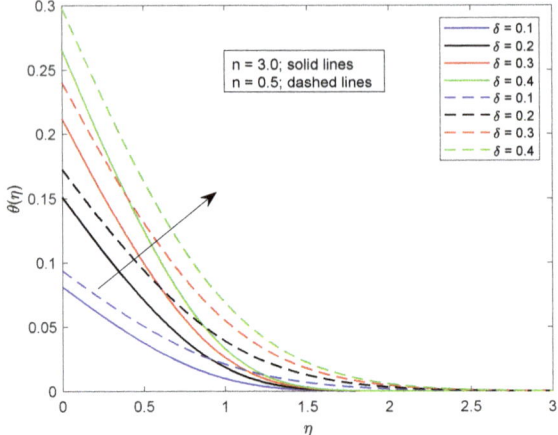

Figure 13. Illustration of δ versus $\theta(\eta)$.

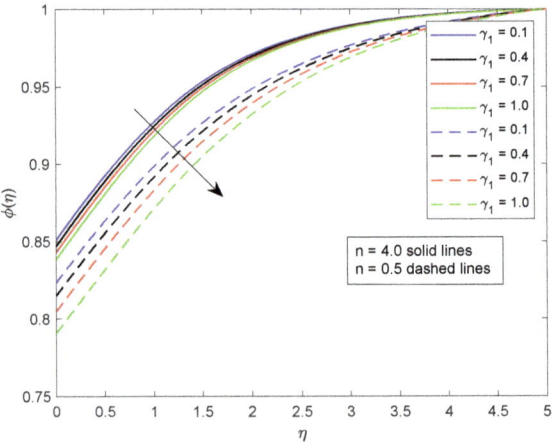

Figure 14. Illustration of γ_1 versus $\phi(\eta)$.

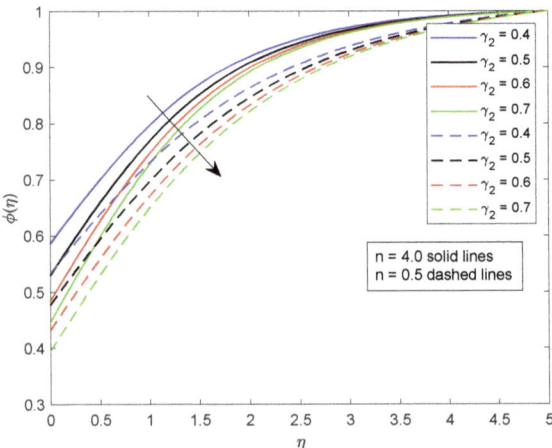

Figure 15. Illustration of γ_2 versus $\phi(\eta)$.

5. Conclusions

In this exploration, impacts of h-h reactions on three-dimensional Carreau fluid flow is witnessed with the presence of temperature dependent thermal conductivity and magneto-hydrodynamic past a bidirectional stretched surface. Furthermore, the impact of C–C heat flux accompanying convective boundary condition is also witnessed. A numerical method is betrothed to find the solution. The notable features of the present study are appended below:

- Strength of homogeneous and heterogeneous reactions show the same decreasing trend on concentration distribution.
- Effects of Prandtl number and Biot number on temperature field are also conflicting.
- The velocity of the fluid is in decline for a stronger magnetic effect.
- Velocity escalates for growing estimates of ratios of stretching rate.
- With an increase in the value of the Schmidt number, the concentration of the fluid is enhanced.

Author Contributions: Formal Analysis, M.R.; Funding Acquisition, M.M., F.H.; Investigation, M.R.; Methodology, D.L.; Project Administration, M.R., M.B.; Software, M.M., M.S.; Validation, M.B.; Writing—Original Draft, M.B.; Writing—Review and Editing, M.S.

Funding: Dr. Mutaz Mohammad and Dr. Fares Howari would like to extend their thanks and appreciation to the UAE Space Agency for funding this research (Grant No. Z01-2016-00).

Acknowledgments: The authors are highly thankful for exceptional support raised by the Zayed University, Abu Dhabi, UAE, Jiangsu University, China, and Bahria University, Islamabad, Pakistan.

Conflicts of Interest: The authors declare no conflict of interest.

Abbreviations

a, b	concentrations of chemical species
A, B	chemical species
a_0	positive dimensional constants
B_0	Magnetic field strength [kg s^{-2} A^{-1}]
C_p	Specific heat [J/kg K]
c, d	stretching constants
Cf_x	Skin friction coefficient
D_A	diffusion coefficient of species A
D_B	diffusion coefficient of species B
f', g'	Dimensionless velocities
h_f	Heat transfer coefficient
h	dimensionless concentration due to heterogeneous reaction
K_1	thermal relaxation time
k_∞	ambient thermal conductivity
k_c	rate constant of chemical species A
k_s	rate constant of chemical species B
k_w	thermal conductivity at wall
M	Magnetic parameter
n	power law index
Pr	Prandtl number
q	heat flux
Sc	Schmidth number
t	time
T_∞	Ambient temperature [K]
T	Temperature of fluid [K]
T_w	Wall temperature [K]
u_w	sheet velocity along x-axis [m/s]
v_w	sheet velocity along y-axis [m/s]

V	Velocity vector
(u, v, w)	Velocity components [m/s]
$u_w(x)$	Stretching velocity along x-axis [m/s]
(x, y, z)	Rectangular coordinate axis [m]
We_1	Weissenberg number
We_2	Weissenberg number
α	variable thermal diffusivity
β^*	ratio of viscosities
γ_1	Thermal Biot number
γ_2	Concentration Biot number
λ	ratio of stretching rates
λ_1	thermal relaxation time coefficient
ν	Kinematic viscosity [m^2/s]
θ	Dimensionless temperature
σ	Electrical conductivity [m^{-3} kg^{-1} s^3 A^2]
μ	Dynamic viscosity [kg/m/s]
η	Similarity variable
ρ	Density of fluid [kg/m^3]
δ	Deborah number
ϕ	dimensionless concentration
ξ	ratio of diffusion coefficients
∇	nibla operator
Γ	material parameter
ϵ	variable thermal conductivity

References

1. Bird, R.; Byron, R.; Armstrong, R.C.; Hassager, O. *Dynamics of Polymeric Liquids. Vol. 1: Fluid Mechanics*; Wiley: London, UK, 1987.
2. Quoc, V.T.; Kim B. Transport phenomena of water in molecular fluidic channels. *Sci. Rep.* **2016**, *6*, 33881.
3. Quoc, V.T.; Park, B.; Park, C.; Kim, B. Nano-scale liquid film sheared between strong wetting surfaces: Effects of interface region on the flow. *J. Mech. Sci. Technol.* **2015**, *29*, 1681–1688.
4. Ghorbanian, J.; Beskok, A. Scale effects in nanochannel liquid Flows. *Microfluid. Nanofluid.* **2016**, *20*, 121. [CrossRef]
5. Ghorbanian, J.; Celebi, A.T.; Beskok, A. A phenomenological continuum model for force-driven nano-channel liquid flows. *J. Chem. Phys.* **2016**, *145*, 184109. [CrossRef]
6. Carreau, P.J. Rheological equations from molecular network Theories. *Trans. Soc. Rheol.* **1972**, *16*, 99–127. [CrossRef]
7. Chhabra, R.P.; Uhlherr P.H.T. Creeping motion of spheres through shear-thinning elastic fluids described by the Carreau viscosity equation. *Rheol. Acta* **1980**, *19*, 187–195. [CrossRef]
8. Bush, M.B.; Phan-Thien, N. Drag force on a sphere in creeping motion throug a Carreau model fluid. *J. Non-Newton. Fluid Mech.* **1984**, *16*, 303–313. [CrossRef]
9. Uddin, J.; Marston, J.O.; Thoroddsen, S.T. Squeeze flow of a Carreau fluid during sphere impact. *Phys. Fluids* **2012**, *24*, 073104. [CrossRef]
10. Tshehla, M.S. The flow of a Carreau fluid down an incline with a free surface. *Int. J. Phys. Sci.* **2011**, *6*, 3896–3910.
11. Khan, M.; Irfan, M.; Khan, W.A.; Alshomrani, A.S. A new modeling for 3D Carreau fluid flow considering nonlinear thermal radiation. *Results Phys.* **2017**, *7*, 2692–2704. [CrossRef]
12. Khan, M.; Ijaz, M.; Kumar, A.; Hayat, T.; Waqas, M.; Singh, R. Entropy generation in flow of Carreau nanofluid. *J. Mol. Liq.* **2019**, *278*, 677–687. [CrossRef]
13. Khan, M.; Irfan, M.; Khan, W.A. Thermophysical properties of unsteady 3D flow of magneto Carreau fluid in the presence of chemical species: A numerical approach. *J. Braz. Soc. Mech. Sci. Eng.* **2018**, *40*, 108. [CrossRef]

14. Irfan, M.; Khan, W.A.; Khan, M.; Gulzar, M. Influence of Arrhenius activation energy in chemically reactive radiative flow of 3D Carreau nanofluid with nonlinear mixed convection. *J. Phys. Chem. Solids* **2019**, *125*, 141–152. [CrossRef]
15. Vasu, B.; Ray, A.K. Numerical study of Carreau nanofluid flow past vertical plate with the Cattaneo–Christov heat flux model. *Int. J. Numer. Methods Heat Fluid Flow* **2019**, *29*, 702–723.
16. Waqas, M.; Farooq, M.; Khan, M.I.; Alsaedi, A.; Hayat, T.; Yasmeen, T. Magnetohydrodynamic (MHD) mixed convection flow of micropolar liquid due to nonlinear stretched sheet with convective condition. *Int. J. Heat Mass Transf.* **2016**, *102*, 766–772. [CrossRef]
17. Ramzan, M.; Farooq, M.; Hayat, T.; Chung, J.D. Radiative and Joule heating effects in the MHD flow of a micropolar fluid with partial slip and convective boundary condition. *J. Mol. Liq.* **2016**, *221*, 394–400. [CrossRef]
18. Besthapu, P.; Haq, R.U.; Bandari, S.; Al-Mdallal, Q.M. Mixed convection flow of thermally stratified MHD nanofluid over an exponentially stretching surface with viscous dissipation effect. *J. Taiwan Inst. Chem. Eng.* **2017**, *71*, 307–314. [CrossRef]
19. Khan, M.; Azam, M. Unsteady heat and mass transfer mechanisms in MHD Carreau nanofluid flow. *J. Mol. Liq.* **2017**, *225*, 554–562. [CrossRef]
20. Turkyilmazoglu, T. Mixed convection flow of magnetohydrodynamic micropolar fluid due to a porous heated/cooled deformable plate: Exact solutions. *Int. J. Heat Mass Transf.* **2017**, *106*, 127–134. [CrossRef]
21. Hayat, T.; Muhammad, T.; Shehzad, S.A.; Alsaedi, A. An analytical solution for magnetohydrodynamic Oldroyd-B nanofluid flow induced by a stretching sheet with heat generation/absorption. *Int. J. Therm. Sci.* **2017**, *111*, 274e288. [CrossRef]
22. Khan, M.I.; Waqas, M.; Hayat, T.; Alsaedi, A. A comparative study of Casson fluid with homogeneous-heterogeneous reactions. *J. Colloid Interface Sci.* **2017**, *498*, 85–90. [CrossRef]
23. Ramzan, M.; Bilal, M.; Chung, J.D. MHD stagnation point Cattaneo–Christov heat flux in Williamson fluid flow with homogeneous–heterogeneous reactions and convective boundary condition—A numerical approach. *J. Mol. Liq.* **2017**, *225*, 856–862. [CrossRef]
24. Ramzan, M.; Farooq, M.; Hayat, T.; Alsaedi, A.; Cao, J. MHD stagnation point flow by a permeable stretching cylinder with Soret-Dufour effects. *J. Cent. South Univ.* **2015**, *22*, 707–716. [CrossRef]
25. Su, X.; Zheng, L.; Zhang, X.; Zhang, J. MHD mixed convective heat transfer over a permeable stretching wedge with thermal radiation and ohmic heating. *Chem. Eng. Sci.* **2012**, *78*, 1–8. [CrossRef]
26. Pal, D.; Chatterjee, S. Soret and Dufour effects on MHD convective heat and mass transfer of a power-law fluid over an inclined plate with variable thermal conductivity in a porous medium. *Appl. Math. Comput.* **2013**, *219*, 7556–7574. [CrossRef]
27. Fourier, J.B.J. *Théorie Analytique de la Chaleur*; Chez Firmin Didot: Paris, France, 1822.
28. Cattaneo, C. Sulla conduzione del calore, Attidel Seminario Matematico e Fisico Dell. *Modena Reggio Emilia* **1948**, *3*, 83–101.
29. Christov, C.I. On frame indifferent formulation of the Maxwell–Cattaneo model of finite-speed heat conduction. *Mech. Res. Commun.* **2009**, *36*, 481–486. [CrossRef]
30. Ramzan, M.; Bilal, M.; Chung, J.D. Effects of MHD homogeneous-heterogeneous reactions on third grade fluid flow with Cattaneo–Christov heat flux. *J. Mol. Liq.* **2016**, *223*, 1284–1290. [CrossRef]
31. Hayat, T.; Khan, M.I.; Farooq, M.; Alsaedi, A.; Khan, M.I. Thermally stratified stretching flow with Cattaneo–Christov heat flux. *Int. J. Heat Mass Transf.* **2017**, *106*, 289–294. [CrossRef]
32. Sui, J.; Zheng, L.; Zhang, X. Boundary layer heat and mass transfer with Cattaneo–Christov double-diffusion in upper-convected Maxwell nanofluid past a stretching sheet with slip velocity. *Int. J. Therm. Sci.* **2016**, *104*, 461–468. [CrossRef]
33. Liu, L.; Zheng, L.; Liu, F.; Zhang, X. Heat conduction with fractional Cattaneo–Christov upper-convective derivative flux model. *Int. J. Therm. Sci.* **2017**, *112*, 421–426. [CrossRef]
34. Kumar, R.; Kumar, R.; Sheikholeslami, M.; Chamkha, A.J. Irreversibility analysis of the three-dimensional flow of carbon nanotubes due to nonlinear thermal radiation and quartic chemical reactions. *J. Mol. Liq.* **2019**, *274*, 379–392. [CrossRef]
35. Xu, H. Homogeneous–Heterogeneous Reactions of Blasius Flow in a Nanofluid. *J. Heat Transf.* **2019**, *141*, 024501. [CrossRef]

36. Sithole, H.; Mondal, H.; Magagula, V.M.; Sibanda, P.; Motsa, S. Bivariate Spectral Local Linearisation Method (BSLLM) for unsteady MHD Micropolar-nanofluids with Homogeneous–Heterogeneous chemical reactions over a stretching surface. *Int. J. Appl. Comput. Math.* **2019**, *5*, 12. [CrossRef]
37. Khan, W.A.; Ali, M.; Sultan, F.; Shahzad, M.; Khan, M.; Irfan, M. Numerical interpretation of autocatalysis chemical reaction for nonlinear radiative 3D flow of cross magnetofluid. *Pramana* **2019**, *92*, 16. [CrossRef]
38. Raees, A.; Wang, R.Z.; Xu. H. A homogeneous-heterogeneous model for mixed convection in gravity-driven film flow of nanofluids. *Int. Commun. Heat Mass Transf.* **2018**, *95*, 19–24. [CrossRef]
39. Lu, D.; Li, Z.; Ramzan, M.; Shafee, M.; Chung, J.D. Unsteady squeezing carbon nanotubes based nano-liquid flow with Cattaneo–Christov heat flux and homogeneous–heterogeneous reactions. *Appl. Nanosci.* **2019**, *9*, 169–178. [CrossRef]
40. Lu, D.; Ramzan, M.; Ahmad, S.; Chung, J.D.; Farooq, U. A numerical treatment of MHD radiative flow of Micropolar nanofluid with homogeneous-heterogeneous reactions past a nonlinear stretched surface. *Sci. Rep.* **2018**, *8*, 12431. [CrossRef]
41. Lu, D.; Ramzan, M.; Bilal, M.; Chung, J.D.; Farooq, U.; Tahir, S. On three-dimensional MHD Oldroyd-B fluid flow with nonlinear thermal radiation and homogeneous–heterogeneous reaction. *J. Braz. Soc. Mech. Sci. Eng.* **2018**, *40*, 387. [CrossRef]
42. Ramzan, M.; Bilal, M.; Chung, J.D. Influence of homogeneous-heterogeneous reactions on MHD 3D Maxwell fluid flow with Cattaneo–Christov heat flux and convective boundary condition. *J. Mol. Liq.* **2017**, *230*, 415–422. [CrossRef]
43. Nadeem, S.; Muhammad, N. Impact of stratification and Cattaneo–Christov heat flux in the flow saturated with porous medium. *J. Mol. Liq.* **2016**, *224*, 423–430. [CrossRef]
44. Hayat, T.; Rashid, M.; Alsaedi, A. Three dimensional radiative flow of magnetite-nanofluid with homogeneous-heterogeneous reactions. *Results Phys.* **2018**, *8*, 268–275. [CrossRef]
45. Merkin, J.H. A model for isothermal homogeneous–heterogeneous reactions in boundary layer flow. *Math. Comput. Model.* **1996**, *24*, 125–136. [CrossRef]
46. Chaudhary, M.A.; Merkin, J.H. A simple isothermal model for homogeneous-heterogeneous reactions in boundary layer flow: I. Equal diffusivities. *Fluid Dyn. Res.* **1995**, *16*, 311–333. [CrossRef]
47. Zargartalebi, H.; Ghalambaz, M.; Noghrehabadi, A.; Chamkha, A. Stagnation-point heat transfer of nanofluids toward stretching sheets with variable thermo-physical properties. *Adv. Powder Technol.* **2015**, *26*, 819–829. [CrossRef]
48. Ahmad, K.; Nazar, R. Magnetohydrodynamic three dimensional flow and heat transfer over a stretching surface in a viscoelastic fluid. *J. Sci. Technol.* **2010**, *3*, 1–14.

© 2019 by the authors. Licensee MDPI, Basel, Switzerland. This article is an open access article distributed under the terms and conditions of the Creative Commons Attribution (CC BY) license (http://creativecommons.org/licenses/by/4.0/).

Article

Integer and Non-Integer Order Study of the GO-W/GO-EG Nanofluids Flow by Means of Marangoni Convection

Taza Gul [1,2], Haris Anwar [1], Muhammad Altaf Khan [1], Ilyas Khan [3,*] and Poom Kumam [4,5,6,*]

1. Department of mathematics, City University of Science and Information Technology, Peshawar 25000, Pakistan; tazagul@cusit.edu.pk (T.G.); harismathe@gmail.com (H.A.); makhan@cusit.edu.pk (M.A.K.)
2. Department of Mathematics, Govt. Superior Science College Peshawar, Khyber Pakhtunkhwa, Peshawar 25000, Pakistan
3. Faculty of Mathematics and Statistics, Ton Duc Thang University, Ho Chi Minh City 72915, Vietnam
4. KMUTT-Fixed Point Research Laboratory, Room SCL 802 Fixed Point Laboratory, Science Laboratory Building, Department of Mathematics, Faculty of Science, King Mongkut's University of Technology Thonburi (KMUTT), 126 Pracha-Uthit Road, Bang Mod, Thrung Khru, Bangkok 10140, Thailand
5. KMUTT-Fixed Point Theory and Applications Research Group, Theoretical and Computational Science Center (TaCS), Science Laboratory Building, Faculty of Science, King Mongkut's University of Technology Thonburi (KMUTT), 126 Pracha-Uthit Road, Bang Mod, Thrung Khru, Bangkok 10140, Thailand
6. Department of Medical Research, China Medical University Hospital, China Medical University, Taichung 40402, Taiwan
* Correspondence: ilyaskhan@tdtu.edu.vn (I.K.); poom.kum@kmutt.ac.th (P.K.)

Received: 7 February 2019; Accepted: 22 March 2019; Published: 7 May 2019

Abstract: Characteristically, most fluids are not linear in their natural deeds and therefore fractional order models are very appropriate to handle these kinds of marvels. In this article, we studied the base solvents of water and ethylene glycol for the stable dispersion of graphene oxide to prepare graphene oxide-water (GO-W) and graphene oxide-ethylene glycol (GO-EG) nanofluids. The stable dispersion of the graphene oxide in the water and ethylene glycol was taken from the experimental results. The combined efforts of the classical and fractional order models were imposed and compared under the effect of the Marangoni convection. The numerical method for the non-integer derivative that was used in this research is known as a predictor corrector technique of the Adams–Bashforth–Moulton method (Fractional Differential Equation-12) or shortly (FDE-12). The impact of the modeled parameters were analyzed and compared for both GO-W and GO-EG nanofluids. The diverse effects of the parameters were observed through a fractional model rather than the traditional approach. Furthermore, it was observed that GO-EG nanofluids are more efficient due to their high thermal properties compared with GO-W nanofluids.

Keywords: integer and non-integer order derivatives; GO-W/GO-EG nanofluids; Marangoni convection; FDE-12 numerical method

1. Introduction

Fractional order models are very useful in the study of nanofluids that contain small nanosized particles at the rate of small intervals rather than the traditional concept of integer order derivatives. A fractional order study has the credibility to explain the actual behavior of the physical parameters and is possible only in the case of the small intervals. The influences of the physical parameters in the classical models are limited and, in some cases, different from the fractional order models near the wall surface. Caputo [1] introduced the idea of fractional derivatives from the modified Darcy's law using the concept of unsteadiness. This idea was further modified by the researchers El Amin [2], Atangana and Alqahtani [3], and Alkahtani [4] by introducing varieties of new fractional derivatives and their applications. The fractional derivative concept can potentially be applied to the study of complicated control system problems. Yilun Shang [5] studied finite-time state consensus problems in continuous multi-agent systems with non-linear particles. Liu et al. [6] investigated the fixed-time event-triggered consensus control problem for multi-agent systems with non-linear uncertainties.

Advanced energy assets are the hot issue amid engineers and researchers as a response to rising energy demands. The base liquids have no sufficient thermal efficiency to fulfill the required demands of the industry. The small size of metal particles are used in common solvents to improve the thermal efficiency of the liquids. Water-, ethylene glycol-, and mineral oil-like convectional heat transfer fluids play an imperative role in many industrial and technological approaches such as heat generation, air-conditioning, chemical production, microelectronics, and transportation. The rate of change at small intervals has been examined by Atangana and Baleanu [7] to investigate the physical constraints of nanofluids for the heat transfer applications.

The physical aspects of the nanofluids and the role of the small sized nanoparticles in the enhancement of heat transfer applications using the traditional concept were introduced by Choi [8] to enhance the thermal efficiency of the nanofluids through nanoparticles.

The carbon family has the tendency to provide rapid cooling and fast thermal productivities. The experimental results demonstrated for carbon materials include the results of graphite nanoparticles, graphene oxides, and carbon nanotubes. Ellahi et al. [9] comprehensively discussed the effect of Carbon Nanotubes (CNT) nanofluid flow along a vertical cone with variable wall temperature. The results of both types of nanofluid can be obtained. Gul et al. [10] discussed effective Prandtl number model influences on $Al_2O_3 - H_2O$ and $Al_2O_3 - C_2H_6O_2$ nanofluids' spray along a stretching cylinder. Ellahi [11] worked on the effects of Magneto Hydrodynamic (MHD) and temperature-dependent viscosity on the flow of a non-Newtonian nanofluid in a pipe, using the analytical solution. Ellahi et al. [12] studied shiny film coating for multi-fluid flows of a rotating disk suspended with nanosized silver and gold particles. Khan et al. [13] worked on the Optimal Homotopy Analysis Method (OHAM) solution of Multi Walled Carbon Nanotubes and Single Walled Carbon Nanotubes (MWCNT/SWCNT) nanofluid thin film flow over a nonlinear extending disc.

Hummers and Offeman [14] developed a speedy and comparatively safe technique for the production of graphitic oxide from graphite in what is basically a crystalline substance of sulfuric acid H_2SO_4, potassium permanganate $KMnO_4$, and sodium nitrate $NaNO_3$.

The high thermal conductivity and characteristic lubricity of graphene make it a perfect claimant for the alteration of functional fluids. The solid particles, having an efficient thermal conductivity, are assorted to the base fluid to enhance the overall thermal conductivity of the fluid, as depicted in Maxwell [15]. Balandin et al. [16] examined the efficient thermal conductivity of single layer graphene in different solvents. Wei et al. [17] were pioneers in expressing the use of graphene oxide in ethylene glycol to enhance the thermal conductivity of ethylene glycol (EG). The graphene oxide nanosheets were set and isolated in EG and water at 5% capacity concentrations to enhance the thermal conductivity up to 60% compared with the base liquid EG.

Recently, Gul and Firdous [18] experimentally examined the stable dispersion of the graphene oxide in water and then analyzed the numerical study of the graphene oxide-water (GO-W) nanofluid between two rotating discs for the thermal applications.

Another type of convection which is used for temperature-dependent situations is called Marangoni convection. The existence of a spontaneous interface was first reported in 1855 by Thomson [19] and later represented in detail in 1865 by Marangoni [20] by spreading an oil droplet on a water surface, revealing that lower surface tension will spread on a liquid with higher surface tension.

In light of the previous meaningful discussion, the aim of this study was to examine the GO-W and graphene oxide and ethylene glycol (GO-EG) nanofluid flow under the effect of Marangoni convection using the classical and fractional order models. The comparison of the two types of nanofluids was conducted to investigate the impacts of the physical parameters. The physical and numerical outputs of the classical and fractional models were also compared and discussed. Sheikholeslami and Ganji [21] examined the Cu–H$_2$O nanofluid flow under the impact of Marangoni convection. The numerical approach to find the solution of a different type of problem was previously discussed [22–27]. The numerical scheme of Runge Kutta method of order 4 (RK-4) was used in their study to determine the impact of the physical parameters and numerical outputs.

The published work of Gul and Kiran [18] was extended by including the GO-EG nanofluid and a comparison of GO-EG and GO-W was made. Furthermore, integer and non-integer models h were compared under the effect of Marangoni convection. The fractional order differential equations were tackled numerically with the help of the Fractional Differential Equation-12 (FDE-12) technique [28–32]. A variety of numerical techniques are used to find the solutions of the classical models [33] and these techniques are further combined for the solutions of fractional order problems. Agarwal et al. [34] studied the neural network models using the GML synchronization and impulsive Caputo fractional differential equations. Morales-Delgado et al. [35] worked on the analytic solution for oxygen diffusion from capillaries to tissues involving external force effects using a fractional calculus approach. Khan et al. [36] researched the dynamics of the Zika virus with the Caputo fractional derivative. The physical configuration of the problem is shown in Figure 1.

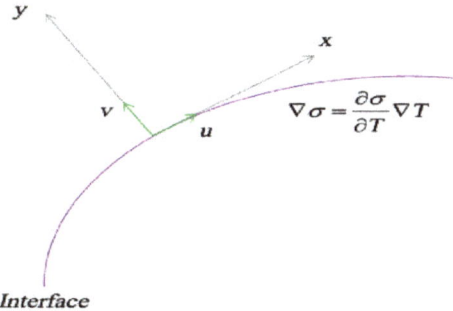

Figure 1. Geometry of the problem.

2. Problem Formulation

The two-dimensional Marangoni boundary layer flow of GO-W and GO-EG nanofluids is considered. The magnetic field is functional to the flow pattern in the transverse direction. The interface temperature is vigilant as a function of x. Assume that both base solvents (water and EG) contain GO nanoplatelets that are present in the thermally stable stage and no slippage. The flow under observation can be put into the following plan for GO nanofluids [14]:

$$\frac{\partial u}{\partial x} + \frac{\partial v}{\partial y} = 0, \qquad (1)$$

$$u\left(\frac{\partial u}{\partial x}\right) + v\left(\frac{\partial v}{\partial y}\right) = v_{nf}\frac{\partial^2 u}{\partial y^2} - \frac{\sigma_{nf}B_0^2}{\rho_{nf}}u, \qquad (2)$$

$$u\left(\frac{\partial T}{\partial x}\right) + v\left(\frac{\partial T}{\partial y}\right) = \frac{k_{nf}}{(\rho c_p)_{nf}}\left(\frac{\partial^2 T}{\partial y^2}\right), \qquad (3)$$

where Equation (1) is the continuity equation, Equation (2) is the momentum equation, and Equation (3) is the energy equation. Exposed boundary conditions are expressed as:

$$v = 0, \ \frac{\mu_{nf}}{\mu_f}\left(\frac{\partial u}{\partial y}\right) = -\frac{\partial \sigma^\otimes}{\partial x} = \sigma_0\left(\gamma\frac{\partial T}{\partial x}\right), \ T = T_\infty + T_0 X^2, X = \frac{x}{L}, \text{ at } y \to 0$$
$$u = 0, \ T = T_\infty, \text{ at } y \to \infty. \quad (4)$$

The Marangoni conditions at the interface are revealed in Equation (4), taking the surface tension $\sigma^\otimes = \sigma[1 - \gamma(T - T_\infty)], \gamma = -\frac{1}{\sigma}\left(\frac{\partial \sigma^\otimes}{\partial T}\right)$, where γ stands for the surface tension temperature coefficient and σ represents the surface tension constant at the origin.

Also, u, v specify velocity components in the x-, y-directions. The interface and external flow of the temperature are represented by T, T_∞ respectively.

The effective $\rho_{nf}, \mu_{nf}, \sigma_{nf}, k_{nf}, (\rho c_p)_{nf}$ indicate the density, dynamic viscosity, electrical conductivity, thermal conductivity, and specific heat capacity of nanoplatelets, respectively, and are defined as:

$$\frac{(\rho c_p)_{nf}}{(\rho c_p)_f} = \left[1 - \phi + \frac{(\rho c_p)_s}{(\rho c_p)_f}\phi\right], \ \frac{\mu_{nf}}{\mu_f} = \frac{1}{(1-\phi)^{2.5}}, \ \frac{\rho_{nf}}{\rho_f} = \left[(1-\phi) + \left(\frac{\rho_s}{\rho_f}\right)\phi\right],$$
$$\frac{\sigma_{nf}}{\sigma_f} = \left(1 + \frac{3(\sigma-1)\phi}{(\sigma+2)-(\sigma-1)\phi}\right). \quad (5)$$

where ϕ is the solid volume fraction and $\sigma_f, \rho_f, (\rho c_p)_f$ are the electrical conductivity, density, and specific heat capacity of the base fluids, respectively.

The similarity transformations are considered as [14]:

$$\eta = \frac{y}{L}, \ \psi = v_f X f(\eta), \ u = \frac{\partial \psi}{\partial y}, \ v = -\frac{\partial \psi}{\partial x}, \ T = T_\infty + T_0 X^2 \Theta(\eta). \quad (6)$$

Using the aforementioned assumption and condition, Equation (1) is verified identically, whereas Equations (2)–(4) are transformed in the following form:

$$\frac{\mu_{nf}}{\mu_f}\frac{\rho_{nf}}{\rho_f}\frac{\partial f^3(\eta)}{\partial \eta^3} + f(\eta)\frac{df^2(\eta)}{d\eta^2} - \left(\frac{df(\eta)}{d\eta}\right)^2 - M^2\frac{\sigma_{nf}}{\sigma_f}\left(\frac{df(\eta)}{d\eta}\right) = 0, \quad (7)$$

$$\left(\frac{k_{nf}}{k_f} + \frac{4}{3}Rd\right)\frac{d\Theta^2(\eta)}{d\eta^2} + \frac{(\rho c_p)_{nf}}{(\rho c_p)_f}Pr\left[f(\eta)\frac{d\Theta(\eta)}{d\eta} - 2\frac{df(\eta)}{d\eta}\Theta(\eta)\right] = 0. \quad (8)$$

$$f = 0, \ \frac{df^2}{d\eta^2} = -2(1-\phi)^{2.5}, \ \Theta = 1, \text{ at } \eta = 0$$
$$\frac{df}{d\eta} = 0, \ \Theta = 0, \text{ at } \eta \to \infty \quad (9)$$

where M, Pr, indicate the transformation of the magnetic parameter and Prandtl number, respectively, and are defined individually as:

$$M^2 = \frac{\sigma_f B_0^2 L^2}{\mu_0 \rho_f}, \ Pr = \frac{(\rho c_p)_f}{k_f}. \quad (10)$$

The local Nusselt number Nu_x is:

$$Nu = -2\frac{k_{nf}}{k_f}\left(\frac{\partial T}{\partial y}\right)_{y=0}. \quad (11)$$

3. Preliminaries on the Caputo Fractional Derivatives

The basic definition and properties related to non-integer or fractional derivatives derived by Caputo are as follows.

3.1. Definition 1

Let $b > 0$, $t > b$; $b, \alpha, t \in R$. The Caputo fractional derivative of order α of function $f \in C^n$ is given by:

$$_b^C D_t^\alpha f(t) = \frac{1}{\Gamma(n-\alpha)} \int_b^t \frac{f^{(n)}(\xi)}{(t-\xi)^{\alpha+1-n}} d\xi, \ n-1 < \alpha < n \in N. \tag{12}$$

3.2. Property 1

Let $f(t), g(t) : [a,b] \to \mathfrak{R}$ be such that $_b^C D_t^\alpha f(t)$ and $_b^C D_t^\alpha g(t)$ exist almost everywhere and let $c_1, c_2 \in \mathfrak{R}$. Then $_b^C D_t^\alpha \{c_1 f(t) + c_2 g(t)\}$ exists almost everywhere and:

$$_b^C D_t^\alpha \{c_1 f(t) + c_2 g(t)\} = c_1 \, _b^C D_t^\alpha f(t) + c_2 \, _b^C D_t^\alpha g(t). \tag{13}$$

3.3. Property 2

The function $f(t) \equiv c$ is constant and therefore the fractional derivative is zero: $_b^C D_t^\alpha c = 0$. The general description of the fractional differential equation is assumed, including the Caputo concept:

$$_b^C D_t^\alpha x(t) = f(t, x(t)), \ \alpha \in (0,1) \tag{14}$$

with the initial conditions $x_0 = x(t_0)$.

4. Solution Methodology

The variables were selected to alter Equations (7)–(9) into the system of the first order differential equations:

$$y_1 = \eta, \ y_2 = f, \ y_3 = f', \ y_4 = f'', \ y_5 = \Theta, \ y_6 = \Theta'. \tag{15}$$

The variables selected in Equation (15) were used for the classical (integer) system and Equations (7)–(9) are settled as:

$$y_1' = 1, \ y_2' = y_3, \ y_3' = y_4, \ y_4' = \left[\frac{\mu_{nf}\rho_{nf}}{\mu_f\rho_f}\right]^{-1}\left[-y_2 y_3 + (y_4)^2 + \frac{\sigma_{nf}}{\sigma_f} M^2 y_3\right], \ y_5' = y_6,$$
$$y_6' = \left[\frac{k_{nf}}{k_f} + \frac{4}{3}Rd\right]^{-1}\left[-\frac{(\rho c_p)_{nf}}{(\rho c_p)_f} \Pr(y_2 y_6 - 2 y_3 y_5)\right] \tag{16}$$

with initial conditions:

$$y_1 = 0, \ y_2 = 0, \ y_3 = u_1, \ y_4 = -2(1-\phi)^{2.5}, \ y_5 = 1, \ y_6 = u_2. \tag{17}$$

The first order ordinary differential equations system (15) is further transformed into the Caputo fractional order derivatives.

The FDE-12 technique was adopted for the fractional order differential equations. The final system and initial conditions are as follows:

$$\begin{pmatrix} D_\eta^\alpha y_1 \\ D_\eta^\alpha y_2 \\ D_\eta^\alpha y_3 \\ D_\eta^\alpha y_4 \\ D_\eta^\alpha y_5 \\ D_\eta^\alpha y_6 \end{pmatrix} = \begin{pmatrix} 1 \\ y_3 \\ y_4 \\ \left(\frac{\mu_{nf}\rho_{nf}}{\mu_f\rho_f}\right)^{-1}\left(-y_2 y_3 + (y_4)^2 + \frac{\sigma_{nf}}{\sigma_f} M^2 y_3\right) \\ y_6 \\ \left(\frac{k_{nf}}{k_f} + \frac{4}{3}Rd\right)^{-1}\left(-\frac{(\rho c_p)_{nf}}{(\rho c_p)_f} \Pr(y_2 y_6 - 2 y_3 y_5)\right) \end{pmatrix}, \ \begin{pmatrix} y_1 \\ y_2 \\ y_3 \\ y_4 \\ y_5 \\ y_6 \end{pmatrix} = \begin{pmatrix} 0 \\ 0 \\ u_1 \\ -2(1-\phi)^{2.5} \\ 1 \\ u_2 \end{pmatrix}. \tag{18}$$

5. Results and Discussions

The GO-W and GO-EG nanofluid flows under the effect of Marangoni convection were analyzed using the classical and fractional models for heat transfer applications. The impact of the physical parameters was obtained through the classical and fractional order models and compared. Moreover, the impact of the embedded parameters, comprising GO-W and GO-EG nanofluids, was compared, and it was observed that due to rich thermophysical properties the GO-EG nanofluid is a comparatively better heat transfer solvent.

In the following figures, an upward arrow shows an increasing effect while a downward arrow shows a decreasing effect.

The effect of the nanofluid volume fraction ϕ using the classical model versus the velocity profile $f(\eta)$ for the GO-W and GO-EG nanofluids is depicted in Figure 2. The rising values of ϕ lead to enhance the velocity field linearly in the classical model. Physically, the larger amount of nanoparticle volume fraction generates the friction force, and this force is more visible near the wall, reducing the flow motion. However, this impact is unclear in the classical model. The increase in the flow motion due to the rising values of ϕ indicates that the thermal efficiency of the nanofluid provides strength to the flow field. Moreover, this impact is comparatively high using the GO-EG nanofluids.

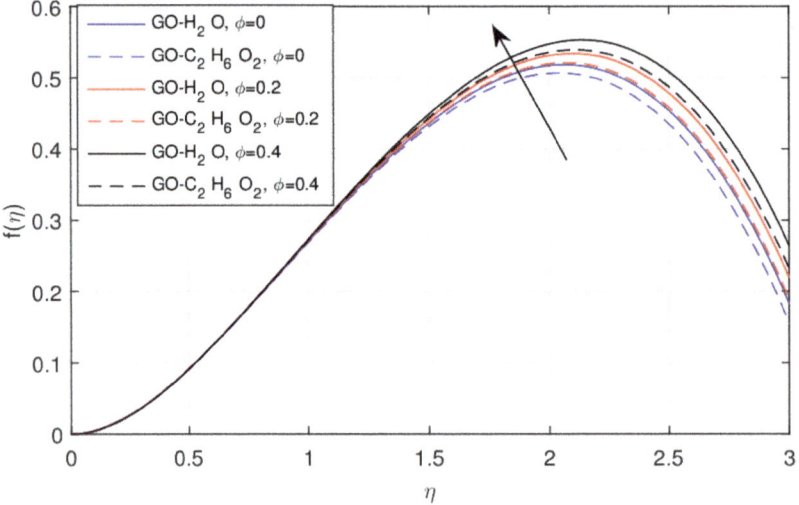

Figure 2. The impact of ϕ versus classical $f(\eta)$, when $M = 0.1$.

The effect of the nanofluid volume fraction ϕ using the fractional model for the same values of ϕ is shown in Figure 3. Near the wall surface the velocity field falls, because near the wall surface friction increases, which retards the velocity and increases after the critical point due to the reduction in friction. Physically, the larger nanoparticle volume fraction generates the friction force and this force is more visible near the wall, reducing the flow motion. This effect is clearer using the fractional model. The impact of the increasing values of ϕ versus the radial velocity field $f'(\eta)$ using the integer model is shown in Figure 4. The same effect as discussed above was observed. The larger amount of ϕ increases the value of $f'(\eta)$ in the integer order model. The impact of the increasing values of ϕ versus the radial velocity field $f'(\eta)$ using the fractional model is shown in Figure 5. The larger amount of ϕ reduces $f'(\eta)$ in the fractional order model near the wall surface and after the point of inflection the velocity enhances, as shown in Figure 5.

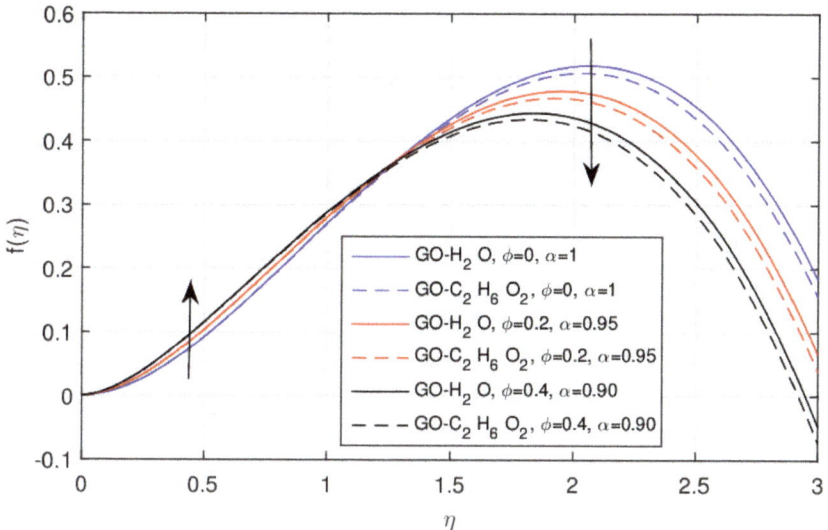

Figure 3. The impact of ϕ versus fractional $f(\eta)$, when $M = 0.1$.

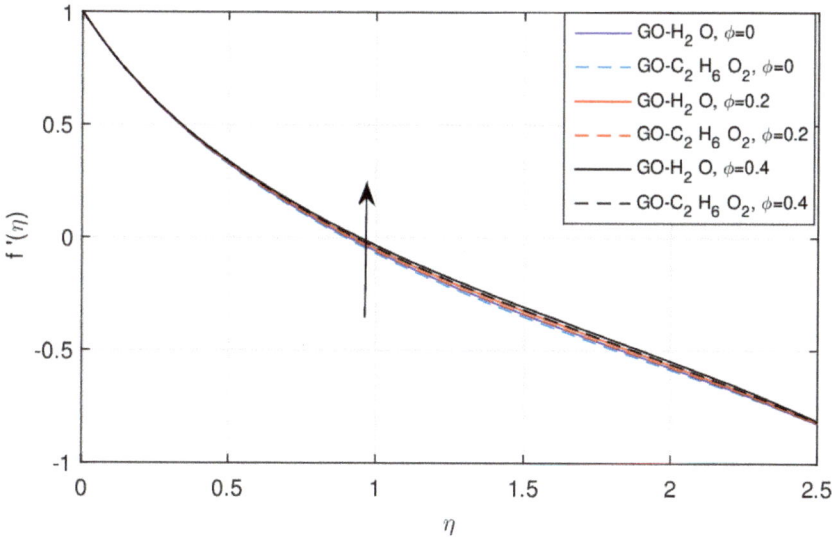

Figure 4. The impact of ϕ versus classical $f'(\eta)$, when $M = 0.1$.

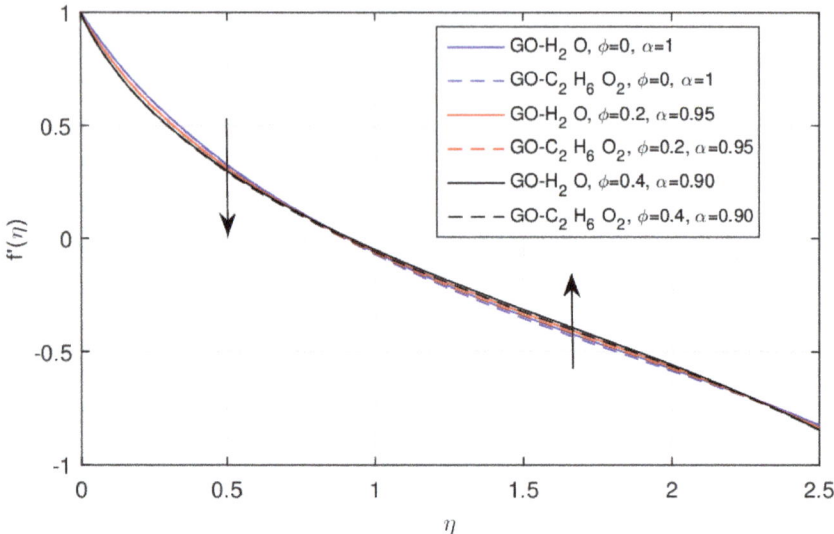

Figure 5. The impact of ϕ versus fractional $f'(\eta)$, when $M = 0.1$.

The influence of the larger values of the magnetic parameter M versus the temperature profile $\Theta(\eta)$ for the integer order and fractional order problems is shown in Figures 6 and 7, respectively. This is due to the Lorentz force, which results in resistance to the transport phenomena. This retarding force controls the GO-W and GO-EG nanofluid velocities, which is useful in numerous industrial and engineering applications such as heat transferring, industrial cooling, and nanofluid coolant. Mathematically, the magnetic parameter represents the ratio of the magnetic induction to the viscous force. Moreover, GO-EG was found to show more dominant results than GO-W.

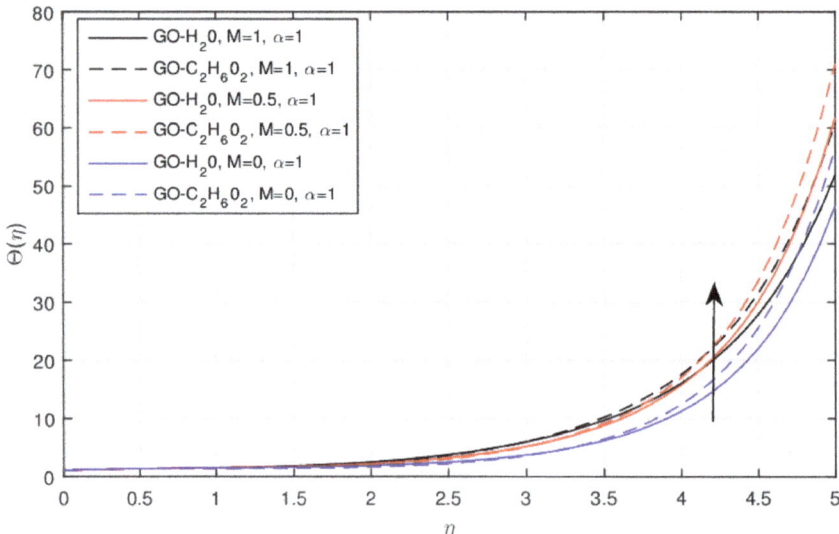

Figure 6. The impact of M versus classical $\Theta(\eta)$, when $\phi = 0.2, \mathrm{Pr} = 6.7$.

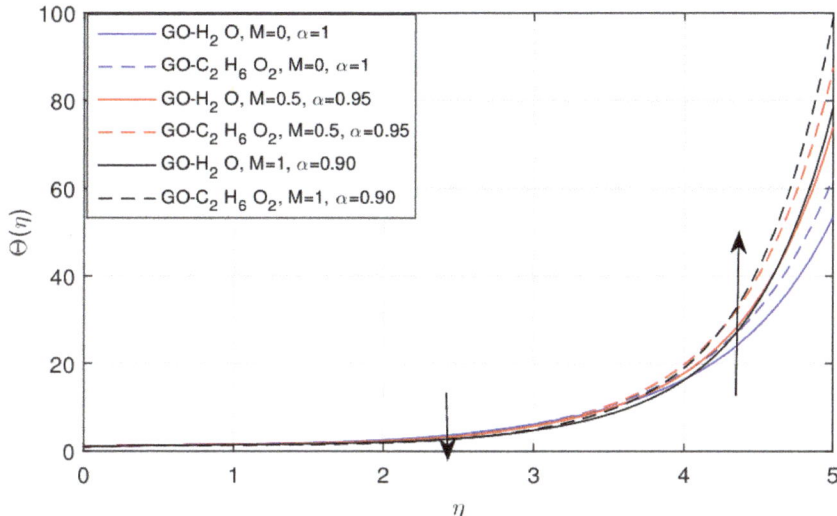

Figure 7. The impact of M versus fractional $\Theta(\eta)$, when $\phi = 0.2, \mathrm{Pr} = 6.7$.

The impact of ϕ using the classical model and the fractional order model versus the temperature profile $\Theta(\eta)$ for the GO-W and GO-EG nanofluids is depicted in Figures 8 and 9. The nanoparticle volume fraction ϕ is basically used as the heat transport agent parameter and its increase boosts up the temperature profile. In fact, the cohesive forces among the liquid molecules release with the increasing amount of ϕ and as a result the thermal boundary layer enhancement. This effect is comparatively efficient in GO-EG nanofluids due to their enriching thermophysical properties.

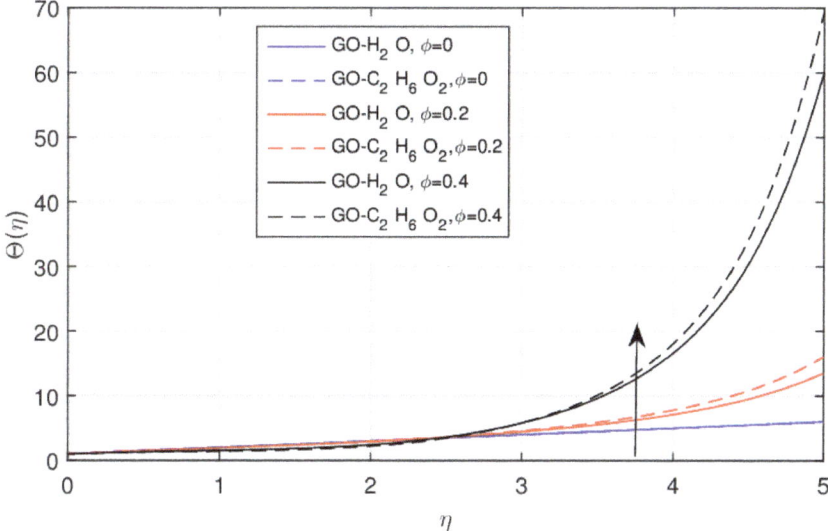

Figure 8. The impact of ϕ versus the integer order of $\Theta(\eta)$, when $M = 0.1, \mathrm{Pr} = 6.7$.

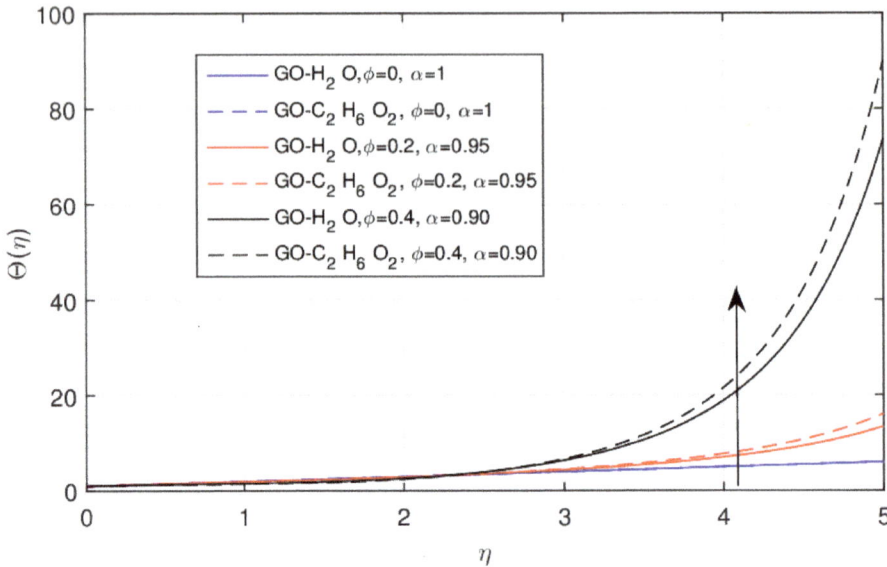

Figure 9. The impact of ϕ versus fractional $\Theta(\eta)$, when $M = 0.1, \mathrm{Pr} = 6.7$.

The impact of the magnetic parameter M versus the Nusselt number Nu of integer and fraction order problems is displayed in Figures 10 and 11, respectively. A higher value of the magnetic parameter enhances the temperature field and reduces the Nusselt number. This effect is slightly clearer using the fractional model for similar values of M, as shown in Figure 11. We noticed that in the cases of GO-EG and GO-W the temperature distribution is dominant and almost completely closed.

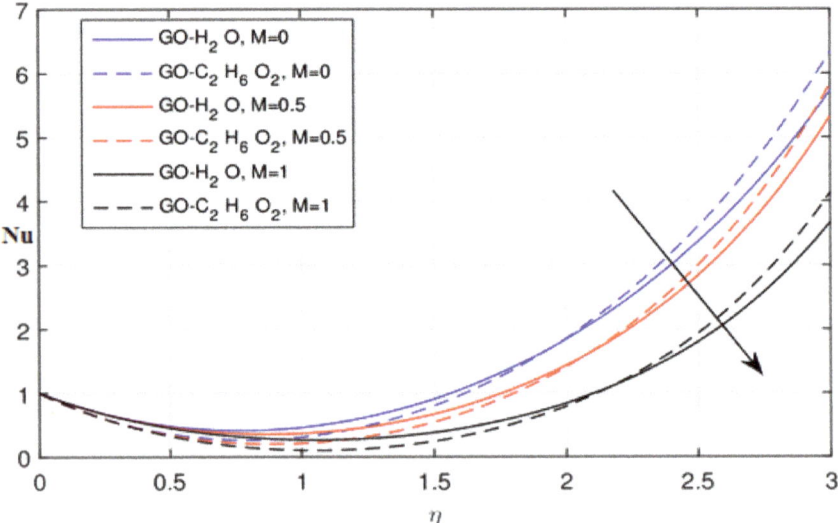

Figure 10. The impact of M versus classical Nu, when $\phi = 0.2, \mathrm{Pr} = 6.7$.

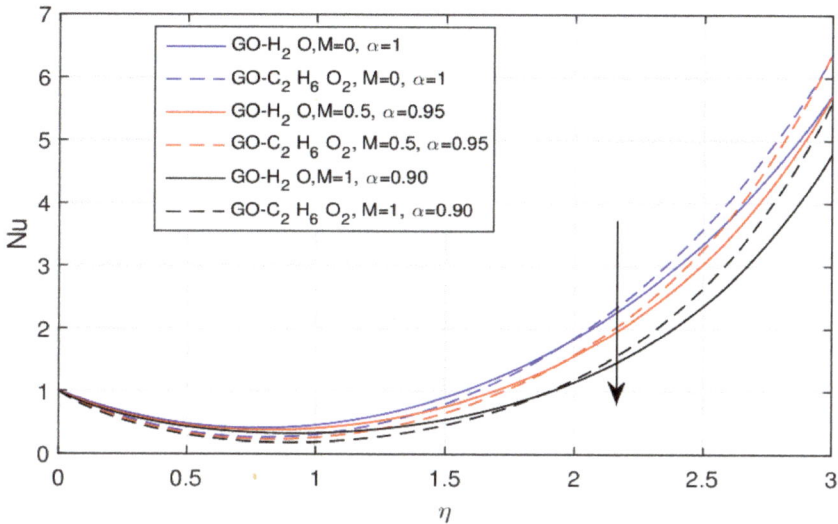

Figure 11. The impact of M versus fractional Nu, when $\phi = 0.2, \Pr = 6.7$.

The impact of the fractional order $\alpha = 1, 0.95, 0.90, 0.85$ versus Nu for both sorts of nanofluids is depicted in Figure 12. It was observed that the heat transfer and cooling efficiency of the GO-EG nanofluid is comparatively higher than the GO-W nanofluid. The Nusselt number increases near the wall surface and declines towards the free surface.

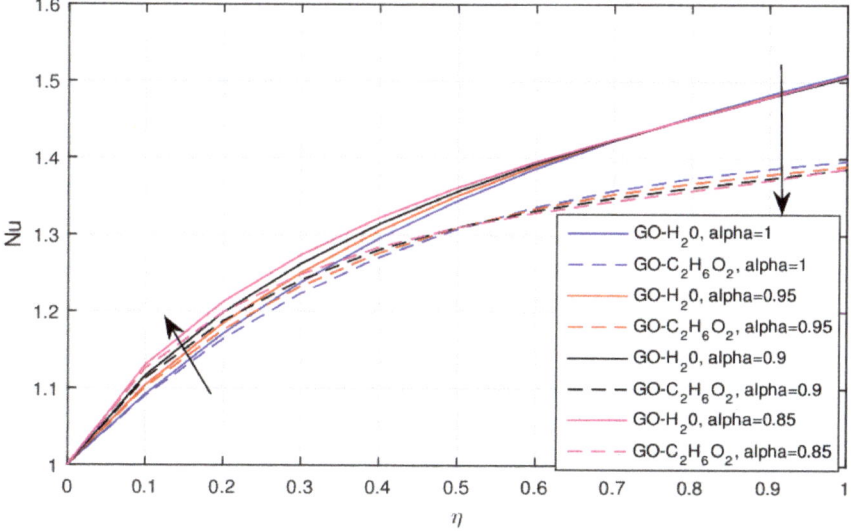

Figure 12. $\alpha = 1, 0.95, 0.90, 0.85$ versus Nu, when $\phi = 0.2, \Pr = 6.7, M = 0.5$.

The thermophysical properties of the two sorts of nanofluids (GO-W and GO-EG) were examined from the experimental results and are displayed in Table 1. These properties of the base fluids were initially calculated at 25 °C. The thermophysical properties were examined at a different temperature level from 25 °C to 40 °C.

Table 1. The experimental values (thermophysical properties) of water, ethylene glycol, and graphene oxide nanoparticles.

Model	ρ (kg/m^3)	C_p (kg^{-1}/k^{-1})	k (Wm^{-1}k^{-1})
Water (W)	997.1	4179	0.613
Graphene oxide (GO)	1800	717	5000
Ethylene glycol (EG)	1.115	0.58	0.1490

The numerical outputs for the heat transfer rate using the fractional order problem are displayed in Table 2. The fractional order $\alpha = 1, 0.95, 0.90, 0.85$ enhances the heat transfer rate in increasing intervals and this effect is relatively high in the GO-W nanofluid.

Table 2. $\alpha = 1, 0.95, 0.90, 0.85$ versus Nu, when $\phi = 0.2, \Pr = 6.7, M = 0.5$.

$\alpha = 1$ η.	$\Theta'(0)$ GO-W	$\Theta'(0)$ GO-EG	$\alpha = 0.95$ η.	$\Theta'(0)$ GO-W	$\Theta'(0)$ GO-EG	$\alpha = 0.9$ η.	$\Theta'(0)$ GO-W	$\Theta'(0)$ GO-EG	$\alpha = 0.85$ η.	$\Theta'(0)$ GO-W	$\Theta'(0)$ GO-EG
0.1	1.0921	1.0903	0.1	1.1039	1.1014	0.1	1.1167	1.1133	0.1	1.1304	1.1259
0.2	1.1708	1.1639	0.2	1.1845	1.1758	0.2	1.1983	1.1875	0.2	1.2121	1.1986
0.3	1.2380	1.2234	0.3	1.2504	1.2328	0.3	1.2624	1.2413	0.3	1.2736	1.2486
0.4	1.2953	1.2706	0.4	1.3051	1.2764	0.4	1.3141	1.2809	0.4	1.3221	1.2840
0.5	1.3442	1.3076	0.5	1.3509	1.3095	0.5	1.3567	1.3101	0.5	1.3617	1.3096
0.6	1.3862	1.3361	0.6	1.3899	1.3344	0.6	1.3928	1.3318	0.6	1.3952	1.3288
0.7	1.4225	1.3578	0.7	1.4235	1.3532	0.7	1.4242	1.3484	0.7	1.4250	1.3441
0.8	1.4544	1.3741	0.8	1.4534	1.3676	0.8	1.4527	1.3619	0.8	1.4527	1.3577
0.9	1.4832	1.3866	0.9	1.4809	1.3794	0.9	1.4796	1.3741	0.9	1.4800	1.3715
1.0	1.5098	1.3966	1.0	1.5072	1.3900	1.	1.5064	1.3866	1.0	1.5083	1.3872

6. Conclusions

The flow of the two types of nanofluids, GO-W and GO-EG, were analyzed for the augmentation of temperature. Numerical and theoretical analyses were carried out under the effect of Marangoni convection. The classical and fractional models were used to investigate the impact of the physical parameters for similar values of the constraint. It was observed that the outputs of the physical parameters over the velocity and temperature profiles in the classical model are limited, but in utilizing the fractional model the effect varies in each interval. The fractional order model specifies the outputs at the small number of intervals, leading to the accurate determination of the physical parameters, which is very necessary for industrial and engineering applications.

The main features of this study are as follows:

- The rising values of ϕ lead to the linear enhancement of the velocity field, which was observed more clearly in the non-integer case compared with the classical model.
- The increasing values of the magnetic parameter increase the temperature field and decrease the Nusselt number. This effect is somewhat better in the fractional case compared to the integer model.
- Due to the rising values of ϕ, the thermal boundary layer increases and this effect is somewhat better in the GO-EG nanofluid rather than the GO-W nanofluid.
- The cooling efficiency and heat transfer of the GO-EG nanofluid is far better than that of the GO-W nanofluid.
- With the Lorentz force, resistance arises in the transport phenomenon. This particular phenomenon controls the GO-W and GO-EG nanofluid velocities. Also, this effect is more visible in GO-EG than in GO-W.
- Due to the fractional order $\alpha = 1, 0.95, 0.90, 0.85$, the heat transfer rate enhances in growing increments and this effect is far better in the GO-W nanofluid compared with the GO-EG nanofluid.

7. Future Work

This mathematical model is extendable for future work considering gold nanoparticles, carbon nanotubes, porous media, variable viscosity, thermal radiation, and hall effects. The fractional ordered derivative scheme is also extendable using the Caputo–Fabrizio and Atangana–Baleanu operators.

Author Contributions: T.G. and H.A.; conceptualization, M.A.K. and T.G.; methodology, I.K.; software, I.K.; and P.K.; validation, T.G., M.A.K. and H.A.; formal analysis, I.K.; P.K.; investigation, T.G.; I.K.; writing—original draft preparation, T.G.; M.A.K. and P.K. I.K.; writing—review and editing.

Funding: This research was funded by the Center of Excellence in Theoretical and Computational Science (TaCS-CoE), KMUTT.

Acknowledgments: This project was supported by the Theoretical and Computational Science (TaCS) Center under Computational and Applied Science for Smart Innovation Research Cluster (CLASSIC), Faculty of Science, KMUTT.

Conflicts of Interest: The authors declare that they have no conflict of interest.

References

1. Caputo, M. Models of flux in porous media with memory. *Water Resour. Res.* **2000**, *36*, 693–705. [CrossRef]
2. El Amin, M.F.; Radwan, A.G.; Sun, S. Analytical solution for fractional derivative gas-flow equation in porous media. *Results Phys.* **2017**, *7*, 2432–2438. [CrossRef]
3. Atangana, A.; Alqahtani, R.T. Numerical approximation of the space-time Caputo-Fabrizio fractional derivative and application to groundwater pollution equation. *Adv. Differ. Equ.* **2016**, *2016*, 156–169. [CrossRef]
4. Alkahtani, B.S.T.; Koca, I.; Atangan, A. A novel approach of variable order derivative: Theory and Methods. *J. Nonlinear Sci. Appl.* **2016**, *9*, 4867–4876. [CrossRef]
5. Shang, Y. Finite-time consensus for multi-agent systems with fixed topologies. *Int. J. Syst. Sci.* **2012**, *43*, 499–506. [CrossRef]
6. Liu, J.; Yu, Y.; Wang, Q.; Sun, C. Fixed-time event-triggered consensus control for multi-agent systems with nonlinear uncertainties. *Neurocomputing* **2017**, *260*, 497–504. [CrossRef]
7. Atangana, A.; Baleanu, D. New fractional derivatives with non-local and non singular kernel: Theory and application to heat transfer model. *arXiv* **2016**, arXiv:1602.03408.
8. Choi, S.U.S. Enhancing thermal conductivity of fluids with nanoparticles, developments and applications of non-Newtonian flows. *FED-231/MD* **1995**, *66*, 99–105.
9. Ellahi, R.; Hassan, M.; Zeeshan, A. Study of Natural Convection MHD Nanofluid by Means of Single and Multi-Walled Carbon Nanotubes Suspended in a Salt-Water Solution. *IEEE Trans. Nanotechnol.* **2015**, *14*, 726–734. [CrossRef]
10. Gul, T.; Nasir, S.; Islam, S.; Shah, Z.; Khan, M.A. Effective prandtl number model influences on the Al_2O_3-H_2O and Al_2O_3-$C_2H_6O_2$ nanofluids spray along a stretching cylinder. *Arab. J. Sci. Eng.* **2019**, *2*, 1601–1616. [CrossRef]
11. Ellahi, R. The effects of MHD and temperature dependent viscosity on the flow of a non-Newtonian nanofluid in a pipe, Analytical solution. *Appl. Math. Model.* **2013**, *37*, 1451–1457. [CrossRef]
12. Ellahi, R.; Zeeshan, A.; Hussain, F.; Abbas, T. Study of Shiny Film Coating on Multi-Fluid Flows of a Rotating Disk Suspended with Nano-Sized Silver and Gold Particles: A Comparative Analysis. *Coatings* **2018**, *8*, 422. [CrossRef]
13. Taza, G.; Waris, K.; Muhammad, S.; Muhammad, A.K.; Ebenezer, B. MWCNTs/SWCNTs Nanofluid Thin Film Flow over a Nonlinear Extending Disc: OHAM solution. *J. Therm. Sci.* **2019**, *28*, 115–122. [CrossRef]
14. Hummers, W.S.; Offeman, R.E. Preparation of graphitic oxide. *J. Am. Chem. Soc.* **1958**, *80*, 1339. [CrossRef]
15. Balandin, A.A.; Ghosh, S.; Bao, W.; Calizo, I.; Teweldebrhan, D.; Miao, F.; Lau, C.N. Superior thermal conductivity of single-layer graphene. *Nano Lett.* **2008**, *8*, 902–907. [CrossRef]
16. Maxwell, J.C. *Treatise on Electricity and Magnetism*; Clarendon Press: Oxford, UK, 1873.
17. Wei, Y.; Huaqing, X.; Dan, B. Enhanced thermal conductivities of nanofluids containing graphene oxide nanosheets. *Nanotechnology* **2010**, *21*, 055705.

18. Gul, T.; Ferdous, K. The experimental study to examine the stable dispersion of the graphene nanoparticles and to look at the GO–H_2O nanofluid flow between two rotating disks. *Appl. Nanosci.* **2018**, *8*, 1711–1728. [CrossRef]
19. Thomson, J. On certain curious motions observable at the surface of wine and other alcoholic liquors. *Philos. Mag.* **1855**, *10*, 330–333. [CrossRef]
20. Marangoni, C. Ueber die Ausbreitung der Tropfeneiner Flussigkeit auf der Oberflache einer anderen. *Ann. Phys.* **1871**, *143*, 337–354. [CrossRef]
21. Sheikholeslami, M.; Ganji, D.D. Influence of magnetic field on $CuOeH_2O$ nanofluid flow considering Marangoni boundary layer. *Int. J. Hydrog. Energy* **2017**, *42*, 2748–2755. [CrossRef]
22. Shirvan, K.M.; Ellahi, R.; Sheikholeslami, T.F.; Behzadmehr, A. Numerical investigation of heat and mass transfer flow under the influence of silicon carbide by means of plasmaenhanced chemical vapor deposition vertical reactor. *Neural Comput. Appl.* **2018**, *30*, 3721–3731. [CrossRef]
23. Barikbin, Z.; Ellahi, R.; Abbasbandy, S. The Ritz-Galerkin method for MHD Couette Fow of non-Newtonian fluid. *Int. J. Ind. Math.* **2014**, *6*, 235–243.
24. Hayat, T.; Saif, R.S.; Ellahi, R.; Muhammad, T.; Ahmad, B. Numerical study of boundary-layer flow due to a nonlinear curved stretching sheet with convective heat and mass conditions. *Results Phys.* **2017**, *7*, 2601–2606. [CrossRef]
25. Hayat, T.; Saif, R.S.; Ellahi, R.; Muhammad, T.; Ahmad, B. Numerical study for Darcy-Forchheimer flow due to a curved stretching surface with Cattaneo-Christov heat flux and homogeneous heterogeneous reactions. *Results Phys.* **2017**, *7*, 2886–2892. [CrossRef]
26. Javeed, S.; Baleanu, D.; Waheed, A.; Khan, M.S.; Affan, H. Analysis of Homotopy Perturbation Method for Solving Fractional Order Differential Equations. *Mathematics* **2019**, *7*, 40. [CrossRef]
27. Srivastava, H.M.; El-Sayed, A.M.A.; Gaafar, F.M. A Class of Nonlinear Boundary Value Problems for an Arbitrary Fractional-Order Differential Equation with the Riemann-Stieltjes Functional Integral and Infinite-Point Boundary Conditions. *Symmetry* **2018**, *10*, 508. [CrossRef]
28. Diethelm, K.; Freed, A.D. The Frac PECE subroutine for the numerical solution of differential equations of fractional order. In *Forschung und Wissenschaftliches Rechnen*; Heinzel, S., Plesser, T., Eds.; 1998 Gessellshaft fur Wissenschaftliche Datenverarbeitung: Gottingen, Germany, 1999; pp. 57–71.
29. Diethelm, K.; Ford, N.J.; Freed, A.D. Detailed error analysis for a fractional Adams method. *Numer. Algorithms* **2004**, *36*, 31–52. [CrossRef]
30. Saifullah Khan, M.A.; Farooq, M. A fractional model for the dynamics of TB virus. *Chaos Solitons Fractals* **2018**, *116*, 63–71.
31. Gul, T.; Khan, M.A.; Khan, A.; Shuaib, M. Fractional-order three-dimensional thin-film nanofluid flow on an inclined rotating disk. *Eur. Phys. J. Plus* **2018**, *133*, 500–5011. [CrossRef]
32. Gul, T.; Khan, M.A.; Noman, W.; Khan, I.; Alkanhal, T.A.; Tlili, I. Fractional Order Forced Convection Carbon Nanotubes Nanofluid Flow Passing Over a Thin Needle. *Symmetry* **2019**, *11*, 312. [CrossRef]
33. Ullah, S.; Khan, M.A.; Farooq, M.; Gul, T.; Hussai, F. A fractional order HBV model with hospitalization. *Discret. Contin. Dyn. Syst.* **2019**, 957–974. [CrossRef]
34. Agarwal, R.; Hristova, S.; O'Regan, D. Global Mittag-Leffler Synchronization for Neural Networks Modeled by Impulsive Caputo Fractional Differential Equations with Distributed Delays. *Symmetry* **2018**, *10*, 473. [CrossRef]
35. Morales-Delgado, V.F.; Gómez-Aguilar, J.F.; Saad, K.M.; Khan, M.A.; Agarwal, P. Analytic solution for oxygen diffusion from capillary to tissues involving external force effects: A fractional calculus approach. *Phys. A Stat. Mech. Its Appl.* **2019**, *523*, 48–65. [CrossRef]
36. Khan, M.A.; Ullah, S.; Farhan, M. The dynamics of Zika virus with Caputo fractional derivative. *AIMS Math.* **2019**, *4*, 134–146. [CrossRef]

© 2019 by the authors. Licensee MDPI, Basel, Switzerland. This article is an open access article distributed under the terms and conditions of the Creative Commons Attribution (CC BY) license (http://creativecommons.org/licenses/by/4.0/).

Article

Two-Phase Couette Flow of Couple Stress Fluid with Temperature Dependent Viscosity Thermally Affected by Magnetized Moving Surface

Rahmat Ellahi [1,2,*], **Ahmed Zeeshan** [2], **Farooq Hussain** [2,3] **and Tehseen Abbas** [4]

1. Center for Modeling & Computer Simulation, Research Institute, King Fahd University of Petroleum & Minerals, Dhahran 31261, Saudi Arabia
2. Department of Mathematics & Statistics, Faculty of Basic and Applied Sciences (FBAS), International Islamic University (IIUI), Islamabad 44000, Pakistan; ahmad.zeeshan@iiu.edu.pk (A.Z.); farooq.hussain@buitms.edu.pk (F.H.)
3. Department of Mathematics, Faculty of Arts and Basic Sciences (FABS), Balochistan University of Information Technology, Engineering, and Management Sciences (BUITEMS), Quetta 87300, Pakistan
4. Department of Mathematics, University of Education Lahore, Faisalabad Campus, Faisalabad 38000, Pakistan; tehseen.abbas@ue.edu.pk
* Correspondence: rellahi@alumni.ucr.edu

Received: 19 March 2019; Accepted: 5 May 2019; Published: 8 May 2019

Abstract: The Couette–Poiseuille flow of couple stress fluid with magnetic field between two parallel plates was investigated. The flow was driven due to axial pressure gradient and uniform motion of the upper plate. The influence of heating at the wall in the presence of spherical and homogeneous Hafnium particles was taken into account. The temperature dependent viscosity model, namely, Reynolds' model was utilized. The Runge–Kutta scheme with shooting was used to tackle a non-linear system of equations. It was observed that the velocity decreased by increasing the values of the Hartman number, as heating of the wall reduced the effects of viscous forces, therefore, resistance of magnetic force reduced the velocity of fluid. However, due to shear thinning effects, the velocity was increased by increasing the values of the viscosity parameter, and as a result the temperature profile also declined. The suspension of inertial particles in an incompressible turbulent flow with Newtonian and non-Newtonian base fluids can be used to analyze the biphase flows through diverse geometries that could possibly be future perspectives of proposed model.

Keywords: couple stress fluid; Hafnium particles; Couette–Poiseuille flow; shooting method; magnetic field

1. Introduction

Diverse forms of flow paths appear when fluid flow is diverted by debris blocking streams. Such multiphase flows take place naturally due to the various factors on plateaus. The physical occurrence of multiphase flows includes chemical processes, pharmaceutical, wastewater management, and power generation. Consequently, the multiphase flows have attracted the attention of scientists and engineers due to the frequently arising issues in industrial and mechanical problems. For instance, couple stress fluid flow under the influence of heat between two parallel walls was examined by Farooq et al. [1]. Mahabaleshwar et al. [2] have investigated the magnetohydrodynamics (MHD) couple stress fluid over the flat sheet affected by the radiation. Exact solutions for the velocity were derived using a power series method for two different models. The First case described the surface temperature while the second case dealt with heat flux. Saad and Ashmway [3] have studied the flow of an unsteady couple stress fluid between two plates. The fluid flows with constant motion of the upper plate which was initially at rest. Influence of lubrication on walls was pondered in such a way that the couple stresses on the boundaries had no impact at all. A suitable transform helps to obtain the velocity of fluid numerically. Akhtar and Shah [4] have presented the exact results for three different types of fundamental flows by taking couple stress fluid as a base fluid. Khan et al. [5] reported an incompressible flow of MHD couple stress in which thermally charged fluid was disturbed by transversely applied magnetic fields. The unsteady Couette flow of non-uniform magnetic field has been investigated by Asghar and Ahamd [6]. Shaowei and Mingyu [7] have devoted their efforts for the study of the Couette flow of Maxwell fluid. Integral and Weber transforms have been used to analyze the physical phenomenon. The Couette flow through a symmetric channel was numerically tackled by Eegunjobi et al. [8]. Few core investigations on Couette flow [9–12] and couple stress fluid [13–15] are listed for those working in the same regimes.

Moreover, Poply et al. [16] have examined the temperature-dependent fluid properties of MHD flow with heat transfer. Ellahi et al. [17] have considered two different viscosity models for their investigations of heated flow. They chose third-grade nanofluid flow through coaxial cylinders. Homotopy analysis method is used to produce a closed form solution. In Reference [18] authors have discussed a temperature dependent thick flow between two opposite walls of uneven configurations. The viscosity of two-dimensional flow was assumed to be decreasing exponentially subject to temperature rise. The study contained the simultaneous effects of radiation and a porous medium. A steady-state flow of fourth-grade fluid in a cylinder was analyzed by Nadeem and Ali [19] and offered a comparative analysis in it. Ellahi et al. [20] studied the thermally charged couple stress fluid suspended with spherically homogenous metallic Hafnium particles for bi-phase flow along slippery walls. The rough surfaces of the walls is tackled with the lubrication effects. Variation in the viscosity of viscoelastic fluid by the Runge–Kutta technique with the shooting technique can be seen in Reference [21]. Makinde [22] focused on the impact of viscosity on the steady fluid flow with gravitational effects. The overhead surface was assumed to be at a constant temperature while the adjunct surface of the plate was heated with some external source. A few core investigations for viscous dissipation can be found in References [23,24].

Furthermore, to enhance the thermal performance, different types of nanoparticles having sizes from 1–100 nm have been utilized in bi-phase fluids. For example, Karimipour et al. [25] have studied the role of miscellaneous nanoparticles for heat transfer flow with MHD. Hosseini et al. [26] repotted a unique model on thermal conductivity of nanofluids. Nasiri et al. [27] have proposed a particle hydrodynamics approach for nano-fluid flows. Safaei et al. [28] have examined nanoplatelets–silver/water nanofluids in fully developed turbulent flows of graphene. All said investigations including References [29–33] end up stating that the presence of nanoparticles always sped up the heat transfer rate.

In the current article, we aim to study the magnetized multiphase Couette-Poiseuille flow of non-Newtonian couple stress fluid suspended by metallic particles of Hafnium with temperature dependent viscosity. The viscosity of the base fluid is exponentially decreasing due to the heating effects at the lower wall of the channel which is at rest. However, the motion of the upper wall causes the multiphase (i.e., solid–liquid) transport. The contribution of the pressure gradient simultaneously distinguishes the investigation further. The humble effort will not only speak about the mechanical and industrial multi-phase flows but would also fill the gap yet not available in the existing literature on the topic under consideration.

2. Mathematical Analysis

Consider a plane Couette flow between two opposite flat plates at $\eta = \pm h$, as shown in Figure 1. Flow is investigated in (ξ, η) plane in such a way that ξ-axis lies in the middle and along the plates. It is a well-established fact [34] that when the flow is generated by the constantly moving upper plate, then only can the unidirectional disturbance in the ξ-direction occur. The axial velocity $[u, 0, 0]$ was along the ξ-direction, whereas lateral velocity was in the η-direction is zero. When the metallic particles of

Hafnium were suspended in couple stress fluid under the influence of higher temperature of the lower wall, then the governing equations in component form [35] can be expressed as:

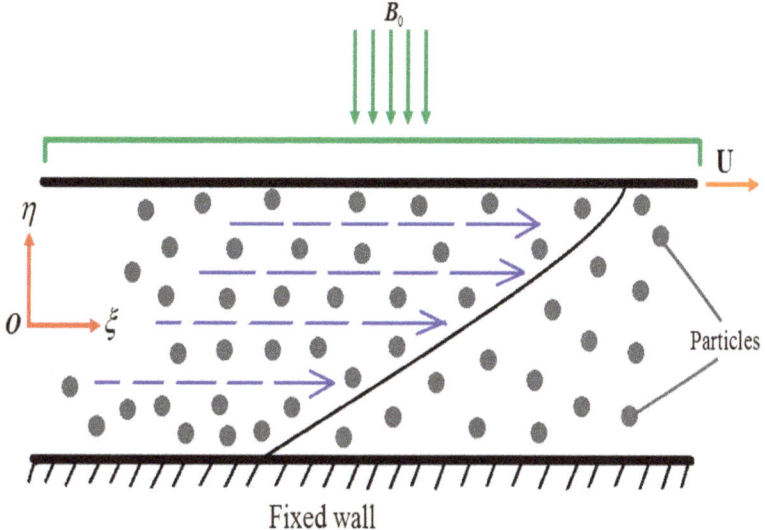

Figure 1. Configuration of the flow.

(i). For fluid phase

$$\frac{\partial u_f}{\partial \xi} = 0, \qquad (1)$$

$$\frac{\partial}{\partial \eta}\left(\mu_s \frac{\partial u_f}{\partial \eta}\right) - \eta_1\left(\frac{\partial^4 u_f}{\partial \eta^4}\right) + \frac{CS}{(1-C)}(u_p - u_f) - \frac{\sigma B_0^2}{(1-C)}u_f = \frac{\partial p}{\partial \xi}. \qquad (2)$$

(ii). For particle phase

$$\frac{\partial u_p}{\partial \xi} = 0, \qquad (3)$$

$$u_f = u_p + \frac{1}{S}\left(\frac{\partial p}{\partial \xi}\right). \qquad (4)$$

(iii). Energy equation

$$\frac{\partial^2 \Theta}{\partial \eta^2} + \frac{\mu_s}{k}\left(\frac{\partial u_f}{\partial \eta}\right)^2 = \frac{\eta_1}{k}\left(\frac{\partial u_f}{\partial \eta}\right)\left(\frac{\partial^3 u_f}{\partial \eta^3}\right). \qquad (5)$$

where, C denotes concentration of the particles, μ_s is the viscosity of solid-liquid, μ_0 viscosity of the base liquid, η_1 is a material constant associated with couple stress fluid, σ is the electric conductivity of the fluid, B_0^2 is the magnetic strength, Θ is temperature, and k is the thermal conductivity of the fluid whereas, ξ and η are, respectively, axial and lateral coordinates. Moreover, S denotes the drag coefficient of interaction for the force exerted by particle on the fluid, and is given by Tam [36]:

$$S = \frac{4.5\,\mu_0}{r}\lambda(C), \qquad (6)$$

$$\lambda(C) = \frac{4 + 3\sqrt{8C - 3C^2} + 3C}{(2 - 3C)^2}. \qquad (7)$$

where, in Equation (6) the radius of the Hafnium particles is denoted by r.

Boundary Conditions

The flow interaction at the surfaces of the parallel plates are denoted by the following:

$$\left.\begin{array}{l}(i).\ u_f(\eta) = 0,\\ (ii).\ \frac{\partial^2 u_f}{\partial \eta^2} = 0,\\ (iii).\Theta(\eta) = \Theta_0.\end{array}\right\}; \text{ When } \eta = -h, \tag{8}$$

$$\left.\begin{array}{l}(iv).\ u_f(\eta) = U,\\ (v).\ \frac{\partial^2 u_f}{\partial \eta^2} = 0,\\ (vi).\Theta(\eta) = \Theta_w.\end{array}\right\}; \text{ When } \eta = h, \tag{9}$$

By using the following appropriate quantities:

$$\frac{u_f}{U} = u_f^*;\ \frac{u_p}{U} = u_p^*;\ \frac{\eta}{h} = \eta^*;\ \frac{\xi}{h} = \xi^*;\ \frac{\mu_s}{\mu_0} = \mu^*;\ \frac{hp}{\mu_0 U} = p^*;\ B_r = \frac{U^2 \mu_0}{k(\Theta_w - \Theta_0)};$$
$$\gamma = \sqrt{\frac{\mu_0}{\eta_1}}h;\ M = \sqrt{\frac{\sigma}{\mu_0}}hB_0;\ m = \frac{\mu_0}{h^2 S};\ \Theta^* = \frac{\Theta - \Theta_0}{(\Theta_w - \Theta_0)}. \tag{10}$$

Equations (1)–(5), in non-dimensional form after neglecting asterisk can be written as

$$\frac{dp}{d\xi} = \frac{d}{d\eta}\left(\mu \frac{du_f}{d\eta}\right) - \frac{1}{\gamma^2}\left(\frac{d^4 u_f}{d\eta^4}\right) + \frac{C(u_p - u_f)}{m(1-C)} - \frac{M^2}{(1-C)}u_f, \tag{11}$$

$$u_p = u_f - m\frac{dp}{d\xi}, \tag{12}$$

$$\frac{d^2\Theta}{d\eta^2} + \mu B_r\left(\frac{du_f}{d\eta}\right)^2 = \frac{B_r}{\gamma^2}\left(\frac{du_f}{d\eta}\right)\left(\frac{d^3 u_f}{d\eta^3}\right). \tag{13}$$

In which, M is the Hartmann number, γ is the couple stress parameter, m is the drag constant and B_r is he Brinkman number.

3. Results and Discussion

3.1. Variable Viscosity

The Reynolds' model for temperature dependent viscosity [37] can be defined as

$$\mu_s(\Theta) = \mu_0 e^{-\alpha(\Theta - \Theta_0)}. \tag{14}$$

In view of expression given in (10), the non-dimensional form of Equation (14), after dropping asterisk is obtained as

$$\mu(\Theta) = e^{-\alpha(\Theta_w - \Theta_0)\Theta} = e^{-\beta\Theta}, \tag{15}$$

where $\beta = \alpha(\Theta_w - \Theta_0)$.

Obviously, for the convergence of Equation (15), $\beta \in [0\ 1]$.

By Walter' lemma, the Maclaurin' series of Equation (15) can be linearized as

$$\mu(\Theta) = 1 - \beta\Theta. \tag{16}$$

In view of Equations (12) and (16), Equations (11) and (13) provide the set of nonlinear coupled differential equations involving the viscosity of the fluid deeply affected by the presence of heat applied at the wall along with a constant pressure gradient at each point of the channel (i.e., $\frac{dp}{d\xi} = P$) as follows

$$\frac{d^4 u_f}{d\eta^4} + \gamma^2 \beta \left(\frac{d\Theta}{d\eta}\right)\left(\frac{du_f}{d\eta}\right) + \gamma^2(\beta\Theta - 1)\frac{d^2 u_f}{d\eta^2} + \frac{M^2 \gamma^2}{(1-C)} u_f + \frac{\gamma^2 P}{(1-C)} = 0, \qquad (17)$$

$$\frac{d^2 \Theta}{d\eta^2} + B_r(1 - \beta\Theta)\left(\frac{du_f}{d\eta}\right)^2 = \frac{B_r}{\gamma^2}\left(\frac{du_f}{d\eta}\right)\left(\frac{d^3 u_f}{d\eta^3}\right). \qquad (18)$$

On the same contrast, Equations (8) and (9), in view of (10), are acquired as

$$\left. \begin{array}{l} (i).\ u_f(\eta) = 0, \\ (ii).\ \frac{\partial^2 u_f}{\partial \eta^2} = 0, \\ (iii).\Theta(\eta) = 0. \end{array} \right\} ;\ \text{When}\ \eta = -1, \qquad (19)$$

$$\left. \begin{array}{l} (iv).\ u_f(\eta) = 1, \\ (v).\ \frac{\partial^2 u_f}{\partial \eta^2} = 0, \\ (vi).\Theta(\eta) = 1. \end{array} \right\} ;\ \text{When}\ \eta = 1. \qquad (20)$$

3.2. Numerical Procedure

The set of non-linear differential Equations (17) and (18) with the boundary conditions (19) and (20) are solved by employing the most efficient numerical procedure consist of Runge–Kutta method and the shooting scheme [38] using MATLAB software. It is an iterative scheme, in which each step possible error can be successively reduced by changing higher order derivatives.

Let:
$$u_f = f_1 \qquad (21)$$

be the velocity of the fluid phase, then the derivatives of u_f, in terms of system of first ordinary differential equations (ODEs) can be expressed as:

$$f_2 = \frac{du_f}{d\eta} = f_1', \qquad (22)$$

$$f_3 = \frac{d^2 u_f}{d\eta^2} = f_2', \qquad (23)$$

$$f_4 = \frac{d^3 u_f}{d\eta^3} = f_3', \qquad (24)$$

$$\Theta = f_5, \qquad (25)$$

$$f_6 = \frac{d\Theta}{d\eta} = f_5', \qquad (26)$$

here the sign of prime (′) at the top indicates the derivative with respect to "η". In view of Equations (22)–(26) the fluid phase differential equation is transformed as:

$$f_4' = \gamma^2(1 - \beta(f_5))f_3 - \gamma^2 \beta(f_2)(f_6) - \left(\frac{\gamma^2 M^2}{1-C}\right)f_1 - \left(\frac{\gamma^2}{1-C}\right)P, \qquad (27)$$

$$f_6' = \frac{B_r}{\gamma^2}(f_2)(f_4) + B_r\ (\beta(f_5) - 1)(f_2)^2. \qquad (28)$$

The transformed set of conditions are given as:

$$\left.\begin{array}{l}(i).\ f_1 = 0,\\ (ii).\ f_2 = k_1,\\ (iii).\ f_3 = 0,\\ (iv).\ f_4 = k_2,\\ (v).\ f_5 = 0,\\ (vi).\ f_6 = k_3.\end{array}\right\};\ \text{When}\ \eta = -1, \qquad (29)$$

$$\left.\begin{array}{l}(i).\ f_1 = 1\\ (ii).\ f_2 = k_4,\\ (iii).\ f_3 = 0,\\ (iv).\ f_4 = k_5,\\ (v).\ f_5 = 1,\\ (vi).\ f_6 = k_6.\end{array}\right\};\ \text{When}\ \eta = 1. \qquad (30)$$

where $k_1, k_2, k_3, k_4, k_5,$ and k_6 can be easily determined during the routine numerical procedure.

3.3. Graphical Illustration

To see the effects of physical parameters for Reynolds model on velocity and temperature, Figures 2–6 have been displayed. The range of all physical parameters available in the existing literatures are as follows: the range of Hartmann number is $0 < M < 1$ [39], the Brinkman number B_r varies from 0.5 to 2.0 [40], the range of couple stress parameter γ is 0.5 to 2.0 [41], the range of concentration of the metallic particles' C is 0 to 0.2 [42], and the range of viscous parameter β lies between 0 to 1. The role of transversely applied magnetic fields can be sighted in Figure 2. It is found that the velocity of fluid decreases by increasing the values of Hartmann number. It is in accordance with the physical expectation, as increased in the Hartmann number, means to strengthen the magnetic field lines which result to impede the flow. Therefore, the obtained results validate the expected outcomes. In Figure 3, addition of some extra metallic particles to the system that expedites the flow is observed. It is found that velocity of fluid escalates for higher values of C. It is very much obvious as the constant movement of the upper wall does not allow the particle to exert an extra drag force to attenuate the base fluid motion. Thus, particle-to-particle interaction and fluid–particle interaction gets meager, which causes the frisky movement of the Hafnium particles in the base fluid. In Figure 4, we show that an increase in the couple stress parameter weakens the rotational field of couple stress fluid particles. It was revealed that the velocity profile increased by increasing the values of the couple stress parameter. It is because of friction force that fails to gain enough strength which can cause enough resistance to slow down the celerity of the flow. Similarly, the application of heat on the lower wall contributes in shear thinning effects which aids the fluid particles to get extra momentum. Hence, increase in the velocity of the fluid flow is vivid in display. Figure 5 shows the impact of decreasing viscous parameter β on the flow dynamics. In Equation (14), it can be inferred that as the temperature difference mounts, the shear thinning effects on the viscosity of the base fluid aggravates. This attenuation of physical property results in the increase of the celerity of the fluid and particles movements. Figure 6 describes the role of Brinkman number B_r on the temperature. It is seen that higher values of Brinkman heats up the fluid by surging the temperature. However, the quite opposite behavior was observed for the case of viscosity parameter as shown in Figure 7. It was revealed that the temperature of the fluid declines for the higher values of β. This temperature decline was in fact due to the rapid movement of the couple stress fluid.

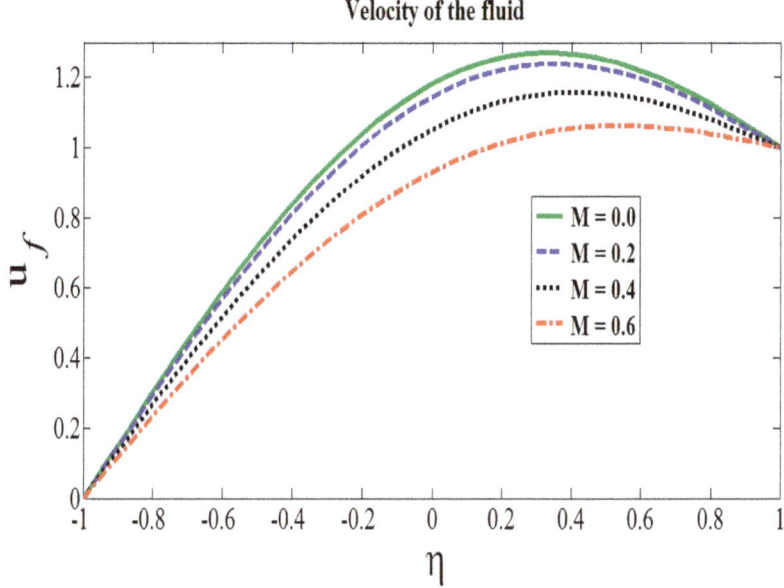

Figure 2. Effects of Hartmann number on the flow.

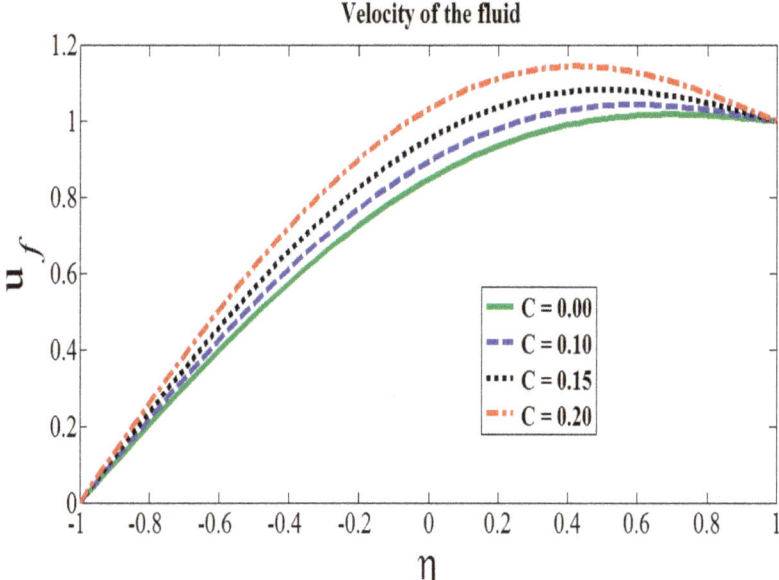

Figure 3. Effects of metallic particle concentration on the flow.

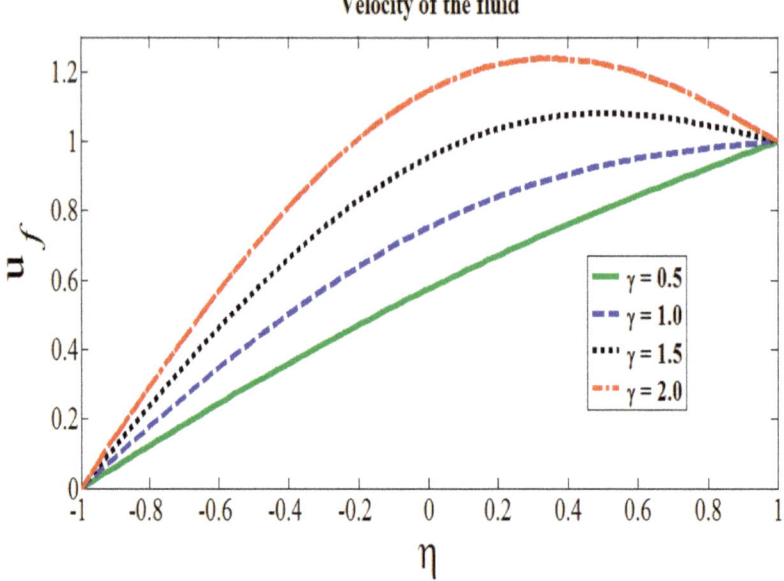

Figure 4. Couples stress parameter affecting the flow.

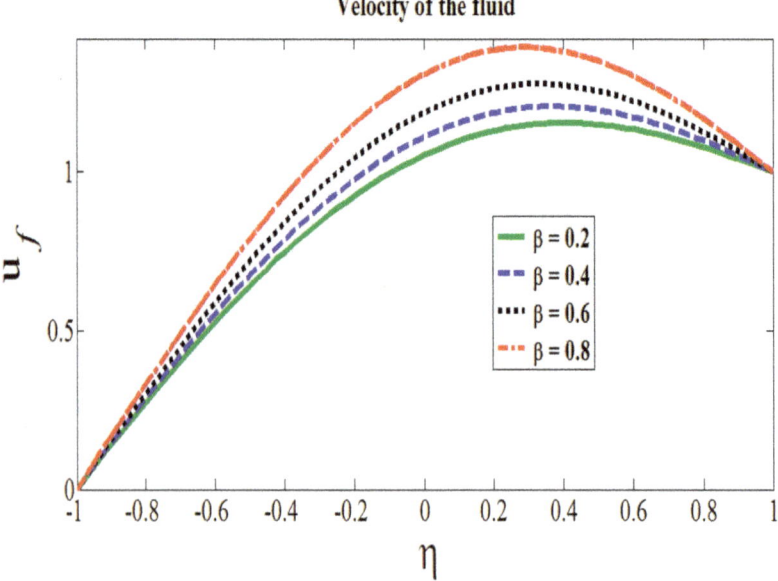

Figure 5. Effects of viscous parameter on the flow.

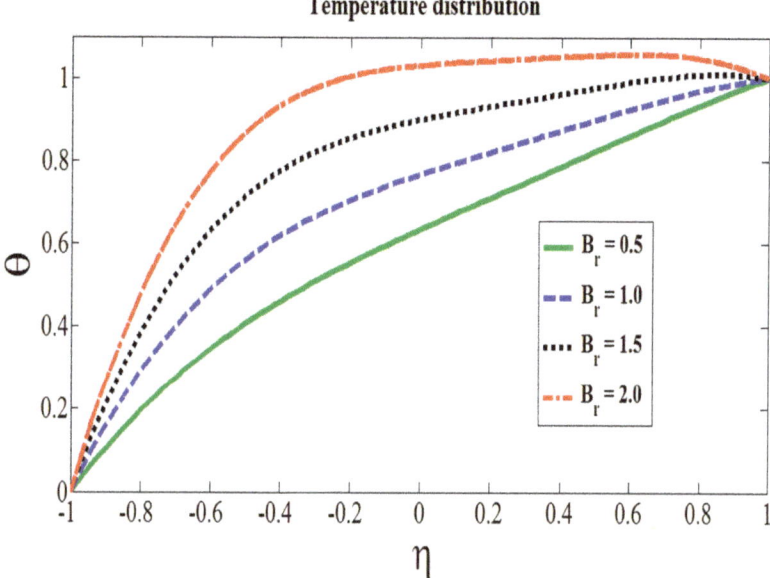

Figure 6. Role of Brinkman number on the temperature.

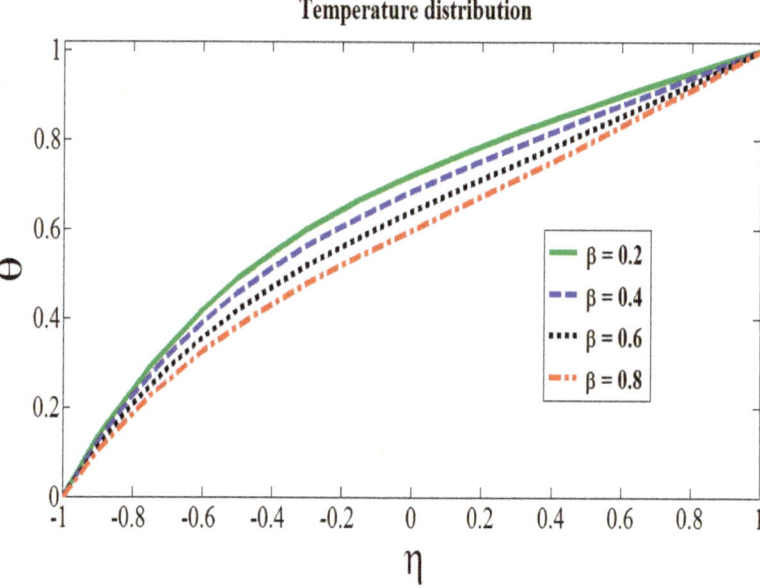

Figure 7. Role of viscous parameter on the temperature.

3.4. Validation

The numerical results are being presented in Tables 1–3. The variation in the velocities of both phases against couple stress parameter when $M = 1.0$, $C = 0.4$, and $B_r = 2.0$ are kept fixed are given in Table 1 whereas the variation in the velocities for single- and two-phase flows at different points of the domain when $M = 1.0$, $\gamma = 2.0$, and $B_r = 2.0$ are specified in Table 2. The thermal variation at the different points of given domain when $M = 1.0$ can be seen in Table 3. In all three table, one can conclude that temperature and velocities for both fluid and nanoparticles were an increasing function of metallic particles concentration C, couple stress parameter γ, and the Brinkman number B_r. The results extracted by numerical computation were found to be in excellent agreement with graphical illustrations and also satisfied all the subjected conditions. This provides a useful check that the presented solutions are correct.

Table 1. Variation in the velocities of both phases for Newtonian case and couple stress fluid.

y	u_p Newtonian Fluid ($\gamma = 0.0$)	u_p Couple Stress Fluid ($\gamma = 2.0$)	u_f Newtonian Fluid ($\gamma = 0.0$)	u_f Couple Stress Fluid ($\gamma = 2.0$)
−1.0	1.0000	1.0000	0.0000	0.0000
−0.6	1.2000	1.3221	0.2000	0.3221
−0.2	1.4000	1.5826	0.4000	0.5826
0.2	1.6000	1.7698	0.6000	0.7698
0.6	1.8000	1.8998	0.8000	0.8998
1.0	2.0000	2.0000	1.0000	1.0000

Table 2. Variation in the velocities for single- and two-phase flows.

y	u_f Single Phase ($C = 0.0$)	u_p Solid–Liquid Phase ($C = 0.4$)	u_f Solid–Liquid Phase ($C = 0.4$)
−1.0	0.0000	1.0000	0.0000
−0.6	0.2741	1.3221	0.3221
−0.2	0.5117	1.5826	0.5826
0.2	0.7047	1.7698	0.7698
0.6	0.8618	1.8998	0.8998
1.0	1.0000	2.0000	1.0000

Table 3. Thermal variation at the different points.

y	Θ $B_r = 0.0$	Θ $B_r = 2.0$	Θ $\gamma = 0.0$	Θ $C = 0.0$
−1.0	0.0000	0.0000	0.0000	0.0000
−0.6	0.2000	0.3916	0.3512	0.3578
−0.2	0.4000	0.6066	0.5629	0.5870
0.2	0.6000	0.7528	0.7095	0.7504
0.6	0.8000	0.8785	0.8446	0.8830
1.0	1.0000	1.0000	1.0000	1.0000

4. Conclusions

The Couette–Poiseuille flow of couple stress fluid in the presence of Hafnium particles was studied. The viscous dissipation effects were also reported. Exponentially decreasing viscosity of base fluid was presented by the Reynolds model. Transversely acting magnetic fields contributed by hindering the bi-phase flow. The key findings are described as:

- ➤ The flow of couple stress fluid resists for increasing values of Hartmann number.
- ➤ The temperature effectively variates the viscosity of the fluid to cause the shear thinning effects.
- ➤ The temperature of the flow mounts in response of higher values of Brinkman number.
- ➤ Attenuation of the viscosity results to expedite the flows.
- ➤ Viscosity parameter brings celerity in the velocity of bi-phase fluid due to high temperature difference.

> Molecules additives of base fluid reduce the force of friction and hence in both phases the velocity is galvanized.
> Due to the immense applications of multiphase flows in industrial and pharmaceutical, the proposed theoretical model is now available to vet relevant experimental investigations.

Author Contributions: Conceptualization, R.E.; Investigation, F.H.; Methodology, T.A.; Visualization, A.Z.

Funding: This research received no external funding.

Acknowledgments: R. Ellahi thanks to Sadiq M. Sait, the Director Office of Research Chair Professors, King Fahd University of Petroleum and Minerals, Dhahran, Saudi Arabia, to honor him with the Chair Professor at KFUPM. F. Hussain is also acknowledged Higher Education Commission Pakistan to provide him indigenous scholar for the pursuance of his Ph.D. studies.

Conflicts of Interest: The authors declare no conflict of interest.

References

1. Farooq, M.; Rahim, M.T.; Islam, S.; Siddiqui, A.M. Steady Poiseuille flow and heat transfer of couple stress fluids between two parallel inclined plates with variable viscosity. *J. Assoc. Arab Univ. Basic Appl. Sci.* **2013**, *14*, 9–18. [CrossRef]
2. Mahabaleshwar, U.S.; Sarris, I.E.; Hill, A.; Lorenzini, G.; Pop, I. An MHD couple stress fluid due to a perforated sheet undergoing linear stretching with heat transfer. *Int. J. Heat Mass. Transf.* **2017**, *105*, 157–167. [CrossRef]
3. Saad, H.; Ashmawy, E.A. Unsteady plane Couette flow of an incompressible couple stress fluid with slip boundary conditions. *Int. J. Med. Health Sci.* **2016**, *3*, 85–92. [CrossRef]
4. Akhtar, S.; Shah, N.A. Exact solutions for some unsteady flows of a couple stress fluid between parallel plates. *Ain Shams Eng. J.* **2018**, *9*, 985–992. [CrossRef]
5. Khan, N.A.; Khan, H.; Ali, S.A. Exact solutions for MHD flow of couple stress fluid with heat transfer. *J. Egypt. Math. Soc.* **2016**, *24*, 125–129. [CrossRef]
6. Asghar, S.; Ahmad, A. Unsteady Couette flow of viscous fluid under a non-uniform magnetic field. *Appl. Math. Lett.* **2012**, *25*, 1953–1958. [CrossRef]
7. Shaowei, W.; Mingyu, X. Exact solution on unsteady Couette flow of generalized Maxwell fluid with fractional derivative. *Acta Mech.* **2006**, *187*, 103–112. [CrossRef]
8. Eegunjobi, A.S.; Makinde, O.D.; Tshehla, M.S.; Franks, O. Irreversibility analysis of unsteady Couette flow with variable viscosity. *J. Hydrodyn. B* **2015**, *27*, 304–310. [CrossRef]
9. Ellahi, R.; Wang, X.; Hameed, M. Effects of heat transfer and nonlinear slip on the steady flow of Couette fluid by means of Chebyshev Spectral Method. *Z. Naturforsch A.* **2014**, *69*, 1–8. [CrossRef]
10. Ellahi, R.; Shivanian, E.; Abbasbandy, S.; Hayat, T. Numerical study of magnetohydrodynamics generalized Couette flow of Eyring-Powell fluid with heat transfer and slip condition. *Int. J. Numer. Method Heat Fluid Flow* **2016**, *26*, 1433–1445. [CrossRef]
11. Zeeshan, A.; Shehzad, N.; Ellahi, R. Analysis of activation energy in Couette-Poiseuille flow of nanofluid in the presence of chemical reaction and convective boundary conditions. *Results Phys.* **2018**, *8*, 502–512. [CrossRef]
12. Shehzad, N.; Zeeshan, A.; Ellahi, R. Electroosmotic flow of MHD Power law Al2O3-PVC nanofluid in a horizontal channel: Couette-Poiseuille flow model. *Commun. Theor. Phys.* **2018**, *69*, 655–666. [CrossRef]
13. Hussain, F.; Ellahi, R.; Zeeshan, A.; Vafai, K. Modelling study on heated couple stress fluid peristaltically conveying gold nanoparticles through coaxial tubes: A remedy for gland tumors and arthritis. *J. Mol. Liq.* **2018**, *268*, 149–155. [CrossRef]
14. Ellahi, R.; Zeeshan, A.; Hussain, F.; Asadollahi, A. Peristaltic blood flow of couple stress fluid suspended with nanoparticles under the influence of chemical reaction and activation energy. *Symmetry* **2019**, *11*, 276. [CrossRef]
15. Ellahi, R.; Bhatti, M.M.; Fetecau, C.; Vafai, K. Peristaltic flow of couple stress fluid in a non-uniform rectangular duct having compliant walls. *Commun. Theor. Phys.* **2016**, *65*, 66–72. [CrossRef]

16. Poply, V.; Singh, P.; Yadav, A.K. A study of Temperature-dependent fluid properties on MHD free stream flow and heat transfer over a non-linearly stretching sheet. *Procedia Eng.* **2015**, *127*, 391–397. [CrossRef]
17. Ellahi, R.; Raza, M.; Vafai, K. Series solutions of non-Newtonian nanofluids with Reynolds model and Vogel's model by means of the homotopy analysis method. *Math. Comput. Model.* **2012**, *55*, 1876–1891. [CrossRef]
18. Disu, A.B.; Dada, M.S. Reynolds model viscosity on radiative MHD flow in a porous medium between two vertical wavy walls. *J. Taibah Univ. Sci.* **2017**, *11*, 548–565. [CrossRef]
19. Nadeem, S.; Ali, M. Analytical solutions for pipe flow of a fourth-grade fluid with Reynolds and Vogel's models of viscosities. *Commun. Nonlin. Sci. Numer. Simulat.* **2009**, *14*, 2073–2090. [CrossRef]
20. Ellahi, R.; Zeeshan, A.; Hussain, F.; Abbas, T. Thermally charged MHD bi-phase flow coatings with non-Newtonian nanofluid and Hafnium particles through slippery walls. *Coatings* **2019**, *9*, 300. [CrossRef]
21. Mahmoud, M.A. Chemical reaction and variable viscosity effects on flow and mass transfer of a non-Newtonian visco-elastic fluid past a stretching surface embedded in a porous medium. *Meccanica* **2010**, *45*, 835–846. [CrossRef]
22. Makinde, O.D. Laminar falling liquid film with variable viscosity along an inclined heated plate. *Appl. Math. Comput.* **2006**, *175*, 80–88. [CrossRef]
23. Jawad, M.; Shah, Z.; Islam, S.; Majdoubi, J.; Tlili, I.; Khan, W.; Khan, I. Impact of nonlinear thermal radiation and the viscous dissipation effect on the unsteady three-dimensional rotating flow of single-wall carbon nanotubes with aqueous suspensions. *Symmetry* **2019**, *11*, 207. [CrossRef]
24. Ellahi, R. A study on the convergence of series solution of non-Newtonian third grade fluid with variable viscosity: By means of homotopy analysis method. *Adv. Math. Phys.* **2012**, *2012*, 634925. [CrossRef]
25. Karimipour, A.; Orazio, A.D.; Shadloo, M.S. The effects of different nano particles of Al_2O_3 and Ag on the MHD nano fluid flow and heat transfer in a microchannel including slip velocity and temperature jump. *Phys. E* **2017**, *86*, 146–153. [CrossRef]
26. Hosseini, S.M.; Safaei, M.R.; Goodarzi, M.; Alrashed, A.A.A.A.; Nguyen, T.K. New temperature, interfacial shell dependent dimensionless model for thermal conductivity of nanofluids. *Int. J. Heat Mass Transf.* **2017**, *114*, 207–210. [CrossRef]
27. Nasiri, H.; Jamalabadi, M.Y.A.; Sadeghi, R.; Safaei, M.R.; Nguyen, T.K.; Shadloo, M.S. A smoothed particle hydrodynamics approach for numerical simulation of nano-fluid flows. *J. Therm. Anal. Calorim.* **2018**, 1–9. [CrossRef]
28. Safaei, M.R.; Ahmadi, G.; Goodarzi, M.S.; Shadloo, M.S.; Goshayeshi, H.R.; Dahari, M. Heat transfer and pressure drop in fully developed turbulent flows of graphene nanoplatelets–silver/water nanofluids. *Fluids* **2016**, *1*, 20. [CrossRef]
29. Sadiq, M.A. MHD stagnation point flow of nanofluid on a plate with anisotropic slip. *Symmetry* **2019**, *11*, 132. [CrossRef]
30. Rashidi, S.; Esfahani, J.A.; Ellahi, R. Convective heat transfer and particle motion in an obstructed duct with two side by side obstacles by means of DPM model. *Appl. Sci.* **2017**, *7*, 431. [CrossRef]
31. Shehzad, N.; Zeeshan, A.; Ellahi, R.; Rashidid, S. Modelling study on internal energy loss due to entropy generation for non-Darcy Poiseuille flow of silver-water nanofluid: An application of purification. *Entropy* **2018**, *20*, 851. [CrossRef]
32. Hassan, M.; Ellahi, R.; Bhatti, M.M.; Zeeshan, A. A comparative study of magnetic and non-magnetic particles in nanofluid propagating over a wedge. *Can. J. Phys.* **2019**, *97*, 277–285. [CrossRef]
33. Zeeshan, A.; Shehzad, N.; Abbas, A.; Ellahi, R. Effects of radiative electro-magnetohydrodynamics diminishing internal energy of pressure-driven flow of titanium dioxide-water nanofluid due to entropy generation. *Entropy* **2019**, *21*, 236. [CrossRef]
34. Ashrafi, N.; Khayat, R.E. A low-dimensional approach to nonlinear plane-Couette flow of viscoelastic fluids. *Phys. Fluids.* **2000**, *12*, 345–365. [CrossRef]
35. Srivastava, L.M.; Srivastava, V.P. Peristaltic transport of a particle-fluid suspension. *J. Biomech. Eng.* **1989**, *111*, 157–165. [CrossRef]
36. Tam, C.K.W. The drag on a cloud of spherical particles in a low Reynolds number flow. *J. Fluid Mech.* **1969**, *38*, 537–546. [CrossRef]
37. Ellahi, R. The effects of MHD and temperature dependent viscosity on the flow of non-Newtonian nanofluid in a pipe: Analytical solutions. *Appl. Math. Model.* **2013**, *37*, 1451–1457. [CrossRef]

38. Hossain, M.A.; Subba, R.; Gorla, R. Natural convection flow of non-Newtonian power-law fluid from a slotted vertical isothermal surface. *Int. J Numer. Methods Heat Fluid Flow* **2009**, *19*, 835–846. [CrossRef]
39. Makinde, O.D.; Onyejekwe, O.O. A numerical study of MHD generalized Couette flow and heat transfer with variable viscosity and electrical conductivity. *J. Magn. Magn. Mater.* **2011**, *323*, 2757–2763. [CrossRef]
40. Coelho, M.P.; Faria, J.S. On the generalized Brinkman number definition and its importance for Bingham fluids. *J. Heat Transf.* **2011**, *133*, 545051–545055. [CrossRef]
41. Swarnalathamma, B.V.; Krishna, M.V. Peristaltic hemodynamic flow of couple stress fluid through a porous medium under the influence of magnetic field with slip effect. *AIP Conf. Proc.* **2016**, *1728*, 0206031–0206039. [CrossRef]
42. Charm, S.E.; Kurland, G.S. *Blood Flow and Microcirculation*; Wiley: New York, NY, USA, 1974.

© 2019 by the authors. Licensee MDPI, Basel, Switzerland. This article is an open access article distributed under the terms and conditions of the Creative Commons Attribution (CC BY) license (http://creativecommons.org/licenses/by/4.0/).

Article

Boundary Layer Flow through Darcy–Brinkman Porous Medium in the Presence of Slip Effects and Porous Dissipation

Muhammad Salman Kausar [1], Abid Hussanan [2,3,*], Mustafa Mamat [1] and Babar Ahmad [4]

1. Faculty of Informatics and Computing, University Sultan Zainal Abidin (Kampus Gong Badak), Kuala Terengganu, Terengganu 21300, Malaysia; salmanrao603@gmail.com (M.S.K.); must@unisza.edu.my (M.M.)
2. Division of Computational Mathematics and Engineering, Institute for Computational Science, Ton Duc Thang University, Ho Chi Minh City 700000, Vietnam
3. Faculty of Mathematics and Statistics, Ton Duc Thang University, Ho Chi Minh City 70000, Vietnam
4. Department of Mathematics, COMSATS University Islamabad (CUI) Park Road, Tarlai Kalan, Islamabad 455000, Pakistan; babar.sms@gmail.com
* Correspondence: abidhussanan@tdtu.edu.vn

Received: 20 March 2019; Accepted: 6 May 2019; Published: 11 May 2019

Abstract: This paper aims to examine the Darcy–Brinkman flow over a stretching sheet in the presence of frictional heating and porous dissipation. The governing equations are modeled and simplified under boundary layer approximations, which are then transformed into system of self-similar equations using appropriate transformations. The resulting system of nonlinear equations was solved numerically under velocity and thermal slip conditions, by fourth-order Runge–Kutta method and built-in routine bvp4c in Matlab. Under special conditions, the obtained results were compared with the results available in the literature. An excellent agreement was observed. The variation of parameters was studied for different flow quantities of interest and results are presented in the form of tables and graphs.

Keywords: Darcy–Brinkman porous medium; viscous dissipation; slip conditions; porous dissipation; permeable sheet

1. Introduction

The porous medium is a continuous solid phase having void spaces/pores in it. The fraction of the void space to the total volume is named as porosity. There are plenty of porous media available naturally and many of them are artificial. Some examples of porous medium are rocks such as limestone, sand stone, beach sand, pumice and dolomite, lathes packed with pebbles, cloth sponge, rye bread, foamed plastics, endothelial surface layer, catalyst pellets, gall bladder with stones, human lung, and drug permeation through human skin. Industrial and engineering applications of flows through porous medium have attracted the attention of researchers. Purification and filtration processes, seepage of water in river beds, migration of pollutants into the soil and aquifers, drying of porous materials in textile industries, the movement of moisture through and under engineering structures, the saturation of porous materials by chemicals, and heat and mass transport in packed bed reactor columns stand among many other applications. "Flow is linearly dependent on the pressure gradient and the gravitational force" is known as Darcy Law. This law is generally accepted as the macroscopic equation of motion for the Newtonian fluids in porous media at small Reynolds numbers and when the medium is close-packed (lower permeability). However, when the pore distribution in the medium is sparse and the pores are large, the porous medium will have large voids, giving rise to viscous shear in addition to Darcy's resistance. In that case, the usual viscous resistance term (Brinkman term) should be considered, along with the Darcy resistance term. This model is known as Darcy–Brinkman model [1–7].

In a stretching flow, an elastic flat sheet that is stretched in its own plane with a velocity changing with the distance from a fixed point. The sheeting material is being produced in numerous manufacturing, industrial and engineering processes. In the manufacture of the polymer sheets, the melt material, when pushed through an extrusion die, cools and solidifies at a distance from the die before reaching the cooling phase. Applications of the boundary layer flow generated by stretching sheet can also be witnessed in procedures such as spinning of fibers, glass blowing, hot rolling, continuous casting, and in thin film flow and many others [8–14]. Boundary layer flow over a stretching sheet in the presence of Darcy porous medium are investigated by several researchers [15–20]. Darcy–Brinkman flow over a stretching sheet was performed by Waqar and Pop [21] and Khan et al. [22]. When fluid is forced to move due to the stretching of sheet, the fluid gains some velocity as well as kinetic energy and this kinetic energy is converted into the heat energy. In the presence of porous medium, viscous dissipation term in energy equation is modified and this phenomenon is called porous dissipation. In [21,22], the authors neglected the viscous dissipation effects. Moreover, even in the case of Darcy flow, the authors neglected the porous dissipation terms in the modeling. From the literature survey and to best of our knowledge, no one has investigated the Darcy–Brinkman flow over a stretching sheet in the presence of frictional heating and porous dissipation.

The phenomenon of slip condition has many industrial and practical applications, especially in microchannels or nanochannels. To study heat transfer flows more accurately, slip conditions are required, which strongly influence fluid motion at the fluid–solid interface. Zhang et al. [23] investigated the heat transfer performance in microchannel under the slip flow regime and constant heat flux boundary condition by considering into account the effects of velocity slip and temperature jump. Hooman and Ejlali [24] showed that the combined effects of temperature jump and velocity slip on forced convection in both parallel plate and circular microchannels for fully developed gas–liquid slip flows. Hussanan et al. [25] studied the Newtonian heating problem with additional effects of velocity slip and free convection on heat transfer flow over a vertical plate. Liu and Guo [26] used second-order slip condition while studying analytical solution of fractional Maxwell flow under magnetic field. Jing et al. [27] investigated the hydraulic resistance and heat transfer rate in elliptical microchannel with the velocity slip for different length ratios. Andersson [28] obtained the analytical solution for the slip flow over a stretching sheet. Turkyilmazoglu [29] performed the heat and mass transfer analysis of MHD flow over a stretching sheet in presence of velocity and thermal slip effects. Yazdi et al. [30] studied the effects of viscous dissipation on MHD flow over a porous stretching sheet in the presence of slip and convective boundary conditions. Hsiao [31] examined the MHD stagnation point flow of nanofluid towards a stretching sheet with slip boundary conditions.

The aim of this paper is to investigate the Darcy–Brinkman flow over a permeable stretching sheet in the presence of viscous and porous dissipation under the velocity and thermal slip conditions. Governing equations are modeled and then transformed into self-similar forms using the suitable similarity transformations. Note that, in the presence of viscous dissipation, similar solutions are very rare. Resulting self-similar equations were solved numerically using shooting method. Comparative study between shooting method and built in routine bvp4c [32] in Matlab was also made, to check the accuracy of our results. Moreover, in special cases, comparison between the existing available results was performed. The variations of pertinent parameters on the dimensionless velocity, temperature, skin friction coefficient and local Nusselt number are illustrated and discussed.

2. Mathematical Formulation

We consider the flow of over a permeable stretching surface embedded in a porous medium. In Cartesian coordinates, x-axis and y-axis are perpendicular to the sheet, which is being stretched with velocity $U_s = ax$. Let $T_s = T_\infty + cx^2$ be the temperature of sheet and T_∞ be the ambient temperature and $T_s > T_\infty$. In the presence of viscous dissipation, governing equations under boundary layer assumption are

$$\frac{\partial u}{\partial x} + \frac{\partial v}{\partial y} = 0, \tag{1}$$

$$u\frac{\partial u}{\partial x} + v\frac{\partial u}{\partial y} = \frac{\varepsilon^2 \mu_e}{\rho}\left(\frac{\partial^2 u}{\partial y^2}\right) - \frac{\mu \varepsilon^2}{\rho K^*}u, \tag{2}$$

$$u\frac{\partial T}{\partial x} + v\frac{\partial T}{\partial y} = \frac{\kappa}{\rho C_p}\left(\frac{\partial^2 T}{\partial y^2}\right) + \frac{\varepsilon^2}{\rho C_p}\left[\mu_e\left(\frac{\partial u}{\partial y}\right)^2 + \frac{\mu u^2}{K^*}\right], \tag{3}$$

along with boundary conditions

$$\begin{array}{c} u = ax + \beta_1\left(\frac{\partial u}{\partial y}\right),\ v = -V_0,\ T = T_s + \delta_1\left(\frac{\partial T}{\partial y}\right) \text{ at } y = 0, \\ u \to 0,\ T \to T_\infty \text{ as } y \to \infty. \end{array} \tag{4}$$

In above equations, u and v are the components of velocity along x and y directions, respectively. Introducing the similarity transformations

$$\xi = \sqrt{\frac{a}{v}}y,\ u = axg'(\xi),\ v = -\sqrt{av}g(\xi),\ \theta(\xi) = \frac{T - T_\infty}{T_s - T_\infty}. \tag{5}$$

Equations (1)–(3) along with boundary conditions Equation (4) are reduced to the dimensionless forms

$$\gamma g''' - g'^2 + gg'' - P_m g' = 0, \tag{6}$$

$$\frac{1}{\Pr}\theta'' + g\theta' - 2g'\theta + Ec\left(\gamma g''^2 + P_m g'^2\right) = 0, \tag{7}$$

$$\begin{array}{c} g(0) = S,\ g'(0) = 1 + \beta g''(0),\ g'(\infty) = 0, \\ \theta(0) = 1 + \delta \theta'(0),\ \theta(\infty) = 0. \end{array} \tag{8}$$

Skin friction coefficients S_{fx} and the local Nusselt number N_{Rx} are defined as

$$S_{fx} = \frac{\mu}{\rho U_s^2}\left(\frac{\partial u}{\partial y}\right)_{y=0},\ N_{Rx} = -\frac{x}{(T_s - T_\infty)}\left(\frac{\partial T}{\partial y}\right)_{y=0}. \tag{9}$$

In dimensionless form, quantities defined in Equation (9) take the form

$$S_{fx}\text{Re}_x^{1/2} = g''(0),\ N_{Rx}\text{Re}_x^{-1/2} = -\theta'(0). \tag{10}$$

In Equations (6)–(10), dimensionless physical parameters are defined as

Dimensionless Physical Parameters	Notations	Definitions
Brinkmann parameter	γ	$\varepsilon^2 \frac{\mu_e}{\mu}$
Porosity parameter	P_m	$\frac{\mu \varepsilon^2}{\rho a K^*}$
Suction/injection parameter	S	$\frac{V_0}{\sqrt{av}}$
Prandtl number	$\Pr$	$\frac{\mu C_p}{\kappa}$
Eckert number	Ec	$\frac{a^2}{cC_p}$
Velocity slip parameter	β	$\beta_1 \sqrt{\frac{a}{v}}$
Thermal slip parameter	δ	$\delta_1 \sqrt{\frac{a}{v}}$

3. Solution Methodologies

The nonlinear differential Equations (6) and (7) subject to the boundary conditions in Equation (8) were solved numerically using an efficient Runge–Kutta fourth-order method along with shooting technique. The asymptotic boundary conditions given by Equation (8) were replaced by using a value of 15 for the similarity variable ξ_{max}. The choice of $\xi_{max} = 15$ and the step size $\Delta \xi = 0.001$, ensured that all numerical solutions approached the asymptotic values correctly. To check the accuracy of computed results, comparison between analytical, exact and shooting method was made for special cases available in the literature. Moreover, for present general case results obtained by shooting method were also compared with the built-in routine bvp4c in MATLAB. The obtained results are in excellent agreement, which confirms the accuracy of our results.

4. Results and Discussion

We analyzed the effects of significant physical parameters on dimensionless velocity $g'(\xi)$, dimensionless temperature $\theta(\xi)$, skin friction coefficient $S_{fx}Re_x^{1/2} = -g''(0)$ and local Nusselt number $N_{Rx}Re_x^{-1/2} = -\theta'(0)$. Table 1 presents different values of velocity slip parameter β when there is no porous medium and sheet is impermeable. Skin friction coefficient decreases by increasing the velocity slip parameter. In the case of no-slip i.e., when $\beta = 0$, Equation (6) admits exact solution of the form [22]

$$g(\xi) = S + \frac{1}{A}\left(1 - e^{-A\xi}\right), \quad A = \frac{S + \sqrt{S^2 + 4\gamma(1 + P_m)}}{2\gamma}. \tag{11}$$

Table 1. Comparison between analytical solution [22] and shooting method for different values of β when $\gamma = 1$ and $P_m = 0.0, S = 0.0$.

Velocity Slip Parameter β	$g'(0)$		$-g''(0)$	
	Andersson [22]	Present	Andersson [22]	Present
0.0	1.0000	1.0000	1.0000	1.0000
0.1	0.9128	0.91278	0.8721	0.87215
0.2	0.8447	0.84471	0.7764	0.77645
0.5	0.7044	0.70436	0.5912	0.59127
1.0	0.5698	0.56974	0.4302	0.43025
2.0	0.4320	0.43183	0.2840	0.28408
5.0	0.2758	0.27530	0.1448	0.14493
10.0	0.1876	0.18670	0.0812	0.08132
20.0	0.1242	0.12285	0.0438	0.04385
50.0	0.0702	0.06801	0.0186	0.01863
100.0	0.0450	0.04225	0.0095	0.00957

Table 2 shows that the numerical solution obtained by shooting method is in good agreement with the exact solution. Moreover, we observed the skin friction coefficient Brinkman viscosity for parameter γ, whereas an oppose behavior is noted with increasing values suction parameter S and porosity parameter P_m. Table 3 shows that the skin friction coefficient is higher for the slip case in comparison with no-slip case. Numerical values of local Nusselt number for different physical parameters are presented in Table 4. Nusselt number increases by increasing Prandtl number Pr and decreases by increasing Eckert number Ec and thermal slip parameter δ.

Table 2. Skin friction coefficient $S_{fx}\text{Re}_x^{1/2} = -g''(0)$ for no slip case $\beta = 0$. Comparison between exact and numerical solution.

Physical Parameters			$S_{fx}\text{Re}_x^{1/2}=-g''(0)$	
γ	S	P_m	Exact (See Equation (11))	Numerical (Shooting Method)
1.0	1.0	0.5	1.82287	1.82287
2.0	1.0	0.5	1.15138	1.15140
3.0	1.0	0.5	0.89314	0.89324
0.5	**0.0**	0.3	1.61245	1.61245
0.5	**1.0**	0.3	2.89736	2.89736
0.5	**2.0**	0.3	4.56904	4.56905
2.0	0.5	**0.0**	0.84307	0.84336
2.0	0.5	**0.4**	0.97094	0.97098
2.0	0.5	**0.8**	1.08188	1.08188

Table 3. Skin friction coefficient $S_{fx}\text{Re}_x^{1/2} = -g''(0)$ for slip case $\beta = 1.0$. Comparison between Shooting method and MATLAB bvp4c.

Physical Parameters			$S_{fx}\text{Re}_x^{1/2}=-g''(0)$	
γ	S	P_m	Shooting Method	bvp4c
1.0	1.0	0.5	0.610511	0.610497
2.0	1.0	0.5	0.500008	0.500006
3.0	1.0	0.5	0.439566	0.439507
0.5	**0.0**	0.3	0.550438	0.550437
0.5	**1.0**	0.3	0.712228	0.712227
0.5	**2.0**	0.3	0.808872	0.808872
2.0	0.5	**0.0**	0.406493	0.406209
2.0	0.5	**0.4**	0.452006	0.451987
2.0	0.5	**0.8**	0.485908	0.485905

Table 4. Local Nusselt number $N_{Rx}\text{Re}_x^{-1/2} = -\theta'(0)$ when $\beta = 1.0$ and $S = 0.5$ Comparison between Shooting method and MATLAB bvp4c.

Physical Parameters					$N_{Rx}\text{Re}_x^{-1/2}=-\theta'(0)$	
Pr	Ec	δ	P_m	γ	Shooting Method	bvp4c
0.7	0.5	1.0	0.4	2.0	0.456141	0.456203
1.2	0.5	1.0	0.4	2.0	0.538161	0.538197
6.8	0.5	1.0	0.4	2.0	0.738928	0.738983
3.0	**0.0**	1.0	0.4	2.0	0.738078	0.738124
3.0	**0.6**	1.0	0.4	2.0	0.642319	0.642382
3.0	**1.2**	1.0	0.4	2.0	0.546560	0.546591
3.0	1.0	**0.0**	0.4	2.0	2.208602	2.208638
3.0	1.0	**0.6**	0.4	2.0	0.820808	0.820821
3.0	1.0	**1.2**	0.4	2.0	0.504071	0.504105
3.0	1.0	1.0	**0.0**	2.0	0.640207	0.640288
3.0	1.0	1.0	**0.5**	2.0	0.566207	0.566224
3.0	1.0	1.0	**1.0**	2.0	0.517044	0.517133
3.0	1.0	1.0	0.4	**1.0**	0.619665	0.619690
3.0	1.0	1.0	0.4	**2.0**	0.578480	0.578501
3.0	1.0	1.0	0.4	**3.0**	0.546450	0.546487

Figure 1 is plotted to see effect of Brinkman viscosity ratio parameter γ on velocity profile. It was observed that velocity increases on increasing the values of Brinkman viscosity ratio number. This also makes the sense because the Brinkman viscosity ratio number appears with the velocity gradient term in the momentum equation, consequently large values of Brinkman viscosity parameter increases the velocity. Figure 2 portrays the effects of porosity parameter P_m on the velocity profile. It was observed that velocity and momentum boundary layer decreases by increasing porosity parameter. In Figures 3 and 4, we can see that, by increasing the suction parameter S and slip parameter β, the velocity of fluid decreases and momentum boundary layer becomes thinner. Figure 5 displays the impact of Brinkman viscosity number on the temperature profile $\theta(\xi)$. Thermal boundary layer is an increasing function of γ. Increase in the suction velocity S and porosity parameter P_m decreases the fluid temperature with in the boundary layer (see Figures 6 and 7). Figure 8 exhibits the influence of Prandtl number on the temperature field $\theta(\xi)$. Temperature inside the boundary layer decreases with increasing Prandtl number Pr. This is true because, by increasing Prandtl number (decreasing thermal conductivity) of fluid, the heat transfer rate from the stretching sheet decreases and therefore thermal boundary layer decreases. Figure 9 depicts the influence of Eckert number Ec on temperature profile. As predictable, it is noticed that the thermal boundary layer increases with increasing values of Ec as Eckert number increases fluid friction between the adjacent layers increases, which results in conversion of the kinetic energy into heat energy. Figure 10 illustrates that temperature and thermal boundary layer reduces by increasing thermal slip parameter δ.

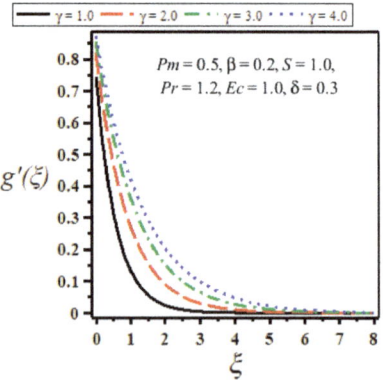

Figure 1. Variation of velocity with viscosity ratio parameter γ.

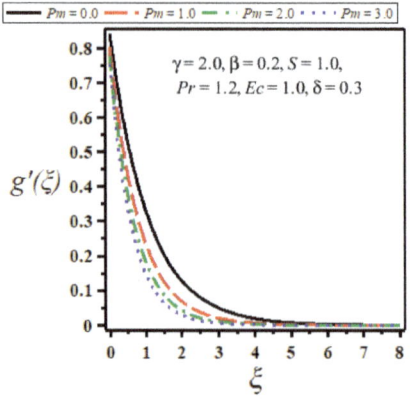

Figure 2. Variation of velocity with porosity parameter P_m.

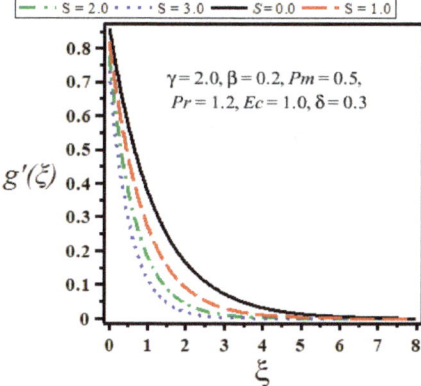

Figure 3. Variation of velocity with suction parameter S.

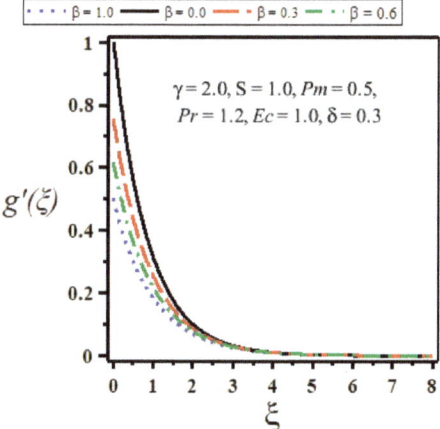

Figure 4. Variation of velocity with velocity slip parameter β.

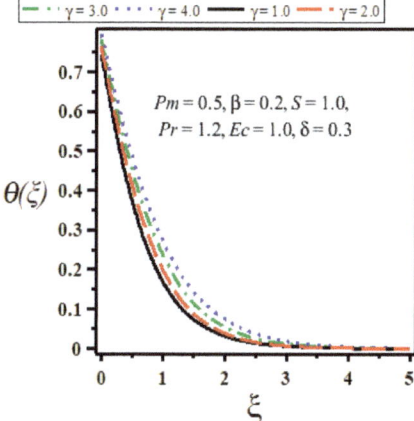

Figure 5. Variation of temperature with Brinkman parameter γ.

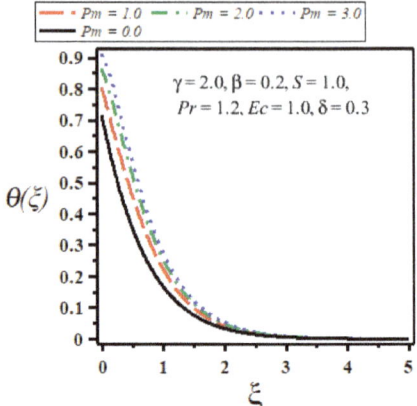

Figure 6. Variation of temperature with porosity parameter P_m.

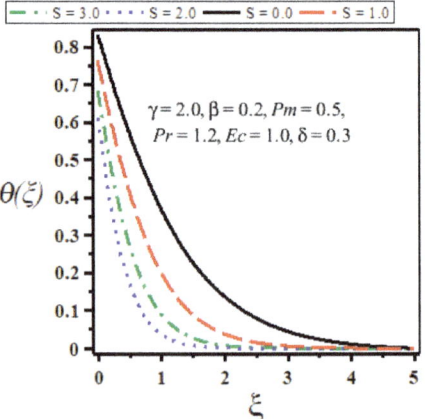

Figure 7. Variation of temperature with suction parameter S.

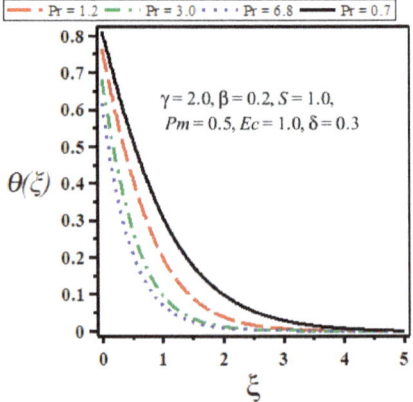

Figure 8. Variation of temperature with Prandtl number Pr.

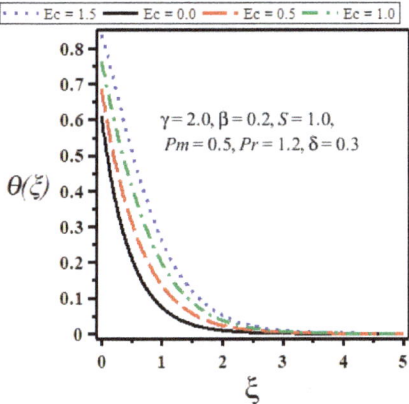

Figure 9. Variation of temperature with Eckert number Ec.

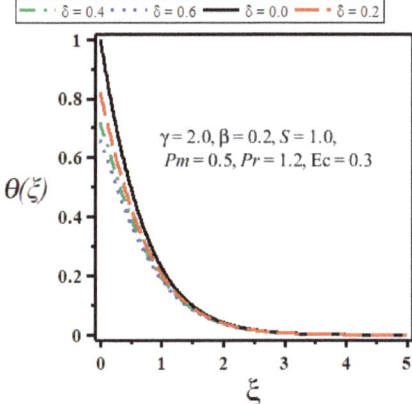

Figure 10. Variation of temperature with thermal slip parameter δ.

5. Conclusions

The present work gives the numerical solutions for Darcy–Brinkman flow over a stretching sheet in the presence of porous dissipation and frictional heating. From the numerical results obtained, some important conclusions are summarized:

(i) Both velocity and temperature decrease with the increase of suction parameter.
(ii) The slip parameter has high impact on skin friction coefficient as compared with no-slip condition.
(iii) Heat transfer rate is reduced due to increase in Eckert number and thermal slip parameter.

Author Contributions: M.S.K. and A.H. formulated the problem. M.S.K. and B.A. solved the problem. M.S.K., A.H., M.M. and B.A. computed and analyzed the results. All the authors equally contributed in writing and proof reading of the paper.

Funding: The APC was funded by Ton Duc Thang University, Ho Chi Minh City, Vietnam.

Conflicts of Interest: The authors declare no conflict of interest.

References

1. Patil, P.R.; Vaidyanathan, G. Effect of variable viscosity on thermohaline convection in a porous medium. *J. Hydrol.* **1982**, *57*, 147–161. [CrossRef]
2. Ingham, D.B.; Pop, I. *Transport Phenomena in Porous Media*; Elsevier: Amsterdam, The Netherlands, 1998.
3. Vafai, K. *Handbook of Porous Media*; CRC Press: Boca Raton, FL, USA, 2005.
4. Makinde, O.D.; Mhone, P.Y. Heat transfer to MHD oscillatory flow in a channel filled with porous medium. *Rom. J. Phys.* **2005**, *50*, 931–938.
5. Yu, L.H.; Wang, C.Y. Darcy-Brinkman flow through a bumpy channel. *Transp. Porous Media* **2013**, *97*, 281–294. [CrossRef]
6. Wang, C.Y. Darcy-Brinkman flow over a grooved surface. *Transp. Porous Media* **2010**, *84*, 219–227. [CrossRef]
7. Liu, H.; Patil, P.R.; Narusawa, U. On Darcy-Brinkman equation: Viscous flow between two parallel plates packed with regular square arrays of cylinders. *Entropy* **2007**, *9*, 118–131. [CrossRef]
8. Crane, L.J. Flow past a stretching plate. *Z. Angew. Math. Phys. Zamp* **1970**, *4*, 645–647. [CrossRef]
9. Ishak, A.; Nazar, R.; Pop, I. Hydromagnetic flow and heat transfer adjacent to a stretching vertical sheet. *Heat Mass Transf.* **2008**, *44*, 921–927. [CrossRef]
10. Chamkha, A.J. Hydromagnetic three-dimensional free convection on a vertical stretching surface with heat generation or absorption. *Int. J. Heat Fluid Flow* **1999**, *20*, 84–92. [CrossRef]
11. Salleh, M.Z.; Nazar, R.; Pop, I. Boundary layer flow and heat transfer over a stretching sheet with Newtonian heating. *J. Taiwan Inst. Chem. Eng.* **2010**, *41*, 651–655. [CrossRef]
12. Hsiao, K.L. Micropolar nanofluid flow with MHD and viscous dissipation effects towards a stretching sheet with multimedia feature. *Int. J. Heat Mass Transf.* **2017**, *112*, 983–990. [CrossRef]
13. Hussanan, A.; Salleh, M.Z.; Khan, I. Microstructure and inertial characteristics of a magnetite ferrofluid over a stretching/shrinking sheet using effective thermal conductivity model. *J. Mol. Liq.* **2018**, *255*, 64–75. [CrossRef]
14. Jamaludin, A.; Nazar, R.; Pop, I. Mixed convection stagnation-point flow of a nanofluid past a permeable stretching/shrinking sheet in the presence of thermal radiation and heat source/sink. *Energies* **2019**, *12*, 788. [CrossRef]
15. Mukhopadhyay, S. Effect of thermal radiation on unsteady mixed convection flow and heat transfer over a porous stretching surface in porous medium. *Int. J. Heat Mass Transf.* **2009**, *52*, 3261–3265. [CrossRef]
16. Chauhan, D.S.; Agrawal, R. MHD flow through a porous medium adjacent to a stretching sheet: Numerical and an approximate solution. *Eur. Phys. J. Plus* **2011**, *126*, 47. [CrossRef]
17. Pal, D.; Mondal, H. Influence of chemical reaction and thermal radiation on mixed convection heat and mass transfer over a stretching sheet in Darcian porous medium with Soret and Dufour effects. *Energy Convers. Manag.* **2012**, *62*, 102–108. [CrossRef]
18. Zheng, L.; Zhang, C.; Zhang, X.; Zhang, J. Flow and radiation heat transfer of a nanofluid over a stretching sheet with velocity slip and temperature jump in porous medium. *J. Frankl. Inst.* **2013**, *350*, 990–1007. [CrossRef]
19. Hussanan, A.; Salleh, M.Z.; Khan, I.; Tahar, R.M. Heat transfer in magnetohydrodynamic flow of a casson fluid with porous medium and Newtonian heating. *J. Nanofluids* **2017**, *6*, 784–793. [CrossRef]
20. Yasin, M.H.M.; Ishak, A.; Pop, I. Boundary layer flow and heat transfer past a permeable shrinking surface embedded in a porous medium with a second-order slip: A stability analysis. *Appl. Therm. Eng.* **2017**, *115*, 1407–1411. [CrossRef]
21. Khan, W.A.; Pop, I. Boundary layer flow past a stretching surface in a porous medium saturated by a nanofluid: Brinkman-Forchheimer model. *PLoS ONE* **2012**, *7*, e47031. [CrossRef]
22. Khan, Z.H.; Qasim, M.; Haq, R.U.; Al-Mdallal, Q.M. Closed form dual nature solutions of fluid flow and heat transfer over a stretching/shrinking sheet in a porous medium. *Chin. J. Phys.* **2017**, *55*, 1284–1293. [CrossRef]
23. Zhang, T.; Jia, L.; Yang, L.; Jaluria, Y. Effect of viscous heating on heat transfer performance in microchannel slip flow region. *Int. J. Heat Mass Transf.* **2010**, *53*, 4927–4934. [CrossRef]
24. Hooman, K.; Ejlali, A. Effects of viscous heating, fluid property variation, velocity slip, and temperature jump on convection through parallel plate and circular microchannels. *Int. Commun. Heat Mass Transf.* **2010**, *37*, 34–38. [CrossRef]

25. Hussanan, A.; Khan, I.; Salleh, M.Z.; Shafie, S. Slip effects on unsteady free convective heat and mass transfer flow with Newtonian heating. *Therm. Sci.* **2016**, *20*, 1939–1952. [CrossRef]
26. Liu, Y.; Guo, B. Effects of second-order slip on the flow of a fractional Maxwell MHD fluid. *J. Assoc. Arab Univ. Basic Appl. Sci.* **2017**, *24*, 232–241. [CrossRef]
27. Jing, D.; Song, S.; Pan, Y.; Wang, X. Size dependences of hydraulic resistance and heat transfer of fluid flow in elliptical microchannel heat sinks with boundary slip. *Int. J. Heat Mass Transf.* **2018**, *119*, 647–653. [CrossRef]
28. Andersson, H.I. Slip flow past a stretching surface. *Acta Mech.* **2002**, *158*, 121–125. [CrossRef]
29. Turkyilmazoglu, M. Analytic heat and mass transfer of the mixed hydrodynamic/thermal slip MHD viscous flow over a stretching sheet. *Int. J. Mech. Sci.* **2011**, *53*, 886–896. [CrossRef]
30. Yazdi, M.H.; Abdullah, S.; Hashim, I.; Sopian, K. Effects of viscous dissipation on the slip MHD flow and heat transfer past a permeable surface with convective boundary conditions. *Energies* **2011**, *4*, 2273–2294. [CrossRef]
31. Hsiao, K.L. Stagnation electrical MHD nanofluid mixed convection with slip boundary on a stretching sheet. *Appl. Therm. Eng.* **2016**, *98*, 850–861. [CrossRef]
32. Shampine, L.F.; Kierzenka, J. Solving boundary value problems for ordinary differential equations in MATLAB with bvp4c. *Tutor. Notes* **2000**, *2000*, 1–27.

© 2019 by the authors. Licensee MDPI, Basel, Switzerland. This article is an open access article distributed under the terms and conditions of the Creative Commons Attribution (CC BY) license (http://creativecommons.org/licenses/by/4.0/).

Article

Significance of Velocity Slip in Convective Flow of Carbon Nanotubes

Ali Saleh Alshomrani and Malik Zaka Ullah *

Department of Mathematics, Faculty of Science, King Abdulaziz University, Jeddah 21589, Saudi Arabia; aszalshomrani@kau.edu.sa
* Correspondence: malikzakas@gmail.com

Received: 7 April 2019; Accepted: 18 April 2019; Published: 17 May 2019

Abstract: The present article inspects velocity slip impacts in three-dimensional flow of water based carbon nanotubes because of a stretchable rotating disk. Nanoparticles like single and multi walled carbon nanotubes (CNTs) are utilized. Graphical outcomes have been acquired for both single-walled carbon nanotubes (SWCNTs) and multi-walled carbon nanotubes (MWCNTs). The heat transport system is examined in the presence of thermal convective condition. Proper variables lead to a strong nonlinear standard differential framework. The associated nonlinear framework has been tackled by an optimal homotopic strategy. Diagrams have been plotted so as to examine how the temperature and velocities are influenced by different physical variables. The coefficients of skin friction and Nusselt number have been exhibited graphically. Our results indicate that the skin friction coefficient and Nusselt number are enhanced for larger values of nanoparticle volume fraction.

Keywords: stretchable rotating disk; CNTs (MWCNTs and SWCNTs); velocity slip; convective boundary condition; OHAM

---

1. Introduction

The investigation of liquid flow by a rotating disk has various applications in aviation science, pivot of hardware, synthetic enterprises and designing, creating frameworks of warm power, rotor-stator frameworks, medicinal contraption, electronic and PC putting away apparatuses, gem developing wonders, machines of air cleaning, nourishment preparing advances, turbo apparatus and numerous others. Von Karman [1] analyzed flow of thick fluid by a rotating disk. Turkyilmazoglu and Senel [2] explored effects of heat and mass transport in thick liquid flow over a permeable rotating disk. Rashidi et al. [3] dissected MHD flow of viscous liquid because of a turn of disk. Turkyilmazoglu [4] exhibited nanoliquid flow by a rotating plate. Hatami et al. [5] examined laminar flow of a thick nanofluid because of the revolution and constriction of disks. Nanoliquid flow because of an extending disk is considered by Mustafa et al. [6]. Sheikholeslami et al. [7] examined nanoliquid flow by a slanted rotatory plate. Recently Hayat et al. [8] analyzed MHD nanoliquid flow over a rotatory disk with slip impacts.

Carbon nanotubes (CNTs) were first discovered by Lijima in 1991. CNTs have long cylindrical pofiles such as frames of carbon atoms with diameter ranges from 0.70–50 nm. CNTs have individual importance in nano-technology, hardwater, air purification systems, structural composite materials, conductive plastics, extra strong fibres, sensors, flat-panel displays, gas storage, biosensors and many others. Thus Choi et al. [9] examined anamolous enhancement of thermal conductivity in nanotubes suspension. Ramasubramaniam et al. [10] examined homogeneous polymer composites/carbon nanotubes for electrical utilizations. Xue [11] proposed a relation for CNT-based composites. Heat transfer enhancement using carbon nanotubes-based-non-Newtonian nanofluids is discussed by Kamali et al. [12]. Wang et al. [13] illustrated laminar flows of nanofluids containing single-walled carbon nanotubes (SWCNT) and multi-walled carbon nanotubes (MWCNTs). Hammouch et al. [14] analyzed squeezed flow of CNTs between parallel disks. Thermal transfer upgrade in front aligned contracting channel by taking FMWCNT nanoliquids is analyzed by Safaei et al. [15]. MHD flow of carbon nanotubes is portrayed by Ellahi et ai. [16]. Karimipour et al. [17] dissected MHD laminar flow of carbon nanotubes in a microchannel with a uniform warmth transition. Hayat et al. [18] represented homogeneous-heterogeneous responses in nanofluid flows over a non-direct extending surface of variable thickness. Unsteady squeezed flow of CNTs with convective surface was contemplated by Hayat et al. [19]. Hayat et al. [20] likewise talked about Darcy Forchheimer flow of CNTs over a turning plate. Further relevant investigations on nanofluids can be seen through the studies [21–25].

Motivated by the aforementioned applications of rotating flows, the underlying objective of this article is to develop a mathematical model for three-dimensional flow of water-based carbon nanotubes because of a stretchable rotating disk considering velocity slip effects. Thermal conductivity of carbon nanotubes is estimated through the well-known Xue model. Such research work was not carried out in the past even in the absence of a convective heating surface. Researchers also found that dispersion of carbon nanotubes in water elevates the thermal conductivity of the resulting nanofluid by 100% (see Choi et al. [9]). Both single-walled carbon nanotubes (SWCNTs) and multi-walled carbon nanotubes (MWCNTs) are considered. Optimal homotopic strategy (OHAM) [26–35] is utilized for solutions of temperature and velocities. Impacts of different flow variables are examined and investigated. Nusselt number and skin friction have been analyzed graphically. Emphasis is given to the role of the main ingredients of the problem, namely volume fraction of carbon nanotubes and a rotating stretchable disk. The benefits of carbon nanotubes towards heat transfer enhancement are also justified via thorough analysis.

2. Mathematical Formulation

Let us assume three-dimensional flow of water-based carbon nanotubes by a stretchable rotating disk. The disk at $z = 0$ rotates subject to constant angular velocity Ω (see Figure 1). Let us assume CNT nanoparticles: SWCNTs and MWCNTs within base liquid (water). Due to axial symmetry, derivatives of φ are neglected. The surface of the disk has temperature T_f, while ambient fluid temperature is T_∞. The velocity components are (u, v, w) in cylindrical coordinate (r, φ, z) respectively. The resulting boundary-layer expressions are [8,20]:

$$\frac{\partial u}{\partial r} + \frac{u}{r} + \frac{\partial w}{\partial z} = 0, \tag{1}$$

$$u\frac{\partial u}{\partial r} - \frac{v^2}{r} + w\frac{\partial u}{\partial z} = \nu_{nf}\left(\frac{\partial^2 u}{\partial r^2} + \frac{\partial^2 u}{\partial z^2} + \frac{1}{r}\frac{\partial u}{\partial r} - \frac{u}{r^2}\right), \tag{2}$$

$$u\frac{\partial v}{\partial r} + \frac{uv}{r} + w\frac{\partial v}{\partial z} = \nu_{nf}\left(\frac{\partial^2 v}{\partial r^2} + \frac{\partial^2 v}{\partial z^2} + \frac{1}{r}\frac{\partial v}{\partial r} - \frac{v}{r^2}\right), \tag{3}$$

$$u\frac{\partial w}{\partial r} + w\frac{\partial w}{\partial z} = \nu_{nf}\left(\frac{\partial^2 w}{\partial r^2} + \frac{\partial^2 w}{\partial z^2} + \frac{1}{r}\frac{\partial w}{\partial r}\right), \tag{4}$$

$$u\frac{\partial T}{\partial r} + w\frac{\partial T}{\partial z} = \alpha_{nf}\left(\frac{\partial^2 T}{\partial r^2} + \frac{\partial^2 T}{\partial z^2} + \frac{1}{r}\frac{\partial T}{\partial r}\right), \tag{5}$$

with subjected boundary conditions [8]:

$$u = rs + L_1\mu_{nf}\frac{\partial u}{\partial z}, \ v = r\Omega + L_1\mu_{nf}\frac{\partial v}{\partial z}, \ w = 0, \ -k_{nf}\frac{\partial T}{\partial z} = h_f\left(T_f - T\right) \text{ at } z = 0, \tag{6}$$

$$u \to 0, \ v \to 0, \ T \to T_\infty \text{ as } z \to \infty. \tag{7}$$

Here u, v and w depict flow velocities in increasing directions of r, φ and z respectively, while $\nu_{nf} = (\mu_{nf}/\rho_{nf})$ stands for kinematic viscosity, $\alpha_{nf} = k_{nf}/(\rho c_p)_{nf}$ for thermal diffusivity, μ_{nf} for dynamic viscosity, L_1 for wall-slip coefficient, T for fluid temperature, k_{nf} for thermal conductivity of nanofluids, ρ_{nf} for effective density, k_{CNT} for thermal conductivity of CNTs and $(\rho c_p)_{nf}$ for effective heat capacitance of nanoparticle material. Xue [11] proposed a theoratical model which is expressed by

$$\begin{aligned}
\rho_{nf} &= \rho_f(1-\phi) + \rho_{CNT}\phi, \quad \mu_{nf} = \frac{\mu_f}{(1-\phi)^{2.5}}, \\
(\rho c_p)_{nf} &= (\rho c_p)_f(1-\phi) + (\rho c_p)_{CNT}\phi, \\
\frac{k_{nf}}{k_f} &= \frac{(1-\phi) + 2\phi\frac{k_{CNT}}{k_{CNT}-k_f}\ln\frac{k_{CNT}+k_f}{2k_f}}{(1-\phi) + 2\phi\frac{k_f}{k_{CNT}-k_f}\ln\frac{k_{CNT}+k_f}{2k_f}},
\end{aligned} \qquad (8)$$

where ϕ represents solid volume fraction of nanoparticles and nf represents thermophysical properties of nanofluid. Table 1 describes thermo-physical features of water and CNT.

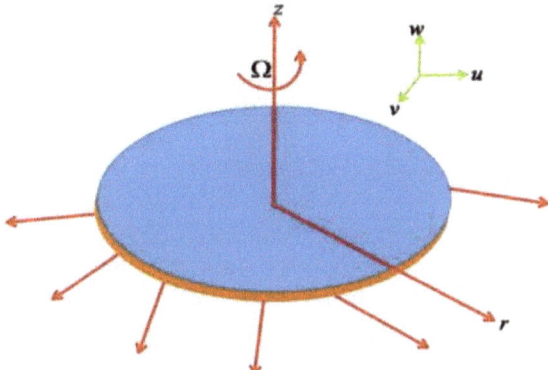

Figure 1. Geometry of the problem.

Table 1. Thermophysical features of water and carbon nanotubes (CNT).

Physical Features	Water	CNT	
		SWCNTs	MWCNTs
ρ (kg/m^3)	997.1	2600	1600
k (W/mK)	0.613	6600	3000
c_p (J/kgK)	4179	425	796

We now introduce the following transformations:

$$\begin{aligned}
u = r\Omega f'(\zeta), \; v = r\Omega g(\zeta), \; w = -\sqrt{2\nu_f \Omega} f(\zeta), \\
\zeta = z\left(\frac{2\Omega}{\nu_f}\right)^{1/2}, \; \theta(\zeta) = \frac{T-T_\infty}{T_f - T_\infty}.
\end{aligned} \qquad (9)$$

Expression (1) is automatically satisfied while Equations (2)–(8) yield

$$\frac{1}{(1-\phi)^{5/2}(1-\phi+\frac{\rho_{CNT}}{\rho_f}\phi)} f'''(\zeta) + f(\zeta)f''(\zeta) - \frac{1}{2}f'^2(\zeta) + \frac{1}{2}g^2(\zeta) = 0, \qquad (10)$$

$$\frac{1}{(1-\phi)^{5/2}(1-\phi+\frac{\rho_{CNT}}{\rho_f}\phi)} g''(\zeta) + f(\zeta)g'(\zeta) - f'(\zeta)g(\zeta) = 0, \qquad (11)$$

$$\frac{1}{\Pr}\frac{k_{nf}}{k_f}\theta''(\zeta) + \left(1-\phi+\phi\frac{(\rho c_p)_{CNT}}{(\rho c_p)_f}\right) f(\zeta)\theta'(\zeta) = 0, \qquad (12)$$

with the boundary conditions

$$f(0) = 0, \ f'(0) = C + \frac{\alpha}{(1-\phi)^{5/2}} f''(0), \ g(0) = 1 + \frac{\alpha}{(1-\phi)^{5/2}} g'(0), \ \theta'(0) = -\frac{k_f}{k_{nf}} Bi(1 - \theta(0)), \quad (13)$$

$$f'(\infty) \to 0, \ g(\infty) \to 0, \ \theta(\infty) \to 0. \quad (14)$$

Here C stands for stretching-strength parameter, α for velocity slip number, Pr for Prandtl number and Bi for the Biot number. These numbers are described by:

$$C = \frac{s}{\Omega}, \ \alpha = L_1 \mu_f \left(\frac{2\Omega}{\nu_f}\right)^{1/2}, \ Bi = \frac{h_f}{k_f} \left(\frac{\nu_f}{2\Omega}\right)^{1/2}, \ \Pr = \frac{\nu_f (\rho c_p)_f}{k_f}. \quad (15)$$

Nusselt number and skin friction are defined by

$$\left. \begin{array}{l} \operatorname{Re}_r^{-1/2} Nu_r = -\frac{k_{nf}}{k_f} \theta'(0), \\ \operatorname{Re}_r^{1/2} C_f = \frac{1}{(1-\phi)^{5/2}} \left(f''(0)^2 + g'(0)^2\right)^{1/2}, \end{array} \right\}, \quad (16)$$

where $\operatorname{Re}_r = 2\Omega r^2 / \nu_f$ depicts the local Reynolds number.

3. Solutions by OHAM

The optimal solutions of expressions (10)–(12) through (13) and (14) have been established by considering optimal homotopic strategy (OHAM). The proper operators and guesses are

$$f_0(\zeta) = \frac{C}{\left(1 + \frac{\alpha}{(1-\phi)^{5/2}}\right)} (1 - e^{-\zeta}), \ g_0(\zeta) = \frac{1}{\left(1 + \frac{\alpha}{(1-\phi)^{5/2}}\right)} e^{-\zeta}, \ \theta_0(\zeta) = \frac{Bi}{\left(\frac{k_{nf}}{k_f} + Bi\right)} e^{-\zeta}, \quad (17)$$

$$\mathcal{L}_g = \frac{d^2 g}{d\zeta^2} - g, \ \mathcal{L}_\theta = \frac{d^2 \theta}{d\zeta^2} - \theta, \ \mathcal{L}_f = \frac{d^3 f}{d\zeta^3} - \frac{df}{d\zeta}. \quad (18)$$

The above operators satisfy

$$\mathcal{L}_f \left[F_1^{****} + F_2^{****} e^\zeta + F_3^{****} e^{-\zeta} \right] = 0, \ \mathcal{L}_g \left[F_4^{****} e^\zeta + F_5^{****} e^{-\zeta} \right] = 0, \ \mathcal{L}_\theta \left[F_6^{****} e^\zeta + F_7^{****} e^{-\zeta} \right] = 0, \quad (19)$$

in which F_i^{****} (i = 1–7) portrays arbitrary constants. The m-th and zero-th order systems are easily established in view of above operators. By using BVPh2.0 of the software Mathematica, the obtained deformation problems have been computed.

4. Optimal Convergence-Control Parameters

In homotopic solutions, the non zero auxiliary variables $\hbar_f$, $\hbar_g$ and $\hbar_\theta$ determine the convergence portion and also rate of homotopy solution. The idea of minimization has been applied by defining averaged squared residuals errors as proposed by Liao [26].

$$\varepsilon_m^f = \frac{1}{k+1} \sum_{j=0}^{k} \left[\mathcal{N}_f \left(\sum_{i=0}^{m} \hat{f}(\zeta), \sum_{i=0}^{m} \hat{g}(\zeta) \right)_{\zeta = j\delta\zeta} \right]^2, \quad (20)$$

$$\varepsilon_m^g = \frac{1}{k+1} \sum_{j=0}^{k} \left[\mathcal{N}_g \left(\sum_{i=0}^{m} \hat{f}(\zeta), \sum_{i=0}^{m} \hat{g}(\zeta) \right)_{\zeta = j\delta\zeta} \right]^2, \quad (21)$$

$$\varepsilon_m^\theta = \frac{1}{k+1} \sum_{j=0}^{k} \left[\mathcal{N}_\theta \left(\sum_{i=0}^{m} \hat{f}(\zeta), \sum_{i=0}^{m} \hat{g}(\zeta), \sum_{i=0}^{m} \hat{\theta}(\zeta) \right)_{\zeta = j\delta\zeta} \right]^2. \quad (22)$$

Following Liao [26]:

$$\varepsilon_m^t = \varepsilon_m^f + \varepsilon_m^g + \varepsilon_m^\theta, \qquad (23)$$

where ε_m^t represents total squared residual error, $\delta\zeta = 0.5$ and $k = 20$. At the second order of deformations, convergence-control parameters for SWCNTs–water have optimal values i.e., $h_f = -0.35923$, $h_g = -0.736096$ and $h_\theta = -0.00105197$ and total averaged squared residuals error is $\varepsilon_m^t = -0.0255367$ while optimal data of convergence-control parameters for MWCNTs–water is $h_f = -0.385385$, $h_g = -0.729057$ and $h_\theta = -0.00232643$ and total averaged squared residuals error is $\varepsilon_m^t = -0.025173$. Figures 2 and 3 display error plots for MWCNTs–water and SWCNTs–water. Tables 2 and 3 show that averaged squared residuals error decreases for higher order deformations.

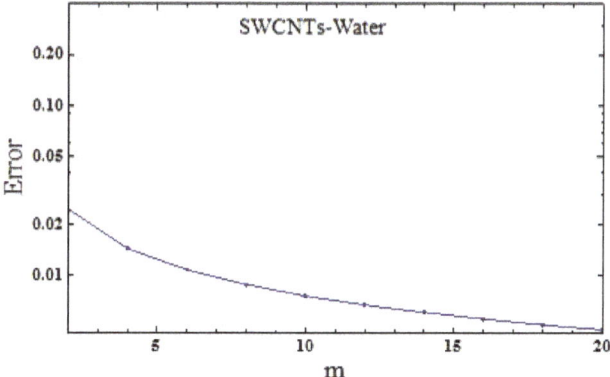

Figure 2. Error sketch for SWCNTs-Water.

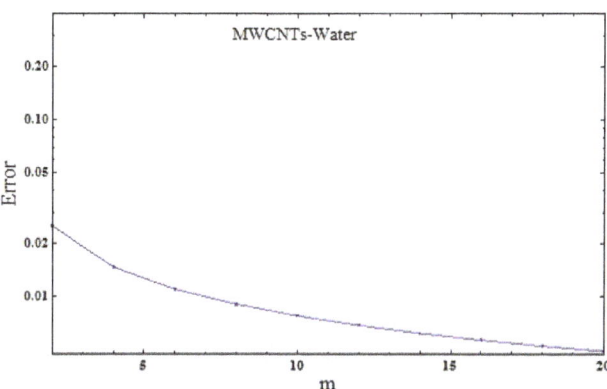

Figure 3. Error sketch for MWCNTs-Water.

Table 2. Individual averaged squared residuals errors for single-walled carbon nanotubes (SWCNTs)–water.

m	ε_m^f	ε_m^g	ε_m^θ
2	9.95225×10^{-5}	2.35341×10^{-2}	7.29447×10^{-7}
6	4.17686×10^{-5}	1.03083×10^{-2}	6.04738×10^{-7}
10	2.95796×10^{-5}	7.24672×10^{-3}	5.69806×10^{-7}
14	2.37325×10^{-5}	5.77429×10^{-3}	5.53122×10^{-7}
18	2.01939×10^{-5}	4.88653×10^{-3}	5.43213×10^{-7}
20	1.88867×10^{-5}	4.55942×10^{-3}	5.39608×10^{-7}

Table 3. Individual averaged squared residuals errors for single-walled carbon nanotubes (MWCNTs)–water.

m	ε_m^f	ε_m^g	ε_m^θ
2	1.0164×10^{-4}	2.40503×10^{-2}	7.29447×10^{-7}
6	4.27165×10^{-5}	1.05447×10^{-2}	6.04522×10^{-7}
10	3.02678×10^{-5}	7.41739×10^{-3}	5.69547×10^{-7}
14	2.42942×10^{-5}	5.91293×10^{-3}	5.52829×10^{-7}
18	2.06785×10^{-5}	5.00567×10^{-3}	5.42892×10^{-7}
20	1.93426×10^{-5}	4.67132×10^{-3}	5.39275×10^{-7}

5. Results and Discussion

The present section presents behaviors of various physical parameters like stretching-strength parameter C, volume fraction ϕ, velocity slip parameter α and Biot number Bi on radial $f'(\zeta)$ and azimuthal $g(\zeta)$ velocities and temperature $\theta(\zeta)$. The results are obtained for both SWCNTs and MWCNTs. Figure 4 shows variation in the radial velocity $f'(\zeta)$ for larger values of α. Radial velocity $f'(\zeta)$ shows reduction for increasing values of α. Figure 5 presents impact of stretching-strength parameter C on radial velocity $f'(\zeta)$. For larger values of C, the radial velocity shows an increasing trend. Figure 6 depicts the effect of nanoparticle volume fraction ϕ on radial velocity $f'(\zeta)$. For higher ϕ, the radial velocity $f'(\zeta)$ is increased. Figure 7 presents that how velocity slip parameter α affects the azimuthal velocity $g(\zeta)$. It is observed that an increment in velocity slip parameter α lead to lower $g(\zeta)$. Figure 8 depicts impact of C on azimuthal velocity $g(\zeta)$. Azimuthal velocity reduces for larger values of streching-strength parameter. Figure 9 depicts the impact of nanoparticles volume fraction ϕ on $g(\zeta)$. The azimuthal velocity $g(\zeta)$ is increased for higher estimations of ϕ. Figure 10 examines that how Biot number Bi affects the temperature profile. For higher values of Bi, the temperature field $\theta(\zeta)$ is enhanced. Higher estimations of Biot number correspond to stronger convection which produces higher temperature field and more associated layer thickness. Figure 11 highlights the impact of stretching-strength parameter C on temperature field $\theta(\zeta)$. Temperature field $\theta(\zeta)$ is reduced for increasing values of C. Figure 12 presents that how volume fraction ϕ affects the temperature field $\theta(\zeta)$. Higher values of ϕ shows an enhancement in temperature $\theta(\zeta)$. Figure 13 shows the effects of volume fraction ϕ and velocity slip parameter α on $Re_r^{1/2}C_f$. Skin friction $Re_r^{1/2}C_f$ is increased for higher estimations of ϕ. Figure 14 displays the behavior of the volume fraction ϕ and Biot number Bi on Nusselt number $Re_r^{-1/2}Nu_r$. The Nusselt number is enhanced for increasing values of ϕ.

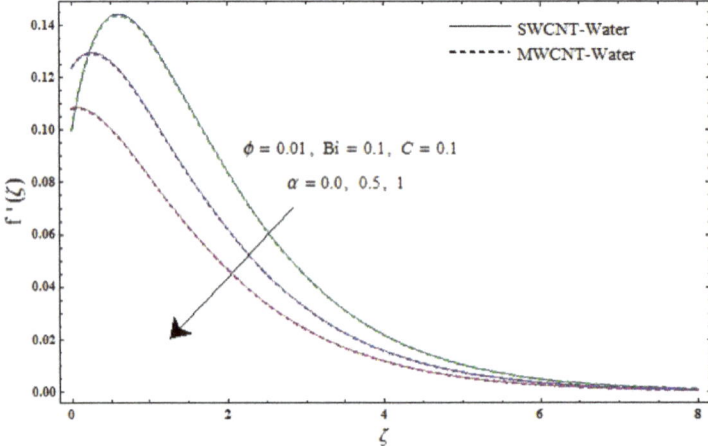

Figure 4. Sketch of $f'(\zeta)$ for α.

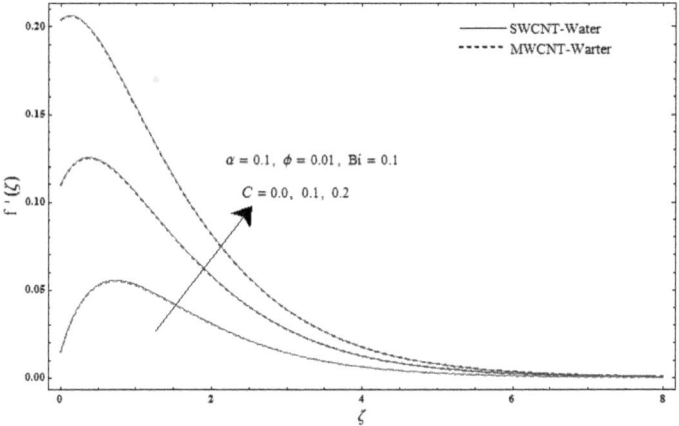

Figure 5. Sketch of $f'(\zeta)$ for C.

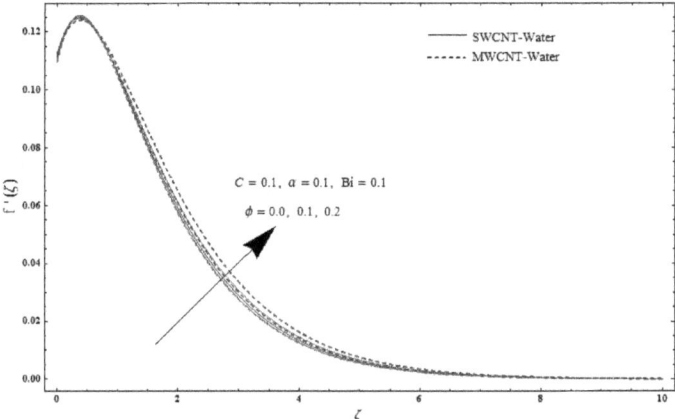

Figure 6. Sketch of $f'(\zeta)$ for ϕ.

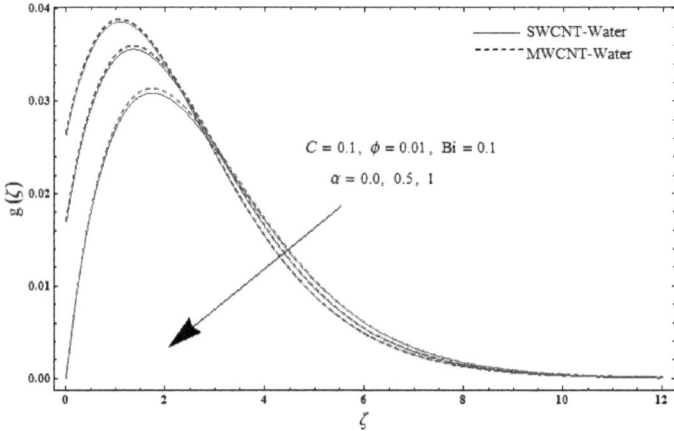

Figure 7. Sketch of $g(\zeta)$ for α.

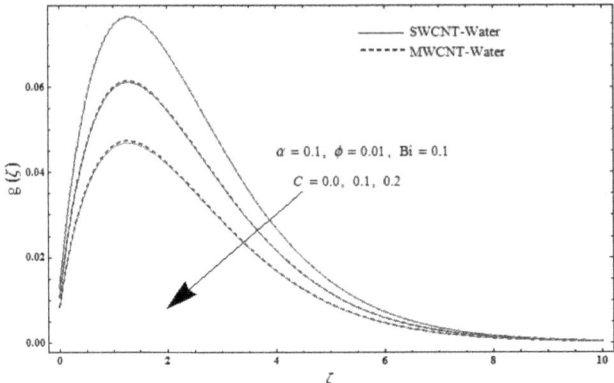

Figure 8. Sketch of $g(\zeta)$ for C.

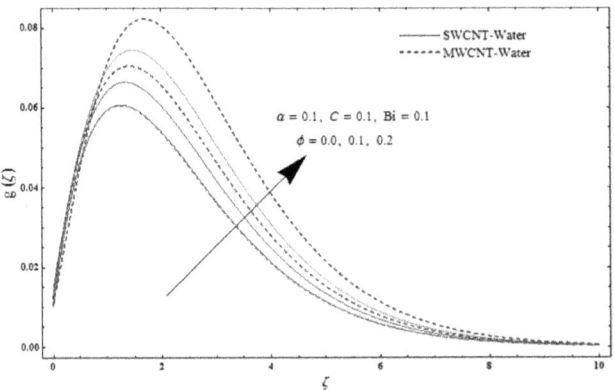

Figure 9. Sketch of $g(\zeta)$ for ϕ.

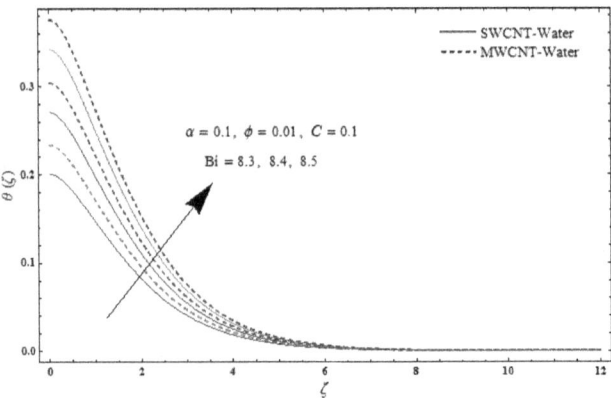

Figure 10. Sketch of $\theta(\zeta)$ for Bi.

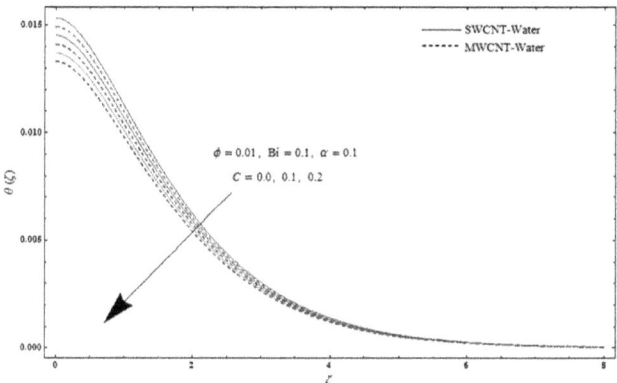

Figure 11. Sketch of $\theta(\zeta)$ for C.

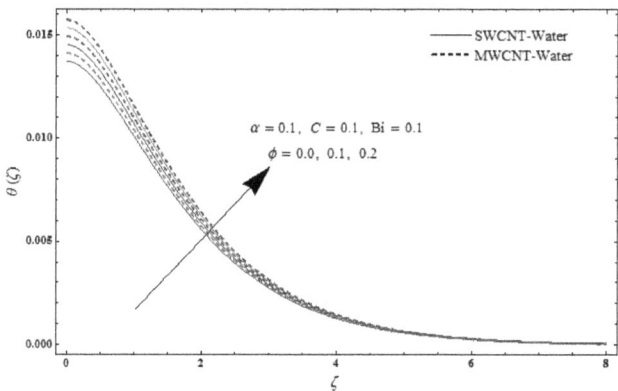

Figure 12. Sketch of $\theta(\zeta)$ for ϕ.

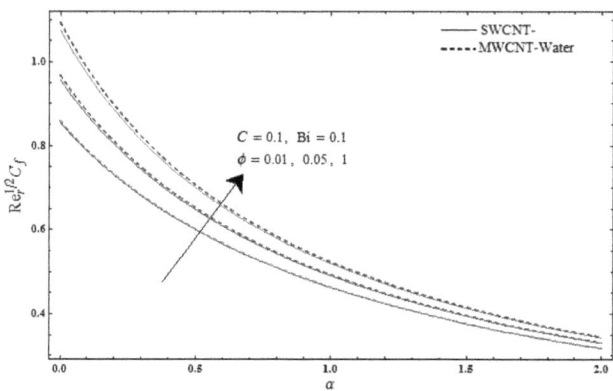

Figure 13. Sketch of $Re_r^{1/2} C_f$ for ϕ and α.

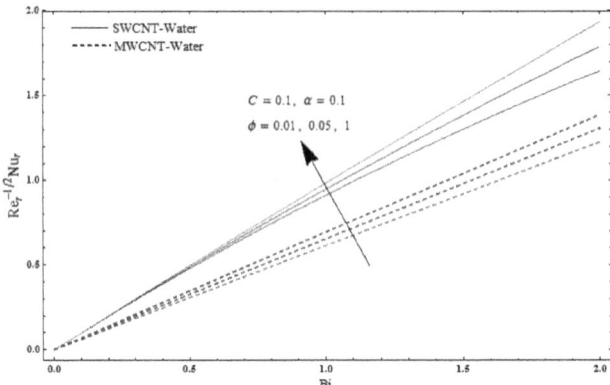

Figure 14. Sketch of $Re_r^{-1/2} Nu_r$ for ϕ and Bi.

6. Conclusions

Three-dimensional flow of carbon nanotubes by a stretchable rotating disk with velocity slip effects is studied. Heat transport is explained by convective heating surface. The key findings of current research are listed below:

- Both velocities $f'(\zeta)$ and $g(\zeta)$ show reduction for higher values of velocity slip parameter α.
- Larger stretching-strength parameter C presents an increase in radial velocity $f'(\zeta)$ while opposite trend is noticed for azimuthal velocity $g(\zeta)$ and temperature $\theta(\zeta)$.
- For higher estimations of the volume fraction ϕ, both the velocity and temperatue field are enhanced.
- Temperature field $\theta(\zeta)$ is enhanced for larger values of the Biot number Bi.
- Nusselt number $Re_r^{-1/2} Nu_r$ is increased for larger values of volume fraction ϕ.
- Coefficient of skin-friction $Re_r^{1/2} C_f$ increases for higher volume fraction ϕ and velocity slip parameter α.
- The used technique for the solution's development has advantages over the other in the sense of the following points:

    a. It is independent of small/large physical parameters.
    b. It provides a simple way to ensure the convergence of series solutions.
    c. It provides a large freedom to choose the base functions and related auxiliary linear operators.

Author Contributions: All the authors contributed equally to the conception of the idea, implementing and analyzing the experimental results, and writing the manuscript.

Funding: This project was funded by the Deanship of Scientific Research (DSR) at King Abdulaziz University, Jeddah, under grant No. G-316-130-38. The authors, therefore, acknowledge with thanks DSR for technical and financial support.

Conflicts of Interest: The authors declare no conflict of interest.

Nomenclature

r, φ, z	space coordinates [m]	u, v, w	velocity components [m·s^{-1}]
ρ_f	fluid density [kg·m^{-3}]	μ_f	fluid dynamic viscosity [Pa·s]
k_{nf}	nanofluids themal conductivity [W·m^{-1}·K^{-1}]	ν_{nf}	kinematic nanofluid viscosity [m^2·s^{-1}]
k_f	basefluid themal conductivity [W·m^{-1}·K^{-1}]	ν_f	kinematic fluid viscosity [m^2·s^{-1}]
α_f	thermal diffusivity of base fluid [m^2·s^{-1}]	α_{nf}	thermal diffusivity of nanofluid [m^2·s^{-1}]
T_f	hot fluid temperature [K]	T_∞	ambient temperature [K]
C	stretching-strength parameter	k_{CNT}	CNTs thermal conductivity [W·m^{-1}·K^{-1}]
α	velocity slip parameter	ϕ	nanomaterial volume fraction
Bi	Biot number	Pr	Prandtl number
f'	dimensionless velocity	Nu_r	Nusselt number
C_f	skin friction coefficient	ζ	dimensionless variable
Re_r	local Reynolds number	θ	dimensionless temperature
CNTs	carbon nanotubes	F_i^{****}	arbitrary constants

References

1. Von Karman, T. Uber laminare and turbulente Reibung. *ZAMM Z. Angew. Math. Mech.* **1921**, *1*, 233–252. [CrossRef]
2. Turkyilmazoglu, M.; Senel, P. Heat and mass transfer of the flow due to a rotating rough and porous disk. *Int. J. Therm. Sci.* **2013**, *63*, 146–158. [CrossRef]
3. Rashidi, M.M.; Kavyani, N.; Abelman, S. Investigation of entropy generation in MHD and slip flow over rotating porous disk with variable properties. *Int. J. Heat Mass Transf.* **2014**, *70*, 892–917. [CrossRef]
4. Turkyilmazoglu, M. Nanofluid flow and heat transfer due to a rotating disk. *Comput. Fluids* **2014**, *94*, 139–146. [CrossRef]
5. Hatami, M.; Sheikholeslami, M.; Gangi, D.D. Laminar flow and heat transfer of nanofluids between contracting and rotating disks by least square method. *Power Technol.* **2014**, *253*, 769–779. [CrossRef]
6. Mustafa, M.; Khan, J.A.; Hayat, T.; Alsaedi, A. On Bödewadt flow and heat transfer of nanofluids over a stretching stationary disk. *J. Mol. Liq.* **2015**, *211*, 119–125. [CrossRef]
7. Sheikholeslami, M.; Hatami, M.; Ganji, D.D. Numerical investigation of nanofluid spraying on an inclined rotating disk for cooling process. *J. Mol. Liq.* **2015**, *211*, 577–583. [CrossRef]
8. Hayat, T.; Muhammad, T.; Shehzad, S.A.; Alsaedi, A. On magnetohydrodynamic flow of nanofluid due to a rotating disk with slip effect: A numerical study. *Comput. Methods Appl. Mech. Eng.* **2017**, *315*, 467–477. [CrossRef]
9. Choi, S.U.S.; Zhang, Z.G.; Yu, W.; Lockwood, F.E.; Grulke, E.A. Anomalous thermal conductivity enhancement in nanotube suspensions. *Appl. Phys. Lett.* **2001**, *79*, 2252. [CrossRef]
10. Ramasubramaniam, R.; Chen, J.; Liu, H. Homogeneous carbon nanotube/polymer composites for electrical applications. *Appl. Phys. Lett.* **2003**, *83*, 2928. [CrossRef]
11. Xue, Q.Z. Model for thermal conductivity of carbon nanotube-based composites. *Phys. B Condens. Matter* **2005**, *368*, 302–307. [CrossRef]
12. Kamali, R.; Binesh, A. Numerical investigation of heat transfer enhancement using carbon nanotube-based non-Newtonian nanofluids. *Int. Commun. Heat Mass Transf.* **2010**, *37*, 1153–1157. [CrossRef]
13. Wang, J.; Zhu, J.; Zhang, X.; Chen, Y. Heat transfer and pressure drop of nanofluids containing carbon nanotubes in laminar flows. *Exp. Therm. Fluid Sci.* **2013**, *44*, 716–721. [CrossRef]
14. Haq, R.U.; Hammouch, Z.; Khan, W.A. Water-based squeezing flow in the presence of carbon nanotubes between two parallel disks. *Therm. Sci.* **2014**, *20*, 148. [CrossRef]

15. Safaei, M.R.; Togun, H.; Vafai, K.; Kazi, S.N.; Badarudin, A. Investigation of heat transfer enhancement in a forward-facing contracting channel using FMWCNT nanofluids. *Numer. Heat Transf. Part A* **2014**, *66*, 1321–1340. [CrossRef]
16. Ellahi, R.; Hassan, M.; Zeeshan, A. Study of natural convection MHD nanofluid by means of single and multi walled carbon nanotubes suspended in a salt water solutions. *IEEE Trans. Nanotechnol.* **2015**, *14*, 726–734. [CrossRef]
17. Karimipour, A.; Taghipour, A.; Malvandi, A. Developing the laminar MHD forced convection flow of water/FMWNT carbon nanotubes in a microchannel imposed the uniform heat flux. *J. Magn. Magn. Mater.* **2016**, *419*, 420–428. [CrossRef]
18. Hayat, T.; Hussain, Z.; Muhammad, T.; Alsaedi, A. Effects of homogeneous and heterogeneous reactions in flow of nanofluids over a nonlinear stretching surface with variable surface thickness. *J. Mol. Liq.* **2016**, *221*, 1121–1127. [CrossRef]
19. Hayat, T.; Muhammad, K.; Farooq, M.; Alsaedi, A. Unsteady squeezing flow of carbon nanotubes with convective boundary conditions. *PLoS ONE* **2016**, *11*, e0152923. [CrossRef]
20. Hayat, T.; Haider, F.; Muhammad, T.; Alsaedi, A. On Darcy-Forchheimer flow of carbon nanotubes due to a rotating disk. *Int. J. Heat Mass Transf.* **2017**, *112*, 248–254. [CrossRef]
21. Akbar, N.S.; Khan, Z.H.; Nadeem, S. The combined effects of slip and convective boundary conditions on stagnation-point flow of CNT suspended nanofluid over a stretching sheet. *J. Mol. Liq.* **2014**, *196*, 21–25. [CrossRef]
22. Arani, A.A.A.; Akbari, O.A.; Safaei, M.R.; Marzban, A.; Alrashed, A.A.A.A.; Ahmadi, G.R.; Nguyen, T.K. Heat transfer improvement of water/single-wall carbon nanotubes (SWCNT) nanofluid in a novel design of a truncated double-layered microchannel heat sink. *Int. J. Heat Mass Transf.* **2017**, *113*, 780–795. [CrossRef]
23. Goodarzi, M.; Javid, S.; Sajadifar, A.; Nojoomizadeh, M.; Motaharipour, S.H.; Bach, Q.V.; Karimipour, A. Slip velocity and temperature jump of a non-Newtonian nanofluid, aqueous solution of carboxy-methyl cellulose/aluminum oxide nanoparticles, through a microtube. *Int. J. Numer. Methods Heat Fluid Flow* **2018**. [CrossRef]
24. Ellahi, R.; Zeeshan, A.; Hussain, F.; Asadollahi, A. Peristaltic blood flow of couple stress fluid suspended with nanoparticles under the influence of chemical reaction and activation energy. *Symmetry* **2019**, *11*, 276. [CrossRef]
25. Suleman, M.; Ramzan, M.; Ahmad, S.; Lu, D.; Muhammad, T.; Chung, J.D. A numerical simulation of silver-water nanofluid flow with impacts of Newtonian heating and homogeneous-heterogeneous reactions past a nonlinear stretched cylinder. *Symmetry* **2019**, *11*, 295. [CrossRef]
26. Liao, S.J. An optimal homotopy-analysis approach for strongly nonlinear differential equations. *Commun. Nonlinear. Sci. Numer. Simul.* **2010**, *15*, 2003–2016. [CrossRef]
27. Dehghan, M.; Manafian, J.; Saadatmandi, A. Solving nonlinear fractional partial differential equations using the homotopy analysis method. *Numer. Meth. Part. Diff. Equ.* **2010**, *26*, 448–479. [CrossRef]
28. Malvandi, A.; Hedayati, F.; Domairry, G. Stagnation point flow of a nanofluid toward an exponentially stretching sheet with nonuniform heat generation/absorption. *J. Thermodyn.* **2013**, *2013*, 764827. [CrossRef]
29. Abbasbandy, S.; Hayat, T.; Alsaedi, A.; Rashidi, M.M. Numerical and analytical solutions for Falkner-Skan flow of MHD Oldroyd-B fluid. *Int. J. Numer. Methods Heat Fluid Flow* **2014**, *24*, 390–401. [CrossRef]
30. Sheikholeslami, M.; Hatami, M.; Ganji, D.D. Micropolar fluid flow and heat transfer in a permeable channel using analytic method. *J. Mol. Liq.* **2014**, *194*, 30–36. [CrossRef]
31. Hayat, T.; Muhammad, T.; Alsaedi, A.; Alhuthali, M.S. Magnetohydrodynamic three-dimensional flow of viscoelastic nanofluid in the presence of nonlinear thermal radiation. *J. Magn. Magn. Mater.* **2015**, *385*, 222–229. [CrossRef]
32. Turkyilmazoglu, M. An effective approach for evaluation of the optimal convergence control parameter in the homotopy analysis method. *Filomat* **2016**, *30*, 1633–1650. [CrossRef]
33. Zeeshan, A.; Majeed, A.; Ellahi, R. Effect of magnetic dipole on viscous ferro-fluid past a stretching surface with thermal radiation. *J. Mol. Liq.* **2016**, *215*, 549–554. [CrossRef]

34. Hayat, T.; Abbas, T.; Ayub, M.; Muhammad, T.; Alsaedi, A. On squeezed flow of Jeffrey nanofluid between two parallel disks. *Appl. Sci.* **2016**, *6*, 346. [CrossRef]
35. Muhammad, T.; Alsaedi, A.; Shehzad, S.A.; Hayat, T. A revised model for Darcy-Forchheimer flow of Maxwell nanofluid subject to convective boundary condition. *Chin. J. Phys.* **2017**, *55*, 963–976. [CrossRef]

© 2019 by the authors. Licensee MDPI, Basel, Switzerland. This article is an open access article distributed under the terms and conditions of the Creative Commons Attribution (CC BY) license (http://creativecommons.org/licenses/by/4.0/).

Article

Numerical Solution of Non-Newtonian Fluid Flow Due to Rotatory Rigid Disk

Khalil Ur Rehman [1,*], M. Y. Malik [2], Waqar A Khan [3], Ilyas Khan [4] and S. O. Alharbi [4]

1. Department of Mathematics, Air University, PAF Complex E-9, Islamabad 44000, Pakistan
2. Department of Mathematics, College of Sciences, King Khalid University, Abha 61413, Saudi Arabia; drmymalik@qau.edu.pk
3. Department of Mechanical Engineering, College of Engineering, Prince Mohammad Bin Fahd University, Al Khobar 31952, Kingdom of Saudi Arabia; wkhan@pmu.edu.sa
4. Department of Mathematics, College of Science Al-Zulfi, Majmaah University, Al-Majmaah 11952, Saudi Arabia; i.said@mu.edu.sa (I.K.); so.alharbi@mu.edu.sa (S.O.A.)
* Correspondence: krehman@math.qau.edu.pk

Received: 12 March 2019; Accepted: 16 April 2019; Published: 22 May 2019

Abstract: In this article, the non-Newtonian fluid model named Casson fluid is considered. The semi-infinite domain of disk is fitted out with magnetized Casson liquid. The role of both thermophoresis and Brownian motion is inspected by considering nanosized particles in a Casson liquid spaced above the rotating disk. The magnetized flow field is framed with Navier's slip assumption. The Von Karman scheme is adopted to transform flow narrating equations in terms of reduced system. For better depiction a self-coded computational algorithm is executed rather than to move-on with build-in array. Numerical observations via magnetic, Lewis numbers, Casson, slip, Brownian motion, and thermophoresis parameters subject to radial, tangential velocities, temperature, and nanoparticles concentration are reported. The validation of numerical method being used is given through comparison with existing work. Comparative values of local Nusselt number and local Sherwood number are provided for involved flow controlling parameters.

Keywords: Casson fluid model; rotating rigid disk; nanoparticles; Magnetohydrodynamics (MHD)

1. Introduction

The examination of non-Newtonian fluids has received remarkable attention from researchers and scientists because of their extensive use in industrial and technological areas. For instance, paints, synthetic lubricants, sugar solutions, certain oils, clay coating, drilling muds, and blood as a biological fluid are common examples of non-Newtonian fluids, just to mention a few. The fundamental mathematical equations given by Navier–Stokes cannot briefly delineate the flow field characteristics of non-Newtonian fluids because of the complex mathematical expression involved in the formulation of flow problem. In addition, the relation between strain rate and shear stress is non-linear so the single constitutive expressions are fruitless to report complete description of flows subject to non-Newtonian fluids. Numerous non-Newtonian fluid models are exposed to explore rheological characteristics, namely Bingham Herschel–Bulkley fluid model, Seely, Carreau Carreau–Yasuda, Sisko, Eyring, Cross, Ellis, Williamson, tangent hyperbolic, Generalized Burgers, Burgers, Oldroyd-8 constants, Oldroyd-A, Oldroyd-B fluid model, Maxwell, Jeffrey, Casson fluid model, etc. Researchers discussed flow characteristics of non-Newtonian fluid models via stretching surfaces by incorporating pertinent physical effects. Among these, Casson fluid model has many advantages as compared to rest of fluid models. This model can be used to approximate the properties of blood and daily life suspensions. One can assessed recent developments in this direction in References [1–15].

The centrifugal filtration, gas turbine rotors, rotating air cleaning machines, food processing, medical equipment, system of electric-power generation, crystal growth processes, and many others are the practical applications of rotational fluids flow. Therefore, analysis of flows due to rotation of solid surfaces is widely recognized by scientists, and researchers like Karman [16] firstly report viscous fluid flow induced by rotating solid disk. A special transformation named as Karman transformation given by him for the first time in this attempt. These transformations are utilized for conversion of fundamental equations termed as Naviers–Stokes equations in terms of ordinary differential system. Later on, a number of studies were given by researchers to depict the flow characteristics of both Newtonian and non-Newtonian fluids model over a rotating disk. Preceding these analyses in 2013, the extension of Karman problem was given by Turkyilmazoglu and Senel [17]. In this attempt they discussed numerical results for heat transfer properties of rotating partial slip fluid flow. In 2014, the magnetized slip flow via porous disk was reported by Rashidi et al. [18]. In addition, they discussed entropy measurements for this case. The flow properties in the presence of nano-size particles were discussed by Turkyilmazoglu [19]. He used numerical algorithm for solution purpose. In fact, he dealt comparative execution to report the impact of various nanoparticles suspended in fluid flow regime. Afterwards, tremendous attempts are given in this direction by way of both analytical and numerical approach. One can find the concern developments on rotating flows in References [20–31].

The present article contains analysis of Casson liquid towards rotating rigid disk. The Casson flow field is magnetized and has nanoparticles. Further, slip effects are also taken into account. The physical model is translated in terms of mathematical model. For solution purposes, the van Karman way of study is adopted. A computational algorithm is applied and the obtained results of involved parameters of concerned quantities are discussed via graphs and tables. Further, the current attempt is compared with existing literature and we found a good agreement which leads to the surety of the present work.

2. Problem Formulation

The Casson liquid is quipped above the disk for $\bar{z} > 0$. The constant frequency $(\bar{\Omega})$ is constant. The semi bounded magnetized flow regime contains suspended nanoparticles. The surface is taken with velocity slip condition. The quantities $(\bar{u},\bar{v},\bar{w})$ are in $(\bar{r},\bar{\phi},\bar{z})$ directions. The ultimate differential system of said problem is:

$$\frac{\partial \bar{w}}{\partial \bar{z}} + \frac{\partial \bar{u}}{\partial \bar{r}} + \frac{\bar{u}}{\bar{r}} = 0, \tag{1}$$

$$\bar{w}\frac{\partial \bar{u}}{\partial \bar{z}} + \bar{u}\frac{\partial \bar{u}}{\partial \bar{r}} - \frac{\bar{v}^2}{\bar{r}} = \nu\left(1 + \frac{1}{\lambda}\right)\left(\frac{\partial^2 \bar{u}}{\partial \bar{z}^2} + \frac{1}{\bar{r}}\frac{\partial \bar{u}}{\partial \bar{r}} + \frac{\partial^2 \bar{u}}{\partial \bar{r}^2} - \frac{\bar{u}}{\bar{r}^2}\right) - \frac{\sigma B_0^2}{\rho_f}\bar{u}, \tag{2}$$

$$\bar{w}\frac{\partial \bar{v}}{\partial \bar{z}} + \bar{u}\frac{\partial \bar{v}}{\partial \bar{r}} + \frac{\bar{u}\bar{v}}{\bar{r}} = \nu\left(1 + \frac{1}{\lambda}\right)\left(\frac{\partial^2 \bar{v}}{\partial \bar{z}^2} + \frac{\partial^2 \bar{v}}{\partial \bar{r}^2} - \frac{\bar{v}}{\bar{r}^2} + \frac{1}{\bar{r}}\frac{\partial \bar{v}}{\partial \bar{r}}\right) - \frac{\sigma B_0^2}{\rho_f}\bar{v}, \tag{3}$$

$$\bar{w}\frac{\partial \bar{w}}{\partial \bar{z}} + \bar{u}\frac{\partial \bar{w}}{\partial \bar{r}} = \nu\left(1 + \frac{1}{\lambda}\right)\left(\frac{\partial^2 \bar{w}}{\partial \bar{z}^2} + \frac{1}{\bar{r}}\frac{\partial \bar{w}}{\partial \bar{r}} + \frac{\partial^2 \bar{w}}{\partial \bar{r}^2}\right), \tag{4}$$

$$\bar{w}\frac{\partial \bar{T}}{\partial \bar{z}} + \bar{u}\frac{\partial \bar{T}}{\partial \bar{r}} = \alpha\left(\frac{\partial^2 \bar{T}}{\partial \bar{z}^2} + \frac{1}{\bar{r}}\frac{\partial \bar{T}}{\partial \bar{r}} + \frac{\partial^2 \bar{T}}{\partial \bar{r}^2}\right) + \frac{(\rho c)_p}{(\rho c)_f}\left[D_B\left(\frac{\partial \bar{T}}{\partial \bar{z}}\frac{\partial \bar{C}}{\partial \bar{z}} + \frac{\partial \bar{T}}{\partial \bar{r}}\frac{\partial \bar{C}}{\partial \bar{r}}\right)\right]$$
$$+ \frac{(\rho c)_p}{(\rho c)_f}\left[\frac{D_T}{T_\infty}\left(\left(\frac{\partial \bar{T}}{\partial \bar{z}}\right)^2 + \left(\frac{\partial \bar{T}}{\partial \bar{r}}\right)^2\right)\right], \tag{5}$$

$$\bar{w}\frac{\partial \bar{C}}{\partial \bar{z}} + \bar{u}\frac{\partial \bar{C}}{\partial \bar{r}} = D_B\left(\frac{\partial^2 \bar{C}}{\partial \bar{z}^2} + \frac{1}{\bar{r}}\frac{\partial \bar{C}}{\partial \bar{r}} + \frac{\partial^2 \bar{C}}{\partial \bar{r}^2}\right) + \frac{D_T}{T_\infty}\left[\frac{\partial^2 \bar{T}}{\partial \bar{z}^2} + \frac{1}{\bar{r}}\frac{\partial \bar{T}}{\partial \bar{r}} + \frac{\partial^2 \bar{T}}{\partial \bar{r}^2}\right], \tag{6}$$

$$\bar{u} = L\frac{\partial \bar{u}}{\partial \bar{z}}, \bar{v} = \bar{r}\bar{\Omega} + L\frac{\partial \bar{v}}{\partial \bar{z}}, \bar{w} = 0, \bar{T} = \bar{T}_w, \bar{C} = \bar{C}_w \text{ at } \bar{z} = 0, \tag{7}$$

$$\bar{u} \to 0, \bar{v} \to 0, \bar{T} \to \bar{T}_\infty, \bar{C} \to \bar{C}_\infty \text{ as } \bar{z} \to \infty, \tag{8}$$

for order reduction one can use the variables [16],

$$\bar{u} = \bar{r}\bar{\Omega}\frac{dF(\xi)}{d\xi}, \bar{v} = G(\xi)\bar{r}\bar{\Omega}, \bar{w} = -F(\xi)\sqrt{2\bar{\Omega}\nu},$$
$$C(\xi) = \frac{\bar{C}-\bar{C}_\infty}{\bar{C}_w-\bar{C}_\infty}, T(\xi) = \frac{\bar{T}-\bar{T}_\infty}{\bar{T}_w-\bar{T}_\infty}, \xi = \sqrt{\frac{2\bar{\Omega}}{\nu}}z.$$

(9)

We get:

$$2\frac{d^3F(\xi)}{d\xi^3}\left(1+\frac{1}{\lambda}\right) + 2F(\xi)\frac{d^2F(\xi)}{d\xi^2} - \left(\frac{dF(\xi)}{d\xi}\right)^2 + (G(\xi))^2 - \gamma\frac{dF(\xi)}{d\xi} = 0,$$ (10)

$$2\frac{d^2G(\xi)}{d\xi^2}\left(1+\frac{1}{\lambda}\right) + 2F(\xi)\frac{dG(\xi)}{d\xi} - 2G(\xi)\frac{dF(\xi)}{d\xi} - \gamma G(\xi) = 0,$$ (11)

$$\frac{d^2T(\xi)}{d\xi^2} + \Pr\left(F(\xi)\frac{dT(\xi)}{d\xi} + N_B\frac{dT(\xi)}{d\xi}\frac{dC(\xi)}{d\xi} + N_T\left(\frac{dT(\xi)}{d\xi}\right)^2\right) = 0,$$ (12)

$$\frac{d^2C(\xi)}{d\xi^2} + Le\Pr F(\xi)\frac{dC(\xi)}{d\xi} + \frac{N_T}{N_B}\frac{d^2T(\xi)}{d\xi^2} = 0,$$ (13)

$$F(\xi) = 0, \frac{dF(\xi)}{d\xi} = \beta\frac{d^2F(\xi)}{d\xi^2}, G(\xi) = 1+\beta\frac{dG(\xi)}{d\xi}, T(\xi) = 1, C(\xi) = 1, \text{at } \xi = 0,$$
$$\frac{dF(\xi)}{d\xi} \to 0, G(\xi) \to 0, T(\xi) \to 0, C(\xi) \to 0, \text{as } \xi \to \infty.$$

(14)

and:

$$\gamma = \sqrt{\frac{\sigma B_0^2}{\rho_f \bar{\Omega}}}, \beta = L\sqrt{\frac{2\bar{\Omega}}{\nu}}, N_B = \frac{(\rho c)_p}{(\rho c)_f}\frac{(\bar{T}_w-\bar{T}_\infty)D_T}{\bar{T}_\infty \nu},$$
$$Le = \frac{\alpha}{D_B}, \Pr = \frac{\nu}{\alpha}, N_T = \frac{(\rho c)_p}{(\rho c)_f}\frac{(\bar{C}_w-\bar{C}_\infty)D_B}{\nu},$$

(15)

the surface quantities are defined as:

$$\sqrt{Re_r}C_F = \left(1+\frac{1}{\lambda}\right)\frac{d^2F(0)}{d\xi^2}, \sqrt{Re_r}C_G = \left(1+\frac{1}{\lambda}\right)\frac{dG(0)}{d\xi},$$
$$\frac{Nu}{\sqrt{Re_r}} = -\frac{dT(0)}{d\xi}, \frac{Sh}{\sqrt{Re_r}} = -\frac{dC(0)}{d\xi},$$

(16)

3. Computational Outline

To transform the system of Equations (10)–(13) into an initial value problem one can use the dummy substitutions:

$Y_2 = F'(\xi), Y_3 = F'_2 = F''(\xi), Y_5 = G'(\xi), Y_7 = T'(\xi), Y_9 = C'(\xi)$, so we have

$$\begin{bmatrix} Y'_1 \\ Y'_2 \\ Y'_3 \\ Y'_4 \\ Y'_5 \\ Y'_6 \\ Y'_7 \\ Y'_8 \\ Y'_9 \end{bmatrix} = \begin{bmatrix} Y_2 \\ Y_3 \\ \frac{\gamma Y_2 + (Y_2)^2 - 2Y_1Y_3 - (Y_4)^2}{2(1+\frac{1}{\lambda})} \\ Y_5 \\ \frac{2Y_2Y_4 + \gamma Y_4 - 2Y_1Y_5}{2(1+\frac{1}{\lambda})} \\ Y_7 \\ -\Pr[Y_1Y_7 + N_BY_7Y_9 + N_TY_7^2] \\ Y_9 \\ -Le\Pr Y_9 + \frac{N_T}{N_B}Y'_7 \end{bmatrix}$$ (17)

$Y_1(\xi) = 0, Y_2(\xi) = \beta F''(\xi) = \beta\alpha_1, Y_3(\xi) = F''(\xi), Y_4(\xi) = 1+\beta G'(\xi) = 1+\beta\alpha_2,$
$Y_5(\xi) = G'(\xi), Y_6(\xi) = 1, Y_7(\xi) = \alpha_3, Y_8(\xi) = 1, Y_9(\xi) = \alpha_4, \text{when } \xi \to 0,$ (18)

with

$$Y_2(\xi) = 0, \ Y_4(\xi) = 0, Y_6(\xi) = 0, \ Y_8(\xi) = 0, \text{ when } \xi \to \infty, \quad (19)$$

here, α_1, α_2, α_3 and α_4 are initial guess values.

4. Analysis

The Casson fluid (CF) flow is considered on a rigid disk. The flow field is magnetized with suspended nanoparticles. The said problem is controlled mathematically and a numerical solution is offered through the shooting method. In detail, Figures 1–6 are used to highlight the variations of both CF velocities ($F'(\xi)$ and $G(\xi)$) via physical parameters, namely λ, γ, and β. Figures 1 and 2 are plotted to examine the impact of λ on CF velocity. It is clear from Figures 1 and 2 that the CF velocity decreases against λ.

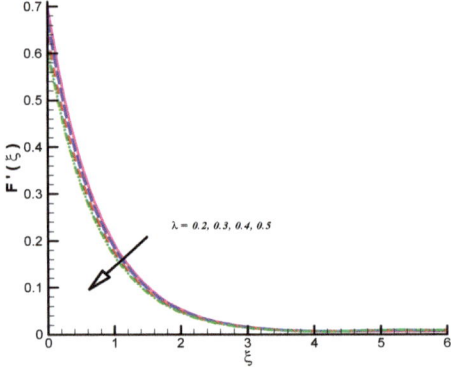

Figure 1. Effect of λ on $F'(\xi)$.

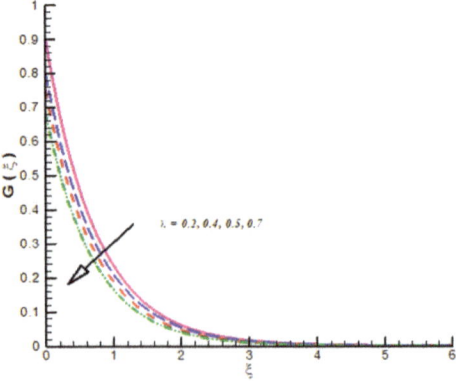

Figure 2. Effect of λ on $G(\xi)$.

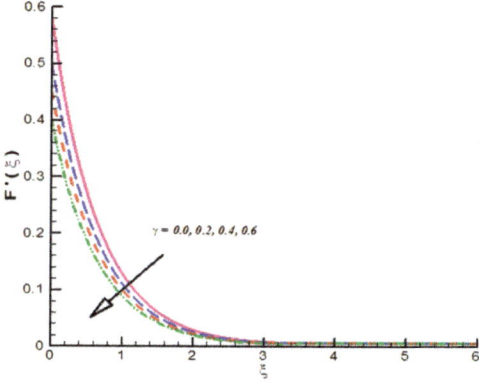

Figure 3. Effect of γ on $F'(\xi)$.

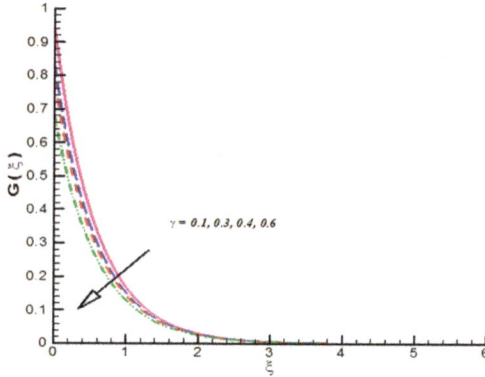

Figure 4. Effect of γ on $G(\xi)$.

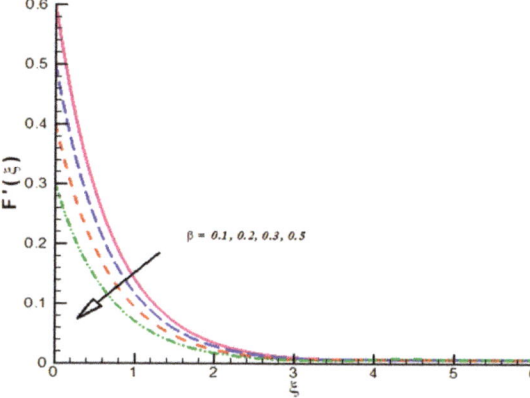

Figure 5. Effect of β on $F'(\xi)$.

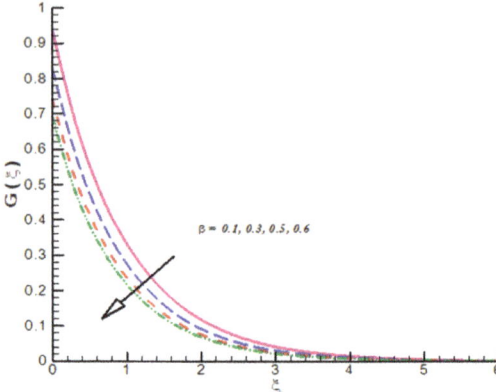

Figure 6. Effect of β on $G(\xi)$.

The impact of γ on CF velocity is examined and provided via Figure 3. The CF velocity decreases for higher values of γ. This is due to activation of Lorentz force via increasing γ. Similarly, the effect of γ on tangential velocity $G(\xi)$ is examined and given by means of Figure 4. It is important to note that the tangential velocity decreases for γ like radial one. The effect of β on radial velocity is offer in Figure 5. It is noticed that the radial velocity reflects a diminishing nature for positive values of β and the corresponding momentum boundary layer is also effected and admits decline values. Figure 6 gives the effect of β on tangential velocity of Casson fluid parameter. It is observed that the tangential velocity decreases for slip parameter. The Casson fluid temperature is examined and provided via Figures 7–9. Particularly, Figure 7 is plotted against N_T while Figure 8 is used to identify the influence of Pr on $T(\xi)$. Figure 9 reports influence of N_B on $T(\xi)$. From these figures we observed that Casson fluid temperature increases towards N_T, N_B but opposite trend is testified for Pr. Figures 10–12 reports the impact of Le, N_B and N_t on $C(\xi)$. In detail, Figure 10 paints the effect of Le on $C(\xi)$. The Casson concertation decreases for positive variations in Le. The $C(\xi)$ effected significantly towards N_B. Figure 11 is evident that the N_B results decline values in $C(\xi)$ for both zero and non-zero values of β. Such decreasing trend is due to higher values of Brownian force. The change in $C(\xi)$ is observed towards N_t and offer in Figure 12. The higher values of N_B corresponds increasing trends in $C(\xi)$ and related momentum boundary layer. In this attempt the MHD Casson nanofluid flow brought by rotating solid disk in the presence of slip conditions is examined. For comparison purpose, when Casson fluid parameter approaches to infinity our problem absolutely match with Hayat et al. [32]. In this work they studied nanoparticle aspects on viscous fluid flow due to rotating disk along with slip effects numerically. We have compared the variation of both Nusselt and Sherwood numbers with their findings as shown in Tables 1 and 2. One can see from these tables our finding match with existing values in a limiting sense. The trifling difference is due to choice of numerical method used in both attempts. Their values are obtained by build in command in Mathematica while we have used self-coded algorithm (shooting method with R-K scheme) subject to Casson nanofluid flow induced by solid rotating disk. Beside this one can extend idea to computational fluid dynamics in context of industrial and standpoints, see References [32–42].

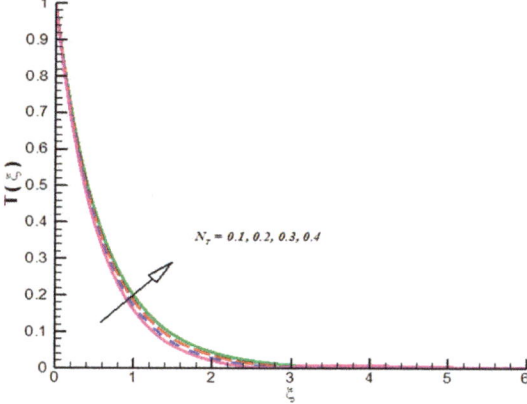

Figure 7. Effect of N_T on $T(\xi)$.

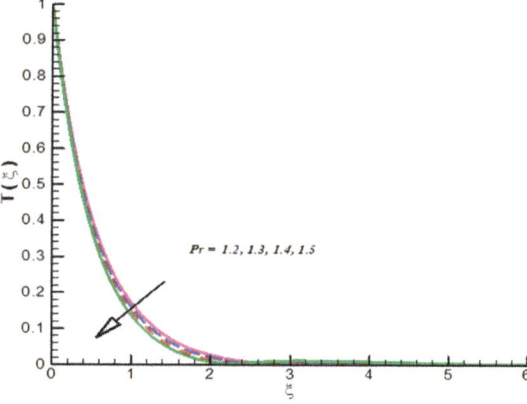

Figure 8. Effect of Pr on $T(\xi)$.

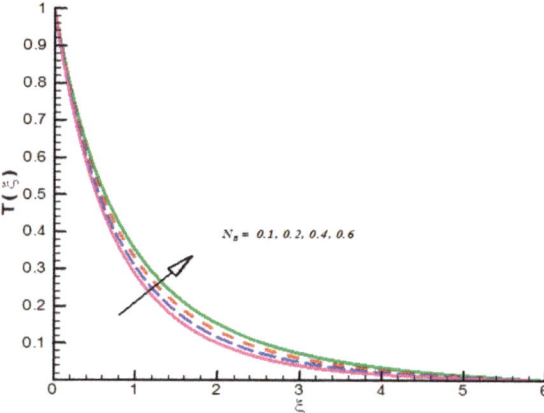

Figure 9. Effect of N_B on $T(\xi)$.

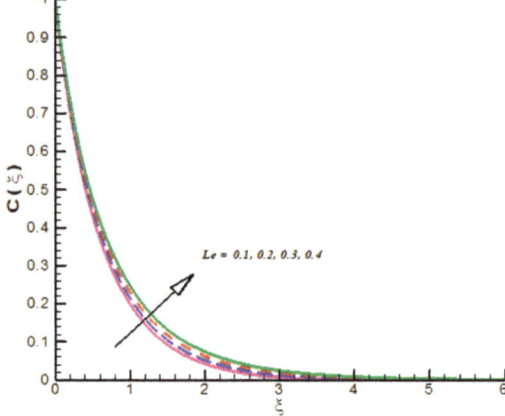

Figure 10. Effect of Le on $C(\xi)$.

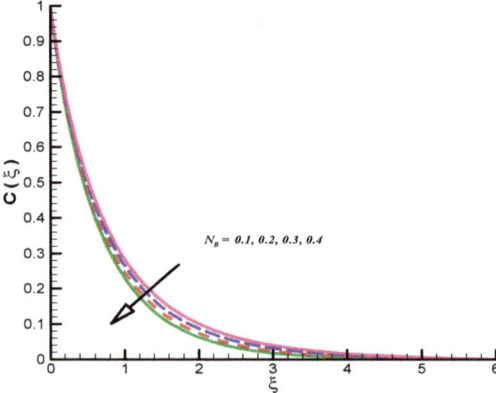

Figure 11. Effect of N_B on $C(\xi)$.

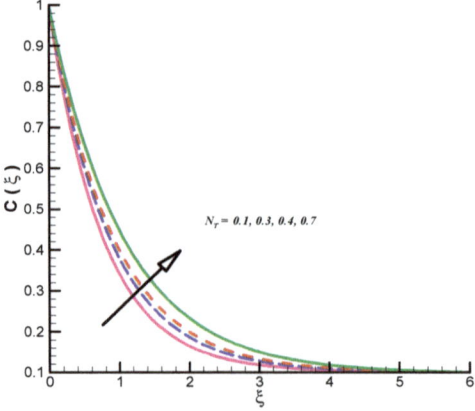

Figure 12. Effect of N_T on $C(\xi)$.

Table 1. Local Nusselt number comparison with Hayat et al. [32].

						$\frac{Nu}{\sqrt{Re_\gamma}} = -\frac{dT(0)}{d\xi}$	
β	γ	N_T	Le	Pr	N_B	Hayat et al. [32]	Present values
0.2	-	-	-	-	-	0.32655	0.326600
0.5	-	-	-	-	-	0.30360	0.30363
0.8	-	-	-	-	-	0.28715	0.28724
-	0.0	-	-	-	-	0.30494	0.30502
-	0.7	-	-	-	-	0.24421	0.24434
-	1.4	-	-	-	-	0.17566	0.17575
-	-	0.5	-	-	-	0.25913	0.25916
-	-	0.7	-	-	-	0.23865	0.23879
-	-	1.0	-	-	-	0.21010	0.21025
-	-	-	0.5	-	-	0.29633	0.29642
-	-	-	1.0	-	-	0.28954	0.28963
-	-	-	1.5	-	-	0.28395	0.28398
-	-	-	-	0.5	-	0.24989	0.24999
-	-	-	-	1.0	-	0.29211	0.29224
-	-	-	-	1.5	-	0.32286	0.32294
-	-	-	-	-	0.5	0.26341	0.26358
-	-	-	-	-	0.7	0.23677	0.23687
-	-	-	-	-	1.0	0.20056	0.20068

Table 2. Local Sherwood number comparison with Hayat et al. [32].

						$\frac{Sh}{\sqrt{Re_\gamma}} = -\frac{dC(0)}{d\xi}$	
β	γ	N_T	Le	Pr	N_B	Hayat et al. [32]	Present values
0.2	-	-	-	-	-	0.27583	0.27593
0.5	-	-	-	-	-	0.26933	0.26945
0.8	-	-	-	-	-	0.26493	0.26498
-	0.0	-	-	-	-	0.27000	0.27012
-	0.7	-	-	-	-	0.25387	0.25394
-	1.4	-	-	-	-	0.23722	0.23735
-	-	0.5	-	-	-	0.22206	0.22215
-	-	0.7	-	-	-	0.22539	0.22564
-	-	1.0	-	-	-	0.22285	0.22288
-	-	-	0.5	-	-	0.21373	0.21380
-	-	-	1.0	-	-	0.30132	0.30145
-	-	-	1.5	-	-	0.38690	0.38696
-	-	-	-	0.5	-	0.22934	0.22944
-	-	-	-	1.0	-	0.26624	0.26636
-	-	-	-	1.5	-	0.31262	0.31276
-	-	-	-	-	0.5	0.30338	0.30342
-	-	-	-	-	0.7	0..31875	0..31887
-	-	-	-	-	1.0	0.32959	0.32978

5. Closing Remarks

A Casson fluid (CF) flow yield by rotating rigid disk is considered. **Both the** Brownian and thermophoresis aspects are entertained by incorporating nanoparticles. The flow characteristics are reported numerically with the support of computational algorithm. The summary is as follows:

- CF velocities which includes $[G(\xi), F'(\xi)]$ reflects decline trend towards β.
- CF velocities are decreasing function of λ and γ.
- CFT $[T(\xi)]$ admits inciting nature towards both N_T and N_B but opposite trend is observed for Pr.
- CFC $[C(\xi)]$ shows decline values for both Le, and N_B.
- CFC $[C(\xi)]$ reflect inciting trend for N_T.
- Comparative values of HTR and MTR are provided for involved flow controlling parameters.

Author Contributions: Conceptualization, K.U.R. and M.Y.M.; methodology, W.A.K.; software, I.K.; validation K.U.R. and M.Y.M. and W.A.K.; formal analysis, I.K.; investigation, W.A.K.; resources, K.U.R. and M.Y.M; writing original draft preparation, W.A.K. and I.K.; writing review and editing, I.K.; visualization, K.U.R. and M.Y.M; supervision, W.A.K.; project administration, W.A.K; funding acquisition, S.O.A.

Funding: This research was funded by Deanship of Scientific Research at King Khalid University, Abha 61413, Saudi Arabia grant number R.G.P-2/29/40.

Acknowledgments: The authors extend their appreciation to the Deanship of Scientific Research at King Khalid University, Abha 61413, Saudi Arabia for funding this work through research groups program under grant number R.G.P-2/29/40.

Conflicts of Interest: The authors declare no conflict of interest.

Nomenclature

$V = (\bar{u}, \bar{v}, \bar{w})$	**Velocity field**
$(\bar{r}, \bar{\phi}, \bar{z})$	Polar coordinates
ν	Kinematic viscosity
λ	Casson fluid parameter
ρ_f	Fluid density
σ	Electrical conductivity
B_0	Uniform applied magnetic field
α	Thermal diffusivity
D_B	Brownian diffusion coefficient
D_T	Thermophoretic diffusion coefficient
$\bar{T}_\infty$	Ambient temperature
L	Velocity slip parameter
$\bar{T}_w$	Surface temperature
$\bar{C}_w$	Surface concentration
$\bar{C}$	Concentration
$F'(\xi), G(\xi)$	Dimensionless velocities
$T(\xi)$	Dimensionless temperature
$C(\xi)$	Dimensionless concentration
γ	Magnetic field parameter
Pr	Prandtl number
N_B	Brownian motion parameter
N_T	Thermophoresis parameter
Le	Lewis number
β	Velocity slip parameter
Re_r	Reynolds number

References

1. Mustafa, M.; Hayat, T.; Pop, I.; Aziz, A. Unsteady boundary layer flow of a Casson fluid due to an impulsively started moving flat plate. *Heat Transf. Asian Res.* **2011**, *6*, 563–576. [CrossRef]
2. Nadeem, S.; Rizwan, U.H.; Lee, C. MHD flow of a Casson fluid over an exponentially shrinking sheet. *Sci. Iran.* **2012**, *19*, 1550–1553. [CrossRef]
3. Mustafa, M.; Tasawar, H.; Pop, I.; Awatif, H. Stagnation-point flow and heat transfer of a Casson fluid towards a stretching sheet. *Z. Naturforsch. A* **2012**, *67*, 70–76. [CrossRef]
4. Mukhopadhyay, S. Casson fluid flow and heat transfer over a nonlinearly stretching surface. *Chin. Phys. B* **2013**, *22*, 074701. [CrossRef]
5. Mukhopadhyay, S.; Iswar, C.M.; Tasawar, H. MHD boundary layer flow of Casson fluid passing through an exponentially stretching permeable surface with thermal radiation. *Chin. Phys. B* **2014**, *23*, 104701. [CrossRef]
6. Mustafa, M.; Junaid, A.K. Model for flow of Casson nanofluid past a non-linearly stretching sheet considering magnetic field effects. *AIP Adv.* **2015**, *5*, 077148. [CrossRef]
7. Ramesh, K.; Devakar, M. Some analytical solutions for flows of Casson fluid with slip boundary conditions. *Ain Shams Eng. J.* **2015**, *6*, 967–975. [CrossRef]
8. Sandeep, N.; Olubode, K.K.; Isaac, L.A. Modified kinematic viscosity model for 3D-Casson fluid flow within boundary layer formed on a surface at absolute zero. *J. Mol. Liq.* **2016**, *221*, 1197–1206. [CrossRef]
9. Qing, J.; Muhammad, M.B.; Munawwar, A.A.; Mohammad, M.R.; Mohamed, E.-S.A. Entropy generation on MHD Casson nanofluid flow over a porous stretching/shrinking surface. *Entropy* **2016**, *18*, 123. [CrossRef]
10. Ali, M.E.; Sandeep, N. Cattaneo-christov model for radiative heat transfer of magnetohydrodynamic Casson-ferrofluid: A numerical study. *Results Phys.* **2017**, *7*, 21–30. [CrossRef]
11. Reddy, J.V.R.; Sugunamma, V.; Sandeep, N. Enhanced heat transfer in the flow of dissipative non-Newtonian Casson fluid flow over a convectively heated upper surface of a paraboloid of revolution. *J. Mol. Liq.* **2017**, *229*, 380–388. [CrossRef]
12. Rehman, K.U.; Aneeqa, A.M.; Malik, M.Y.; Sandeep, N.; Saba, N.U. Numerical study of double stratification in Casson fluid flow in the presence of mixed convection and chemical reaction. *Results Phys.* **2017**, *7*, 2997–3006. [CrossRef]
13. Kumaran, G.; Sandeep, N. Thermophoresis and Brownian moment effects on parabolic flow of MHD Casson and Williamson fluids with cross diffusion. *J. Mol. Liq.* **2017**, *233*, 262–269. [CrossRef]
14. Ali, F.; Nadeem, A.S.; Ilyas, K.; Muhammad, S. Magnetic field effect on blood flow of Casson fluid in axisymmetric cylindrical tube: A fractional model. *J. Magn. Magn. Mater.* **2017**, *423*, 327–336. [CrossRef]
15. Raju, C.S.K.; Mohammad, M.H.; Sivasankar, T. Radiative flow of Casson fluid over a moving wedge filled with gyrotactic microorganisms. *Adv. Powder Technol.* **2017**, *28*, 575–583. [CrossRef]
16. Kármán, T.V. Über laminare und turbulente Reibung. *ZAMM* **1921**, *1*, 233–252. [CrossRef]
17. Turkyilmazoglu, M.; Senel, P. Heat and mass transfer of the flow due to a rotating rough and porous disk. *Int. J. Therm. Sci.* **2013**, *63*, 146–158. [CrossRef]
18. Rashidi, M.M.; Kavyani, N.; Abelman, S. Investigation of entropy generation in MHD and slip flow over a rotating porous disk with variable properties. *Int. J. Heat Mass Transf.* **2014**, *70*, 892–917. [CrossRef]
19. Turkyilmazoglu, M. Nanofluid flow and heat transfer due to a rotating disk. *Comput. Fluids* **2014**, *94*, 139–146. [CrossRef]
20. Griffiths, P.T.; Stephen, J.G.; Stephen, S.O. The neutral curve for stationary disturbances in rotating disk flow for power-law fluids. *J. Non Newton. Fluid Mech.* **2014**, *213*, 73–81. [CrossRef]
21. Mustafa, M.; Junaid, A.K.; Hayat, T.; Alsaedi, A. On Bödewadt flow and heat transfer of nanofluids over a stretching stationary disk. *J. Mol. Liq.* **2015**, *211*, 119–125. [CrossRef]
22. Sheikholeslami, M.; Hatami, M.; Ganji, D.D. Numerical investigation of nanofluid spraying on an inclined rotating disk for cooling process. *J. Mol. Liq.* **2015**, *211*, 577–583. [CrossRef]
23. Xun, S.; Zhao, J.; Zheng, L.; Chen, X.; Zhang, X. Flow and heat transfer of Ostwald-de Waele fluid over a variable thickness rotating disk with index decreasing. *Int. J. Heat Mass Transf.* **2016**, *103*, 1214–1224. [CrossRef]
24. Latiff, N.A.; Uddin, M.J.; Ismail, A.M. Stefan blowing effect on bioconvective flow of nanofluid over a solid rotating stretchable disk. *Propuls. Power Res.* **2016**, *5*, 267–278. [CrossRef]
25. Ming, C.; Zheng, L.; Zhang, X.; Liu, F.; Anh, V. Flow and heat transfer of power-law fluid over a rotating disk with generalized diffusion. *Int. Commun. Heat Mass Transf.* **2016**, *79*, 81–88. [CrossRef]

26. Imtiaz, M.; Tasawar, H.; Ahmed, A.; Saleem, A. Slip flow by a variable thickness rotating disk subject to magnetohydrodynamics. *Results Phys.* **2017**, *7*, 503–509. [CrossRef]
27. Doh, D.H.; Muthtamilselvan, M. Thermophoretic particle deposition on magnetohydrodynamic flow of micropolar fluid due to a rotating disk. *Int. J. Mech. Sci.* **2017**, *130*, 350–359. [CrossRef]
28. Hayat, T.; Madiha, R.; Maria, I.; Ahmed, A. Nanofluid flow due to rotating disk with variable thickness and homogeneous-heterogeneous reactions. *Int. J. Heat Mass Transf.* **2017**, *113*, 96–105. [CrossRef]
29. Devi, M.; Chitra, L.; Rajendran, A.B.Y.; Fernandez, C. Non-linear differential equations and rotating disc electrodes: Padé approximationtechnique. *Electrochim. Acta* **2017**, *243*, 1–6. [CrossRef]
30. Guha, A.; Sayantan, S. Non-linear interaction of buoyancy with von Kármán's swirling flow in mixed convection above a heated rotating disc. *Int. J. Heat Mass Transf.* **2017**, *108*, 402–416. [CrossRef]
31. Ellahi, R.; Ahmed, Z.; Farooq, H.; Tehseen, A. Study of shiny film coating on multi-fluid flows of a rotating disk suspended with nano-sized silver and gold particles: A comparative analysis. *Coatings* **2018**, *8*, 422. [CrossRef]
32. Hayat, T.; Taseer, M.; Sabir, A.S.; Ahmed, A. On magnetohydrodynamic flow of nanofluid due to a rotating disk with slip effect: A numerical study. *Comput. Methods Appl. Mech. Eng.* **2017**, *315*, 467–477. [CrossRef]
33. Vo, T.Q.; Park, B.S.; Park, C.H.; Kim, B.H. Nano-scale liquid film sheared between strong wetting surfaces: Effects of interface region on the flow. *J. Mech. Sci. Technol.* **2015**, *29*, 1681–1688. [CrossRef]
34. Kherbeet, A.; Sh, H.A.; Mohammed, B.H.; Salman, H.E.; Ahmed, O.; Alawi, A.; Rashidi, M.M. Experimental study of nanofluid flow and heat transfer over microscale backward-and forward-facing steps. *Exp. Therm. Fluid Sci.* **2015**, *65*, 13–21. [CrossRef]
35. Abbas, T.; Muhammad, A.; Muhammad, B.; Mohammad, R.; Mohamed, A. Entropy generation on nanofluid flow through a horizontal Riga plate. *Entropy* **2016**, *18*, 223. [CrossRef]
36. Bhatti, M.M.; Abbas, T.; Rashidi, M.M. Numerical study of entropy generation with nonlinear thermal radiation on magnetohydrodynamics non-Newtonian nanofluid through a porous shrinking sheet. *J. Magn.* **2016**, *21*, 468–475. [CrossRef]
37. Ghorbanian, J.; Alper, T.C.; Beskok, A. A phenomenological continuum model for force-driven nano-channel liquid flows. *J. Chem. Phys.* **2016**, *145*, 184109. [CrossRef]
38. Bao, L.; Priezjev, N.V.; Hu, H.; Luo, K. Effects of viscous heating and wall-fluid interaction energy on rate-dependent slip behavior of simple fluids. *Phys. Rev. E* **2017**, *96*, 033110. [CrossRef]
39. Ghorbanian, J.; Beskok, A. Temperature profiles and heat fluxes observed in molecular dynamics simulations of force-driven liquid flows. *Phys. Chem. Chem. Phys.* **2017**, *19*, 10317–10325. [CrossRef]
40. Mohebbi, R.; Rashidi, M.M.; Mohsen, I.; Nor, A.C.S.; Hong, W.X. Forced convection of nanofluids in an extended surfaces channel using lattice Boltzmann method. *Int. J. Heat Mass Transf.* **2018**, *117*, 1291–1303. [CrossRef]
41. Rehman, K.U.; Malik, M.Y.; Iffat, Z.; Alqarni, M.S. Group theoretical analysis for MHD flow fields: A numerical result. *J. Braz. Soc. Mech. Sci. Eng.* **2019**, *41*, 156. [CrossRef]
42. Rehman, K.U.; Malik, M.Y.; Mahmood, R.; Kousar, N.; Zehra, I. A potential alternative CFD simulation for steady Carreau–Bird law-based shear thickening model: Part-I. *J. Braz. Soc. Mech. Sci. Eng.* **2019**, *41*, 176. [CrossRef]

© 2019 by the authors. Licensee MDPI, Basel, Switzerland. This article is an open access article distributed under the terms and conditions of the Creative Commons Attribution (CC BY) license (http://creativecommons.org/licenses/by/4.0/).

Article

Heat Transfer of Oil/MWCNT Nanofluid Jet Injection Inside a Rectangular Microchannel

Esmaeil Jalali [1], Omid Ali Akbari [2], M. M. Sarafraz [3], Tehseen Abbas [4] and Mohammad Reza Safaei [5,6,*]

1. Department of Mechanical Engineering, Najafabad Branch, Islamic Azad University, Najafabad, Iran; esmaiil.j66@yahoo.com
2. Young Researchers and Elite Club, Khomeinishahr Branch, Islamic Azad University, Khomeinishahr, Iran; Akbariomid11@gmail.com
3. Centre for Energy Technology, School of Mechanical Engineering, The University of Adelaide, South Australia, Australia; Mohammadmohsen.sarafraz@adelaide.edu.au
4. Department of Mathematics, University of Education Lahore, Faisalabad Campus, Faisalabad, Pakistan; tehseen.abbas@ue.edu.pk
5. Division of Computational Physics, Institute for Computational Science, Ton Duc Thang University, Ho Chi Minh City, Vietnam
6. Faculty of Electrical and Electronics Engineering, Ton Duc Thang University, Ho Chi Minh City, Vietnam
* Correspondence: cfd_safaei@tdtu.edu.vn

Received: 20 March 2019; Accepted: 27 May 2019; Published: 4 June 2019

Abstract: In the current study, laminar heat transfer and direct fluid jet injection of oil/MWCNT nanofluid were numerically investigated with a finite volume method. Both slip and no-slip boundary conditions on solid walls were used. The objective of this study was to increase the cooling performance of heated walls inside a rectangular microchannel. Reynolds numbers ranged from 10 to 50; slip coefficients were 0.0, 0.04, and 0.08; and nanoparticle volume fractions were 0–4%. The results showed that using techniques for improving heat transfer, such as fluid jet injection with low temperature and adding nanoparticles to the base fluid, allowed for good results to be obtained. By increasing jet injection, areas with eliminated boundary layers along the fluid direction spread in the domain. Dispersing solid nanoparticles in the base fluid with higher volume fractions resulted in better temperature distribution and Nusselt number. By increasing the nanoparticle volume fraction, the temperature of the heated surface penetrated to the flow centerline and the fluid temperature increased. Jet injection with higher velocity, due to its higher fluid momentum, resulted in higher Nusselt number and affected lateral areas. Fluid velocity was higher in jet areas, which diminished the effect of the boundary layer.

Keywords: Oil/MWCNT nanofluid; heat transfer; finite volume method; laminar flow; slip coefficient; microchannel

---

1. Introduction

In recent years, industrial developments have led scientists to search for methods to improve heat transfer in heat exchangers and industrial equipment. Therefore, a new generation of cooling fluids, called nanofluids, is used in industrial and commercial applications. Cooling fluid jet is used in turbine blade cooling and indirect access surfaces. Metal and non-metal nanoparticles have a higher thermal conductivity coefficient than water and lead to higher conductive heat transfer coefficients of fluid, as well as improve the temperature distribution of the nanofluid. Experimental results show that adding nanoparticles to the base fluid increases the heat transfer coefficient of nanofluids. Using nanofluids is one of the novel heat transfer improvement methods with high efficiency [1–4].

Numerous researchers have investigated the thermal or hydrodynamic performance of nanofluids in different microchannel heat sinks [5–7]. Hang et al. [8] numerically investigated the heat transfer performance of a microchannel heat sink with different nanofluids. Akbari et al. [9] studied laminar flow and heat transfer parameters of water/Al_2O_3 nanofluid with different nanoparticle volume fractions inside a rectangular microchannel and found that using rough surfaces in microchannel leads to higher heat transfer. Behnampour et al. [10] numerically investigated laminar flow and heat transfer parameters of water/AgO nanofluid with different nanoparticle volume fractions in a rectangular microchannel and showed that by increasing fluid velocity, an optimized trade-off can be obtained between heat transfer, hydrodynamic behavior of nanofluid, and the performance evaluation criteria (PEC) variations. Geravandian et al. [11] numerically simulated the laminar heat transfer of nanofluid flow in a rectangular microchannel and revealed that by increasing TiO_2 nanoparticles, heat transfer, friction coefficient, PEC, and pressure drop increase. Studies on the effect of using cooling fluid jet injection on heated surfaces [12] and other methods to increase heat transfer, such as using dimples, rough surfaces [13,14], and twisted tapes [15,16], have been conducted for different industrial and experimental geometries. These studies show that by creating vortexes, uniform temperature distribution can be obtained. Fluid jet plays an important role in cooling technologies and by creating better mixtures of cooling fluid flow, thermal performance can be enhanced [17–19]. Chen et al. [20] numerically and experimentally investigated the forced convection heat transfer inside a rectangular channel for determining fluid flow and heat transfer properties. They also compared the performance of heat sinks with solid and perforated pins and showed that by increasing the number of perforations and their diameter, pressure drop decreases and the Nusselt number increases. In their experiments, thermal performance of the heat sink with perforated pins was better than with solid pins. Nafon et al. [21] experimentally studied the effects of inlet temperature, Reynolds number, and heat flux on heat transfer properties of a water/TiO_2 nanofluid jet in a semi-rectangular heat sink, and showed that the average heat transfer coefficient of nanofluid is higher than base fluid, and the pressure drop increases by increasing the nanoparticle volume fraction. Jasperson et al. [22] studied the thermal and hydrodynamic performance of a copper microchannel and a pin fin microchannel and showed that by increasing the volume rate of flow, the thermal resistance of a pin fin heat sink decreases. Zhuwang et al. [23] studied heat transfer inside a microchannel with fluid jet and different coolants and showed that using fluid jet results in higher heat transfer compared with ordinary parallel flows.

Few studies have investigated the thermal and hydrodynamic performance of laminar nanofluid flow in a rectangular microchannel with cooling fluid jet injection. Therefore, the main purpose of this study is to investigate heat transfer improvement methods, such as using nanofluid and direct fluid jet injection. We also investigated the effect of the hydrodynamic velocity boundary condition on the cooling performance of the microchannel. In the present numerical study, laminar flow and heat transfer of oil/MWCNT nanofluid inside a two-dimensional microchannel with nanofluid jet injection were simulated using a finite volume method. Cooling nanofluid jet injection in a microchannel disturbs the thermal boundary layer and increases the heat transfer rate. The use of fluid jet injection in various sections of the microchannel is one of the novelties of this study. Fluid flow and heat transfer behavior are separately simulated in cases with no jets and with one, two, and three jets. The effect of applying slip and no-slip boundary conditions on solid walls of microchannel on the flow and heat transfer were studied.

2. Problem Definition

In the present numerical study, oil/MWCNT nanofluid flow was simulated in 0–4% nanoparticle volume fractions and Reynolds numbers of 10 and 50. We considered different numbers of fluid jets on the insulated bottom wall of the microchannel. The top heated wall of the microchannel had a constant temperature of T_h = 303 K. The inlet cold fluid entered from the left side of the microchannel with the temperature of T_c = 293 K. Figure 1 shows the microchannel in the present study.

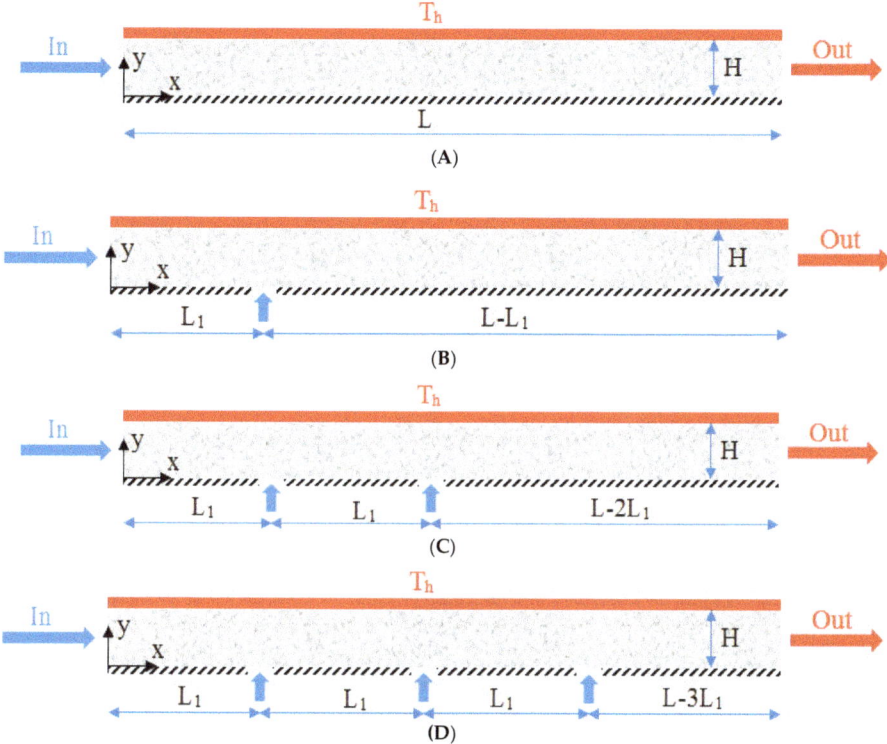

Figure 1. Schematic of the studied geometry in the present numerical study. (**A**) Case 1; (**B**) Case 2; (**C**) Case 3; (**D**) Case 4.

The length of studied microchannel was L = 2500 μm, the width of inlet jet entrance was 10 μm, the height of microchannel was H = 50 μm, and the jet pitch was L_1 = 500 μm. The velocity of the inlet nanofluid jet at all studied Reynolds numbers was constant (Re = 10). Fluid flow and heat transfer inside the microchannel were separately simulated in cases with no jet and with 1, 2, and 3 jets. Nanofluids with high velocity through the lower wall were injected into the micro-channel. As the cooling fluid jet flowed on the surface, the temperature of this surface decreased. Also, in this investigation, the effect of applying slip and no-slip boundary conditions on solid walls of the microchannel in slip coefficients of B = 0.0, 0.04, and 0.08 were investigated. The flow was laminar, forced, Newtonian, single-phase, and incompressible; the nanofluid was homogeneous and uniform; and also, the radiation effects were neglected. The properties of the base fluid and nanofluid in different nanoparticle volume fractions are presented in Table 1 [24].

Table 1. Thermophysical properties of base fluid and solid nanoparticles.

	Oil	MWCNT	ϕ = 0.02	ϕ = 0.04
c_p (J/kg K)	2032	1700	2012.9	1995.1
ρ (kg/m^3)	867	2600	901.66	936.32
k (W/m K)	0.133	3000	0.5255	0.7912
μ (Pa s)	0.0289	-	0.0305	0.0321

3. Governing Equations

Dimensionless Navier–Stokes equations for forced, laminar, steady, and single-phase nanofluid are presented for 2-D space, which are continuity, momentum, and energy equations [25]:

$$\frac{\partial U}{\partial X} + \frac{\partial V}{\partial Y} = 0 \tag{1}$$

$$U\frac{\partial U}{\partial X} + V\frac{\partial U}{\partial Y} = -\frac{\partial P}{\partial X} + \frac{\mu_{nf}}{\rho_{nf}\nu_f}\frac{1}{Re}\left(\frac{\partial^2 U}{\partial X^2} + \frac{\partial^2 U}{\partial Y^2}\right) \tag{2}$$

$$U\frac{\partial V}{\partial X} + V\frac{\partial V}{\partial Y} = -\frac{\partial P}{\partial Y} + \frac{\mu_{nf}}{\rho_{nf}\nu_f}\frac{1}{Re}\left(\frac{\partial^2 V}{\partial X^2} + \frac{\partial^2 V}{\partial Y^2}\right) \tag{3}$$

$$U\frac{\partial \theta}{\partial X} + V\frac{\partial \theta}{\partial Y} = \frac{\alpha_{nf}}{\alpha_f}\frac{1}{RePr}\left(\frac{\partial^2 \theta}{\partial X^2} + \frac{\partial^2 \theta}{\partial Y^2}\right) \tag{4}$$

The dimensionless equations used this study are [26]:

$$X = \frac{x}{H},\ Y = \frac{y}{H},\ U = \frac{u}{u_c},\ V = \frac{v}{u_c},\ P = \frac{\bar{P}}{\rho_{nf}u_c^2}$$
$$\theta = \frac{T-T_c}{\Delta T},\ Re = \frac{u_c \times H}{\nu_f},\ Pr = \frac{\nu_f}{\alpha_f} \tag{5}$$

For calculating local and average Nusselt number along the microchannel walls, the following equations are used:

$$Nu(X) = \frac{k_{eff}}{k_f}\left(\frac{\partial \theta}{\partial Y}\right)_{Y=0} \tag{6}$$

$$Nu_{ave} = \frac{1}{L}\int_0^L Nu_s(X)\,dX \tag{7}$$

The dimensionless boundary conditions are:

$$U = 1, V = 0 \quad \text{and} \quad \theta = 0 \quad \text{for} \quad X = 0 \quad \text{and} \quad 0 \leq Y \leq 1 \tag{8}$$

$$V = 0 \quad \text{and} \quad \frac{\partial U}{\partial X} = 0 \quad \text{for} \quad X = 50 \quad \text{and} \quad 0 \leq Y \leq 1 \tag{9}$$

$$U_w = -B\frac{\partial U}{\partial Y},\ V = 0 \quad \text{and} \quad \theta = 1 \quad \text{for} \quad Y = 1 \quad \text{and} \quad 0 \leq X \leq 50 \tag{10}$$

$$U_w = B\frac{\partial U}{\partial Y},\ V = 0 \quad \text{and} \quad \frac{\partial \theta}{\partial Y} = 0 \quad \text{for} \quad Y = 0 \quad \text{and} \quad 0 \leq X \leq 50 \tag{11}$$

4. Numerical Details

Finite volume method [27,28] and the second-order upwind discretization method with maximum residual of 10^{-6} [29,30] was used. For coupling velocity–pressure equations, the SIMPLEC [31,32] algorithm was used. Also, the effect of nanoparticle volume fraction, Reynolds number, and jet number on flow and heat transfer parameters were investigated; and flow parameters, temperature, and streamlines contours are the presented.

The meshes in this study were rectangular and regular grids ranging from 25,000 to 100,000 grids were used for Reynolds numbers of 10 and 50, 2% volume fraction, and no-slip boundary conditions (Table 2). In order to ensure the independence of flow and heat transfer parameters from mesh number, Nusselt number and average velocity on flow centerline were studied for different mesh numbers. The 850 × 50 mesh number has better performance than 900 × 90, therefore, the former was used in the present simulations.

Table 2. Grid independence test.

Grid Size	500 × 50	750 × 75	850 × 80	900 × 90	1000 × 100
Re = 10					
Nu_{ave}	0.3698	0.3401	0.3202	0.2922	0.3401
$U_{out(Y=H/2)}$	3.806	3.7740	3.7532	3.6178	3.7740
Re = 50					
Nu_{ave}	0.69719	0.6561	0.6178	0.5541	0.6561
$U_{out(Y=H/2)}$	3.972	3.9616	3.9578	3.9257	3.9616

Figures 2 and 3 show the validation of velocity and dimensionless flow temperature at the outlet section of microchannel against the results of Raisi et al. [33], who investigated laminar and forced flow of water/CuO nanofluid with different nanoparticle volume fractions in a two-dimensional rectangular microchannel using a finite volume method. The results of Raisi et al. [33] for 3% nanoparticle volume fraction, Re = 100, and different slip velocity coefficients were used for the validation. There is good agreement between the results of the present study and Raisi et al. [33]. Therefore, the boundary conditions and assumptions of this study are correct.

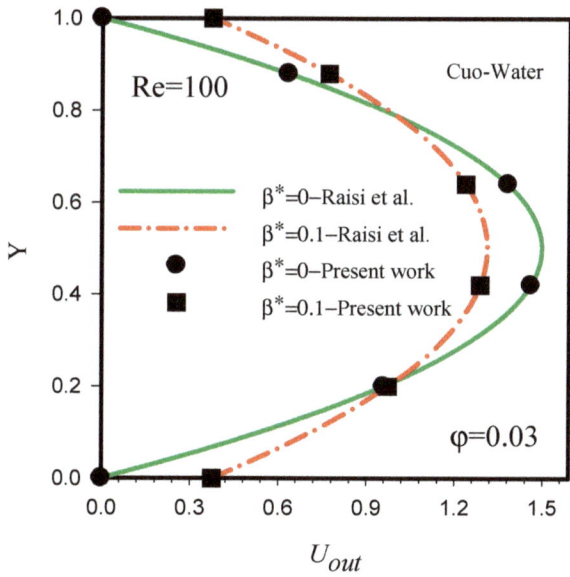

Figure 2. Validation of the present numerical study with Raisi et al. [33].

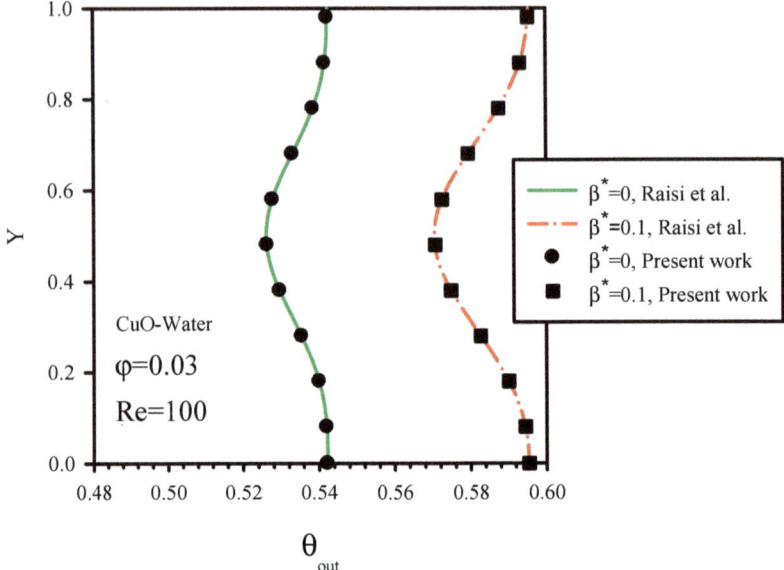

Figure 3. Validation of the present study with Raisi et al. [33].

5. Results and Discussion

In the present numerical study, the effect of nanofluid jet injection on laminar flow and heat transfer of oil/MWCNT nanofluid for 0–4% nanoparticle volume fractions, at Reynolds numbers of 10 and 50, and different slip velocity coefficients was numerically studied. The effects of the variations of jet number, nanoparticle volume fractions, and velocity boundary layer on solid walls of microchannel were also studied. Reynolds number, nanoparticle volume fraction, and slip velocity coefficient were input variables and their effect on Nusselt number, temperature domain, velocity domain, flow parameters, and temperature contours were investigated and compared based on their local and average values.

5.1. Streamlines and Isothermal Contours

In Figure 4, streamline contours (right side) and constant temperature contours (left side) are shown for Re = 10, B = 0.04, and ϕ = 2% for cases 1, 2, 3, and 4. In these contours, the effect of fluid motion with minimum velocity along the microchannel at different nanofluid jet injection on the heated surface was investigated. The temperature difference between inlet cold fluid and the heated wall improves heat transfer and creates temperature gradients. Fluid motion and its impact with the solid walls of the microchannel creates velocity gradients in areas close to the wall. In general, fluid motion along the microchannel leads to the creation of velocity and thermal boundary layers. Nanofluid jet injection on fluid direction results in the elimination of velocity and thermal boundary layers. In all contours, by increasing the number of fluid jets, areas with eliminated boundary layers become larger. Areas with a high number of jet injections have smaller velocity boundary layers. In the cases with more jets, due to the high volume of crossing fluid, axial velocity parameters of fluid are higher than the cases with fewer jets.

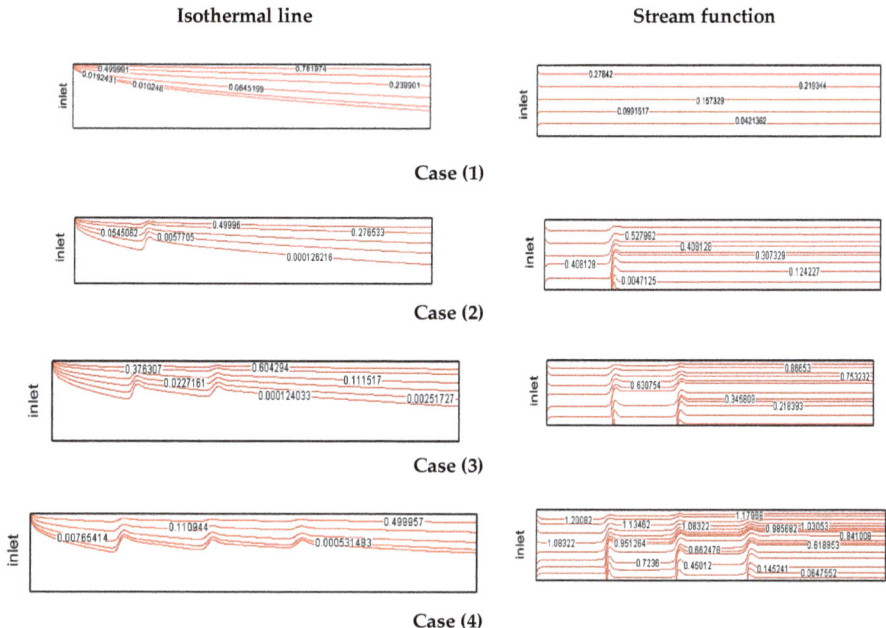

Figure 4. Constant temperature and streamline contours at Re = 10 and B = 0.04 for ϕ = 2%.

Streamline contours (right side) and constant temperature contours (left side) at Re = 50, B = 0.04, and ϕ = 2% are presented in Figure 5. The effect of increasing jet number on the heated surface on these contours is shown for cases 1, 2, 3, and 4. In the inlet areas of the microchannel and jet injection areas is 5 times higher, contrary to Figure 4. Increasing fluid velocity decreases the thermal boundary layer thickness. Increasing the fluid momentum results in higher crossing fluid velocity. At Re = 50, by increasing the number of fluid jets, due to the better mixture of fluid and reduction of temperature curve slope, a significant reduction of the thermal boundary layer thickness is observed. According to the streamline contours, nanofluid jet injection with higher velocity results in better mixture of flow and creation of local vortices. By increasing Reynolds number, especially in the last jet, the effect of vertical jet injection decreases due to the higher of volumetric rate of crossing fluid and creation of a smaller vortex behind each jet.

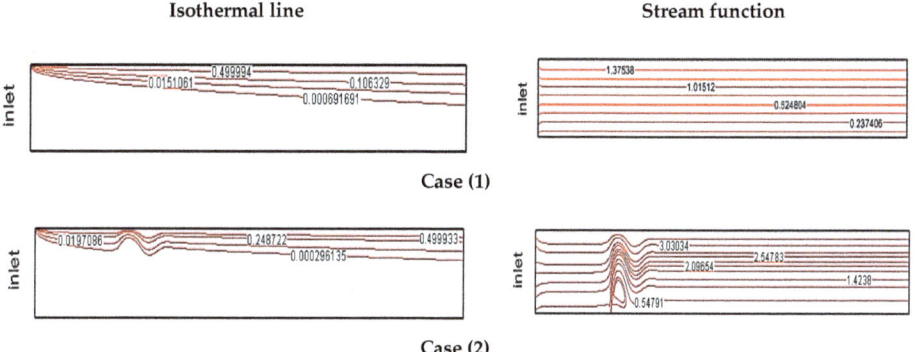

Figure 5. *Cont.*

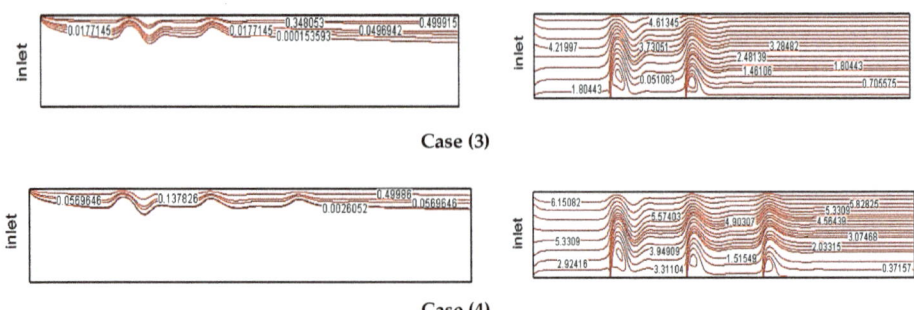

Figure 5. Streamlines and constant temperature contours at Re = 50 and B = 0.04 for ϕ = 2%.

5.2. Local Nusselt Number

In Figure 6, the local Nusselt number on the heated wall of the microchannel is shown for Re = 10, B = 0.04, and different nanoparticles volume fractions. In this figure, we have different jet numbers with minimum fluid velocity. Nanoparticle volume fractions were 0–4%. Heat transfer occurs because of temperature differences between fluid and the top heated wall. The value of heat transfer has a direct relationship with Nusselt number and temperature line slope. Uniform temperature distribution and removal of temperature gradients are desirable and lead to higher Nusselt numbers. The local jump in the Nusselt number in Figure 6 was because of the low temperature of the injected fluid jet on the heated surface. In these areas, the heat transfer distribution was more uniform and the Nusselt number increased. In all curves for Figure 6, by adding nanoparticle volume fraction, the Nusselt number and heat transfer were significantly increased.

In Figure 7, the local Nusselt numbers at Re = 50, B = 0.04, and different nanoparticle volume fractions are shown. In this figure, the effect of the increasing fluid velocity on Nusselt number is compared with Figure 6. Increasing ϕ and the number of jets resulted in the higher Nusselt number graphs in Figure 7. Figures 6 and 7 show the elimination of thermal and velocity boundary layers due to the fluid jet injection, which is more significant at higher Reynolds numbers. Jet injection with higher velocity leads to the increase of the Nusselt number due to the higher fluid momentum which affects the lateral regions. Increasing the nanoparticle volume fraction at Re = 50, compared to Re = 10, resulted in a significant rise of the Nusselt number. In general, the local Nusselt number on heated walls showed that using fluid jet injection with low temperature can increase the heat transfer rate. Compared to the primary jets, the heat transfer performance of the last jets was low, because by adding more jets, the effects of fluid momentum and crossing fluid become more significant, which diminishes the lateral flow by vertical jets as they need higher momentum. Therefore, in cases 2 and 3, the lateral cooling areas were narrower.

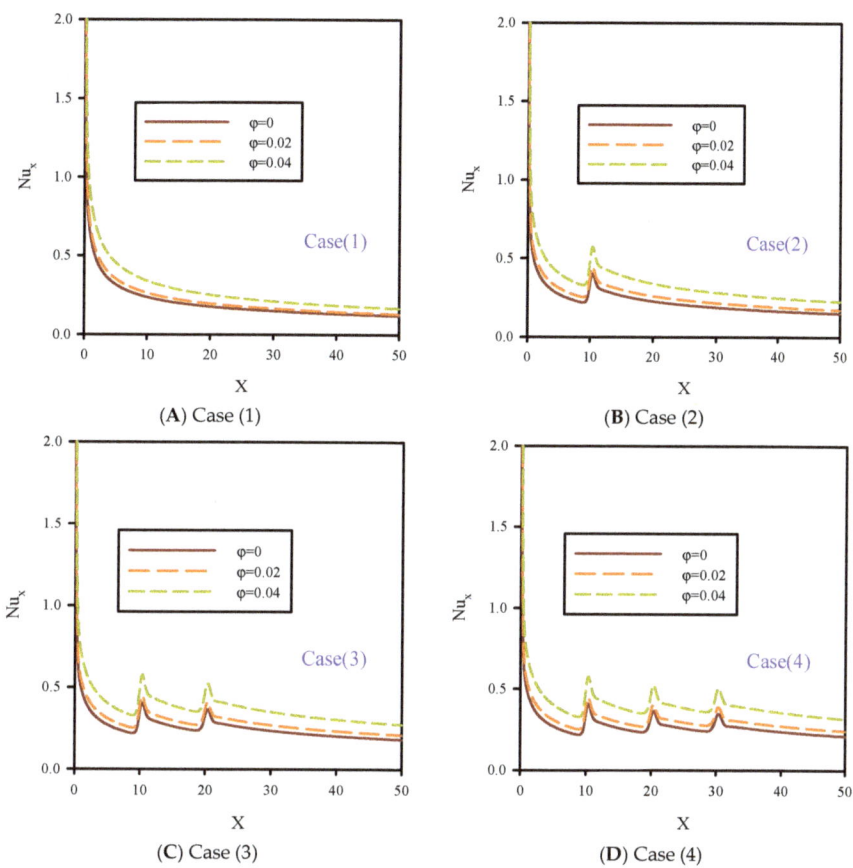

Figure 6. Local Nusselt number graphs at Re = 10 and B = 0.04 for different nanoparticle volume fractions.

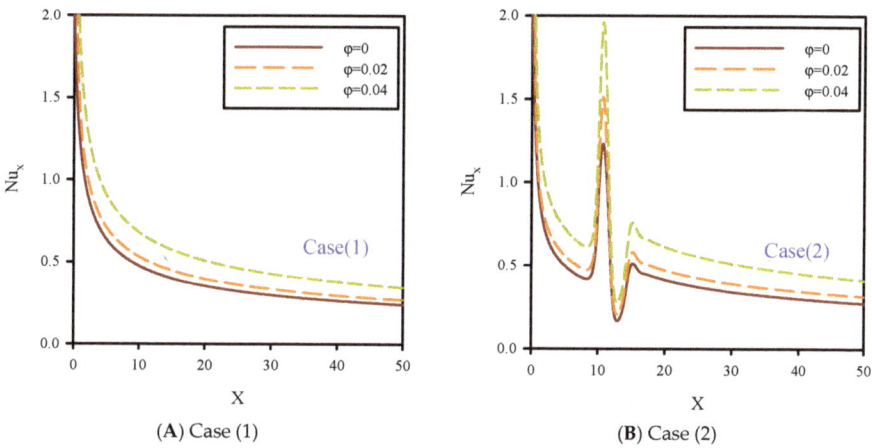

Figure 7. *Cont.*

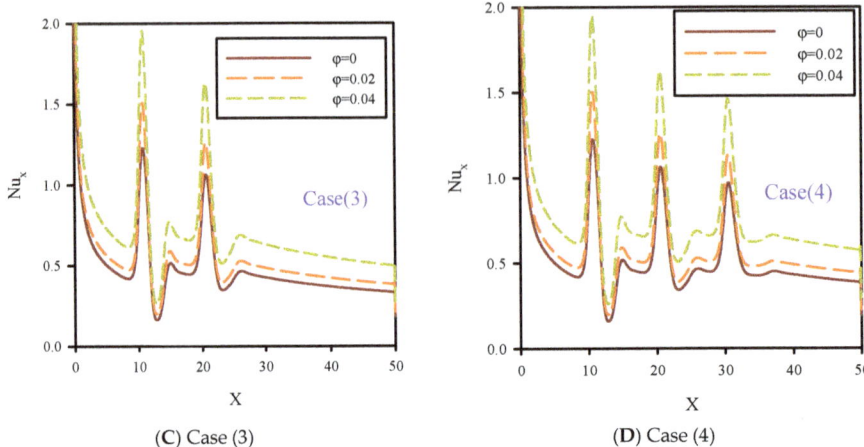

(C) Case (3) (D) Case (4)

Figure 7. Local Nusselt number at Re = 50 and B = 0.04 for different nanoparticle volume fractions.

5.3. Temperature along the Symmetry Plane

In Figure 8, local dimensionless temperature distribution on flow centerline for Re = 10, B = 0.04, and volume fractions of 0–4% is presented. The dimensionless temperature distribution in the central section of the microchannel was influenced by the reduction of fluid velocity and jet number for cases 1, 2, 3, and 4, and different nanoparticle volume fractions. Because of the temperature difference between the fluid and heated surface, the penetration of the thermal boundary layer affected all areas of the microchannel, especially the areas in the central line of flow. Figure 8 shows that increasing φ resulted in higher thermal conductivity of the microchannel and uniform temperature distribution in all areas of the microchannel, especially in the heated areas. Due to the higher thermal conductivity of cooling fluid in higher nanoparticle volume fractions, temperature gradients in heated areas were diminished, and also, the conductivity of the fluid layers was higher. Therefore, by increasing φ, heat transfer penetrated the flow centerline, and because of the higher conductivity of the fluid layers, the fluid temperature rose.

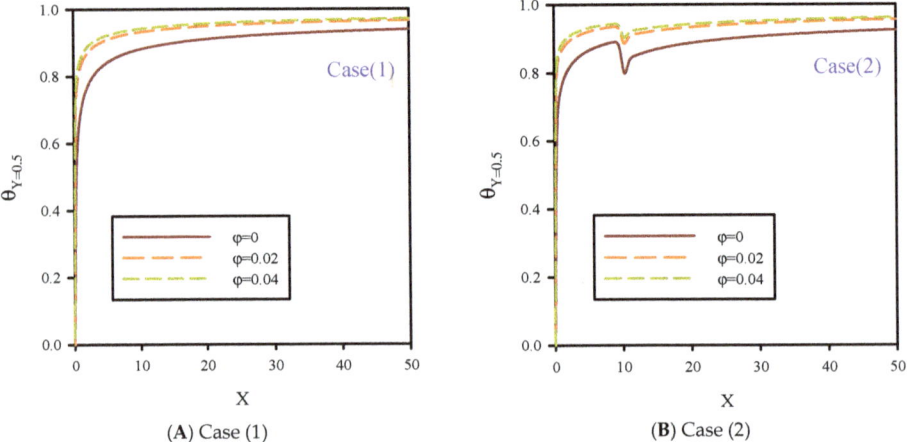

(A) Case (1) (B) Case (2)

Figure 8. *Cont.*

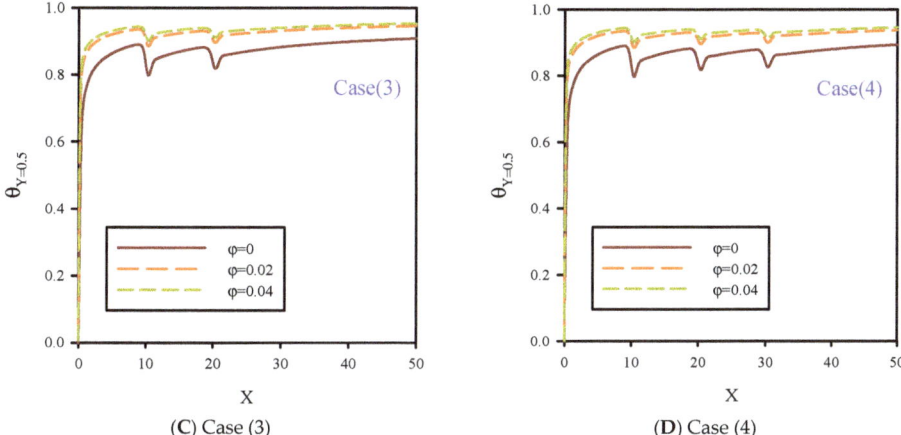

(C) Case (3)   (D) Case (4)

Figure 8. Temperature distribution at Re = 10, B = 0.04, and different nanoparticle volume fractions.

In Figure 9, local dimensionless temperature distribution at Re = 50, B = 0.04, and 0–4% nanoparticle volume fractions along the flow centerline are shown for cases 1, 2, 3, and 4. Increasing the number of jets resulted in lower temperature of the fluid, and thus the dimensionless temperature graphs in the microchannel centerlines were decreased, which was due to the reduction of temperature associated with increased fluid velocity and transferred temperature to the flow centerline. By increasing the velocity of injected cold fluid jet, the cold and hot fluids were better mixed. Consequently, the fluid temperature decreased significantly. According to Figures 8 and 9, by decreasing the fluid velocity, the fluid becomes thermally better developed, which is due to the penetration of heat to the inlet section of the microchannel. By increasing the fluid velocity, the undeveloped length of flow at the inlet section of microchannel increases.

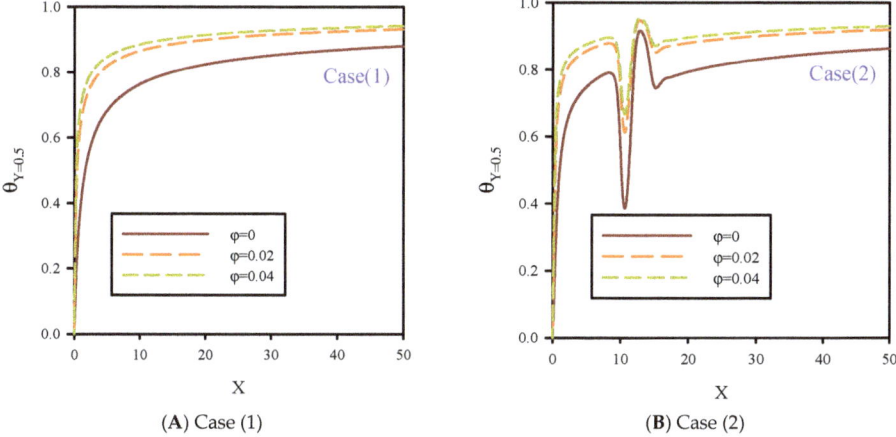

(A) Case (1)   (B) Case (2)

Figure 9. Cont.

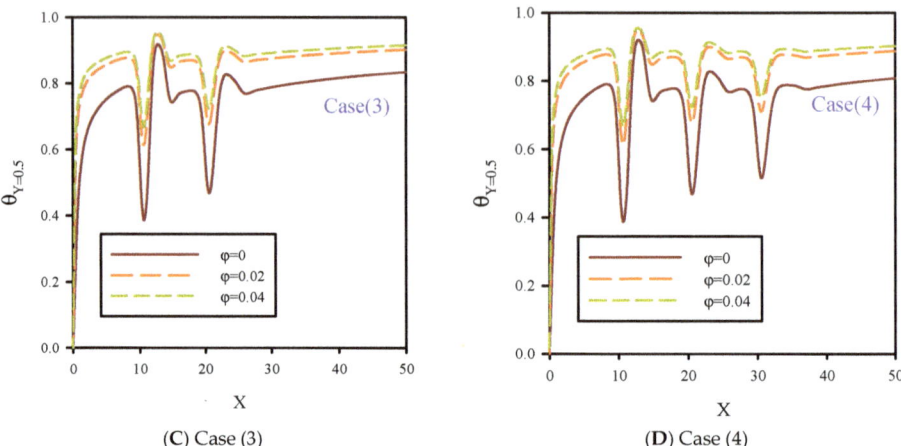

Figure 9. Temperature distribution at R e= 50, B = 0.04, and different nanoparticle volume fractions.

5.4. Axial Velocity along the Symmetric Plate

In Figure 10, dimensionless axial velocity in the flow centerline at Reynolds numbers of 10 and 50, B = 0.0, and ϕ=2% were investigated. Fluid motion along the microchannel was affected by velocity boundary layer caused by the solid walls. Contrary to the inlet velocity, fluid velocity was higher in case 1, which was due to the effects of the velocity boundary layer along the microchannel walls. In case 1, fluid velocity in developed areas at Re = 50 was lower than Re = 10. Fluid jet increased the volume rate of the crossing fluid downstream of the jets compared to areas upstream of the jets. Therefore, in cases 1, 2, and 3, fluid velocity increased in areas after the jets. In fact, high fluid velocity decreased the size of the velocity boundary layer. The velocity curves for fluid with Re = 50 were higher than fluid with Re = 10. In general, fluid jet eliminates the velocity boundary layer and prevents local flow development.

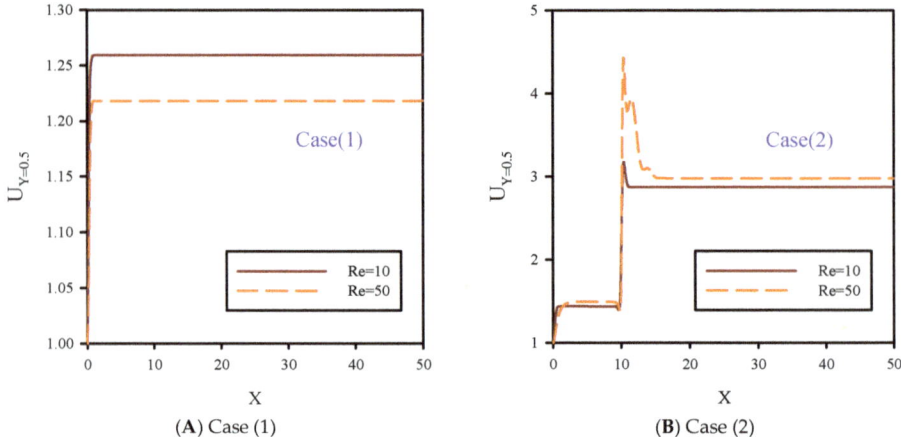

Figure 10. Cont.

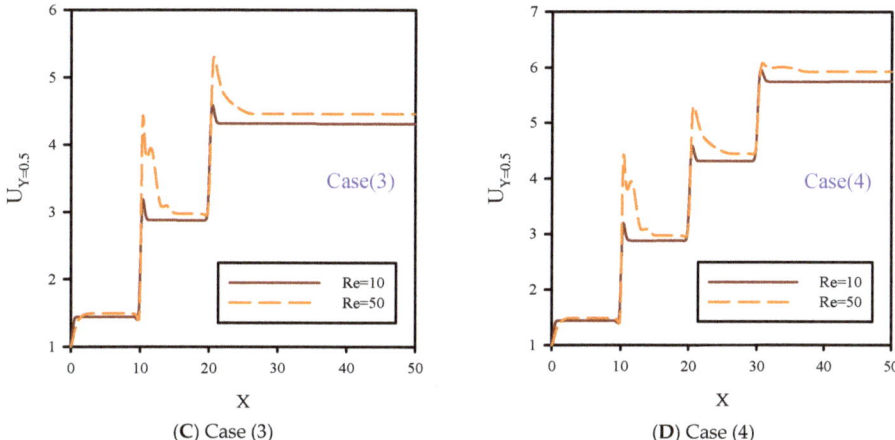

(C) Case (3) (D) Case (4)

Figure 10. Axial velocity along the symmetry plane for Re = 10 and 50, B = 0, and ϕ = 2%.

In Figure 11, dimensionless axial velocity along the flow centerline at Reynolds numbers of 10 and 50, B = 0.08, and ϕ = 2% are shown. Unlike Figure 10, the slip boundary condition was applied to the solid walls which facilitated fluid motion and reduced the velocity boundary layer on the solid walls of the microchannel. Therefore, the velocity profile in the flow centerline for all cases of 1, 2, 3, and 4 were lower than Figure 10. According to Figures 10 and 11, increasing the slip velocity coefficient on the solid walls of the microchannel at Re = 10 had less effect than Re = 50. At Re = 50, the lateral mixture of flow covered more areas in the case with slip boundary conditions (BCs) compared with the no-slip BC case. According to Figure 11, this factor resulted in the microchannel contraction which increased the effect of jet injection areas and local axial velocity.

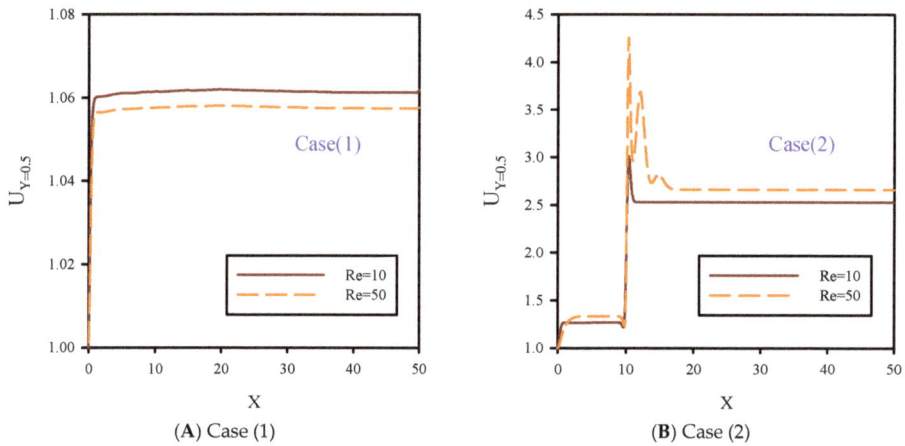

(A) Case (1) (B) Case (2)

Figure 11. Cont.

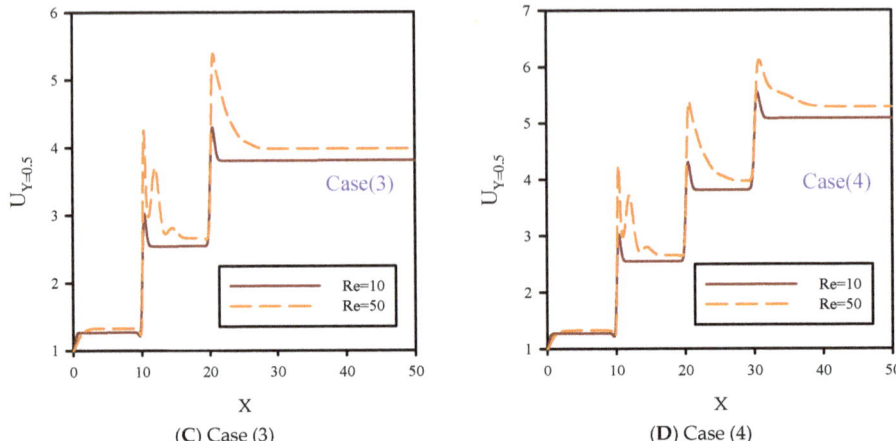

Figure 11. Axial velocity along the symmetry plane for Re = 10 and 50, B = 0.08, and φ = 2%.

5.5. Average Nusselt Number on the Heated Surface

In Figure 12, the average Nusselt number at Reynolds numbers of 10 and 50, volume fractions of 0, 2, and 4%, and B = 0.04 are shown. Using fluid jet injection with low temperature and adding solid nanoparticles to the base fluid, heat transfer was improved. Increasing the number of jets resulted in a better mixture of cold and hot flows in the microchannel, reduction of total temperature of the microchannel, and better cooling performance of the nanofluid. Adding solid nanoparticles to the base fluid with higher volume fractions resulted in better temperature distribution, and the Nusselt number increased as well. Also, increasing the Reynolds number resulted in higher fluid velocity and better temperature distribution, especially in heated areas, and also, the average Nusselt number increased. Among investigated cases, case 4 had the highest value of Nusselt number due to the maximum number of cold fluid jet injection. In case 1, due to the lack of nanofluid jet injection, the effects of thermal and velocity boundary layers became significant, and this case had the lowest heat transfer and Nusselt number.

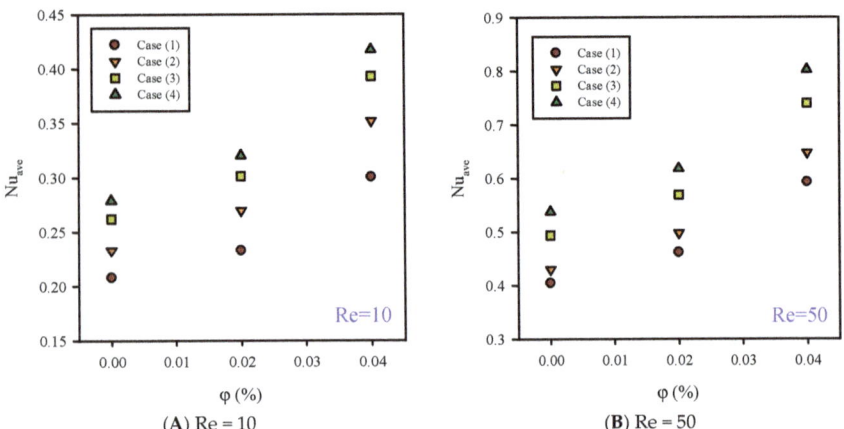

Figure 12. Average Nusselt number for Re = 10, 50; φ = 0, 2%, 4%; and B = 0.04.

5.6. Effect of Slip Coefficient on Axial Velocity

Dimensionless axial velocity in the central line of the microchannel at Re = ϕ = 2% and different slip coefficients was demonstrated in Figure 13. As it is seen, nanofluid jet injection at Re = 10 leads to some changes in jet areas. At this Reynolds number, increasing slip velocity coefficient results in lower fluid momentum dissipation and higher axial velocity. By increasing the slip velocity coefficient, because of the smaller velocity boundary layer, fluid moves on the microchannel walls with less friction. By increasing the jet number with a higher fluid velocity and volumetric rate of crossing fluid, the heat transfer was increased.

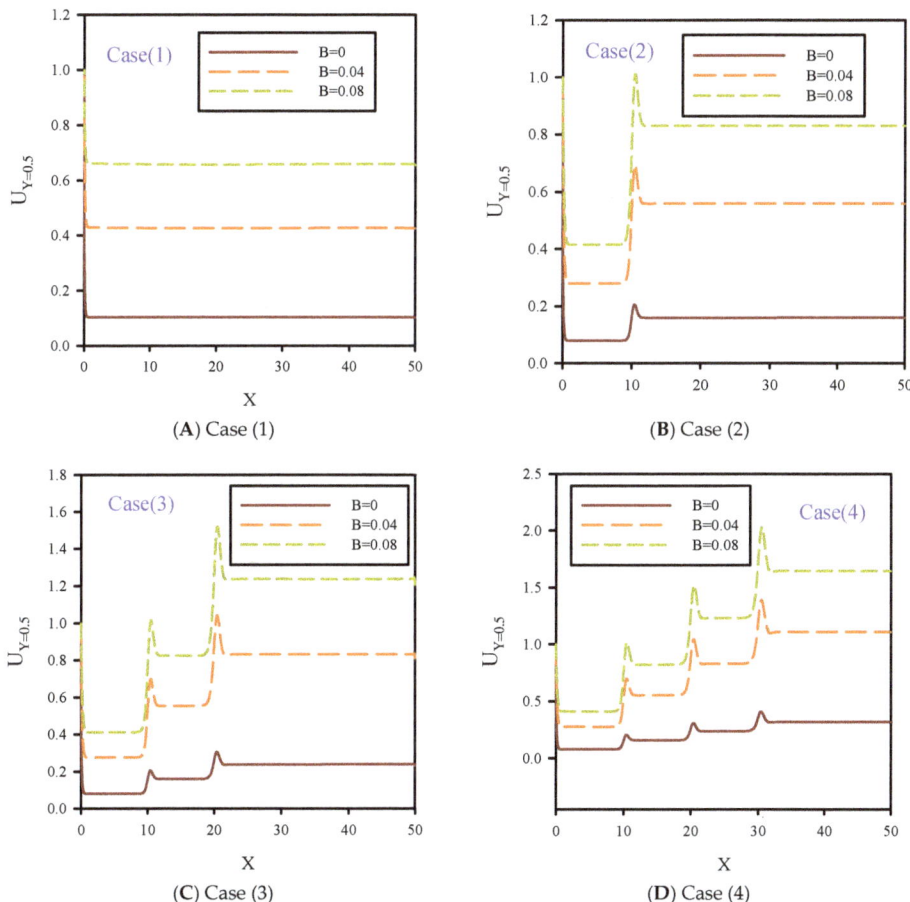

Figure 13. Dimensionless axial velocity at Re = 10 and ϕ = 2% with different slip coefficients on solid walls.

In Figure 14, dimensionless axial velocity in the flow centerline at Re = 50 and ϕ = 2% for cases 1, 2, 3, and 4 are shown. Increasing the slip coefficient on the solid surfaces parallel with the fluid direction, especially at higher Reynolds numbers, leads to the higher of fluid momentum. By increasing the slip velocity coefficient, due to the local flow contraction and significant changes in axial fluid velocity, fluid jet injection creates larger areas in the longitudinal direction. In Figure 14, because of the higher value of Reynolds number, the level of graphs are higher than those in Figure 13.

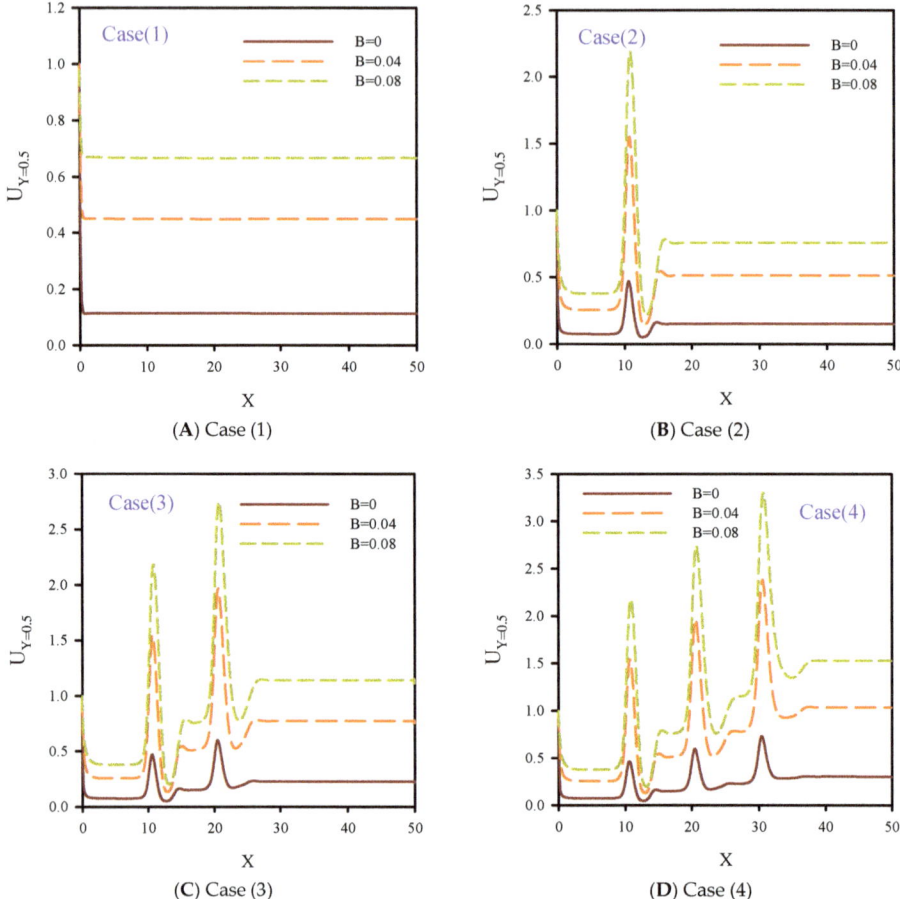

Figure 14. Dimensionless axial velocity at Re = 50 and ϕ = 2% with different slip coefficients on solid walls.

In Figure 15, dimensionless axial velocity profiles at the outlet section of the microchannel at Re = 10, ϕ = 4%, and slip coefficients of 0.0 and 0.08 are shown. Applying the slip coefficient on the solid walls resulted in a slight reduction of the axial velocity. By increasing the slip velocity coefficient, the velocity increased and the size of the velocity boundary layer on the heated surface decreased. According to the axial velocity profiles with the no-slip boundary condition, the effects of the velocity boundary layer was significant for the layers further away from the solid walls, and the velocity profile was stronger in the flow centerline. Therefore, from solid walls to flow centerline, the slope of the axial velocity variations was higher than the velocity profile with B = 0.08. Applying the slip velocity coefficient on the solid walls resulted in lower axial velocity gradients. The slope of the velocity curve from the walls to the flow centerline indicated lower gradients than the case with the no-slip boundary condition. The velocity profile in the central areas of the flow confirmed this issue and the corresponding curve had a milder slope.

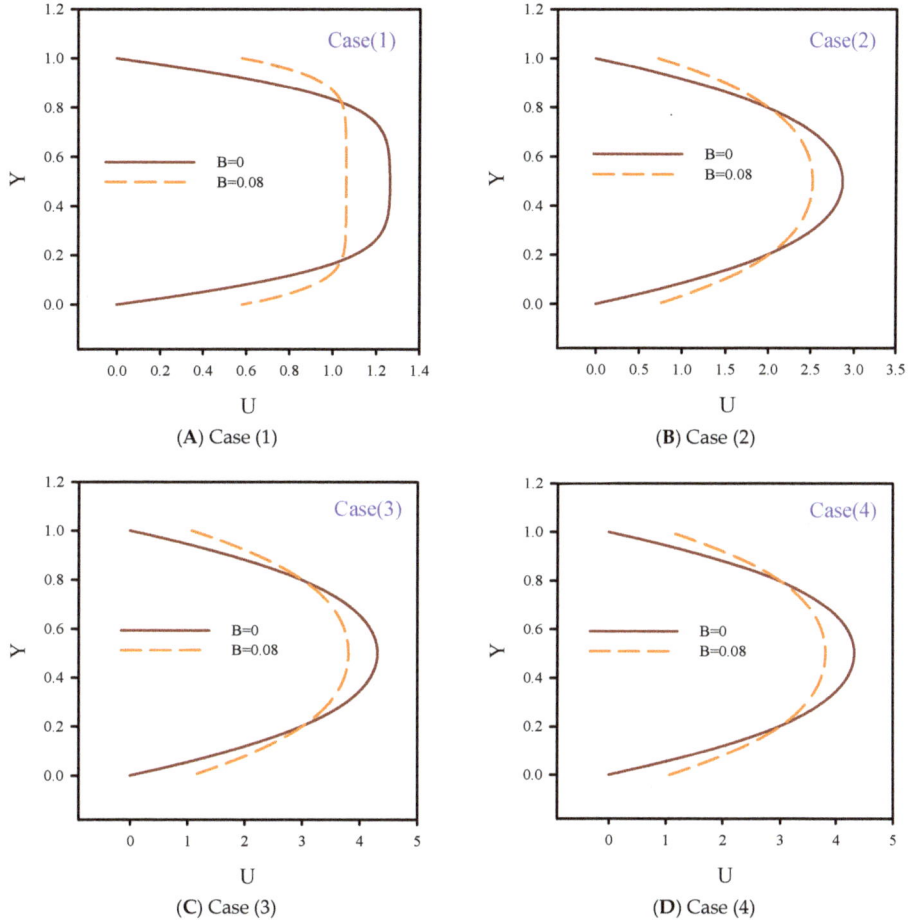

Figure 15. Effect of the slip coefficient on the outlet axial velocity for Re = 10 and ϕ = 4.

6. Conclusions

In the present numerical study, laminar flow of oil/MWCNT nanofluid inside a two-dimensional microchannel with nanofluid jet injection was investigated. This simulation was carried out in a two-dimensional domain for laminar and Newtonian flow and heat transfer of oil/MWCNT nanofluid, with volume fractions of 0–4%, at Reynolds numbers of 10 and 50. Various jet numbers were used on the insulated bottom wall of the microchannel. Results showed that nanofluid jet injection with higher velocity leads to better flow mixture through the formation of local vortexes. In Re = 50, by increasing the number of fluid jets, the thermal boundary layer thickness becomes smaller because of the better mixture of fluid. Also, the temperature line slope was milder. Heat transfer occurs due to the temperature difference between fluid and the top heated wall. Heat transfer had a direct relationship with Nusselt number and the slope of the temperature line. Fluid jets eliminated the thermal and velocity boundary layers, which was more significant in higher Reynolds numbers. Because of the temperature differences between fluid and heated surfaces, penetration of the thermal boundary layer affected all sections of the microchannel, especially areas on the flow centerline. Because of higher conductivity of the fluid layer, by increasing the nanoparticle volume fraction, the heat transfer reached the flow centerline and the fluid temperature increased in that area. By decreasing fluid velocity, due to the higher temperature penetration, the developed length of flow became shorter and the fluid developed faster. Higher fluid velocity in areas with jets eliminated the velocity boundary layer.

Unlike the case with the no-slip boundary condition, by applying the slip coefficient on the solid walls of the microchannel, the lateral mixture of flow at Re = 50 covered larger areas. Increasing the Reynolds number resulted in higher fluid velocity and better temperature distribution, especially in heated areas, and this led to higher average Nusselt number. In fluid with higher velocity, applying the slip boundary condition on solid walls significantly increases the crossing fluid momentum.

Author Contributions: All authors contributed equally.

Funding: This research received no external funding.

Conflicts of Interest: The authors declare no conflict of interest.

Nomenclature

B	dimensionless slip velocity coefficient
C_p	heat capacity, J kg^{-1} K^{-1}
d	diameter, m
H, L	microchannel height and length, m
k	thermal conductivity coefficient, Wm^{-1} K^{-1}
Nu	Nusselt number
P	fluid pressure, Pa
$Pr = \nu_f/f$	Prandtl number
$Re = \rho_f u_c H / \mu_f$	Reynolds number
T	temperature, K
u, v	velocity components in the x-, y-directions, ms^{-1}
u_c	inlet flow velocity, ms^{-1}
$(U, V) = (u/U_0, v/U_0)$	dimensionless flow velocity in the x-, y-direction
x, y	Cartesian coordinates, m
$(X, Y = x/H, y/H)$	dimensionless coordinates
Greek symbols	
α	thermal diffusivity, m^{2}s^{-1}
$\beta *$	dimensionless slip velocity coefficient
ϕ	nanoparticle volume fraction
μ	dynamic viscosity, Pa s
$\theta = (T - T_C)/(T_H - T_C)$	dimensionless temperature
ρ	density, kg m^{-3}
ν	kinematics viscosity m^{2}s^{-1}
Super- and Subscripts	
c	Cold
eff	Effective
f	base fluid (pure water)
h	Hot
m	Mean
nf	Nanofluid
s	solid nanoparticles

References

1. Gorla, R.S.R.; Chamkha, A. Natural convective boundary layer flow over a vertical plate embedded in a porous medium saturated with a non-Newtonian nanofluid. *Int. J. Microscale Nanoscale Therm. Fluid Transp. Phenom.* **2011**, *3*, 1–20.
2. Dogonchi, A.S.; Chamkha, A.J.; Seyyedi, S.M.; Ganji, D.D. Radiative nanofluid flow and heat transfer between parallel disks with penetrable and stretchable walls considering Cattaneo–Christov heat flux model. *Heat Transf. Asian Res.* **2018**, *47*, 735–753. [CrossRef]

3. Goshayeshi, H.R.; Safaei, M.R.; Goodarzi, M.; Dahari, M. Particle Size and Type Effects on Heat Transfer Enhancement of Ferro-nanofluids in a Pulsating Heat Pipe under Magnetic Field. *Powder Technol.* **2016**, *301*, 1218–1226. [CrossRef]
4. Khanafer, K.; Vafai, K. A critical synthesis of thermophysical characteristics of nanofluids. *Int. J. Heat Mass Transf.* **2012**, *54*, 4410–4428. [CrossRef]
5. Kosar, A.; Peles, Y. TCPT-2006-096. R2: Micro Scale Pin Fin Heat Sinks—Parametric Performance Evaluation Study. *IEEE Trans. Compon. Packag. Technol.* **2007**, *30*, 855–865. [CrossRef]
6. Sivasankaran, H.; Asirvatham, G.; Bose, J.; Albert, B. Experimental Analysis of Parallel Plate and Crosscut Pin Fin Heat Sinks for Electronic Cooling Applications. *Therm. Sci.* **2010**, *14*, 147–156. [CrossRef]
7. Mital, M. Analytical Analysis of Heat Transfer and Pumping Power of Laminar Nanofluid Developing Flow in Microchannels. *Appl. Therm. Eng.* **2012**, *50*, 429–436. [CrossRef]
8. Hung, T.C.; Yan, W.M.; Wang, X.D.; Chang, C.Y. Heat Transfer Enhancement in Microchannel Heat Sinks using Nanofluids. *Int. J. Heat Mass Transf.* **2012**, *55*, 2559–2570. [CrossRef]
9. Akbari, O.A.; Toghraie, D.; Karimipour, A. Impact of ribs on flow parameters and laminar heat transfer of water–aluminum oxide nanofluid with different nanoparticle volume fractions in a three-dimensional rectangular microchannel. *Adv. Mech. Eng.* **2015**, *7*, 1–11. [CrossRef]
10. Behnampour, A.; Akbari, O.A.; Safaei, M.R.; Ghavami, M.; Marzban, A.; Ahmadi Sheikh Shabani, G.R.; zarringhalam, M.; Mashayekhi, R. Analysis of heat transfer and nanofluid fluid flow in microchannels with trapezoidal, rectangular and triangular shaped ribs. *Physica E* **2017**, *91*, 15–31. [CrossRef]
11. Gravndyan, Q.; Akbari, O.A.; Toghraie, D.; Marzban, A.; Mashayekhi, R.; Karimi, R.; Pourfattah, F. The effect of aspect ratios of rib on the heat transfer and laminar water/TiO$_2$ nanofluid flow in a two-dimensional rectangular microchannel. *J. Mol. Liq.* **2017**, *236*, 254–265. [CrossRef]
12. Lee, D.Y.; Vafai, K. Comparative analysis of jet impingement and microchannel cooling for high heat flux applications. *Int. J. Heat Mass Transf.* **1999**, *42*, 1555–1568. [CrossRef]
13. Chang, S.W.; Chiang, K.F.; Chou, T.C. Heat transfer and pressure drop in hexagonal ducts with surface dimples. *Exp. Therm. Fluid Sci.* **2010**, *34*, 1172–1181. [CrossRef]
14. Elyyan, M.A.; Ball, K.S.; Diller Thomas, E.; Paul Mark, R.; Ragab Saad, A. Heat Transfer Augmentation Surfaces Using Modified Dimples/Protrusions. Ph.D. Thesis, Virginia Tech, Blacksburg, VA, USA, 2008.
15. Guo, J.; Fan, A.; Zhang, X.; Liu, W. A numerical study on heat transfer and friction factor characteristics of laminar flow in a circular tube fitted with centercleared twisted tap. *Int. J. Therm. Sci.* **2011**, *50*, 1263–1270. [CrossRef]
16. Zhang, X.; Liu, Z.; Liu, W. Numerical studies on heat transfer and flow characteristics for laminar flow in a tube with multiple regularly spaced twisted tapes. *Int. J. Therm. Sci.* **2012**, *58*, 157–167. [CrossRef]
17. Zheng, N.; Liu, W.; Liu, Z.; Liu, P.; Shan, F. A numerical study on heat transfer enhancement and the flow structure in a heat exchanger tube with discrete double inclined ribs. *Appl. Therm. Eng.* **2015**, *90*, 232–241. [CrossRef]
18. Shan, F.; Liu, Z.; Liu, W.; Tsuji, Y. Effects of the orifice to pipe diameter ratio on orifice flows. *Chem. Eng. Sci.* **2016**, *152*, 497–506. [CrossRef]
19. Zheng, N.; Liu, P.; Shan, F.; Liu, Z.; Liu, W. Effects of rib arrangements on the flow pattern and heat transfer in an internally ribbed heat exchanger tube. *Int. J. Therm. Sci.* **2016**, *101*, 93–105. [CrossRef]
20. Chin, S.B.; Foo, J.J.; Lai, Y.L.; Yong, T.K.K. Forced Convective Heat Transfer Enhancement with Perforated Pin Fins. *Heat Mass Transf.* **2013**, *49*, 1447–1458. [CrossRef]
21. Naphon, P.; Nakharintr, L. Heat Transfer of Nanofluids in the Mini-rectangular Fin Heat Sinks. *Int. Commun. Heat Mass Transf.* **2012**, *40*, 25–31. [CrossRef]
22. Jasperson, B.A.; Jeon, Y.; Turner, K.T.; Pfefferkorn, F.E.; Qu, W. Comparison of Micro-pin-fin and Microchannel Heat Sinks Considering Thermal-hydraulic Performance and Manufacturability. *IEEE Trans. Compon. Packag. Technol.* **2010**, *33*, 148–160. [CrossRef]
23. Zhuang, Y.; Ma, C.F.; Qin, M. Experimental study on local heat transfer with liquid impingement flow in two-dimensional micro-channels. *Int. J. Heat Mass Transf.* **1997**, *40*, 4055–4059. [CrossRef]
24. Gholami, M.R.; Akbari, O.A.; Marzban, A.; Toghraie, D.; Ahmadi Sheikh Shabani, G.H.R.; Zarringhalam, M. The effect of rib shape on the behavior of laminar flow of Oil/MWCNT nanofluid in a rectangular microchannel. *J. Therm. Anal. Calorim.* **2017**, 1–18. [CrossRef]

25. Raisi, A.; Aminossadati, S.M.; Ghasemi, B. An innovative nanofluid-based cooling using separated natural and forced convection in low Reynolds flows. *J. Taiwan Inst. Chem. Eng.* **2016**, 1–5. [CrossRef]
26. Aminossadati, S.M.; Raisi, A.; Ghasemi, B. Effects of magnetic field on nanofluid forced convection in a partially heated microchannel. *Int. J. Non-Linear Mech.* **2011**, *46*, 1373–1382. [CrossRef]
27. Bahmani, M.H.; Sheikhzadeh, G.; Zarringhalam, M.; Akbari, O.A.; Alrashed, A.A.A.A.; Ahmadi Sheikh Shabani, G.; Goodarzi, M. Investigation of turbulent heat transfer and nanofluid flow in a double pipe heat exchanger. *Adv. Powder Technol.* **2018**, *29*, 273–282. [CrossRef]
28. Arani, A.A.A.; Akbari, O.A.; Safaei, M.R.; Marzban, A.; Alrashed, A.A.A.A.; Ahmadi, G.R.; Nguyen, T.K. Heat transfer improvement of water/single-wall carbon nanotubes (SWCNT) nanofluid in a novel design of a truncated double layered microchannel heat sink. *Int. J. Heat Mass Transf.* **2017**, *113*, 780–795. [CrossRef]
29. Khodabandeh, E.; Rahbari, A.; Rosen, M.A.; Najafian Ashrafi, Z.; Akbari, O.A.; Anvari, A.M. Experimental and numerical investigations on heat transfer of a water-cooled lance for blowing oxidizing gas in an electrical arc furnace. *Energy Conversat. Manag.* **2017**, *148*, 43–56. [CrossRef]
30. Safaiy, M.R.; Saleh, S.R.; Goudarzi, M. Numerical studies of laminar natural convection in a square cavity with orthogonal grid mesh by finite volume method. *Int. J. Adv. Des. Manuf. Technol.* **2011**, *1*, 13–21.
31. Akbari, O.A.; Goodarzi, M.; Safaei, M.R.; Zarringhalam, M.; Ahmadi Sheikh Shabani, G.R.; Dahari, M. A modified two-phase mixture model of nanofluid flow and heat transfer in 3-D curved microtube. *Adv. Powd. Technol.* **2016**, *27*, 2175–2185. [CrossRef]
32. Safaei, M.R.; Goodarzi, M.; Akbari, O.A.; Safdari Shadloo, M.; Dahari, M. Performance Evaluation of Nanofluids in an Inclined Ribbed Microchannel for Electronic Cooling Applications. *Electron. Cool.* **2016**. [CrossRef]
33. Raisi, A.; Ghasemi, B.; Aminossadati, S.M. A Numerical Study on the Forced Convection of Laminar Nanofluid in a Microchannel with Both Slip and No-Slip Conditions. *Numer. Heat Transf. Part A* **2011**, *59*, 114–129. [CrossRef]

© 2019 by the authors. Licensee MDPI, Basel, Switzerland. This article is an open access article distributed under the terms and conditions of the Creative Commons Attribution (CC BY) license (http://creativecommons.org/licenses/by/4.0/).

Article

Modified MHD Radiative Mixed Convective Nanofluid Flow Model with Consideration of the Impact of Freezing Temperature and Molecular Diameter

Umar Khan [1], Adnan Abbasi [2], Naveed Ahmed [3], Sayer Obaid Alharbi [4] Saima Noor [5], Ilyas Khan [6,*], Syed Tauseef Mohyud-Din [3] and Waqar A. Khan [7]

1. Department of Mathematics and Statistics, Hazara University, Mansehra 21120, Pakistan; umar_jadoon4@yahoo.com
2. Department of Mathematics, Mohi-ud-Din Islamic University Nerian Sharif, Azad Jammu & Kashmir 12080, Pakistan; adnan_abbasi89@yahoo.com
3. Department of Mathematics, Faculty of Sciences, HITEC University Taxila Cantt, Punjab 47080, Pakistan; nidojan@gmail.com (N.A.); syedtauseefs@hotmail.com (S.T.M.-D.)
4. Department of Mathematics, College of Science Al-Zulfi, Majmaah University, Al-Majmaah 11952, Saudi Arabia; so.alharbi@mu.edu.sa
5. Department of Mathematics, COMSATS University Islamabad, Abbottabad 22010, Pakistan; saimanoor@ciit.net.pk
6. Faculty of Mathematics and Statistics, Ton Duc Thang University, Ho Chi Minh City 72915, Vietnam
7. Department of Mechanical Engineering, College of Engineering, Prince Mohammad Bin Fahd University, Al Khobar 31952, Saudi Arabia; wkhan1956@gmail.com
* Correspondence: ilyaskhan@tdtu.edu.vn

Received: 13 March 2019; Accepted: 16 April 2019; Published: 25 June 2019

Abstract: Magnetohydrodynamics (MHD) deals with the analysis of electrically conducting fluids. The study of nanofluids by considering the influence of MHD phenomena is a topic of great interest from an industrial and technological point of view. Thus, the modified MHD mixed convective, nonlinear, radiative and dissipative problem was modelled over an arc-shaped geometry for Al_2O_3 + H_2O nanofluid at 310 K and the freezing temperature of 273.15 K. Firstly, the model was reduced into a coupled set of ordinary differential equations using similarity transformations. The impact of the freezing temperature and the molecular diameter were incorporated in the energy equation. Then, the Runge–Kutta scheme, along with the shooting technique, was adopted for the mathematical computations and code was written in Mathematica 10.0. Further, a comprehensive discussion of the flow characteristics is provided. The results for the dynamic viscosity, heat capacity and effective density of the nanoparticles were examined for various nanoparticle diameters and volume fractions.

Keywords: arched surface; nonlinear thermal radiation; molecular diameter; Al_2O_3 nanoparticles; streamlines; isotherms; RK scheme

1. Introduction

The liquids regularly used in heat transfer applications, such as water, propylene glycol, ethylene glycol, kerosene oil, engine oil and transformer oil, are extensively used in industry and in thermal power plants. Due to their reduced thermal conductivity, these liquids do not have effective heat transfer characteristics. However, for a great deal of industrial production, remarkable amounts of heat are required. The thermal conductivity of the solid materials, such as different metals and oxides, is very high in comparison with regular liquids. Thus, scientists and engineers hypothesized that the heat transfer in working fluids could be enhanced by mixing in the nanoparticles in above regular liquids. Finally, Choi [1] developed a colloidal composition that has effective heat transfer characteristics, as compared to regular fluids. Choi [1] unlocked a new innovative research area and researchers, scientists and engineers focused on the analysis of nanofluids. Before the development of nanofluids, heat transfer was a major problem from an industrial point of view, since considerable amounts of heat transfer were required for a great deal of technological and industrial production, and regular liquids failed to provide the desired amount of heat. Thermal conductivity plays a major role in the heat transfer rate of nanofluids. Thus, several theoretical models were proposed for thermal conductivity. The thermal conductivity model was developed by Maxwell in 1873 [2], who considered nanosized particles, and this can be considered the origin of the concept of the nanofluids.

Several theoretical models based on nanoparticle characteristics that take into account the effects of temperature and the shape and diameter of nanoparticles, as well as Brownian motion, have been presented. In 1935, Bruggemann [3] constructed a thermal conductivity correlation for spherical nanoparticles that was limited to high concentration patterns. The behavior of thermal conductivity was developed by Hamilton [4] in 1962, who explored the effects of nanoparticle shape. In 1996, Lu and Lin [5] proposed a model incorporating the effects of Brownian dynamics. The thermal conductivity model for the interaction between the nanoparticles and their surrounding liquid was developed by Koo and Kleinstreuer [6,7]. Xue [8] developed a thermal conductivity model for carbon nanotubes. In 2005, Prasher et al. [9] found a correlation by considering the influence of convection on the thermal conductivity of nanoparticles. In 2006, Li [10] made apparent the influence of temperature on thermal conductivity and outlined a correlation for Al_2O_3/H_2O and CuO/H_2O nanofluids. In 2011, Corcione [11] developed a model for $Al_2O_3 + H_2O$ nanofluids by incorporating the effects of freezing temperatures. By incorporating thermal conductivity models, the above researchers presented various models and described the heat transfer enhancement due to thermal conductivity. Some useful studies for nanofluids are described in [12–16].

Similar to nanoparticles, carbon nanotubes also have high thermal conductivity and unique mechanical and chemical properties. Carbon nanotubes are subcategorized as either single or multiple walled carbon nanotubes. The concept of colloidal suspension in relation to carbon nanotubes was presented by Iijima [17]. After the development of the thermal conductivity correlation for carbon nanotubes, many studies were presented outlining thermal enhancement due to suspended carbon nanotubes. Recently, Ahmed et al. [18] explored the influence of thermal radiation and viscous dissipation on the flow of water suspended by carbon nanotubes. The effect of thermophysical characteristics of the nanotubes on the heat transfer enhancement water over a curved surface and non-parallel walls was described in [19,20], respectively.

Recently, the flow over an arc-shaped geometry has become a point of interest. Reddy et al. [21] recently modified the curve-shaped flow model for nonlinear radiative heat flux. They also examined the impact of the cross-diffusion phenomenon on heat and mass transfer. Another useful mechanism to enhance the fluid temperature is ohmic heating, which produces extra heat in the conductor, with electrons supplying energy to the atoms of the conductor through collisions. In 2017, Hayat et al. [22] examined the effects of resistive heating on the curve surface flow.

A literature review revealed that there have not yet been any studies of the impact of freezing temperatures and the diameter of nanoparticles on the flow of incompressible fluids due to the effects of nonlinear radiative heat flux, viscous dissipation, mixed convection and Lorentz forces. This study is presented to cover this significant gap. The nanofluids $Al_2O_3 + H_2O$ were used to study the characteristics of the flow and other effective thermophysical properties, such as effective density, heat capacity and thermal conductivity. The results for shear stress and local heat transfer are also described and discussed comprehensively. Finally, major findings of the study is presented.

2. Model Formulation

We considered the laminar time independent and the incompressible flow of the $Al_2O_3 + H_2O$ nanofluid by taking into account the influence of a nonlinear, radiative heat flux and the imposed variable magnetic field over an arc geometry situated in the curvilinear frame r and s. Further, the r-axis was perpendicular, and the arc was placed in the direction of s. The velocity and magnetic field were functions of s and mathematically described as below:

$$\hat{u}_w(s) = \frac{b}{s^{-m}}, \hat{B}(s) = \frac{B_0}{\left(s^{0.5(m-1)}\right)^{-1}},$$

It was assumed that the induced magnetic field was inconsequential, and therefore was not taken into consideration. The temperatures at the arched and the free surface were $\hat{T}_w$ and $\hat{T}_\infty$, respectively. The value of $m > 1$ represented the flow over an arc shaped, which was nonlinearly stretched. $m = 1$ was for the flow of a linearly stretching geometry. Figure 1 presents the flow description in a curvilinear frame. The flow chart of the study presented in Figure 2.

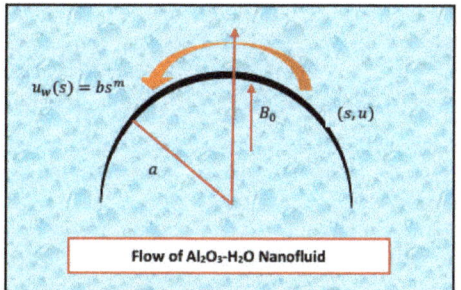

Figure 1. Flow description in a curvilinear frame.

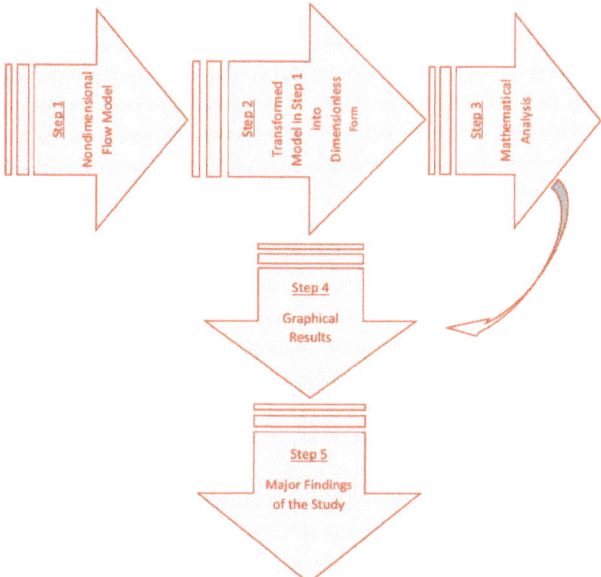

Figure 2. Flowchart of the study.

The Al$_2$O$_3$ + H$_2$O nanofluid flow, incorporating the phenomena of Lorentz forces, viscous dissipation and nonlinear radiative heat flux, is described by the following system [21]:

$$\frac{\partial}{\partial r}(\hat{v}(r+a)) = 0, \tag{1}$$

$$\hat{u}^2 = \frac{\partial p}{\partial r}(r+a)(\rho_{nf})^{-1}, \tag{2}$$

$$\rho_{nf}\left(\hat{v}\frac{\partial \hat{u}}{\partial r} + \left(\frac{a}{r+a}\right)\hat{u}\frac{\partial \hat{u}}{\partial s} + \hat{u}\frac{\hat{v}}{(r+a)}\right) = -\frac{a}{(r+a)}\frac{\partial p}{\partial s} + \mu_{nf}\left(\frac{\partial^2 \hat{u}}{\partial r^2} + \left(\frac{1}{r+a}\right)\frac{\partial \hat{u}}{\partial r} - \left(\frac{1}{r+a}\right)^2 \hat{u}\right) - \sigma_{nf}B_0^2\hat{u} + g(\rho\beta)_{nf}(T - T_\infty), \tag{3}$$

$$\frac{a}{(r+a)}\hat{u}\frac{\partial \hat{T}}{\partial s} + \hat{v}\frac{\partial \hat{T}}{\partial r} = \frac{k_{nf}}{(\rho C_p)_{nf}}\left(\frac{1}{(r+a)}\frac{\partial \hat{T}}{\partial r} + \frac{\partial^2 \hat{T}}{\partial r^2}\right) + \frac{\mu_{nf}}{(\rho C_p)_{nf}}\left(\frac{\partial \hat{u}}{\partial r} - \frac{\hat{u}}{(r+a)}\right)^2 - \frac{1}{(\rho C_p)_{nf}(r+a)}\frac{\partial}{\partial r}(q_R(r+a)), \tag{4}$$

with boundary conditions at the curve and far from the curve of:

At $r = 0$:

$$\left.\begin{array}{l} \hat{u} = bs^m \\ \hat{v} = 0 \\ \hat{T} = T_w \end{array}\right\}. \tag{5}$$

At $r \to \infty$:

$$\left.\begin{array}{l} \hat{u} \to 0 \\ \frac{\partial \hat{u}}{\partial r} \to 0 \\ \hat{T} \to T_\infty \end{array}\right\}. \tag{6}$$

The expression for nonlinear radiative heat flux is described as:

$$q_R = -4\frac{\hat{\sigma}}{3\hat{k}}\frac{\partial}{\partial r}(\hat{T}^4) = -\frac{16\hat{\sigma}}{3\hat{k}}\hat{T}^3\frac{\partial \hat{T}}{\partial r}, \tag{7}$$

Equations (1)–(4) are the conservation of mass, momentum and energy, respectively. Further, the Stefan–Boltzmann law and adsorption coefficient are $\hat{\sigma}$ and $\hat{k}$, respectively. To enhance the thermal and physical characteristics of the model, the following effective models for thermophysical characteristics were used [23]:

$$\rho_{nf} = \left[(1-\phi) + \frac{\phi \rho_p}{\rho_f}\right]\rho_f, \tag{8}$$

$$(\rho C_p)_{nf} = \left[(1-\phi) + \frac{\phi(\rho C_p)_p}{(\rho C_p)_f}\right](\rho C_p)_f, \tag{9}$$

$$\mu_{nf} = \mu_f\left(1 - 34.87\left(\frac{d_{particle}}{d_{fluid}}\right)^{-0.3}\phi^{1.03}\right)^{-1}. \tag{10}$$

$$k_{nf} = k_f\left(1 + 4.4Re_b^{0.4}Pr^{0.66}\left(\frac{T}{T_{freezing}}\right)^{10}\left(\frac{k_p}{k_f}\right)^{0.03}\phi^{0.66}\right), \tag{11}$$

$$(\rho\beta)_{nf} = \left[(1-\phi) + \frac{\phi(\rho\beta)_p}{(\rho\beta)_f}\right](\rho\beta)_f, \tag{12}$$

$$\sigma_{nf} = \sigma_f\left[1 + \frac{3(\Theta-1)\phi}{(\Theta+2)-(\Theta-1)\phi}\right], \text{ where, } \Theta = \frac{\sigma_p}{\sigma_f}. \tag{13}$$

In Equation (11), Re_b shows the Reynolds number due to Brownian motion, and is described in the following pattern:

$$Re_b(\mu_f) = d_p \rho_f u_b, \tag{14}$$

In Equation (14), the velocity of Brownian motion is calculated by the formula:

$$u_b = 2Tk_b(\pi d_p^2 \mu_f), \tag{15}$$

Here, k_b is the Stefan–Boltzmann coefficient and its value is 1.380648×10^{-23} (JK^{-1}). d_p represents the molecular diameter which is calculated by the expression [24]:

$$d_f = 6M^*(N^* \rho_f \pi)^{-1}, \tag{16}$$

The molecular weight of the regular liquid and Avogadro number are denoted by M^* and N^*, respectively. Further, the value of d_f is calculated as:

$$d_f = \left(\frac{6 \times 0.01801528}{998.62 \times (6.022 \times 10^{23}) \times \pi}\right)^{\frac{1}{3}} = 3.85 \times 10^{-10} \text{ nm}, \tag{17}$$

Table 1 shows the data for thermal conductivity, effective density, thermal expansion coefficient, and effective dynamic viscosity [23]:

Table 1. Thermal and physical characteristics of the fluid phase and nanoparticles at $T = 310$ K [23].

Properties	d_p (nm)	ρ (kg/m^3)	β (1/k)	c_p (J/Kg K)	μ_f (kg/ms)	k (W/mk)	σ (S/m)
H$_2$O	0.385	993	36.2×10^5	4178	695×10^6	0.628	0.005
Al$_2$O$_3$	33	3970	0.85×10^5	765	-	40	0.05×10^6

The similarity transformations are described in the following set of equations [21]:

$$\left.\begin{array}{l} \eta = \sqrt{\frac{b}{v_{bf}}} s^{0.5(m-1)} r \\ \hat{u} = b s^m L' \\ p = \rho_{bf} b^2 s^{2m} P' \\ \hat{v} = -\frac{a}{(r+a)} \sqrt{bv_{bf}} s^{0.5(m-1)} \{0.5(m+1)L + 0.5(m-1)\eta L'\} \\ N = \frac{T - T_\infty}{T_w - T_\infty} \end{array}\right\} \tag{18}$$

To analyze the phenomena of nonlinear radiative heat flux, the following expression was used:

$$\hat{T} = T_\infty(1 + (\beta_w - 1)N) \tag{19}$$

The ratio of wall and free surface temperature was denoted by β_w.

The following model was attained after incorporating the similarity transformations and partial derivatives in Equations (1)–(4):

$$\frac{1}{\left(1 - 34.87 \left(\frac{d_p}{d_f}\right)^{-0.3} \phi^{1.03}\right)} \left[(\eta + K)^3 L'''' + 2(\eta + K)^2 L''' - (\eta + K)L'' + L'\right] - \\ \left[1 + \frac{3(\Theta - 1)\phi}{(\Theta + 2) - (\Theta - 1)\phi}\right] M\left[(\eta + K)^3 L'' + (\eta + K)^2 L'\right] + \left((1 - \phi) + \frac{\phi(\rho\beta)_p}{(\rho\beta)_f}\right) \\ \alpha\left((\eta + K)^3 N' + (\eta + K)^2 L\right) + \\ \left\{(1 - \phi) + \frac{\phi \rho_p}{\rho_f}\right\} K \left[\begin{array}{c} 0.5(1 - 3m)(\eta + K)(L')^2 + 0.5(m + 1)(\eta + K)LL'' \\ -0.5(m + 1)LL' + 0.5(m + 1)(\eta + K)^2 LL''' + \\ (\eta + K)^2 0.5(1 - 3m)L'L'' \end{array}\right] = 0, \tag{20}$$

$$Rd\left[3(N')^3(1+(\beta_w-1)N)^2(\beta_w-1)(\eta+K)+(1+(\beta_w-1)N)^3((\eta+K)N''+N')\right]+$$
$$\left[1+4.4Re_b^{0.4}Pr^{0.66}\left(\frac{T}{T_{freezing}}\right)^{10}\left(\frac{k_p}{k_f}\right)^{0.03}\phi^{0.66}\right]((\eta+K)N''+N')+$$
$$\frac{PrEc}{\left(1-34.87\left(\frac{d_p}{d_f}\right)^{-0.3}\phi^{1.03}\right)(\eta+K)}((\eta+K)L''-L')^2=0. \tag{21}$$

In Equations (20) and (21), L and N are the functions of η. After solving, L and N provide the velocity and temperature distributions, respectively.

$$L(\eta)=0,\ L'(\eta)=1,\ N(\eta)=1, \tag{22}$$

At $\eta\to\infty$:
$$L'(\eta)\to 0, L''(\eta)\to 0, N(\eta)\to 0, \tag{23}$$

Dimensionless quantities were described by the following formulas:

$$K=a\sqrt{\frac{b}{v_{bf}}},\ Ec=\frac{b^2s^{2m}}{(C_p)_f(T_w-T_\infty)},\ Rd=16\delta\frac{T_\infty^3}{3\bar{k}k_f},\ M=\frac{\sigma_f B_0^2}{\rho_f b},\ Pr=\frac{(c_p)_f\mu_f}{k_f}, \tag{24}$$
$$\alpha=Gr_s\left(Re_s^2\right)^{-1},\ Re_s=bs^2\left(v_f\right)^{-1},\ Gr_s=s^3(T_w-T_\infty)g\beta\left(v_f\right)^{-2}.$$

Moreover, the dimensional formula for skin friction and local Nusselt number were defined as:

$$C_f=\tau_{rs}\left(\rho_f\hat{u}_w^2\right)^{-1},\ Nu_s=sq_w\left(k_f(T_w-T_\infty)\right)^{-1}, \tag{25}$$

where,
$$\tau_{rs}=\left(\frac{\partial\hat{u}}{\partial r}-\frac{\hat{u}}{(a+r)}\right)\downarrow r=0, q_w=-k_{nf}\frac{\partial\hat{T}}{\partial r}\downarrow_{r=0}, \tag{26}$$

Drawing the values from Equation (26) into Equation (25), the following dimensionless formulas were obtained:

$$C_f(Re_s)^{\frac{1}{2}}=\frac{1}{\left(1-34.87\left(\frac{d_p}{d_f}\right)^{-0.3}\phi^{1.03}\right)}\left(L''(0)-L'(0)K^{-1}\right), \tag{27}$$

$$Nu_s(Re_s)^{-\frac{1}{2}}=-\left(\left(1+4.4Re_b^{0.4}Pr^{0.66}\left(\frac{T}{T_{freezing}}\right)^{10}\left(\frac{k_p}{k_f}\right)^{0.03}\phi^{0.66}\right)+Rd\beta_w^3\right)N'(0). \tag{28}$$

3. Mathematical Analysis

Highly nonlinear and coupled systems of differential equations usually possess no closed-form solution. Our flow model for $Al_2O_3 + H_2O$ is a highly nonlinear fourth-order model defined at a semi-infinite domain. Therefore, the model was tackled numerically using the Runge-Kutta scheme [13,16,25], as the RK scheme is used for the first order initial value problem (IVP). First the following substitutions were made and the model was reduced into first order IVP.

$$h_1=L,\ h_2=L',\ h_3=L'',\ h_4=L''',\ h_5=N,\ h_6=N'. \tag{29}$$

After the successful transformation into the first order IVP, Mathematica 10.0 was used and the system was solved successfully.

4. Graphical Results and Discussion

This section emphasizes the flow and thermophysical characteristics of the fluid phase and the nanoparticles of Al_2O_3. The values for the thermophysical characteristics were calculated at 310 K [23]. The results for the shear stress and heat transfer rate are elaborated using bar charts and are discussed comprehensively.

4.1. Velocity and Temperature Distribution

Magnetic field phenomena are of a great significance from an industrial point of view. Many industrial productions contain impurities that need to be removed. However, the magnetic parameter opposes the fluid motion, and the impurities remain at the bottom and the nanofluid velocity drops. The influence of Lorentz forces on the velocity distribution of $Al_2O_3 + H_2O$ nanofluids is elaborated in Figure 3a. It was shown that the applied magnetic field opposed the nanofluid motion, and the velocity of the $Al_2O_3 + H_2O$ nanofluid dropped. The velocity declined more slowly for a weaker magnetic field, and a rapid decrement in the nanofluid velocity was observed for a stronger magnetic field. Near the arched surface, variations in the velocity $(L'(\eta))$ were almost negligible. This behavior of the velocity distribution was due to the friction between the surface and the nanolayer of $Al_2O_3 + H_2O$. In the successive nanolayers, the velocity field was altered significantly. These influences became negligible far from the curve and showed an asymptotic pattern of velocity distribution at the free surface. Figure 3b shows the velocity distribution of the parameter m. The velocity of the $Al_2O_3 + H_2O$ nanofluid dropped rapidly for m in comparison with M. As the values for parameter m became larger, the velocity decreased promptly.

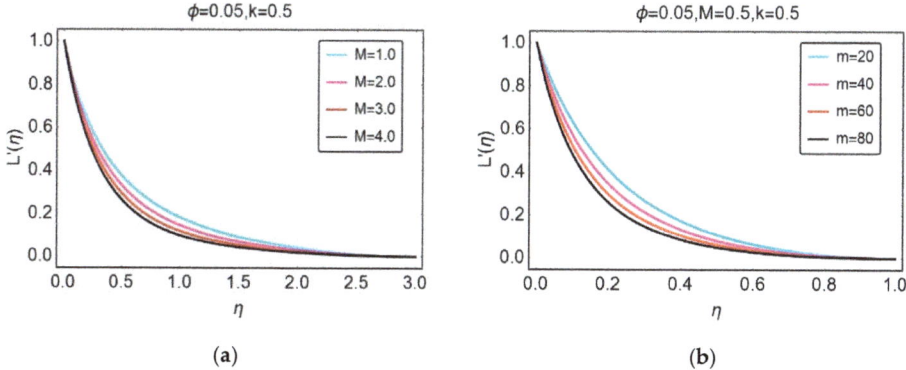

Figure 3. Impacts of (a) M and (b) m on the velocity distribution $(L'(\eta))$.

The effects of surface curvature on the velocity $(L'(\eta))$ are elucidated in Figure 4a. Altering the surface curvature caused the velocity to increase. For a smaller curvature, the velocity increased slowly and then vanished asymptotically far away from the surface. Similar behavior of the velocity of the $Al_2O_3 + H_2O$ nanofluid is depicted in Figure 4b. For α, a prominent behavior of the velocity was noticed in $0.5 \leq \eta \leq 1.5$. Besides this, the velocity $(L'(\eta))$ was almost inconsequential.

The temperature distribution $(N(\eta))$ for the radiation parameter (Rd) and Eckert number is highlighted in Figure 5a,b, respectively. In Figure 5a, we can see rapid drops in the temperature that were investigated by altering the radiation parameter (Rd). The temperature $(N(\eta))$ dropped rapidly for a stronger radiation parameter, however at the free surface this was almost negligible and vanished asymptotically. The Eckert number, which appeared due to viscous dissipation, played a vibrant role in the heat transfer enhancement. These effects are elucidated in Figure 5b. It was obvious that the temperature of $Al_2O_3 + H_2O$ nanofluid grew for the more dissipative nanofluid. For the larger Ec, the temperature distribution rose rapidly.

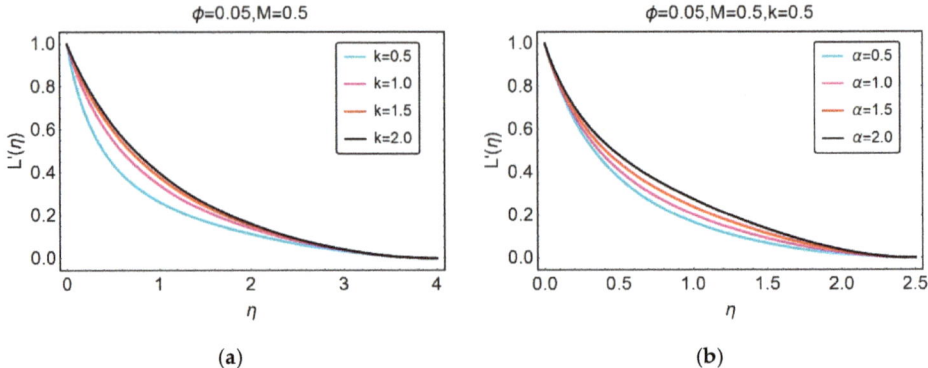

Figure 4. Impacts of (**a**) K and (**b**) α on the velocity distribution ($L'(\eta)$).

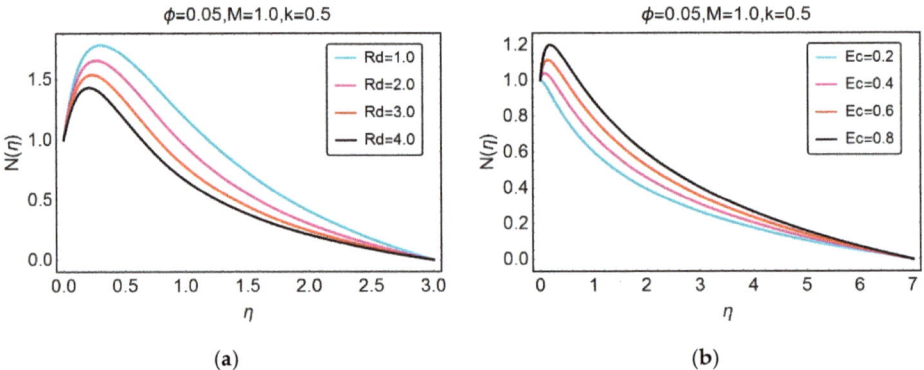

Figure 5. Impacts of (**a**) Rd and (**b**) Ec on the temperature distribution ($N(\eta)$).

4.2. Streamlines and Isotherms

This subsection is devoted to analyzing the behavior of the streamlines and isotherm patterns by altering different pertinent flow parameters. Figure 6 presents the streamline pattern by a varying magnetic parameter (M). For a smaller magnetic parameter, the streamlines were more curved near the surface and for a stronger M, the streamlines assumed a less curved shape. At the free stream, these streamlines became straight, since, as already mentioned for the parameter $m > 1$, it showed that it was a nonlinear stretching curved surface. The streamline pattern versus m is elaborated in Figure 7. It was noticed that by decreasing the parameter m, the streamlines assumed a more curved pattern. The curvature parameter showed a fascinating pattern for the streamlines, in comparison with M and m. These alterations are illustrated in Figure 8. Figure 9 depicts the flow pattern for varying α. For a higher α, the streamlines shrank and became almost straight at the top. Figure 10 elaborates the isotherm pattern for the radiation parameter (Rd). When there was more radiative nanofluid, the isotherms increased, and vice versa. Further, a 3D scenario of the isotherms is depicted in Figure 11.

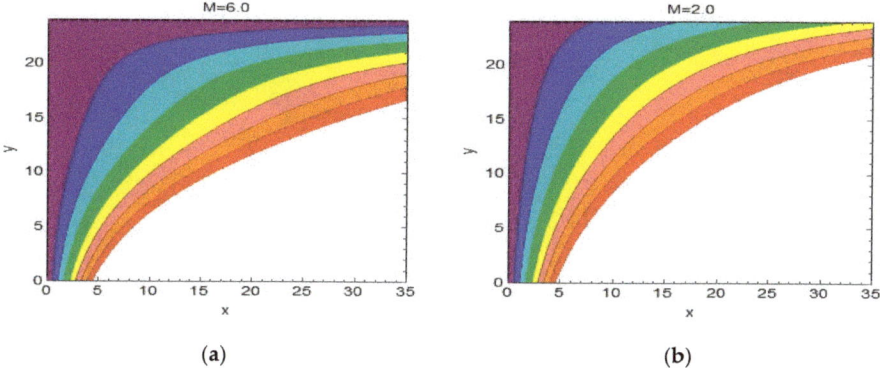

Figure 6. Streamlines for (**a**) $M = 6$ (**b**) $M = 2$.

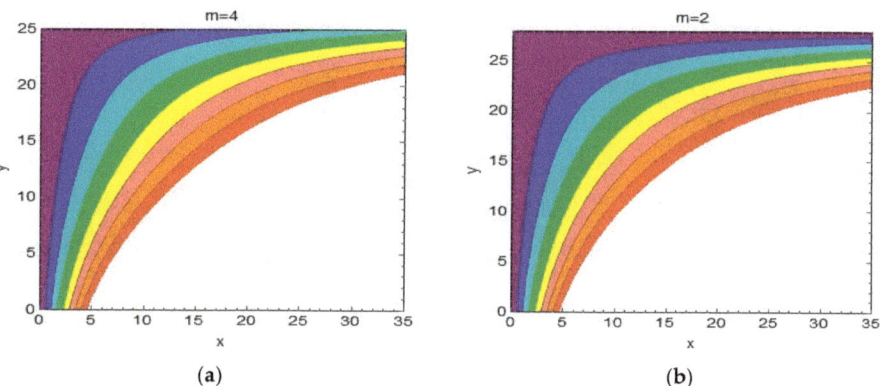

Figure 7. Streamlines for (**a**) $m = 4$ and (**b**) $m = 2$.

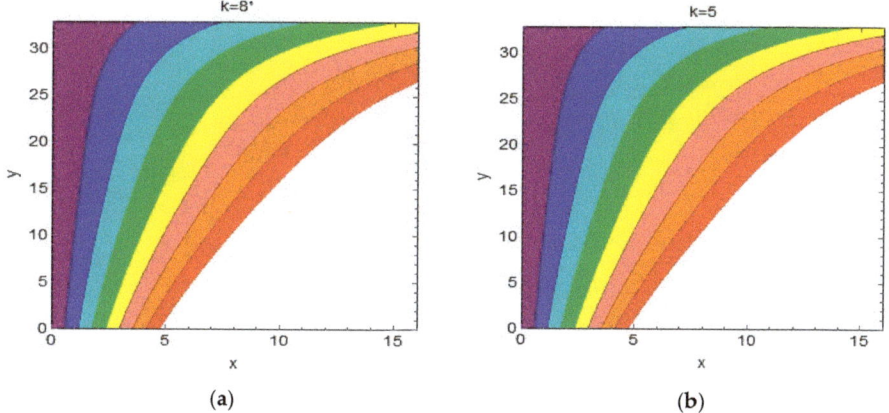

Figure 8. Streamlines for (**a**) $k = 8$ and (**b**) $k = 5$.

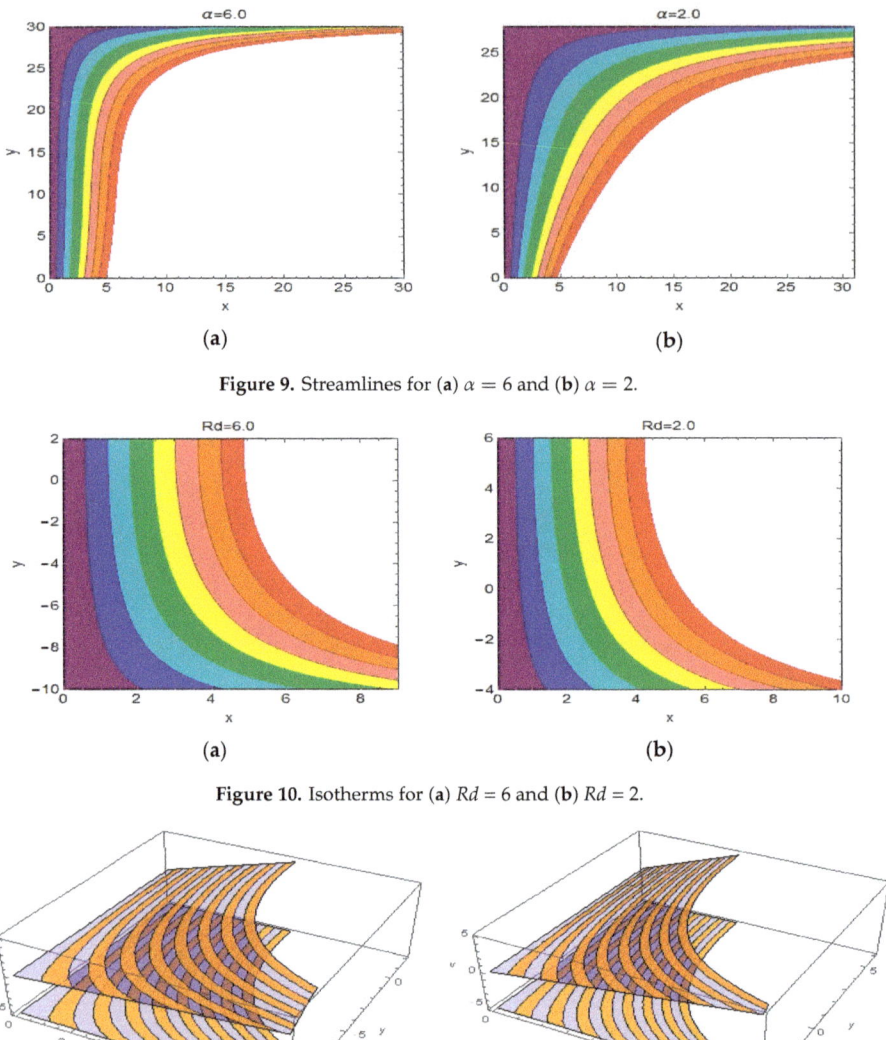

Figure 9. Streamlines for (**a**) $\alpha = 6$ and (**b**) $\alpha = 2$.

Figure 10. Isotherms for (**a**) $Rd = 6$ and (**b**) $Rd = 2$.

Figure 11. 3D scenario of Figure 10; (**a**) $Rd = 6$ and (**b**) $Rd = 2$.

4.3. Thermophysical Characteristics

This subsection describes the impacts of the volume fraction factor (ϕ) on the effective characteristics of the nanofluids, and the behavior of the shear stress and local heat transfer rate by varying embedded flow parameters.

Figure 12 describes the effects of the volume fraction (ϕ,) and diameter of the nanoparticles on the effective dynamic viscosity of $Al_2O_3 + H_2O$. The volume fraction of Al_2O_3 showed a vibrant role in enhancing the dynamic viscosity of the nanofluid. The observed high dynamic viscosity corresponded to a greater volume fraction. On the other hand, the nanoparticle diameter (d_p) induced inverse variations in the dynamic viscosity. Increasing the diameter of the nanoparticles caused the dynamic viscosity to drop. This means that nanoparticles with a smaller diameter are important to enhance the dynamic viscosity of nanofluids.

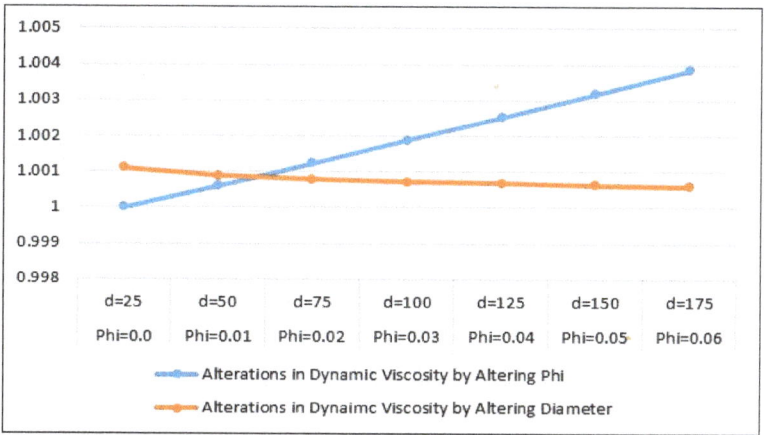

Figure 12. The impact of the volume fraction (ϕ) and the nanoparticle diameter (d_p) on the dynamic viscosity.

Figure 13 highlights the effective density (ρ_{nf}) and heat capacity ($(\rho c_p)_{nf}$) of the nanofluid versus ϕ. ϕ and ρ_{nf} were in direct proportion to each other, and the effective heat capacity dropped when ϕ increased. Therefore, smaller values of ϕ enhanced the effective heat capacity. Due to the high volume fraction, the colloidal suspension $Al_2O_3 + H_2O$ became denser, which enhanced the effective density (ρ_{nf}). Figure 14 highlights that the volume fraction (ϕ) and the effective electrical conductivity were in inverse proportion to each other.

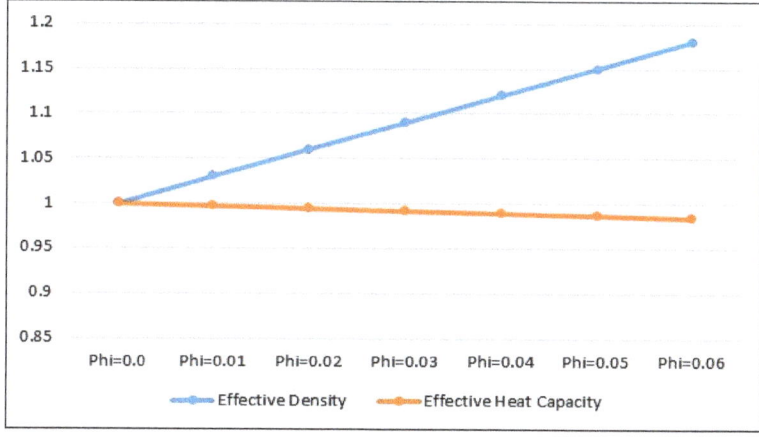

Figure 13. The impact of ϕ on the effective density and heat capacity.

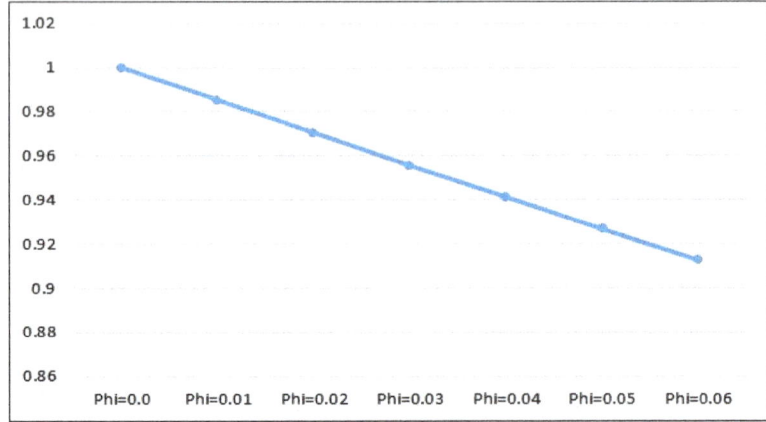

Figure 14. The impact of ϕ on the effective electrical conductivity.

4.4. Skin Fraction and Heat Transfer Rate

The shear stress and the local rate of heat transfer are very interesting and have attained great significance from an industrial point of view. The radiation parameter, Eckert number, and the curvature parameter all play a significant role in the shear stress and local Nusselt number. Figure 15 describes the heat transfer behavior for Rd, Ec, K and the volume fraction of the nanoparticles. It was noted that the more radiative nanofluids favored the heat transfer. On the other hand, a smaller amount of the heat transfer was noticed for a higher Eckert number. Therefore, the less dissipative fluids similarly favored the heat transfer. For a more curved surface, a large curvature worked against the heat transfer. At greater volume concentrations the heat transfer rate grew slowly.

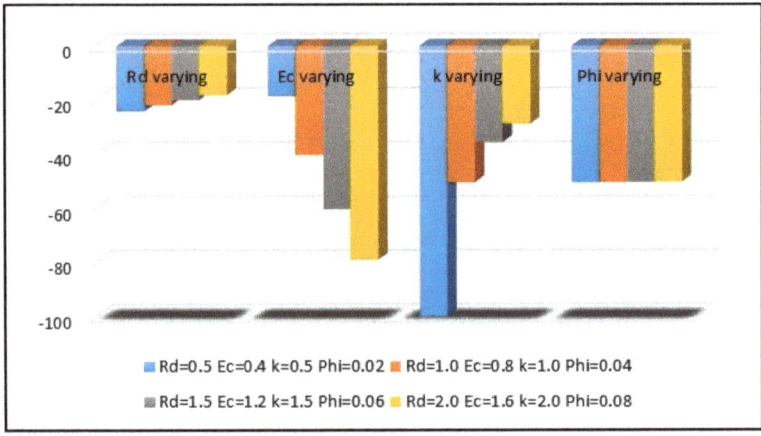

Figure 15. The impact of various flow parameters on the local heat transfer.

Figure 16 elucidates the shear stress behavior for the mixed convection parameter (α), magnetic parameter (M), curvature of the surface, and m. For a more convective fluid, the heat transfer increased rapidly at the surface. The magnetic parameter (M) highlighted the reverse behavior of the shear stress. With an increase of the parameter M, the shear stress dropped quickly, and an almost negligible influence of m on the shear stresses was also observed. Significant alterations were pointed out for surface curvature, as the curve was along the circle loop of radius a. Therefore, for a smaller radius, shear stress declined promptly. Increasing the radius of the loop caused the surface curvature to become larger, which favored the shear stress.

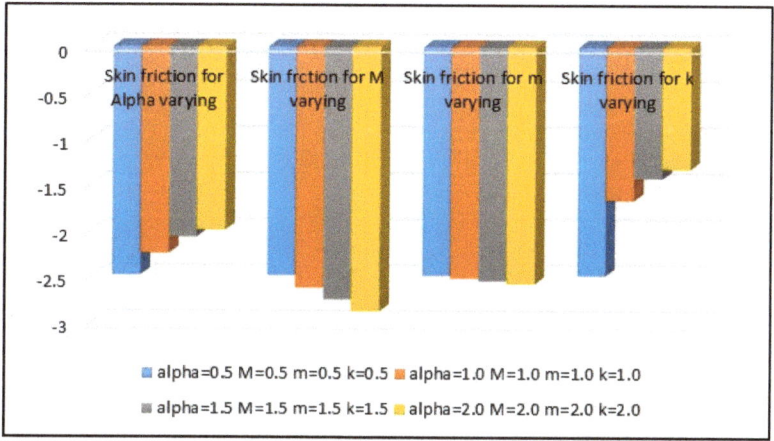

Figure 16. The impact of various parameters on the shear stress.

5. Conclusions

The mixed convective laminar flow of water, composed by Al_2O_3 nanofluids in the presence of Lorentz forces and nonlinear radiative heat flux, was examined over an arc-shaped geometry. To enhance the heat transfer rate, a thermal conductivity model that considered the impact of freezing temperature and molecular diameter was used. It was found that the nanofluid velocity ($L'(\eta)$) dropped for a stronger magnetic parameter (M), which is very significant from an industrial point of view. Further, the velocity of the nanofluid increased due to the mixed convection and larger curvature of the surface. The temperature $N(\eta)$ intensified for more dissipative fluid, and an inverse relationship between the temperature and the thermal radiation parameter was found. The dynamic viscosity of the nanofluid increased with the volume fraction, and the diameter of the nanoparticles showed reverse alterations for dynamic viscosity. The nanofluid became denser for a high volume fraction, and the electrical conductivity dropped. It was found that the heat transfer reduced the surface of a smaller curvature and intensified for larger curvatures. A more convective fluid and a larger curvature better opposed the shear stress. Finally, when considering the influence of the freezing temperature and the molecular diameter, the nanofluid flow model is very useful for heat transfer in comparison with existing studies.

Author Contributions: Conceptualization, U.K. and A.A.; methodology, Adnan, N.A.; software, S.N.; validation, writing—review and editing, I.K. and W.A.K.; supervision, S.T.M.-D; Funding acquisition, S.O.A. All the authors contributed to the manuscript equally.

Funding: This research received no external funding.

Conflicts of Interest: The authors declare no conflict of interest.

Nomenclature

u	Component of the velocity
a	Radius
T	Temperature
T_∞	Temperature far from the surface
k_f	Thermal conductivity of the host fluid
k_s	Thermal conductivity of the nanoparticles
k_{nf}	Effective thermal conductivity of the nanofluid
$(C_p)_f$	Heat capacity of the host fluid
σ_s	Electrical conductivity of the nanoparticles
σ_{nf}	Electrical conductivity of the nanofluid
$\mu)\_f$	Dynamic viscosity of the fluid
M^*	Molecular weight
d_f	Molecular diameter
β_w	Temperature ratio parameter
Pr	Prandtl number
M	Hartmann number
$L(\eta)$	Dimensionless velocity
Nu	Nusselt number
v	Component of the velocity
p	Pressure
T_w	Temperature at the surface
ρ_f	Density of the host fluid
ρ_s	Density of the nanoparticles
ρ_{nf}	Effective density of the nanofluid
$(C_p)\_s$	Heat capacity of the nanoparticles
$(C_p)_{nf}$	Heat capacity of the nanofluid
σ_f	Electrical conductivity of the host fluid
μ_{nf}	Effective dynamic viscosity
ϕ	Volume fraction of the nanoparticles
N^*	Avogadro number
k_b	Stefan Boltzmann constant
Rd	Radiation parameter
Ec	Eckert number
K	Curvature parameter
$N(\eta)$	Dimensionless temperature
C_f	Skin fraction coefficient

References

1. Choi, S. Enhancing thermal conductivity of fluids with nanoparticles in developments and applications of non-newtonians flows. *ASME J. Heat Transf.* **1995**, *66*, 99–105.
2. Clerk, M.J. *Treatise on Electricity and Magnetism*; Oxford University Press: Oxford, UK, 1873.
3. Bruggeman, D.A.G. Berechnung verschiedener physikalischer konstanten von heterogenen substanzen, I—dielektrizitatskonstanten und leitfahigkeiten der mischkorper aus isotropen substanzen. *Ann. Phys. Leipz.* **1935**, *24*, 636–679. [CrossRef]
4. Hamilton, H.L.; Crosser, O.K. Thermal conductivity of heterogeneous two-component systems. *Ind. Eng. Chem. Fundam.* **1962**, *1*, 187–191. [CrossRef]
5. Lu, S.; Lin, H. Effective conductivity of composites containing aligned spherical inclusions of finite conductivity. *J. Appl. Phys.* **1996**, *79*, 6761–6769. [CrossRef]
6. Koo, J.; Kleinstreuer, C. A new thermal conductivity model for nanofluids. *J. Nanopart. Res.* **2004**, *6*, 577–588. [CrossRef]

7. Koo, J.; Kleinstreuer, C. Laminar nanofluid flow in micro-heat sinks. *Int. J. Heat Mass Transf.* **2005**, *48*, 2652–2661. [CrossRef]
8. Xue, Q.Z. Model for thermal conductivity of carbon nanotube-based composites. *Phys. B Phys. Condens. Matter.* **2005**, *368*, 302–307. [CrossRef]
9. Prasher, R.; Bhattacharya, P.; Phelan, P.E. Thermal conductivity of nanoscale colloidal solutions (nanofluids). *Phys. Rev. Lett.* **2005**, *92*, 25901. [CrossRef]
10. Li, C.H.; Peterson, G.P. Experimental investigation of temperature and volume fraction variations on the effective thermal conductivity of nanoparticle suspensions (nanofluids). *J. Appl. Phys.* **2006**, *99*. [CrossRef]
11. Corcione, M. Rayleigh–Bénard convection heat transfer in nanoparticle suspensions. *Int. J. Heat Fluid Flow* **2011**, *32*, 65–77. [CrossRef]
12. Sheikholeslami, M.; Li, Z.; Shamlooei, M. Nanofluid MHD natural convection through a porous complex shaped cavity considering thermal radiation. *Phys. Lett. A* **2018**, *382*, 1615–1632. [CrossRef]
13. Ahmed, N.; Khan, A.U.; Mohyud-Din, S.T. Influence of an effective prandtl number model on squeezed flow of $\gamma Al\_2 O\_3$-$H\_2 O$ and $\gamma Al\_2 O\_3$-$C\_2 H\_6 O\_2$ nanofluids. *J. Mol. Liq.* **2017**, *238*, 447–454. [CrossRef]
14. Sheikholeslami, M.; Zia, Q.M.Z.; Ellahi, R. Influence of induced magnetic field on free convection of nanofluid considering Koo-Kleinstreuer-Li (KKL) correlation. *Appl. Sci.* **2016**, *6*, 324. [CrossRef]
15. Asadullah, A.M.; Khan, U.; Naveed, A.; Mohyud-Din, S.T. Analytical and numerical investigation of thermal radiation effects on flow of viscous incompressible fluid with stretchable convergent/divergent channels. *J. Mol. Liq.* **2016**, *224*, 768–775.
16. Khan, U.; Naveed, A.A.; Mohyud-Din, S.T. 3D squeezed flow of $\gamma Al\_2 O\_3$-$H\_2 O$ and $\gamma Al\_2 O\_3$-$C\_2 H\_6 O\_2$ nanofluids: A numerical study. *Int. J. Hydrog. Energy* **2017**, *42*, 24620–24633. [CrossRef]
17. Iijima, S. Helical microtubules of graphitic carbon. *Nature* **1991**, *354*, 56–58. [CrossRef]
18. Naveed, A.A.; Khan, U.; Mohyud-Din, S.T. Influence of thermal radiation and viscous dissipation on squeezed flow of water between two riga plates saturated with carbon nanotubes. *Colloids Surf. A Physciochem. Eng. Asp.* **2017**, *522*, 389–398.
19. Saba, F.; Naveed, A.; Hussain, S.; Khan, U.; Mohyud-Din, S.T.; Darus, M. Thermal analysis of nanofluid flow over a curved stretching surface suspended by carbon nanotubes with internal heat generation. *Appl. Sci.* **2018**, *8*, 395. [CrossRef]
20. Khan, U.; Naveed, A.; Mohyud-Din, S.T. Heat transfer effects on carbon nanotubes suspended nanofluid flow in a channel with non-parallel walls under the effect of velocity slip boundary condition: A numerical study. *Neural Comput. Appl.* **2017**, *28*, 37–46. [CrossRef]
21. Reddy, J.V.R.; Sugunamma, V.; Sandeep, N. Dual solutions for nanofluid flow past a curved surface with nonlinear radiation, soret and dufour effects. *J. Phys. Conf. Ser.* **2018**, *1000*, 12152. [CrossRef]
22. Hayat, T.; Qayyum, S.; Imtiaz, M.; Alsaedi, A. Double stratification in flow by curved stretching sheet with thermal radiation and joule heating. *J. Therm. Sci. Eng. Appl.* **2017**, *10*. [CrossRef]
23. Alsabery, A.I.; Sheremet, M.A.; Chamkha, A.J.; Hashim, I. MHD convective heat transfer in a discretely heated square cavity with conductive inner block using two-phase nanofuid model. *Sci. Rep.* **2018**, *8*, 1–23. [CrossRef]
24. Corcione, M. Empirical correlating equations for predicting the efective thermal conductivity and dynamic viscosity of nanofuids. *Energy Convers. Manag.* **2011**, *52*, 789–793. [CrossRef]
25. Naveed, A.A.; Khan, U.; Mohyud-Din, S.T. Unsteady radiative flow of chemically reacting fluid over a convectively heated stretchable surface with cross-diffusion gradients. *Int. J. Therm. Sci.* **2017**, *121*, 182–191.

© 2019 by the authors. Licensee MDPI, Basel, Switzerland. This article is an open access article distributed under the terms and conditions of the Creative Commons Attribution (CC BY) license (http://creativecommons.org/licenses/by/4.0/).

Article

Peristaltic Pumping of Nanofluids through a Tapered Channel in a Porous Environment: Applications in Blood Flow

J. Prakash [1], Dharmendra Tripathi [2,*], Abhishek Kumar Tiwari [3], Sadiq M. Sait [4] and Rahmat Ellahi [5,6]

1. Department of Mathematics, Avvaiyar Government College for Women, Karaikal 609602, Puducherry-U.T., India
2. Department of Mathematics, National Institute of Technology, Uttarakhand 246174, India
3. Department of Applied Mechanics, MNNIT Allahabad, Prayagraj, Uttar Pradesh 211004, India
4. Center for Communications and IT Research, Research Institute, King Fahd University of Petroleum & Minerals, Dhahran 31261, Saudi Arabia
5. Center for Modeling & Computer Simulation, Research Institute, King Fahd University of Petroleum & Minerals, Dhahran 31261, Saudi Arabia
6. Department of Mathematics & Statistics, Faculty of Basic and Applied Sciences (FBAS), International Islamic University (IIUI), Islamabad 44000, Pakistan
* Correspondence: dtripathi@nituk.ac.in

Received: 22 May 2019; Accepted: 25 June 2019; Published: 3 July 2019

Abstract: In this study, we present an analytical study on blood flow analysis through with a tapered porous channel. The blood flow was driven by the peristaltic pumping. Thermal radiation effects were also taken into account. The convective and slip boundary conditions were also applied in this formulation. These conditions are very helpful to carry out the behavior of particle movement which may be utilized for cardiac surgery. The tapered porous channel had an unvarying wave speed with dissimilar amplitudes and phase. The non-dimensional analysis was utilized for some approximations such as the proposed mathematical modelling equations were modified by using a lubrication approach and the analytical solutions for stream function, nanoparticle temperature and volumetric concentration profiles were obtained. The impacts of various emerging parameters on the thermal characteristics and nanoparticles concentration were analyzed with the help of computational results. The trapping phenomenon was also examined for relevant parameters. It was also observed that the geometric parameters, like amplitudes, non-uniform parameters and phase difference, play an important role in controlling the nanofluids transport phenomena. The outcomes of the present model may be applicable in the smart nanofluid peristaltic pump which may be utilized in hemodialysis.

Keywords: peristaltic transport; tapered channel; porous medium; smart pumping for hemodialysis; thermal radiation

1. Introduction

Peristaltic motion [1–6] is a fundamental physiological mechanism which has many applications in bio-mechanical and engineering sciences where transport phenomena at micro/macro level occur. This mechanism is also applicable in transporting the nanofluids without any contaminations. The nanofluid term was first invented by Choi and Eastman [7] with reference to a conventional heat transfer liquid retention distribution of a nanosized particle. The behavior of nanoliquid in thermal conductivity enhancement has been observed by Masuda et al. [8] and experimental results of nanoparticle into pure fluid may conduct to reduce in heat transfer. The closed form model for convective transport in nanofluids studying the thermophoresis and Brownian diffusion has been studied by Buongiorno and Hu [9] and Buongiorno [10].

In the peristaltic pumping models, nanoliquid was introduced by Akbar and Nadeem [11]. They investigated endoscopic influences on the peristaltic motion of a nanofluid. Akbar [12] further presented the peristaltic transport of a Sisko nanoliquid in an asymmetric channel. It was noticed that enhances in the Sisko nanoliquid parameter axial pressure rise in the peristaltic pumping region. The effect of nanoliquid features on peristaltic heat transfer in a two-dimensional axisymmetric channel was discussed by Tripathi and Beg [13]. They examined that the nanoliquids incline to suppress backflow equated with Newtonian fluids. Akbar et al. [14] discussed the magnetohydrodynamic (MHD) peristaltic motion of a Carreau nanoliquid in an asymmetric channel. Furthermore, Beg and Tripathi [15] introduced the double diffusion process in peristaltic pumping. They discussed the salute and nanoparticle concentrations in their analysis. The effects of nanoparticle geometry on peristaltic motion has been analyzed by Akbar et al. [16]. MHD peristaltic pumping with viscoelastic nanofluids have been studied by Reddy and Makinde [17]. The velocity and slip influences on peristaltic pumping of nanoliquids have been examined by the Akbar et al. [18]. Heat and mass transfer analysis on peristaltic pumping through the rectangular duct was presented by Nadeem et al. [19]. Peristaltic transport of Prandtl nanofluid through the rectangular duct was studied by Ellahi et al. [20] and with magnetic field [21]. Hyperbolic tangent nanofluid with peristaltic pumping was implemented by Kothandapani and Prakash [22] in the presence of a radiation parameter and inclined magnetic field. Peristaltic pumping by eccentric cylinders has been discussed by the Nadeem et al. [23]. In similar directions, many more investigations [24–37] on peristaltic pumping, nanofluids and non-Newtonian nanofluids with various physical constraints and various flow geometries had been described in the literature.

The analysis of fluid flow through porous channels or tubes had gained attention recently because of its several applications in biomedical engineering and many other engineering areas like the flow of blood oxygenators, gall bladder with stones, in small blood vessels, the design of filters, in transpiration cooling boundary layer control, the flow of blood in the capillaries, the dialysis of blood in artificial kidney, gaseous diffusion in the spreading of fatty cholesterol and artery-clogging blood clots in the lumen of a coronary artery [38–47]. The steady laminar incompressible free convective flow of a nanofluid over a permeable upward facing horizontal plate located in a porous medium in an existence of thermal convective boundary condition was considered numerically by Uddin et al. [48]. Chamkha et al. [49] studied the mixed convection boundary layer flow in the existence of laminar and isothermal vertical porous medium. The onset of convection in a horizontal layer of a porous medium by a nanofluid was analytically studied by Kuznetsov and Nield [50]. Akbar [51] investigated the double-diffusive peristaltic transport of Jeffrey nanoliquids in a porous region in the presence of natural convective. Double-diffusive natural convective peristaltic flow of a Jeffrey nanofluid in a porous channel has been analyzed by Nadeem et al. [52] and investigated the peristaltic flow of nanofluid eccentric tubes which comprises a porous medium. Two-phase flow driven by the peristaltic pumping through porous medium was studied by Bhatti et al. [53]. Perturbation solutions have been obtained and it is observed that chemical reaction and Soret numbers oppose the particle concentration. The applications of porosity can be deeply studied by using nanofluid model in [54,55].

Moreover, the no-slip condition is inadequate when a fluid revealing macroscopic wall slip is considered and that, in general, is governed by the relation between the slip velocity and grip. The slip condition plays significant role in shear skin, spurt and hysteresis belongings. The nanofluids that exhibit boundary slip have vital technological purposes such as in shining valves of the artificial heart and internal holes. The proposed mathematical geometry is very similar to the blood vessel models. The blood vessels can be classified into three types: the largest vessels, small vessels and intermediate blood vessels. The largest vessels are identified in the aorta and vena cava and also experience very little heat transfer with the tissue. In addition, there are also more blood vessels that fall into this category. The smallest vessels are noticed in place of arterioles, capillaries and venules which basically experience ideal heat transfer with the blood departure at tissue temperature. The intermediate blood vessels fall in a relatively narrow band with uniformly distributed temperature. These classifications are dependent on the amplitude of vessels and width of channel. Hence, the main purpose of this paper is to study a theoretical analysis of peristaltic transport of a Newtonian nanofluid with slip through a porous medium in the tapered wavy channel subject to convective boundary conditions. The long wavelength and low Reynolds number assumptions are considered. The exact solutions are found in the form of axial velocity from which temperature and volumetric concentration are deduced. Computational results are illustrated and discussed in detail.

2. Mathematical Formulation

Consider an incompressible viscous nanofluid filling the porous space in the tapered wavy channel. The heat transfer between the blood network and living tissues which passes through the channel depends on the geometry of the blood vessel and it is important to understand the behavior of the blood flow and the neighboring tissue nature. Let $\bar{\eta} = \bar{H}_1$ and $\bar{\eta} = \bar{H}_2$ be, correspondingly, the lower and upper blood vessel boundaries of the channel. The sinusoidal waves propagating along the wavy walls of the tapered channel are demonstrated in Figure 1 and mathematically shown as:

$$\begin{aligned}\bar{H}_2(\bar{\xi}, t') &= \bar{d} + \bar{m}\bar{\xi} + a_2 \sin\left[\tfrac{\pi}{\lambda}(\bar{\xi} - ct')\right]\cos\left[\tfrac{\pi}{\lambda}(\bar{\xi} - ct')\right], \\ \bar{H}_1(\bar{\xi}, t') &= -\bar{d} - \bar{m}\bar{\xi} - a_1 \sin\left[\tfrac{\pi}{\lambda}(\bar{\xi} - ct') + \phi\right]\cos\left[\tfrac{\pi}{\lambda}(\bar{\xi} - ct') + \phi\right]\end{aligned} \qquad (1)$$

here $2\bar{d}, a_1, a_2, \bar{m}(<<1), \lambda, c, \phi$, are the width of the channel at the inlet, amplitudes of lower wavy wall, amplitude of upper wavy wall, dimensional non–uniform parameter, wave length, phase speed of the wave and phase difference varies in the range $0 \le \phi \le \pi$, $\phi = 0$ which corresponds to tapered symmetric channel i.e., together walls move towards inward or outward concurrently.

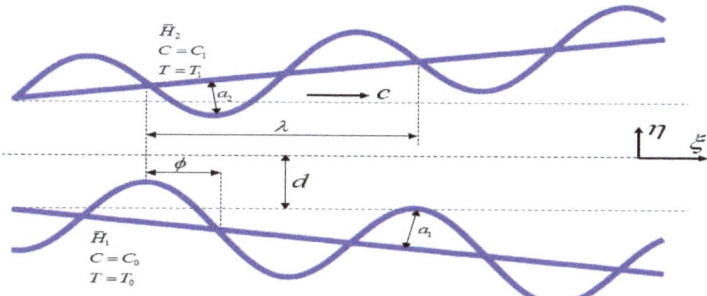

Figure 1. Geometry for peristaltic pumping of nanofluids through a tapered microchannel.

For an incompressible viscous nanofluid the balance of mass, momentum, nanoparticle temperature and volumetric concentration are presented as [56–59]:

$$\frac{\partial \bar{U}}{\partial \bar{\xi}} + \frac{\partial \bar{V}}{\partial \bar{\eta}} = 0, \qquad (2)$$

$$\rho_f\left[\frac{\partial}{\partial t'} + \bar{U}\frac{\partial}{\partial \bar{\xi}} + \bar{V}\frac{\partial}{\partial \bar{\eta}}\right]\bar{U} = -\frac{\partial \bar{P}}{\partial \bar{\xi}} + \mu\left[\frac{\partial^2}{\partial \bar{\xi}^2} + \frac{\partial^2}{\partial \bar{\eta}^2}\right]\bar{U} - \frac{\mu}{k}\bar{U}, \qquad (3)$$

$$\rho_f\left[\frac{\partial}{\partial t'} + \bar{U}\frac{\partial}{\partial \bar{\xi}} + \bar{V}\frac{\partial}{\partial \bar{\eta}}\right]\bar{V} = -\frac{\partial \bar{P}}{\partial \bar{\eta}} + \mu\left[\frac{\partial^2}{\partial \bar{\xi}^2} + \frac{\partial^2}{\partial \bar{\eta}^2}\right]\bar{V} - \frac{\mu}{k}\bar{V}, \qquad (4)$$

$$(\rho c')_f \left[\frac{\partial}{\partial t'} + \overline{U}\frac{\partial}{\partial \xi} + \overline{V}\frac{\partial}{\partial \eta}\right]\overline{T} = \kappa\left[\frac{\partial^2}{\partial \xi^2} + \frac{\partial^2}{\partial \eta^2}\right]\overline{T} - \frac{\partial q_r}{\partial \eta} + \mu\left[4\left(\frac{\partial \overline{U}}{\partial \xi}\right)^2 + \left(\frac{\partial \overline{V}}{\partial \xi} + \frac{\partial \overline{U}}{\partial \eta}\right)^2\right]$$
$$+ (\rho c')_p\left[D_B\left(\frac{\partial \overline{C}}{\partial \xi}\frac{\partial \overline{T}}{\partial \xi} + \frac{\partial \overline{C}}{\partial \eta}\frac{\partial \overline{T}}{\partial \eta}\right) + \frac{D_T}{T_m}\left[\left(\frac{\partial \overline{T}}{\partial \xi}\right)^2 + \left(\frac{\partial \overline{T}}{\partial \eta}\right)^2\right]\right], \quad (5)$$

$$\left[\frac{\partial}{\partial t'} + \overline{U}\frac{\partial}{\partial \xi} + \overline{V}\frac{\partial}{\partial \eta}\right]\overline{C} = D_B\left[\frac{\partial^2}{\partial \xi^2} + \frac{\partial^2}{\partial \eta^2}\right]\overline{C} + \frac{D_T}{T_m}\left[\frac{\partial^2}{\partial \xi^2} + \frac{\partial^2}{\partial \eta^2}\right]\overline{T}, \quad (6)$$

in which $\overline{U}$, $\overline{V}$ are the components of axial velocity along $\overline{\xi}$ and $\overline{\eta}$ directions correspondingly, t', d/dt', $\overline{P}$, ρ_f, ρ_p, $\overline{T}$, κ, k, D_B, D_T, T_m, $\overline{C}$ and $\tau\left(=\frac{(\rho c')_p}{(\rho c')_f}\right)$ are the dimensional time, material time derivative, dimensional pressure, density of the fluid, density of the particle, nanoparticle temperature, thermal conductivity, permeability of porous medium, Brownian diffusion coefficient, themophoretic diffusion coefficient, mean temperature, nanoparticle volumetric volume fraction and the ratio of the effective heat capacity of nanoparticle material and heat capacity of the fluid with ρ being the density. Additionally, T_0, T_1, C_0 and C_1 are the temperature and nanoparticle volume fraction at the lower and upper walls.

3. Convective Boundary Conditions

The convective boundary conditions [60,61] are utilized using Newton's cooling law as:

$$\overline{U} = \frac{\sqrt{\overline{k}}}{\overline{\alpha}}\frac{\partial \overline{U}}{\partial \eta}, \quad -\overline{k}_h\frac{\partial \overline{T}}{\partial \eta} = \overline{h}_h(T_0 - \overline{T}) \text{ and } -\overline{k}_m\frac{\partial \overline{C}}{\partial \eta} = \overline{h}_m(C_0 - \overline{C}) \text{ at } \overline{\eta} = \overline{H}_1, \quad (7)$$

$$\overline{U} = -\frac{\sqrt{\overline{k}}}{\overline{\alpha}}\frac{\partial \overline{U}}{\partial \eta}, \quad -\overline{k}_h\frac{\partial \overline{T}}{\partial \eta} = \overline{h}_h(\overline{T} - T_1) \text{ and } -\overline{k}_m\frac{\partial \overline{C}}{\partial \eta} = \overline{h}_m(\overline{C} - C_1) \text{ at } \overline{\eta} = \overline{H}_2, \quad (8)$$

where $\overline{k}$, $\overline{\alpha}$, $\overline{h}_h$, $\overline{h}_m$, $\overline{k}_h$ and $\overline{k}_m$ are the permeability of the porous walls (Darcy number), slip coefficient at the surface of the porous walls, the heat transfer coefficients, mass transfer coefficients respectively, the thermal conductivity and the mass conductivity.

4. Non-Dimensional Analysis

In order to depict the nanoliquid flow in the following non-dimensional measures are introduced in Equations (1)–(8). $\left(u = \frac{\overline{U}}{c}, u = \frac{\partial \psi}{\partial y}, v = \frac{\overline{V}}{c}, v = -\delta\frac{\partial \psi}{\partial y}\right)$ are the velocity components in direction of $\left(x = \frac{\overline{\xi}}{\lambda}, y = \frac{\overline{\eta}}{d}\right)$, ψ is the stream function, $t = \frac{ct'}{\lambda}$ is the dimensionless time, $h_1 = \frac{\overline{H}_1}{d}$ and $h_2 = \frac{\overline{H}_2}{d}$ represent the dimensionless form of the lower and upper channel, and $R = \frac{\rho_f c d}{\mu}$ is the Reynolds number, $p = \frac{d^2 \overline{P}}{c\lambda\mu}$ is the dimensionless pressure, $a = \frac{a_1}{d}$ is the amplitude of lower wavy wall, $b = \frac{a_2}{d}$ is the amplitude of upper wavy wall, $m = \frac{\lambda \overline{m}}{d}$ is the dimensionless non-uniform parameter, $\delta = \frac{d}{\lambda}$ is the wave number, $Sc = \frac{\nu}{D_B}$ is the Schmidt number, $K = \frac{\overline{k}}{d^2}$ is the Permeability parameter, $\theta = \frac{\overline{T}-T_0}{T_1-T_0}$ is the dimensionless nanoparticle temperature, $\sigma = \frac{\overline{C}-C_0}{C_1-C_0}$ is the nanoparticle volumetric concentration or dimensionless rescaled nanoparticle volume fraction, $Pr = \frac{\mu c_f}{\kappa}$ is the Prandtl number, $N_b = \frac{\tau D_B(C_1-C_0)}{\nu}$ is the Brownian motion parameter $N_t = \frac{\tau D_T(T_1-T_0)}{T_m \nu}$ is the thermophoresis parameter, $Ec = \frac{c^2}{c_f T_m}$ is the Eckert number, $Br = PrEc$ is the Brinkman number, and $R_n = \frac{16\overline{\sigma}T_0^3}{3\overline{k}\mu c_f}$ is the thermal radiation parameter and also, applying the long wavelength and low Reynolds number approximations, we attain:

$$\frac{\partial p}{\partial x} = \frac{\partial^3 \psi}{\partial y^3} - \left(\frac{1}{K}\right)\frac{\partial \psi}{\partial y}, \quad (9)$$

$$\frac{\partial p}{\partial y} = 0, \tag{10}$$

$$(1 + R n \Pr)\frac{\partial^2 \theta}{\partial y^2} + Br\left(\frac{\partial^2 \psi}{\partial y^2}\right)^2 + (Nb\Pr)\left(\frac{\partial \sigma}{\partial y}\frac{\partial \theta}{\partial y}\right) + (Nt\Pr)\left(\frac{\partial \theta}{\partial y}\right)^2 = 0, \tag{11}$$

$$\frac{\partial^2 \sigma}{\partial y^2} + \left(\frac{Nt}{Nb}\right)\frac{\partial^2 \theta}{\partial y^2} = 0. \tag{12}$$

Equation (10) shows that p is not dependent on x. Reducing the pressure gradient term from Equations (9) and (10), it yields:

$$\frac{\partial^4 \psi}{\partial y^4} - \left(\frac{1}{K}\right)\frac{\partial^2 \psi}{\partial y^2} = 0. \tag{13}$$

Additionally, it is noticed that the continuity equation is routinely fulfilled.

$$h_1 = -1 - mx - a\cos(\pi(x-t) + \phi)\sin(\pi(x-t) + \phi) \text{ and}$$
$$h_2 = 1 + mx + b\cos(\pi(x-t))\sin(\pi(x-t))$$

$$\psi = -\frac{F}{2}, \frac{\partial \psi}{\partial y} = L\frac{\partial^2 \psi}{\partial y^2}, \frac{\partial \theta}{\partial y} = B_h \theta \text{ and } \frac{\partial \sigma}{\partial y} = B_m \sigma \text{ at } y = h_1,$$
$$\psi = \frac{F}{2}, \frac{\partial \psi}{\partial y} = -L\frac{\partial^2 \psi}{\partial y^2}, \frac{\partial \theta}{\partial y} = B_h(1-\theta) \text{ and } \frac{\partial \sigma}{\partial y} = B_m(1-\sigma) \text{ at } y = h_2. \tag{14}$$

where $L = \frac{\sqrt{k}}{d\bar{a}}$ is the velocity slip parameter, $B_h = \frac{\bar{h}_h \bar{d}}{k_h}$ is the heat transfer Biot number and $B_m = \frac{\bar{h}_m \bar{d}}{k_m}$ is the mass transfer Biot number.

5. Analytical Solution

The solution of the Equation (13) subject to the conditions in Equation (14) is obtained as:

$$\psi(y) = \begin{array}{l} -(F(\cosh Ny - \sinh Ny)(2(\cosh 2Ny - \sinh 2Ny) - 2(\cosh(N(h_1+h_2)) + \sinh(N(h_1+h_2)))) \\ +(h_1+h_2-2y)N(\cosh Ny + \sinh Ny)(\cosh Nh_1 + \sinh Nh_1 + \cosh Nh_2 + \sinh Nh_2) \\ -LN^2(\cosh Ny + \sinh Ny)(h_1+h_2-2y)(\cosh Nh_1 + \sinh Nh_1 - \cosh Nh_2 - \sinh Nh_2))) \\ \Big/ \left(2\Big((2+LN^2h_1-LN^2h_2)(\cosh Nh_1 + \sinh Nh_1 - \cosh Nh_2 - \sinh Nh_2) \\ -(Nh_1-Nh_2)(\cosh Nh_1 + \sinh Nh_1 + \cosh Nh_2 + \sinh Nh_2)\Big)\right) \end{array} \tag{15}$$

The integration of Equation (12) with respect to y, we obtain

$$\frac{\partial \sigma}{\partial y} + \frac{Nt}{Nb}\frac{\partial \theta}{\partial y} = f(x). \tag{16}$$

Solving Equations (11) and (12) and substituting in Equation (16) subject to boundary conditions of Equations (14), the dimensionless nanoparticle temperature field is attained as

$$\theta(y) = A_8 + A_9(\cosh(A_1 A_5 N_b y) - \sinh(A_1 A_5 N_b y)) - \frac{A_2^3 A_4(\cosh(2Ny) + \sinh(2Ny))}{4N^2 + (\cosh(2A_1 A_5 N_b N) + \sinh(2A_1 A_5 N_b N))} \\ - \frac{A_3^2 A_4(\cosh(2Ny) - \sinh(2Ny))}{4N^2 - (\cosh(2A_1 A_5 N_b N) + \sinh(2A_1 A_5 N_b N))} - \frac{(A_6 \beta + 2A_2 A_3 A_4)y}{A_1 A_5 N_b}, \tag{17}$$

and the nanoparticle volumetric concentration is obtained as:

$$\sigma(y) = A_{10} + A_{11}y - \frac{A_9 N_t(\cosh(A_1 A_5 N_b y) - \sinh(A_1 A_5 N_b y))e^{-A_1 A_5 N_b y}}{N_b} \\ + \frac{A_4 A_3^2 N_t(\cosh(2Ny) - \sinh(2Ny))}{4N^2 N_b - 2A_1 A_5 N N_b^2} + \frac{A_4 A_2^2 N_t(\cosh(2Ny) + \sinh(2Ny))}{4N^2 N_b + 2A_1 A_5 N N_b^2}. \tag{18}$$

The coefficient of nanoparticle heat transfer at the lower wall is specified by

$$Z = h_{1x}\theta_y. \tag{19}$$

The above mentioned constants are elaborated in the Appendix A.

6. Computational Results and Discussion

In general, exact solutions for temperature, nanoparticle volumetric concentrations and coefficient of nanoparticle temperature depend on the value of $f(x)$. First of all, $f(x)$ can be influenced by hiring for θ and σ from Equation (16). It ought to be noticed that observing the value of $f(x)$ analytically from Equation (16) in terms of the other parameters set is a very difficult task and it may be impossible. Nevertheless, with the help of MATHEMATICA/MATLAB software, the numerical solutions are still available. The numerical value of $f(x)$ plays an important role in plotting the graphs for variation of the streamlines, nanoparticle temperature distribution, nanoparticle volumetric concentration and coefficient of nanoparticle temperature.

To analyze the results, instantaneous volume rate $F(x,t)$ is considered as varying exponentially with the relation (Kikuchi [62])

$$F = \Theta e^{-At}, \tag{20}$$

where Θ is the mean flow rate or flow constant, A is the blood flow constant. Figures 2–20 were plotted to examine the stream function $(\psi(x,y))$, nanoparticle temperature $(\theta(y))$, nanoparticle concentration $(\sigma(y))$ and heat transfer coefficient $(Z(x))$. Additionally, it is noticed that the flow rate for the non-positive and positive flow rate $F < 0$ or $F > 0$ may be according to $\Theta < 0$ or $\Theta > 0$. It was detected through an experiment performed by Kikuchi [62] that the flow rate decreases exponentially with time however mean flow rate does not depend on the structural details of the channel.

6.1. Thermal and Concentration Profiles

Effects of permeability parameter (K), slip parameter (L), mean flow rate (Θ), non-uniform parameter (m), Brinkman number (Br), Prandtl number (Pr), heat transfer Biot number (B_h), mass transfer Biot number (B_m), thermal radiation (R_n) and thermophoresis parameter (N_t) on temperature profile are analyzed through Figures 2–11. In accordance with physical laws, the energy fluency requires to destroy cancer cells greatly depends on the number of nanoparticles temperature within the cell. Additionally, the role of this study is to improve correlations and estimation methods for calculating magnitudes of upper and lower tapered wavy wall boundaries of heat transfer in and around the individual blood vessels. The analysis did not consider vessel size and any experimental values because flow oscillations due to the heartbeat are not present in these small vessels. The aim of this study is to improve correlations and estimation methods for scheming magnitudes and upper and lower limits of heat and mass transfer in and around individual blood vessels.

The nanoparticle temperature and concentration profiles resulted in the vessel exit are shown in Figure 2 for three different permeability parameter values such as $(K \to 0, K = 0.2, K \to \infty)$. It is noticed, theoretically, the absence of a permeability parameter shows very few heat exchanges in the blood vessel, but the particles movement is raised in the blood vessel. In Figure 3, we noticed the effects of slip parameter (L) on the nanoparticle temperature and concentration profiles for fixed values of other parameters. Three different slip parameters are used in the nanoparticle temperature and concentration distribution such as $L = 0$, $L = 0.1$ and $L = 0.2$. It is important to note with the enhancement in the velocity slip parameter, the nanoparticle temperature and concentration at any point in the flow medium enhances, but the behavior of temperature profile decreases and at the same time nanoparticle concentration increases when the velocity slip parameter rises. The effects of flow constant (Θ) on nanoparticle temperature and concentration profiles are shown in Figure 4. It is noticed that presence of a flow constant increases the nanoparticle temperature and also enhances uniformly in the boundaries of the channel. However, the nanoparticle concentration shows the revised behavior in nature of temperature distribution. From Figure 5, which elucidates the effect of the non-uniform parameter (m) on the nanoparticle temperature and concentration profiles, it is exposed that when the non-uniform parameter increases, the nanoparticle temperature of blood flow consistently reduces with the flow medium. The nanoparticle displacement increases with increasing of the non-uniform parameter. These physical changes play crucial roles in the treatment of thermotherapy. In Figure 6, the causes of Brinkman number (Br) on nanoparticle temperature and concentration are captured. It is noticed that the temperature of the fluid increases with the increase of Brinkman number. It is well known about nanofluids that when the nanoparticle temperature rises, the distance between molecules increases due to cohesive force decreases. Therefore, viscosity of nanofluids decreases when the nanoparticle temperature increases. On the flip side, absence of Brinkman number shows the maximum displacement of the particles. The effect of the Prandtl number (Pr) on nanoparticle temperature and concentration are depicted in Figure 7. It is observed that with an increase in Pr, the temperature of the fluid increases. It indicates that nanofluids can have significantly better heat transfer characteristics than the base fluids. Additionally, it is noticed that the nanoparticle concentration decreases with increasing the Prandtl number. This indicates that enhances in Prandtl number is accompanied by an enrichment of the heat transfer rate at the tapered wavy wall of the blood vessel. The fundamental physics behind this can be depicted as follows. When the blood achieves a higher Prandtl number, its thermal conductivity is dropped down and so its heat conduction capacity is reduced. Simultaneously, the heat transfer rate at the vessel wall is enhanced. We noticed from Figure 8 that the temperature enhances with rise of heat transfer Biot number (B_h) at the upper portion of the channel, but the influence is reversed at the lower portion of the channel. Further, it can be noted that the temperature at the upper wall is maximum and it reduces slowly towards the lower wall. The small value of heat transfer Biot number shows the conduction nature, while high values of heat transfer Biot number indicates that the convection is the main heat transfer mechanism. Any rate of nanoparticle concentration reduces with increase of the heat transfer Biot number. Figure 9 reveals that the nanoparticle temperature and concentration enhance as mass transfer Biot number increases. Figure 10 illuminates the influence of the thermal radiation on nanoparticle temperature and concentration distribution. This figure highlights that thermal radiation enhances during blood flow in the channel, thereby the nanoparticle temperature of the tapered asymmetric wavy channel is reduced by increase of thermal radiation. Additionally, the converse situation occurred in the nanoparticle concentration profile. It shows that the external radiation dilutes the temperature, and at the same time movement of the particle increases. This concept may be very useful in the treatment of heart transfer mechanism. Figure 11 illuminates a very significant influence of the thermophoresis parameter on the nanoparticle temperature and concentration profiles. It is well known that the strength of thermophoresis rises due to temperature gradient enhancement, which increases the blood flow in the channel. At the same time, the nanoparticle concentration of the particle displacement reduces with increases of the thermophoresis parameter.

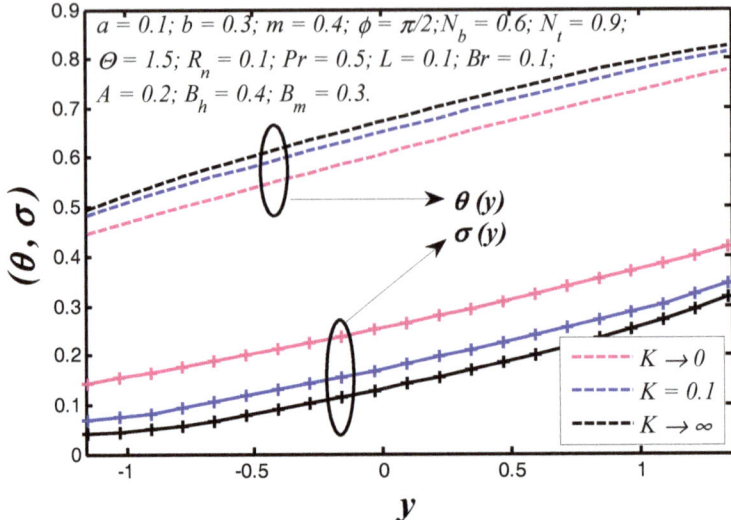

Figure 2. Nanoparticle temperature and concentration profiles $\theta(y)$ for K.

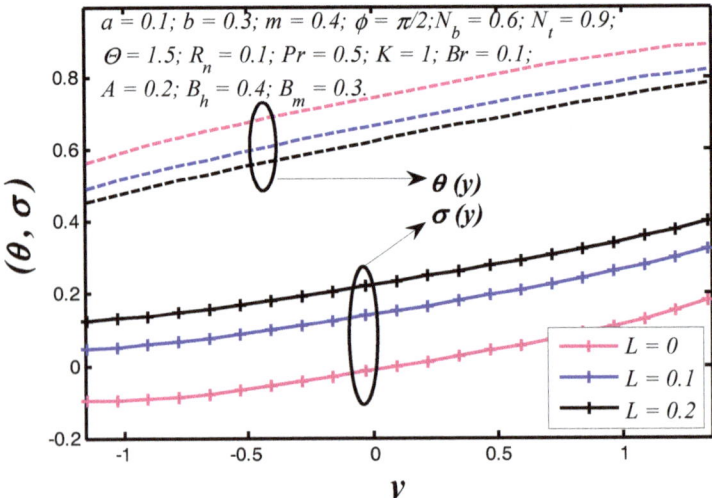

Figure 3. Nanoparticle temperature and concentration profiles $\theta(y)$ for L.

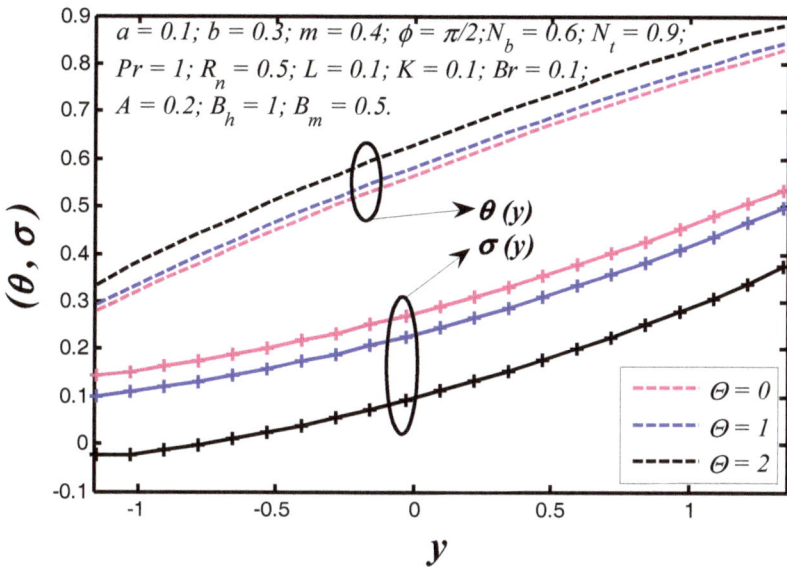

Figure 4. Nanoparticle temperature and concentration profiles $\theta(y)$ for Θ.

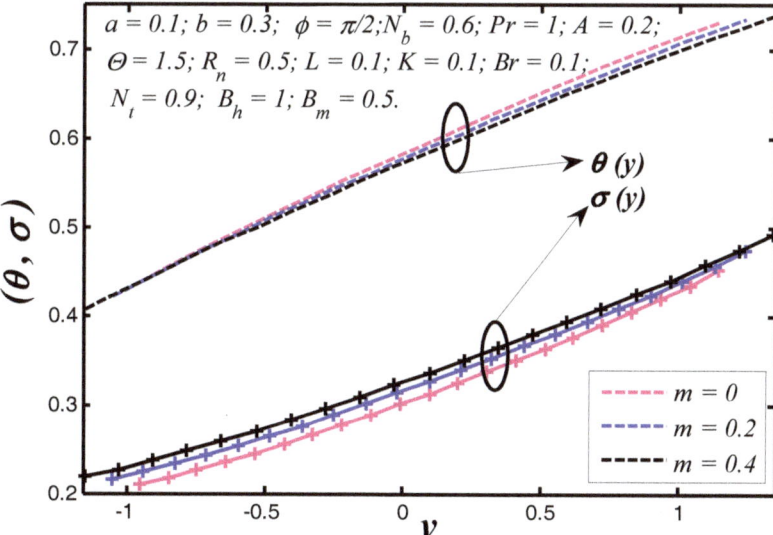

Figure 5. Nanoparticle temperature and concentration profiles $\theta(y)$ for m.

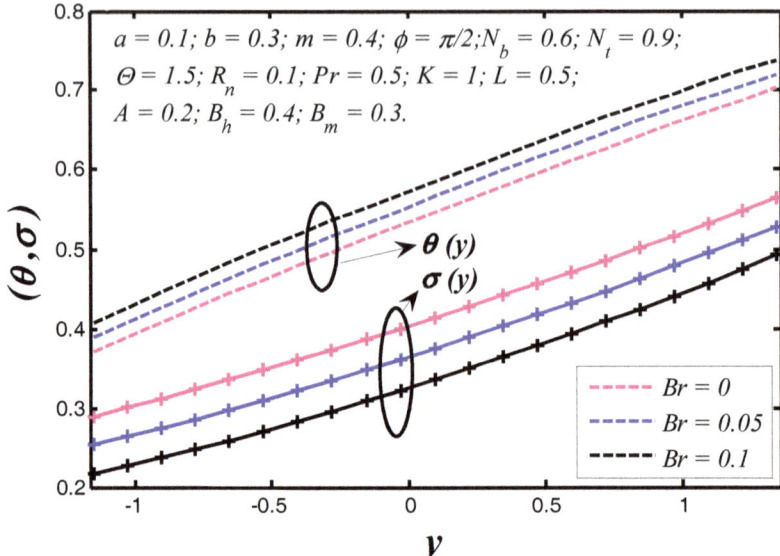

Figure 6. Nanoparticle temperature and concentration profiles $\theta(y)$ for Br.

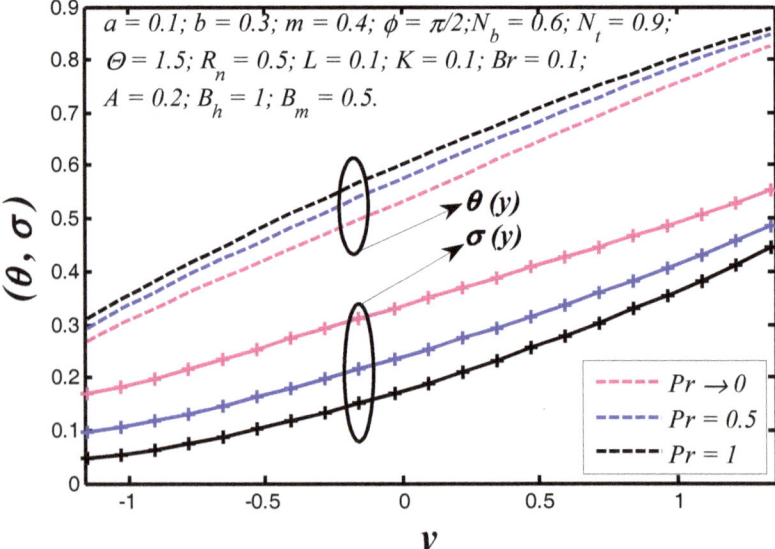

Figure 7. Nanoparticle temperature and concentration profiles $\theta(y)$ for Pr.

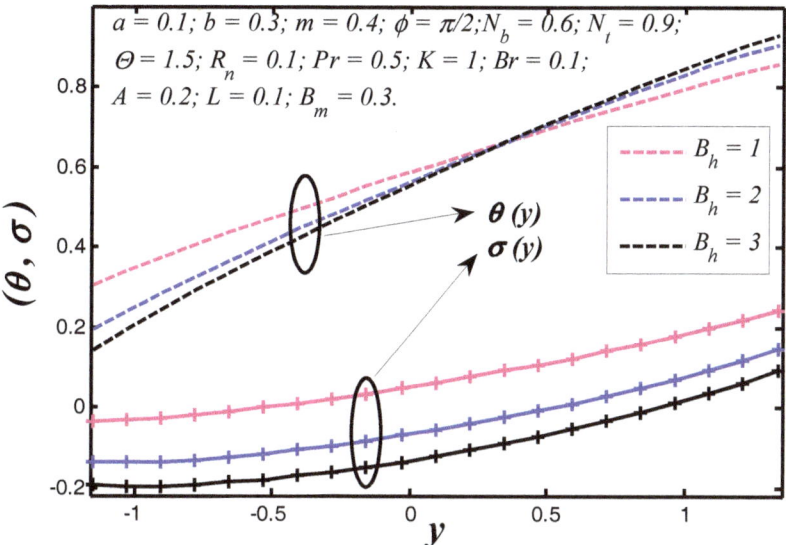

Figure 8. Nanoparticle temperature and concentration profiles $\theta(y)$ for B_h.

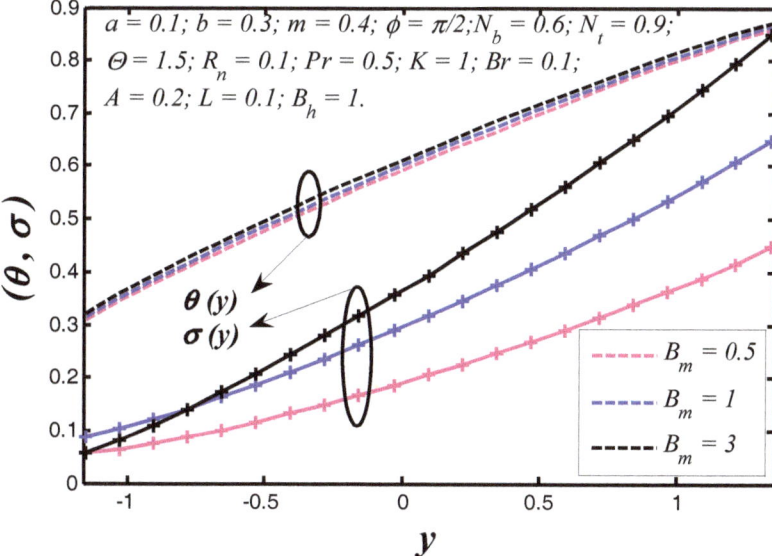

Figure 9. Nanoparticle temperature and concentration profiles $\theta(y)$ for B_m.

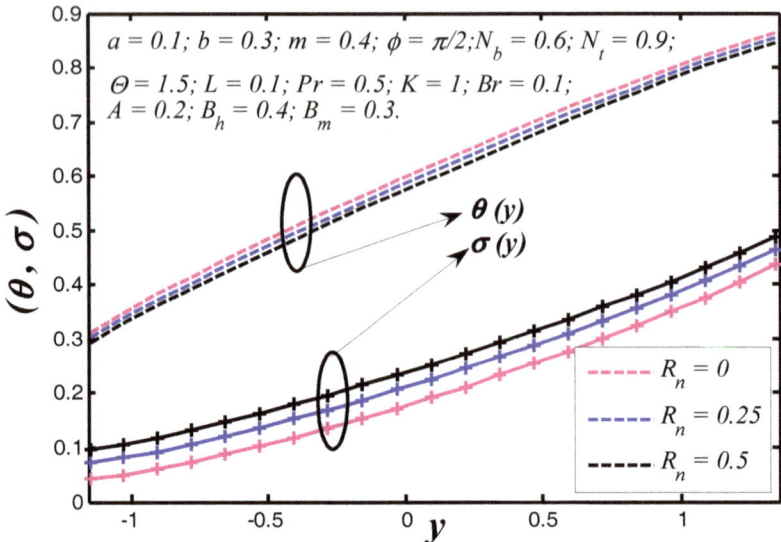

Figure 10. Nanoparticle temperature and concentration profiles $\theta(y)$ for R_n.

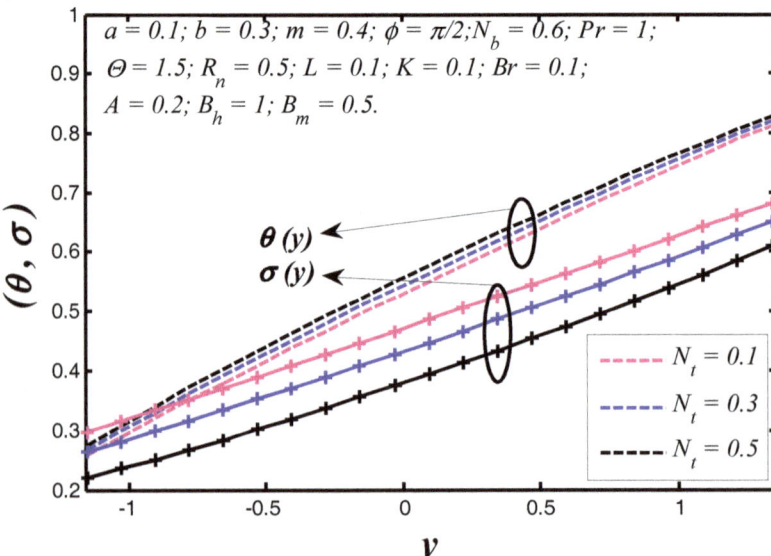

Figure 11. Nanoparticle temperature and concentration profiles $\theta(y)$ for N_t.

6.2. Nanoparticle Heat Transfer Coefficient

The effects of various parameters on the nanoparticle heat transfer coefficient at the upper wall are represented in Figures 12–18. The nanoparticle heat transfer coefficients for a viscous nanofluid in the tapered wavy channel depends on many physical quantities related to the fluid or the geometry of the system through which the fluid is flowing. It is observed that the heat transfer coefficient is in oscillatory behavior which may be due to contraction and equation of walls. The absolute value of heat transfer coefficient increases with the increase of L and Br while it decreases with increasing m, B_h, B_m, N_b and R_n.

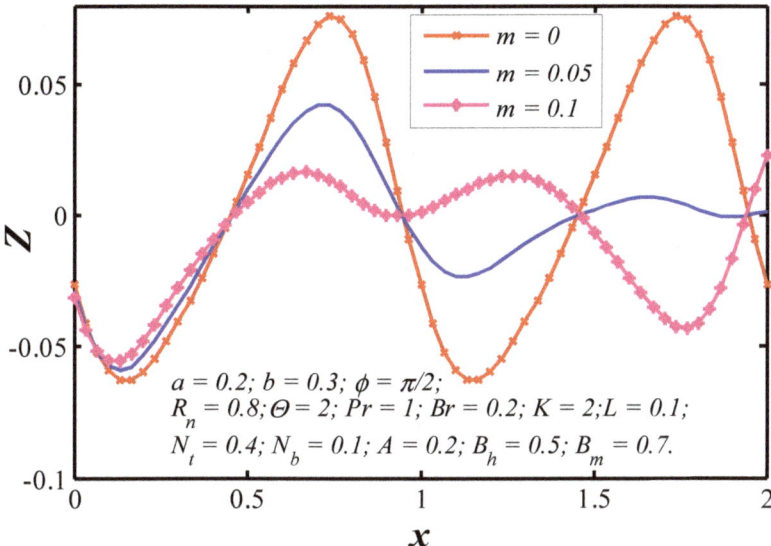

Figure 12. Nanoparticle heat transfer coefficient $Z(x)$ profiles for various values of m.

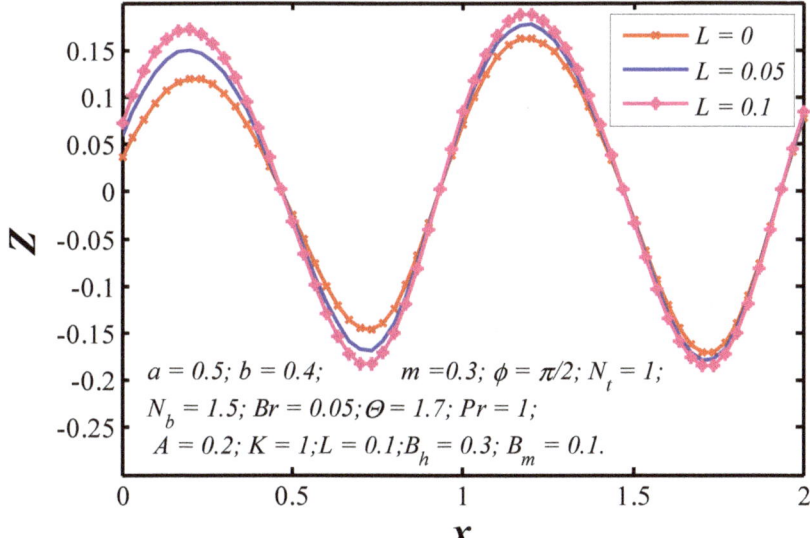

Figure 13. Nanoparticle heat transfer coefficient $Z(x)$ profiles for various values of L.

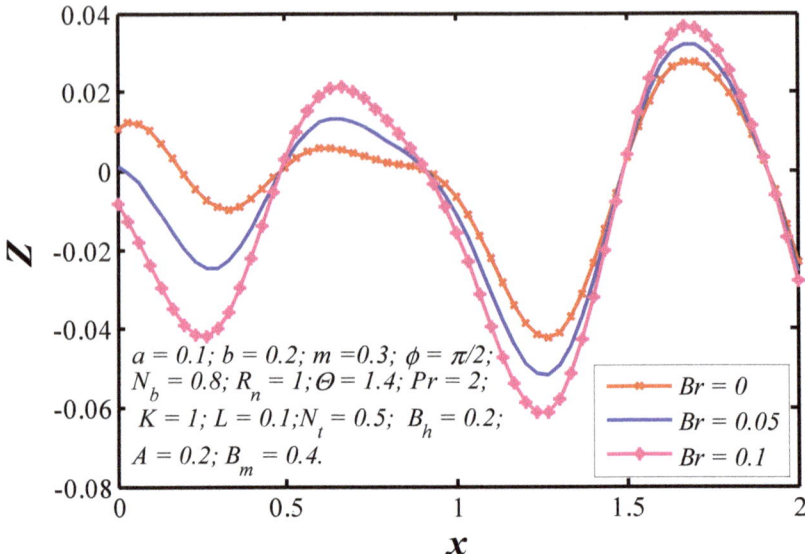

Figure 14. Nanoparticle heat transfer coefficient $Z(x)$ profiles for various values of Br.

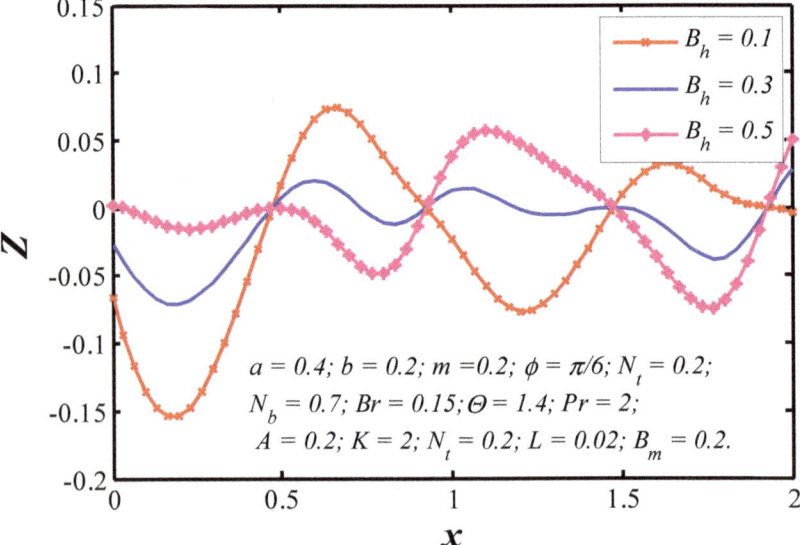

Figure 15. Nanoparticle heat transfer coefficient $Z(x)$ profiles for various values of B_h.

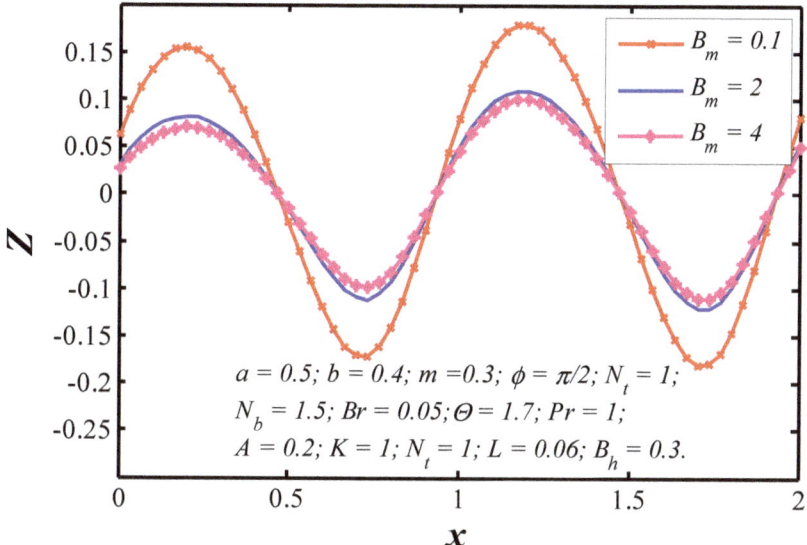

Figure 16. Nanoparticle heat transfer coefficient $Z(x)$ profiles for various values of B_m.

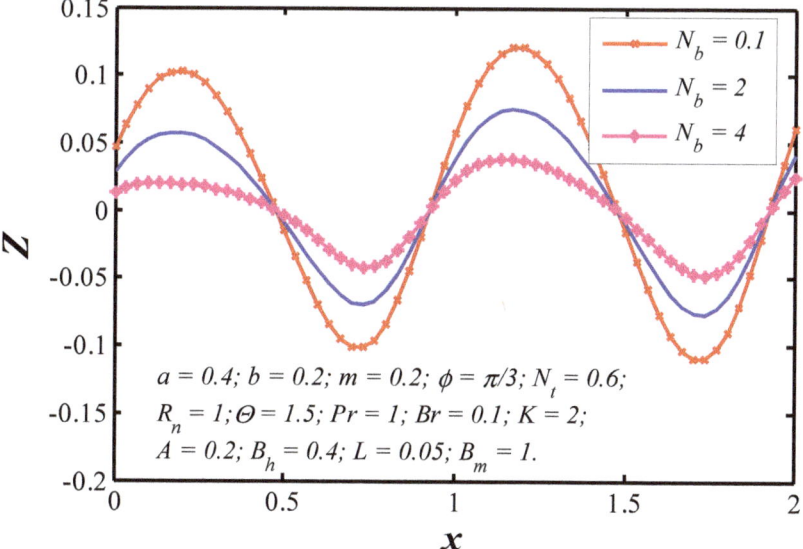

Figure 17. Nanoparticle heat transfer coefficient $Z(x)$ profiles for various values of N_b.

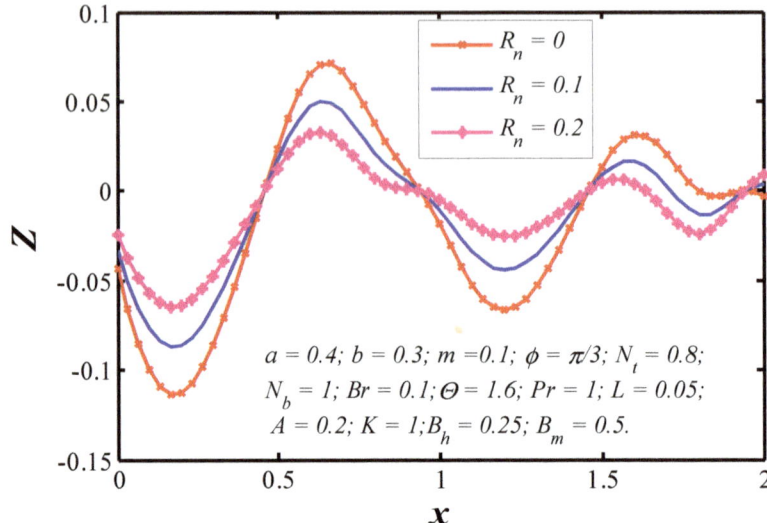

Figure 18. Nanoparticle heat transfer coefficient $Z(x)$ profiles for various values of R_n.

6.3. Trapping

In this subsection, the streamlines for the tapered asymmetry channels are shown in Figure 19. The effects of the non-uniform parameter m on trapping are presented in Figure 19a,b. One can observe that the size of the trapped bolus increases with an increase in m. The effect of slip parameter L on trapping can be seen in Figure 19b,c. It is observed that by increasing the value of velocity slip parameter L, the circulation of trapped bolus increases, at the same time size of the bolus is reduced. To see the effects of permeability parameter K on trapping, Figure 19c,d was illustrated. It is noted that an increase in the permeability parameter increases the size of channel, but the size of the trapped bolus decreases. The streamline patterns in the wave frame for viscous nanofluid for different values of Blood flow rate parameter Θ are shown in Figure 19d,e. It is observed that for small values of Θ only one trapped bolus is formed. It is also observed that the bolus near the upper and lower wavy walls increase eventually with the tapered micro channel.

6.4. Validation

The results of present mathematical model obtained by direct analytical approach were authenticated with the numerical solutions computed by MATLAB through BVP4c command. A validation was completed in Figure 20 and it is portrayed for nanoparticle temperature and concentration distribution at fixed values of pertinent parameters. Additionally, it is noticed that the analytical solution for the entire values of the tapered wavy channel width has a good correlation with the numerical solution computed by the MATLAB. The proposed mathematical formulation has very good correlation in axial velocity with Mishra and Rao [63] in the absence of $a = 0, b = 0$ and $K \to \infty$.

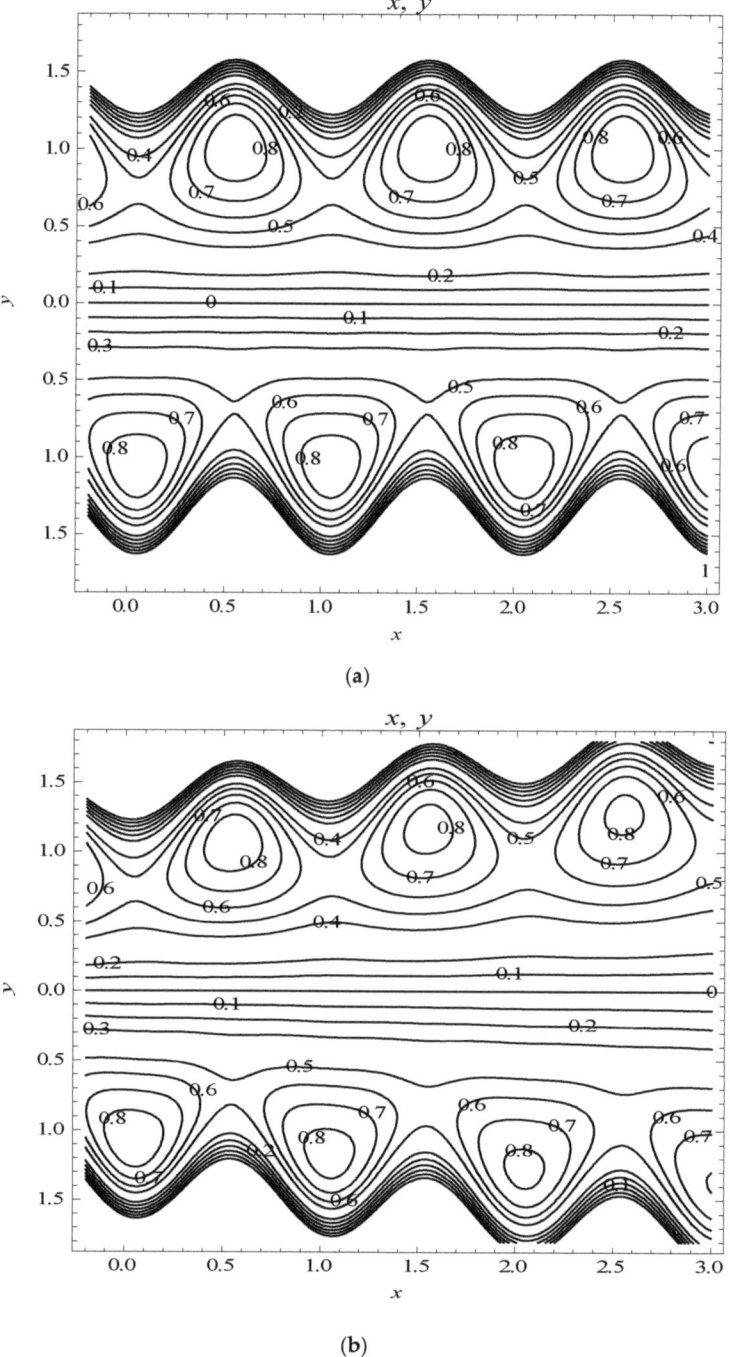

(a)

(b)

Figure 19. *Cont.*

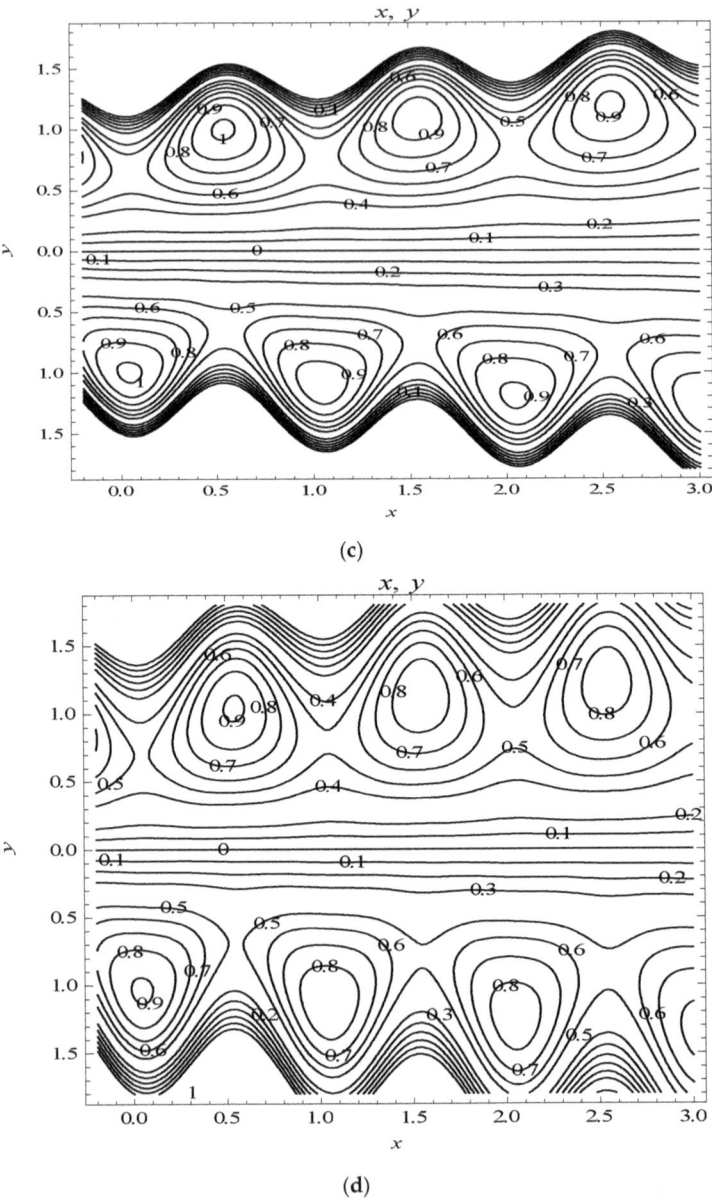

Figure 19. Cont.

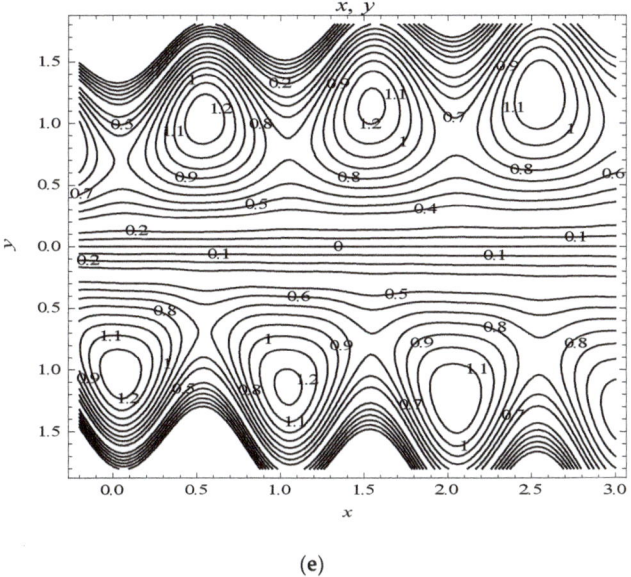

(e)

Figure 19. Streamlines when $a = 0.4$, $b = 0.3$, $\phi = \pi/2$, $t = 0.3$, $A = 0.2$, (**a**) $m = 0$, $L = 0.1$, $K = 0.1$, $\Theta = 1.25$, (**b**) $m = 0.1$, $L = 0.1$, $K = 0.1$, $\Theta = 1.25$, (**c**) $m = 0.1$, $L = 0.15$, $K = 0.1$, $\Theta = 1.25$, (**d**) $m = 0.1$, $L = 0.15$, $K \to \infty$, $\Theta = 1.25$, (**e**) $m = 0.1$, $L = 0.15$, $K \to \infty$, $\Theta = 1.5$.

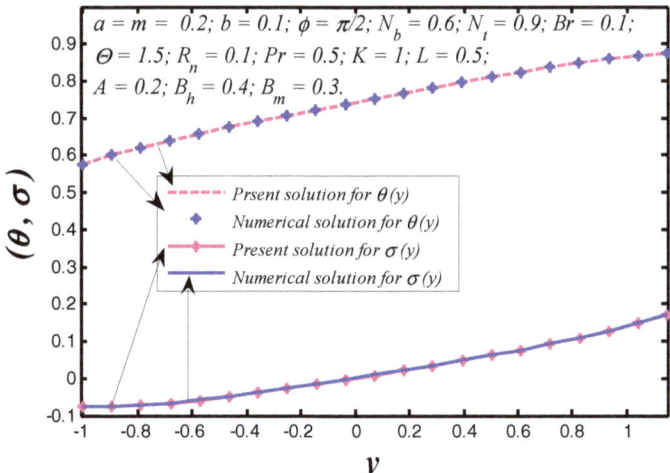

Figure 20. Comparison between numerical and present solutions for temperature and nanoparticle volume fraction profiles.

7. Conclusions

The analytical and numerical solutions of Equations (9)–(12) were estimated for stream function, nanoparticle temperature and concentration distribution. The expressions for nanoparticle temperature, volumetric fraction, heat transfer coefficient profile and stream function were discussed graphically. The following observations were noticed.

- Nanoparticle heat transfer between the tapered walls strongly depends on Brinkman number because the tissue presents the chief resistance to heat flow.
- Thermal radiation contains the potential to contribute a significant change in the nanoparticle temperature distribution.
- With increasing the radiation parameter, the nanoparticle temperature and heat transfer coefficient enhance.
- The nanoparticle temperature reduces with enhancing the Prandtl number, however, reverse behavior is noticed for nanoparticle concentration.
- Heat transfer coefficient depends on the flow, thermal and geometrical nature of flow regime.
- The trapping phenomenon also alters with changing the magnitude of slip and permeability parameters.
- The findings of the present models can be utilized to engineer smart peristaltic pumps which can be applicable for transporting drugs and delivery of nanoparticles.

Author Contributions: Conceptualization, D.T.; Investigation, J.P.; Methodology, A.K.T.; Visualization, R.E.; Writing—review and editing, S.M.S.

Funding: This research received no external funding.

Acknowledgments: R. Ellahi thanks Sadiq M. Sait, the Director Office of Research Chair Professors, King Fahd University of Petroleum and Minerals, Dhahran, Saudi Arabia, to honor him with the Chair Professor at KFUPM.

Conflicts of Interest: The authors declare no conflict of interest.

Nomenclature

Symbol	description	Unit
(a_1, a_2)	Dimensional amplitude of the lower and upper walls	m
c	Wave speed	m/s
$\overline{C}$	Nanoparticle volumetric volume fraction	Kg/m^3
C_0, C_1	Nanoparticle concentration at the lower and upper walls	Kg/m^3
D_B	Brownian diffusion coefficient	m/s
D_T	Themophoretic diffusion coefficient	m^2/s
$\overline{d}$	Dimensionless half width of the channel	m
$\overline{h}_h$	Heat transfer coefficient	W/m^{2}K (or) kg/s^{3}K
$\overline{h}_m$	Mass transfer coefficient	m/s
k	Permeable of porous medium	H/m
$\overline{k}$	Permeability of the porous wavy wall	Darcy (or) m^2
$\overline{k}_h$	Thermal conductivity of wavy wall	W/mK
$\overline{k}_m$	Mass conductivity of wavy wall	W/mK
$\overline{m}$	Dimensional non-uniform parameter	m
$\overline{P}$	Dimensional pressures	Pa (or) N/m^2 (or) kg/ms^2
$\overline{q}_r$	Uni-directional thermal radiative flux	kg/s^3 (or) W/m^2
t'	Dimensional time	s
$\overline{T}$	Nanoparticle temperature	K
T_m	Mean temperature	K
(T_0, T_1)	Temperature at the lower and upper walls	K
$\overline{U}, \overline{V}$	Velocity components in the wave frame	m/s
$\overline{\xi}, \overline{\eta}$	Rectangular coordinates	m
ρ_f	Density of the fluid	Kg/m^3
ρ_p	Density of the particle	Kg/m^3
μ	Dynamic Viscosity	kg/m.s
κ	Thermal conductivity of the fluid	m^2/s
λ	Wave length	m

Dimensionless parameters:

$\overline{\alpha}$	Slip coefficient at the surface of the porous walls
A	Blood flow constant
(a, b)	Dimensionless amplitude of the lower and upper walls
B_h	Heat transfer Biot number
B_m	Mass transfer Biot number
Br	Brinkman number
Ec	Eckert number
F	Dimensionless flow rate
$(\overline{H}_1, \overline{H}_2)$	Lower and upper wall boundaries of the micro-asymmetric channel
(h_1, h_2)	Dimensionless lower and upper wall shapes in wave frame
L	Slip parameter
m	Dimensionless non-uniform parameter
N_t	Thermophoresis parameter
N_b	Brownian motion parameter
p	Dimensionless pressure
Pr	Prandtl number
R	Reynolds number
R_n	Thermal radiation
Sc	Schmidt number
t	Dimensionless time
(u, v)	Velocity components in the wave frame (x, y)
Θ	Constant flow rate
σ	Dimensionless rescaled nanoparticle volume fraction
θ	Dimensionless nanoparticle temperature
ψ	Stream function
ϕ	Phase difference
K	Permeability parameter
δ	Wave number

Appendix A

The following constants are utilized in the solution of the manuscript.

$$N = 1/\sqrt{K}, \quad A_1 = f(x),$$

$$A_2 = \frac{-F}{(LN^2h_1 + 2 - LN^2h_2)\begin{pmatrix} \cosh(Nh_1) + \sinh(Nh_1) \\ -\cosh(Nh_2) - \sinh(Nh_2) \end{pmatrix} + (Nh_2 - Nh_1)\begin{pmatrix} \cosh(Nh_1) + \sinh(Nh_1) \\ +\cosh(Nh_2) + \sinh(Nh_2) \end{pmatrix}},$$

$$A_3 = \frac{F(\cosh(N(h_1 + h_2)) + \sinh(N(h_1 + h_2)))}{(LN^2h_1 + 2 - LN^2h_2)\begin{pmatrix} \cosh(Nh_1) + \sinh(Nh_1) \\ -\cosh(Nh_2) - \sinh(Nh_2) \end{pmatrix} + (Nh_2 - Nh_1)\begin{pmatrix} \cosh(Nh_1) + \sinh(Nh_1) \\ +\cosh(Nh_2) + \sinh(Nh_2) \end{pmatrix}},$$

$$A_4 = \frac{BrN^4}{1 + R_n Pr}, \quad A_5 = \frac{Pr}{1 + R_n Pr},$$

$$A_6 = \frac{(B_h h_1 - 1)2A_2 A_3 A_5}{A_1 A_5 N_b} + \frac{A_2^2 A_4 (\cosh(2Nh_1) + \sinh(2Nh_1))(B_h - 2N)}{4N^2 + 2A_1 A_5 N_b N} + \frac{A_3^2 A_4 (\cosh(2Nh_1) - \sinh(2Nh_1))(B_h + 2N)}{4N^2 - 2A_1 A_6 N_b N},$$

$$A_7 = -B_h - \frac{(B_h h_2 + 1)(A_5 \beta + 2A_2 A_3 A_5)}{A_1 A_5 N_b} - \frac{A_2^2 A_4 (B_h + 2N)(\cosh(2Nh_2) + \sinh(2Nh_2))}{4N^2 + 2A_1 A_5 N_b N} + \frac{A_3^2 A_4 (\cosh(2Nh_2) - \sinh(2Nh_2))(B_h - 2N)}{4N^2 - 2A_1 A_6 N_b N},$$

$$A_8 = \frac{A_7 + A_8(\cosh(A_1 A_5 N_b h_2) - \sinh(A_1 A_5 N_b h_2))(B_h - A_1 A_5 N_b)}{B_h},$$

$$A_9 = \frac{A_6+A_7}{(\cosh(A_1A_5N_bh_1)-\sinh(A_1A_5N_bh_1))(B_h+A_1A_5N_b)-(\cosh(A_1A_5N_bh_2)-\sinh(A_1A_5N_bh_2))(B_h-A_1A_5N_b)}$$

$$A_{10} = -\frac{A_{13}}{B_m} - \frac{(B_mh_2+1)(A_{12}+A_{13})}{B_m(B_m(h_1-h_2)-2)}, \quad A_{11} = \frac{A_{12}+A_{13}}{B_m(h_1-h_2)-2},$$

$$\begin{aligned}A_{12} = \ & A_9\Big(A_1A_5N_t + \tfrac{B_mN_t}{N_b}\Big)(\cosh(A_1A_5N_bh_1) - \sinh(A_1A_5N_bh_1)) \\ & - \tfrac{A_5^2A_4N_t(B_m-2N)(\cosh(2Nh_1)+\sinh(2Nh_1))}{4N^2N_b+2A_1A_5NN_b^2} \\ & - \tfrac{A_5^2A_4N_t(\cosh(2Nh_1)-\sinh(2Nh_1))(B_m+2N)}{4N^2N_b-2A_1A_5NN_b^2},\end{aligned}$$

$$\begin{aligned}A_{13} = \ & A_9(\cosh(A_1A_5N_bh_2) - \cosh(A_1A_5N_bh_2))\Big(A_1A_5N_t - \tfrac{B_mN_t}{N_b}\Big) \\ & -B_m + \tfrac{A_5^2A_4N_t(B_m+2N)(\cosh(2Nh_2)-\cosh(2Nh_2))}{4N^2N_b+2A_1A_5NN_b^2} \\ & + \tfrac{A_5^2A_4N_t(\cosh(2Nh_2)-\cosh(2Nh_2))(B_m-2N)}{4N^2N_b-2A_1A_5NN_b^2},\end{aligned}$$

References

1. Burns, J.C.; Parkes, T. Peristaltic motion. *J. Fluid Mech.* **1967**, *29*, 731–743. [CrossRef]
2. Zien, T.F.; Ostrach, S. A long wave approximation to peristaltic motion. *J. Biomech.* **1970**, *3*, 63–75. [CrossRef]
3. Raju, K.K.; Devanathan, R. Peristaltic motion of a non-Newtonian fluid. *Rheol. Acta* **1972**, *11*, 170–178. [CrossRef]
4. Ellahi, R.; Zeeshan, A.; Hussain, F.; Asadollahi, A. Peristaltic blood flow of couple stress fluid suspended with nanoparticles under the influence of chemical reaction and activation energy. *Symmetry* **2019**, *11*, 276. [CrossRef]
5. Zeeshan, A.; Ijaz, N.; Abbas, T.; Ellahi, R. The sustainable characteristic of Bio-bi-phase flow of peristaltic transport of MHD Jeffery fluid in human body. *Sustainability* **2018**, *10*, 2671. [CrossRef]
6. Hussain, F.; Ellahi, R.; Zeeshan, A.; Vafai, K. Modelling study on heated couple stress fluid peristaltically conveying gold nanoparticles through coaxial tubes: A remedy for gland tumors and arthritis. *J. Mol. Liq.* **2018**, *268*, 149–155. [CrossRef]
7. Choi, S.U.S.; Eastman, J.A. Enhancing Thermal Conductivity of Fluids with Nanoparticles. In *Proceedings of Enhancing Thermal Conductivity of Fluids with Nanoparticles, San Francisco, CA, USA*; American Society of Mechanical Engineers, FED: New York, NY, USA, 1995; Volume 231, pp. 99–105. Available online: https://www.osti.gov/biblio/196525-enhancing-thermal-conductivity-fluids-nanoparticles (accessed on 21 June 2019).
8. Masuda, H.; Ebata, A.; Teramae, K.; Hishinuma, N. Alteration of thermal conductivity and viscosity of liquids by dispersing ultra-fine particles. *Netsu Bussei.* **1993**, *7*, 227–233. [CrossRef]
9. Buongiorno, J.; Hu, W. Nanofluid Coolants for Advanced Nuclear Power Plants. In Proceedings of the nternational Congress on Advances in Nuclear Power Plants (ICAPP'05), Seoul, Korea, 15–19 May 2005.
10. Buongiorno, J. Convective transport in nanofluids. *J. Heat Transf.* **2005**, *128*, 240–250. [CrossRef]
11. Akbar, N.S.; Nadeem, S. Endoscopic effects on the peristaltic flow of a nanofluid. *Commun. Theor. Phys.* **2011**, *56*, 761–768. [CrossRef]
12. Akbar, N.S. Peristaltic Sisko nanofluid in an asymmetric channel. *Appl. Nanosci.* **2014**, *4*, 663–673. [CrossRef]
13. Tripathi, D.; Beg, O.A. A study on peristaltic flow of nanofluids: Application in drug delivery systems. *Int. J. Heat Mass Transf.* **2014**, *70*, 61–70. [CrossRef]
14. Akbar, N.S.; Nadeem, S.; Khan, Z.H. Numerical simulation of peristaltic flow of a Carreau nanofluid in an asymmetric channel. *Alexandria Eng. J.* **2013**, *53*, 191–197. [CrossRef]
15. Bég, O.A.; Tripathi, D. Mathematica simulation of peristaltic pumping with double-diffusive convection in nanofluids: A bio-nano-engineering model. *Proc. Inst. Mech. Eng. Part N J Nanoeng. Nanosyst.* **2012**, *225*, 99–114. [CrossRef]
16. Akbar, N.S.; Tripathi, D.; Bég, A.O. Modeling nanoparticle geometry effects on peristaltic pumping of medical magnetohydrodynamic nanofluids with heat transfer. *J. Mechan. Med. Bio.* **2016**, *16*, 1650088. [CrossRef]

17. Reddy, M.G.; Makinde, O.D. Magnetohydrodynamic peristaltic transport of Jeffrey nanofluid in an asymmetric channel. *J. Mol. Liq.* **2016**, *223*, 1242–1248. [CrossRef]
18. Akbar, N.S.; Huda, A.B.; Tripathi, D. Thermally developing MHD peristaltic transport of nanofluids with velocity and thermal slip effects. *Eur. Phys. J. Plus.* **2016**, *131*, 332. [CrossRef]
19. Nadeem, S.; Riaz, A.; Ellahi, R.; Akbar, N.S.; Zeeshan, A. Heat and mass transfer analysis of peristaltic flow of nanofluid in a vertical rectangular duct by using the optimized series solution and genetic algorithm. *J. Comput. Theor. Nanosci.* **2014**, *11*, 1133–1149. [CrossRef]
20. Ellahi, R.; Riaz, A.; Nadeem, S. A theoretical study of Prandtl nanofluid in a rectangular duct through peristaltic transport. *Appl. Nanosci.* **2014**, *4*, 753–760. [CrossRef]
21. Ellahi, R.; Bhatti, M.M.; Riaz, A.; Sheikholeslami, M. Effects of magnetohydrodynamics on peristaltic flow of Jeffrey fluid in a rectangular duct through a porous medium. *J. Por. Med.* **2014**, *17*, 143–157. [CrossRef]
22. Kothandapani, M.; Prakash, J. Influence of heat source, thermal radiation and inclined magnetic field on peristaltic flow of a hyperbolic tangent nanofluid in a tapered asymmetric channel. *IEEE Trans. NanoBiosci.* **2015**, *14*, 385–392. [CrossRef]
23. Nadeem, S.; Riaz, A.; Ellahi, R.; Akbar, N.S. Effects of heat and mass transfer on peristaltic flow of a nanofluid between eccentric cylinders. *Appl. Nanosci.* **2014**, *4*, 393–404. [CrossRef]
24. Prakash, J.; Sharma, A.; Tripathi, D. Thermal radiation effects on electroosmosis modulated peristaltic transport of ionic nanoliquids in biomicrofluidics channel. *J. Mol. Liq.* **2018**, *249*, 843–855. [CrossRef]
25. Tripathi, D.; Shashi, B.; Bég, O.A.; Akbar, N.S. Transient peristaltic diffusion of nanofluids: A model of micropumps in medical engineering. *J. Hydrodyn.* **2018**, *30*, 1001–1011. [CrossRef]
26. Tripathi, D.; Sharma, A.; Bég, O.A. Joule heating and buoyancy effects in electro-osmotic peristaltic transport of aqueous nanofluids through a microchannel with complex wave propagation. *Adv. Powder Technol.* **2018**, *29*, 639–653. [CrossRef]
27. Prakash, J.; Siva, E.P.; Tripathi, D.; Kuharat, S.; Bég, O.A. Peristaltic pumping of magnetic nanofluids with thermal radiation and temperature-dependent viscosity effects: Modelling a solar magneto-biomimetic nanopump. *Renew. Energ.* **2019**, *133*, 1308–1326. [CrossRef]
28. Prakash, J.; Tripathi, D. Electroosmotic flow of Williamson ionic nanoliquids in a tapered microfluidic channel in presence of thermal radiation and peristalsis. *J. Mol. Liq.* **2018**, *256*, 352–371. [CrossRef]
29. Prakash, J.; Jhorar, R.; Tripathi, D.; Azese, M.N. Electroosmotic flow of pseudoplastic nanoliquids via peristaltic pumping. *J. Braz. Soc. Mech. Sci. Eng.* **2019**, *41*, 61. [CrossRef]
30. Mosayebidorcheh, S.; Hatami, M. Analytical investigation of peristaltic nanofluid flow and heat transfer in an asymmetric wavy wall channel (Part I: Straight channel). *Int. J. Heat Mass Transf.* **2018**, *126*, 790–799. [CrossRef]
31. Abbasi, F.M.; Gul, M.; Shehzad, S.A. Hall effects on peristalsis of boron nitride-ethylene glycol nanofluid with temperature dependent thermal conductivity. *Physica E Low Dimens. Syst Nanostruct.* **2018**, *99*, 275–284. [CrossRef]
32. Ranjit, N.K.; Shit, G.C.; Tripathi, D. Joule heating and zeta potential effects on peristaltic blood flow through porous micro vessels altered by electrohydrodynamic. *Microvasc. Res.* **2018**, *117*, 74–89. [CrossRef]
33. Sadiq, M.A. MHD stagnation point flow of nanofluid on a plate with anisotropic slip. *Symmetry* **2019**, *11*, 132. [CrossRef]
34. Ellahi, R. The effects of MHD and temperature dependent viscosity on the flow of non-Newtonian nanofluid in a pipe: Analytical solutions. *Appl. Math. Model.* **2013**, *37*, 1451–1457. [CrossRef]
35. Zeeshan, A.; Shehzad, N.; Abbas, A.; Ellahi, R. Effects of radiative electro-magnetohydrodynamics diminishing internal energy of pressure-driven flow of titanium dioxide-water nanofluid due to entropy generation. *Entropy* **2019**, *21*, 236. [CrossRef]
36. Hussain, F.; Ellahi, R.; Zeeshan, A. Mathematical models of electro magnetohydrodynamic multiphase flows synthesis with nanosized hafnium particles. *Appl. Sci.* **2018**, *8*, 275. [CrossRef]
37. Ellahi, R.; Zeeshan, A.; Hussain, F.; Abbas, T. Study of shiny film coating on multi-fluid flows of a rotating disk suspended with nano-sized silver and gold particles: A comparative analysis. *Coatings* **2018**, *8*, 422. [CrossRef]
38. Harvey, R.W.; Metge, D.W.; Kinner, N.; Mayberry, N. Physiological considerations in applying laboratory-determined buoyant densities to predictions of bacterial and protozoan transport in groundwater, Results of in-situ and laboratory tests. *Enviorn. Sci. Technol.* **1997**, *31*, 289–295. [CrossRef]

39. Mishra, M.; Rao, A.R. Peristaltic transport in a channel with a porous peripheral layer: Model of a flow in gastrointestinal tract. *J. Biomech.* **2005**, *38*, 779–789. [CrossRef]
40. Mekheimer, K.S. Nonlinear peristaltic transport through a porous medium in an inclined planar channel. *J. Por. Med.* **2003**, *6*, 13. [CrossRef]
41. Siddiqui, A.M.; Ansari, A.R. A note on the swimming problem of a singly flagellated microorganism in a fluid flowing through a porous medium. *J. Porous Med.* **2005**, *8*, 551–556. [CrossRef]
42. Wernert, V.; Schäf, O.; Ghobarkar, H.; Denoyel, R. Adsorption properties of zeolites for artificial kidney applications. *Microporous Mesoporous Mat.* **2005**, *83*, 101–113. [CrossRef]
43. Jafari, A.; Zamankhan, P.; Mousavi, S.M.; Kolari, P. Numerical investigation of blood flow part II: In capillaries. *Commun. Nonlinear Sci. Numeri. Simulat.* **2009**, *14*, 1396–1402. [CrossRef]
44. Goerke, A.R.; Leung, J.; Wickramasinghe, S.R. Mass and momentum transfer in blood oxygenators. *Che. Eng. Sci.* **2002**, *57*, 2035–2046. [CrossRef]
45. Mneina, S.S.; Martens, G.O. Linear phase matched filter design with causal real symmetric impulse response. *AEU Int. J. Electron. Commun.* **2009**, *63*, 83–91. [CrossRef]
46. Andoh, Y.H.; Lips, B. Prediction of porous walls thermal protection by effusion or transpiration cooling. An analytical approach. *Appl. Thermal. Eng.* **2003**, *23*, 1947–1958. [CrossRef]
47. Runstedtler, A. On the modified Stefan–Maxwell equation for isothermal multi component gaseous diffusion. *Chemical Eng. Sci.* **2006**, *61*, 5021–5029. [CrossRef]
48. Uddin, M.J.; Khan, W.A.; Ismail, A.I.M. Free convection boundary layer flow from a heated upward facing horizontal flat plate embedded in a porous medium filled by a nanofluid with convective boundary condition. *Transp. Porous Med.* **2012**, *92*, 867–881. [CrossRef]
49. Chamkha, A.J.; Abbasbandy, S.; Rashad, A.M.; Vajravelu, K. Radiation effects on mixed convection over a wedge embedded in a porous medium filled with a nanofluid. *Transp. Porous Med.* **2011**, *91*, 261–279. [CrossRef]
50. Kuznetsov, A.V.; Nield, D.A. Effect of local thermal non-equilibrium on the onset of convection in a porous medium layer saturated by a nanofluid. *Transp. Porous Med.* **2010**, *83*, 425–436. [CrossRef]
51. Akbar, N.S. Double-diffusive natural convective peristaltic flow of a Jeffrey nanofluid in a porous channel. *Heat Trans. Res.* **2014**, *45*, 293–307. [CrossRef]
52. Nadeem, S.; Riaz, A.; Ellahi, R.; Akbar, N.S. Mathematical model for the peristaltic flow of nanofluid through eccentric tubes comprising porous medium. *Appl. Nanosci.* **2014**, *4*, 733–743. [CrossRef]
53. Bhatti, M.M.; Zeeshan, A.; Ellahi, R.; Shit, G.C. Mathematical modeling of heat and mass transfer effects on MHD peristaltic propulsion of two-phase flow through a Darcy-Brinkman-Forchheimer porous medium. *Adv. Powder Technol.* **2018**, *29*, 1189–1197. [CrossRef]
54. Alamri, S.Z.; Ellahi, R.; Shehzad, N.; Zeeshan, A. Convective radiative plane Poiseuille flow of nanofluid through porous medium with slip: An application of Stefan blowing. *J. Mol. Liq.* **2019**, *273*, 292–304. [CrossRef]
55. Shehzad, N.; Zeeshan, A.; Ellahi, R.; Rashidid, S. Modelling study on internal energy loss due to entropy generation for non-Darcy Poiseuille flow of silver-water nanofluid: An application of purification. *Entropy* **2018**, *20*, 851. [CrossRef]
56. Kothandapani, M.; Prakash, J. The peristaltic transport of Carreau nanofluids under effect of a magnetic field in a tapered asymmetric channel: Application of the cancer therapy. *J. Mech. Med. Bio.* **2015**, *15*, 1550030. [CrossRef]
57. Hayat, T.; Abbasi, F.M.; Al-Yami, M.; Monaquel, S. Slip and Joule heating effects in mixed convection peristaltic transport of nanofluid with Soret and Dufour effects. *J. Mol. Liq.* **2014**, *194*, 93–99. [CrossRef]
58. Kothandapani, M.; Prakash, J. Effects of thermal radiation parameter and magnetic field on the peristaltic motion of Williamson nanofluids in a tapered asymmetric channel. *Int. J. Heat Mass Transf.* **2015**, *51*, 234–245. [CrossRef]
59. Hayat, T.; Yasmin, H.; Ahmad, B.; Chen, B. Simultaneous effects of convective conditions and nanoparticles on peristaltic motion. *J. Mol. Liq.* **2014**, *193*, 74–82. [CrossRef]
60. Makinde, O.D. Thermal stability of a reactive viscous flow through a porous-saturated channel with convective boundary conditions. *Appl. Therm. Eng.* **2009**, *29*, 1773–1777. [CrossRef]
61. Parti, M. Mass transfer Biot numbers. *Periodica Polytechnica Mech. Eng.* **1994**, *38*, 109–122.

62. Kikuchi, Y. Effect of leukocytes and platelets on blood flow through a parallel array of microchannels: Micro-and Macroflow relation and rheological measures of leukocytes and platelate acivities. *Microvasc. Res.* **1995**, *50*, 288–300. [CrossRef]
63. Mishra, M.; Rao, A.R. Peristaltic transport of a Newtonian fluid in an asymmetric channel. *Z. Angew. Math. Phys.* **2003**, *54*, 532–550. [CrossRef]

© 2019 by the authors. Licensee MDPI, Basel, Switzerland. This article is an open access article distributed under the terms and conditions of the Creative Commons Attribution (CC BY) license (http://creativecommons.org/licenses/by/4.0/).

Article

A Particle Method Based on a Generalized Finite Difference Scheme to Solve Weakly Compressible Viscous Flow Problems

Yongou Zhang [1,2] and Aokui Xiong [1,2,*]

1. Key Laboratory of High Performance Ship Technology (Wuhan University of Technology), Ministry of Education, Wuhan 430074, China
2. School of Transportation, Wuhan University of Technology, Wuhan 430074, China
* Correspondence: xiong_ak@163.com

Received: 26 July 2019; Accepted: 26 August 2019; Published: 29 August 2019

Abstract: The Lagrangian meshfree particle-based method has advantages in solving fluid dynamics problems with complex or time-evolving boundaries for a single phase or multiple phases. A pure Lagrangian meshfree particle method based on a generalized finite difference (GFD) scheme is proposed to simulate time-dependent weakly compressible viscous flow. The flow is described with Lagrangian particles, and the partial differential terms in the Navier-Stokes equations are represented as the solution of a symmetric system of linear equations through a GFD scheme. In solving the particle-based symmetric equations, the numerical method only needs the kernel function itself instead of using its gradient, i.e., the approach is a kernel gradient free (KGF) method, which avoids using artificial parameters in solving for the viscous term and reduces the limitations of using the kernel function. Moreover, the order of Taylor series expansion can be easily improved in the meshless algorithm. In this paper, the particle method is validated with several test cases, and the convergence, accuracy, and different kernel functions are evaluated.

Keywords: compressible viscous flow; symmetric linear equations; generalized finite difference scheme; kernel gradient free; Lagrangian approach

1. Introduction

Problems of weakly compressible flows have attracted much attention in aerospace and oceanic applications, such as wind engineering problems, turbine flow, blood flow, and water wave motion. Accurate predictions of such flows are important in computational fluid dynamics. For fluids at low Mach numbers, the ratio between the speed of flow and the speed of sound is extremely small, and therefore, density fluctuations are not obvious. As a result, such a situation can be called weakly compressible flow. Generally, there are three numerical ways to model weakly compressible flow, namely, the Eulerian approach, the Lagrangian approach, and the hybrid approach. The Eulerian approach solves for quantities at fixed locations in space, and the Lagrangian approach uses individual particles that move through both space and time and have their own physical properties, such as density, velocity, and pressure, to represent the dynamically evolving fluid flow. The flow is described by recording the time history of each fluid particle. In the present work, we propose a pure Lagrangian meshfree particle-based method based on a meshless finite difference scheme to solve weakly compressible flow problems.

The Lagrangian meshfree method is rapidly advancing and has been widely used in recent years because it can be easily adapted to modeling problems with complex or time-evolving boundaries for single or multiple phases, such as numerical simulations of dam break flow [1], hydraulic jumps [2], rising bubbles [3] and coalescing [4]. The ability of this method to model non-Newtonian fluid and large scale diffuse fluids has been demonstrated in some recent works [5,6] by introducing different symmetric models. Moreover, because the computations are based on the support domain, which is much smaller than the complete computational region, the ill conditioned system problem is rarely encountered. Among all Lagrangian meshfree methods, the smoothed particle hydrodynamics (SPH) method was one of the earliest methods developed and has been widely applied in different fields. The SPH method was first pioneered independently by Lucy [7] and Gingold and Monaghan [8] to solve astrophysical problems in 1977. Details of the SPH method as a computational fluid dynamics method can be found in recent reviews [9–12] and Liu and Liu's book [13]. Some successful applications of this method include coastal engineering, nuclear engineering, ocean engineering, and bioengineering. However, the accuracy of the conventional SPH method is unsatisfactory, and it is not easy to achieve an accurate high-order SPH approach.

As a meshfree Lagrangian method, the particle distribution generally tends to be irregular in the computations, which leads to inconsistency and low accuracy [14,15]. For that reason, in some cases, only the first-order term of the fluid dynamics equations, the Navier-Stokes equations, is solved, and the viscous term, which contains the second-order differential, is obtained through the artificial viscosity with artificial parameters in the SPH method. This issue can also occur in the incompressible SPH (ISPH) method [16]. To improve the consistency and accuracy of these methods, different modifications have been developed. After using Taylor series expansion to normalize the kernel function, the corrective smoothed particle method (CSPM) [17,18] and the modified smoothed particle method (MSPH) [19] were proposed. Both methods have better accuracy than the conventional SPH method. Nevertheless, it should be noted that both methods improve accuracy by improving the particle approximation of the kernel gradient term, which leads to more strict requirements on the kernel function. These requirements are related to the compact condition, normalization condition, and delta function behavior [20] and limit the selection of the kernel function, especially when the second-order gradient of the kernel function is required.

To avoid the solution of the gradient of the kernel function, a method with kernel gradient free (KGF) features can be developed, as discussed in detail in [21–23]; notably, a KGF-SPH method was proposed in 2015. When a particle method only involves the kernel function itself in kernel and particle approximation, the kernel gradient is not necessary in the computation, and this approach is thus referred to as a KGF method. The KGF-SPH method is used to solve for the viscous term directly without using the artificial viscosity, and the results are good for 2D models. Another KGF method is the consistent particle method (CPM) [24,25] for incompressible flow simulation. In the CPM, Poisson's equation is used in the same way as the moving-particle semi-implicit (MPS) method based on the particle number density and the difference algorithm.

The purpose of our work is to combine a finite difference scheme and the particle method for solving weakly compressible viscous flow problems. In the method, the flow is described with Lagrangian particles, and the partial differential terms in the Navier-Stokes equations are represented as the solution of a symmetric system of linear equations through a generalized finite difference (GFD) scheme. It should be noted that this method is not a completely new method, but we will simply refer to it as the finite difference particle method (FDPM) to simplify the description in the subsequent sections.

Meshless finite difference approximation was first discussed for fully arbitrary meshes by Jensen [26] in 1972. Perrone and Kao [27] also contributed to the development of this method at that time. Subsequently, a variation using the moving least squares method was proposed by Lizska and Orkisz [28], and some recent works have been published [29–31]. The meshless finite difference scheme or GFD approximation we used came from Benito, Urena and Gavete [32], and they provided a discussion of the influence of several factors in the GFD scheme. A comparison between the GFD method and the element-free Galerkin method (EFGM) in solving the Laplace equation was presented in [33,34]. The GFD method was shown to be more accurate than the EFGM and the GFD scheme was used as a Eulerian meshfree method. In the present work, the GFD scheme is utilized to build a Lagrangian meshfree particle-based method, namely, the FDPM.

The FDPM has several advantages compared with the conventional SPH. First, the method is KGF. Only the kernel function itself and the positions of each particle are used to compute the spatial differential through a set of symmetric linear equations. Second, the method can be easily extended to high orders because it is based on Taylor series expansion. We show a fourth-order scheme for the FDPM. Additionally, only a few lines of code need to change to obtain a high-order FDPM, which is simple for users, especially when the users want to start with a low-order but fast computation. Third, the second-order differential term can be obtained without additional limitations on the kernel function. Thus, the viscous term in the Navier-Stokes equations can be computed directly without introducing any artificial parameters. Fourth, the FDPM is characterized by good compatibility. Most boundary conditions in the existing Lagrangian particle-based methods, such as the SPH and MPS methods, can be used directly. In the present work, we focus on the evaluation of the convergence, efficiency, and effects of the kernel function. The method is tested by modeling flow in a pipeline, Poiseuille flow, Couette flow and flow in porous media. These classical flows are used in different ways to solve fluid dynamics problems [35–37].

The present paper is organized as follows. In Section 2, the FDPM is given to solve the Navier-Stokes equations. In Section 3, applications of the particle method are shown. Section 4 summarizes the results of this work.

2. Finite Difference Particle Method for Weakly Compressible Flow

2.1. Lagrangian Form of the Governing Equations for Weakly Compressible Viscous Flow

The Lagrangian form of the Navier-Stokes equations, i.e., the continuity equation and the momentum equation, including viscous and external forces, are defined by Equations (1) and (2), respectively. The Lagrangian form of governing equations is as follows:

$$\frac{D\rho}{Dt} = -\rho \nabla \cdot \boldsymbol{u}, \qquad (1)$$

$$\frac{D\boldsymbol{u}}{Dt} = -\frac{1}{\rho}\nabla p + v_k \nabla^2 \boldsymbol{u} + \boldsymbol{F}, \qquad (2)$$

where ρ is the density, t is the time, $\boldsymbol{u}$ is the particle velocity, p is the pressure, v_k is the kinematic viscosity, and $\boldsymbol{F}$ is an external body force, such as gravity. All these variables are related to the physical properties of fluid particles that can move in both space and time, rather than remain at a fixed position.

The material derivative is written as follows

$$\frac{D}{Dt} = \frac{\partial}{\partial t} + \boldsymbol{u}\cdot\nabla. \qquad (3)$$

The equation of state for weakly compressible fluid flow is

$$\frac{Dp}{Dt} = c^2 \frac{D\rho}{Dt}, \qquad (4)$$

where c is the speed of sound.

2.2. Generalized Finite Difference Scheme

In the FDPM, flow is described with Lagrangian particles, and the GFD approximation [32] is utilized to solve for the spatial differential terms in the governing equations.

Consider a particle i surrounded by particles $j = 1, 2, \ldots, N$, with all $N + 1$ of the particles in a compact support domain, as shown in Figure 1. For a circular support domain, r_s represents the radius of the support domain, which is called the smoothing length in the SPH method. Particles j are white circles around particle i, which is the orange circle, and Ω represents the computational domain. The closest nodes to particle i are selected as j particles, and these particles should be in the support domain at the same time.

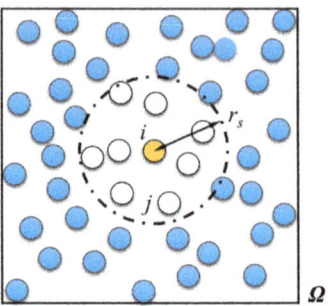

Figure 1. Computational domain Ω, support domain of particle i (point circle line), radius of the support domain r_s, and fluid particles (circles).

The values of an infinitely differentiable function F at the positions of particles i and j are defined as F_i and F_j, respectively. This F can be the pressure, velocity or density of particles in the computation. Let us expand this term as the Taylor series of F for particle i:

$$F_j = F_i + h_j \frac{\partial F_i}{\partial x} + k_j \frac{\partial F_i}{\partial y} + \frac{h_j^2}{2} \frac{\partial^2 F_i}{\partial x^2} + \frac{k_j^2}{2} \frac{\partial^2 F_i}{\partial y^2} + h_j k_j \frac{\partial^2 F_i}{\partial x \partial y} + \frac{h_j^3}{6} \frac{\partial^3 F_i}{\partial x^3} + \frac{k_j^3}{6} \frac{\partial^3 F_i}{\partial y^3} + \frac{h_j^2 k_j}{2} \frac{\partial^3 F_i}{\partial x^2 \partial y} + \frac{h_j k_j^2}{2} \frac{\partial^3 F_i}{\partial x \partial y^2} \\ + \frac{h_j^4}{24} \frac{\partial^4 F_i}{\partial x^4} + \frac{k_j^4}{24} \frac{\partial^4 F_i}{\partial y^4} + \frac{h_j^3 k_j}{6} \frac{\partial^4 F_i}{\partial x^3 \partial y} + \frac{h_j^2 k_j^2}{4} \frac{\partial^4 F_i}{\partial x^2 \partial y^2} + \frac{h_j k_j^3}{6} \frac{\partial^4 F_i}{\partial x \partial y^3} + \cdots ,$$ (5)

where Equation (5) is for two dimensions. x and y are the spatial coordinates of the particles, and $h_j = x_j - x_i$, $k_j = y_j - y_i$.

For the fourth-order FDPM, ignoring the high-order terms, the approximation of F is denoted by f:

$$f_j = f_i + h_j \frac{\partial f_i}{\partial x} + k_j \frac{\partial f_i}{\partial y} + \frac{h_j^2}{2} \frac{\partial^2 f_i}{\partial x^2} + \frac{k_j^2}{2} \frac{\partial^2 f_i}{\partial y^2} + h_j k_j \frac{\partial^2 f_i}{\partial x \partial y} + \frac{h_j^3}{6} \frac{\partial^3 f_i}{\partial x^3} + \frac{k_j^3}{6} \frac{\partial^3 f_i}{\partial y^3} + \frac{h_j^2 k_j}{2} \frac{\partial^3 f_i}{\partial x^2 \partial y} + \frac{h_j k_j^2}{2} \frac{\partial^3 f_i}{\partial x \partial y^2} \\ + \frac{h_j^4}{24} \frac{\partial^4 f_i}{\partial x^4} + \frac{k_j^4}{24} \frac{\partial^4 f_i}{\partial y^4} + \frac{h_j^3 k_j}{6} \frac{\partial^4 f_i}{\partial x^3 \partial y} + \frac{h_j^2 k_j^2}{4} \frac{\partial^4 f_i}{\partial x^2 \partial y^2} + \frac{h_j k_j^3}{6} \frac{\partial^4 f_i}{\partial x \partial y^3}.$$ (6)

After rearranging these equations and multiplying by a kernel function W on both sides of the equation, the sum of these expressions for all particles j is obtained:

$$\sum_{j=1}^{N} \begin{pmatrix} f_i - f_j + h_j \frac{\partial f_i}{\partial x} + k_j \frac{\partial f_i}{\partial y} + \frac{h_j^2}{2} \frac{\partial^2 f_i}{\partial x^2} + \frac{k_j^2}{2} \frac{\partial^2 f_i}{\partial y^2} + h_j k_j \frac{\partial^2 f_i}{\partial x \partial y} \\ + \frac{h_j^3}{6} \frac{\partial^3 f_i}{\partial x^3} + \frac{k_j^3}{6} \frac{\partial^3 f_i}{\partial y^3} + \frac{h_j^2 k_j}{2} \frac{\partial^3 f_i}{\partial x^2 \partial y} + \frac{h_j k_j^2}{2} \frac{\partial^3 f_i}{\partial x \partial y^2} \\ + \frac{h_j^4}{24} \frac{\partial^4 f_i}{\partial x^4} + \frac{k_j^4}{24} \frac{\partial^4 f_i}{\partial y^4} + \frac{h_j^3 k_j}{6} \frac{\partial^4 f_i}{\partial x^3 \partial y} + \frac{h_j^2 k_j^2}{4} \frac{\partial^4 f_i}{\partial x^2 \partial y^2} + \frac{h_j k_j^3}{6} \frac{\partial^4 f_i}{\partial x \partial y^3} \end{pmatrix} W(h_j, k_j, r_s) = 0,$$ (7)

where $W(h_j, k_j, r_s)$ is a kernel function in 2D and r_s represents the size of the support domain. For W, different kernel functions, including Gaussian, cubic spline, and quintic spline functions, can be found in [13]. In the following equations, we use W for the kernel function.

Function G can be defined in 2D as

$$G(f) = \sum_{j=1}^{N} \left[\begin{pmatrix} f_i - f_j + h_j \frac{\partial f_i}{\partial x} + k_j \frac{\partial f_i}{\partial y} + \frac{h_j^2}{2} \frac{\partial^2 f_i}{\partial x^2} + \frac{k_j^2}{2} \frac{\partial^2 f_i}{\partial y^2} + h_j k_j \frac{\partial^2 f_i}{\partial x \partial y} \\ + \frac{h_j^3}{6} \frac{\partial^3 f_i}{\partial x^3} + \frac{k_j^3}{6} \frac{\partial^3 f_i}{\partial y^3} + \frac{h_j^2 k_j}{2} \frac{\partial^3 f_i}{\partial x^2 \partial y} + \frac{h_j k_j^2}{2} \frac{\partial^3 f_i}{\partial x \partial y^2} \\ + \frac{h_j^4}{24} \frac{\partial^4 f_i}{\partial x^4} + \frac{k_j^4}{24} \frac{\partial^4 f_i}{\partial y^4} + \frac{h_j^3 k_j}{6} \frac{\partial^4 f_i}{\partial x^3 \partial y} + \frac{h_j^2 k_j^2}{4} \frac{\partial^4 f_i}{\partial x^2 \partial y^2} + \frac{h_j k_j^3}{6} \frac{\partial^4 f_i}{\partial x \partial y^3} \end{pmatrix} W \right]^2 = 0.$$ (8)

According to Equation (7), the norm of G equals 0, so we obtain:

$$\frac{\partial G(f)}{\partial \left(\frac{\partial f_i}{\partial x}\right)} = 2\sum_{j=1}^{N} \Phi h_j W^2 = 0, \quad \frac{\partial G(f)}{\partial \left(\frac{\partial f_i}{\partial y}\right)} = 2\sum_{j=1}^{N} \Phi k_j W^2 = 0, \quad \frac{\partial G(f)}{\partial \left(\frac{\partial^2 f_i}{\partial x^2}\right)} = \sum_{j=1}^{N} \Phi h_j^2 W^2 = 0,$$

$$\frac{\partial G(f)}{\partial \left(\frac{\partial^2 f_i}{\partial y^2}\right)} = \sum_{j=1}^{N} \Phi k_j^2 W^2 = 0, \quad \frac{\partial G(f)}{\partial \left(\frac{\partial^2 f_i}{\partial x \partial y}\right)} = 2\sum_{j=1}^{N} \Phi h_j k_j W^2 = 0, \quad \frac{\partial G(f)}{\partial \left(\frac{\partial^3 f_i}{\partial x^3}\right)} = \frac{1}{3}\sum_{j=1}^{N} \Phi k_j^3 W^2 = 0,$$

$$\frac{\partial G(f)}{\partial \left(\frac{\partial^3 f_i}{\partial y^3}\right)} = \frac{1}{3}\sum_{j=1}^{N} \Phi k_j^3 W^2 = 0, \quad \frac{\partial G(f)}{\partial \left(\frac{\partial^3 f_i}{\partial x^2 \partial y}\right)} = \sum_{j=1}^{N} \Phi h_j^2 k_j W^2 = 0, \quad \frac{\partial G(f)}{\partial \left(\frac{\partial^3 f_i}{\partial x \partial y^2}\right)} = \sum_{j=1}^{N} \Phi h_j k_j^2 W^2 = 0, \quad (9)$$

$$\frac{\partial G(f)}{\partial \left(\frac{\partial^4 f_i}{\partial x^4}\right)} = \frac{1}{12}\sum_{j=1}^{N} \Phi h_j^4 W^2 = 0, \quad \frac{\partial G(f)}{\partial \left(\frac{\partial^4 f_i}{\partial y^4}\right)} = \frac{1}{12}\sum_{j=1}^{N} \Phi k_j^4 W^2 = 0, \quad \frac{\partial G(f)}{\partial \left(\frac{\partial^4 f_i}{\partial x^3 \partial y}\right)} = \frac{1}{3}\sum_{j=1}^{N} \Phi h_j^3 k_j W^2 = 0,$$

$$\frac{\partial G(f)}{\partial \left(\frac{\partial^4 f_i}{\partial x^2 \partial y^2}\right)} = \frac{1}{2}\sum_{j=1}^{N} \Phi h_j^2 k_j^2 W^2 = 0, \quad \frac{\partial G(f)}{\partial \left(\frac{\partial^4 f_i}{\partial x \partial y^3}\right)} = \frac{1}{3}\sum_{j=1}^{N} \Phi h_j k_j^3 W^2 = 0,$$

where

$$\Phi = f_i - f_j + h_j \frac{\partial f_i}{\partial x} + k_j \frac{\partial f_i}{\partial y} + \frac{h_j^2}{2}\frac{\partial^2 f_i}{\partial x^2} + \frac{k_j^2}{2}\frac{\partial^2 f_i}{\partial y^2} + h_j k_j \frac{\partial^2 f_i}{\partial x \partial y} + \frac{h_j^3}{6}\frac{\partial^3 f_i}{\partial x^3} + \frac{k_j^3}{6}\frac{\partial^3 f_i}{\partial y^3} + \frac{h_j^2 k_j}{2}\frac{\partial^3 f_i}{\partial x^2 \partial y} + \frac{h_j k_j^2}{2}\frac{\partial^3 f_i}{\partial x \partial y^2}$$
$$+ \frac{h_j^4}{24}\frac{\partial^4 f_i}{\partial x^4} + \frac{k_j^4}{24}\frac{\partial^4 f_i}{\partial y^4} + \frac{h_j^3 k_j}{6}\frac{\partial^4 f_i}{\partial x^3 \partial y} + \frac{h_j^2 k_j^2}{4}\frac{\partial^4 f_i}{\partial x^2 \partial y^2} + \frac{h_j k_j^3}{6}\frac{\partial^4 f_i}{\partial x \partial y^3}.$$
(10)

Equation (9) gives us the following equation:

$$\mathbf{AD} = \mathbf{B}, \tag{11}$$

where

$$\mathbf{A} = \begin{bmatrix} \sum_{j=1}^{N} h_j^2 W^2 & \sum_{j=1}^{N} h_j k_j W^2 & \sum_{j=1}^{N} \frac{1}{2}h_j^3 W^2 & \sum_{j=1}^{N} \frac{1}{2}h_j k_j^2 W^2 & \cdots & \sum_{j=1}^{N} \frac{1}{6}h_j^2 k_j^3 W^2 \\ & \sum_{j=1}^{N} k_j^2 W^2 & \sum_{j=1}^{N} \frac{1}{2}h_j^2 k_j W^2 & \sum_{j=1}^{N} \frac{1}{2}k_j^3 W^2 & \cdots & \sum_{j=1}^{N} \frac{1}{6}h_j k_j^4 W^2 \\ & & \sum_{j=1}^{N} \frac{1}{4}h_j^4 W^2 & \sum_{j=1}^{N} \frac{1}{4}h_j^2 k_j^2 W^2 & \cdots & \sum_{j=1}^{N} \frac{1}{12}h_j^3 k_j^3 W^2 \\ & & & \sum_{j=1}^{N} \frac{1}{4}k_j^4 W^2 & \cdots & \sum_{j=1}^{N} \frac{1}{12}h_j k_j^5 W^2 \\ & & & & \ddots & \vdots \\ \text{symmetric} & & & & & \sum_{j=1}^{N} \frac{1}{36}h_j^2 k_j^6 W^2 \end{bmatrix}, \tag{12}$$

$$\mathbf{D} = \left\{ \frac{\partial f_i}{\partial x} \quad \frac{\partial f_i}{\partial y} \quad \frac{\partial^2 f_i}{\partial x^2} \quad \frac{\partial^2 f_i}{\partial y^2} \quad \cdots \quad \frac{\partial^4 f_i}{\partial x \partial y^3} \right\}^T, \tag{13}$$

$$\mathbf{B} = \left\{ \begin{array}{c} \sum_{j=1}^{N}(f_j - f_i)h_j W^2 \\ \sum_{j=1}^{N}(f_j - f_i)k_j W^2 \\ \sum_{j=1}^{N} \frac{1}{2}(f_j - f_i)h_j^2 W^2 \\ \sum_{j=1}^{N} \frac{1}{2}(f_j - f_i)k_j^2 W^2 \\ \vdots \\ \sum_{j=1}^{N} \frac{1}{6}(f_j - f_i)h_j^2 k_j^3 W^2 \end{array} \right\}. \tag{14}$$

Since matrix A is symmetrical, Equation (11) can be solved, and the solution gives the values of the spatial derivatives in matrix **D**. Thus, the spatial derivatives in Equations (1) and (2) can be obtained by solving a set of symmetric linear equations, and the material derivatives in the equations can be integrated using a time integration scheme.

2.3. Particle Representation for Governing Equations

Taking particle i as an example, this section gives the particle representation of the governing equations and the solution to Equation (11). The solution includes the values of the spatial derivatives needed in the governing equations.

The coefficients of **D** and **B** are denoted by $D_m(f_i)$ and $B_m(f_i)$, respectively, with $m = 1, 2, \ldots, 5$. For example, $D_2(f_i) = \frac{\partial f_i}{\partial y}$ (the second coefficient in Equation(13)), and $B_2(f_i) = \sum_{j=1}^{N}(f_j - f_i)k_j W^2$ (the second coefficient in Equation (14)). In addition, the symmetric matrix **A** can be decomposed into the upper and lower triangular matrices $\mathbf{A} = \mathbf{L}\mathbf{L}^T$. The coefficients of the matrix **L** are denoted by $L(m, n)$, with m and $n = 1, 2, 3, 4, 5$.

By using the GFD scheme and Cholesky factorization to solve Equation (11), we obtain the solutions for the Lagrangian derivative terms in Equations (1) and (2) in two-dimensional form

$$\frac{D\rho_i}{Dt} = -\rho_i D_1(u_i) - \rho_i D_2(v_i), \tag{15}$$

$$\frac{Du_i}{Dt} = -\frac{1}{\rho_i} D_1(p_i) + v_k D_3(u_i), \tag{16}$$

$$\frac{Dv_i}{Dt} = -\frac{1}{\rho_i} D_2(p_i) + v_k D_4(v_i) + g, \tag{17}$$

where u_i and v_i are the velocity of particle i in two directions and

$$D_m(f) = \begin{cases} \frac{1}{L(m,m)} \left[Y_m(f) - \sum_{n=m+1}^{N} L(n,m) D_n(f) \right] & m = 1, 2, 3, 4 \\ \frac{Y(f)}{L(m,m)} & m = 5 \end{cases}, \tag{18}$$

where

$$Y_m(f) = \begin{cases} \frac{b_m(f)}{L(m,m)} & m = 1 \\ \frac{1}{L(m,m)} \left[b_m(f) - \sum_{n=1}^{m-1} L(m,n) Y_n(f) \right] & m = 2, 3, 4, 5 \end{cases}. \tag{19}$$

This method considers changes in density and is able to simulate flow at low Mach numbers, so it is used to solve weakly compressible viscous flow problems. Calculations of particle motion and time integration are performed based on second-order leapfrog integration. The equations for updating the position and velocity of particles are

$$v_i\left(t + \frac{1}{2}\Delta t\right) = v_i\left(t - \frac{1}{2}\Delta t\right) + \Delta t \frac{Dv_i(t)}{Dt}, \tag{20}$$

$$r_i(t + \Delta t) = r_i(t) + \Delta t v_i\left(t + \frac{1}{2}\Delta t\right), \tag{21}$$

where $v_i(t + \frac{1}{2}\Delta t)$ is the velocity of fluid particle i at time $t + \frac{1}{2}\Delta t$, and Δt is the time step.

2.4. Artificial Particle Displacement

In simulations of flow in porous media (Section 3.5), artificial particle displacement is suggested as a particle motion correction to avoid particles in the vicinity of the stagnation points of fluid flow [38] and to avoid poor particle distributions [39]. Artificial particle displacement can be expressed as

$$\delta r_i = \alpha \bar{r}_i^2 v_{\max} \Delta t \sum_{j=1}^{N} \frac{r_{ij}}{r_{ij}^3}, \tag{22}$$

where r_i is the position of particle i, α is a problem-dependent parameter that is usually set between 0.01 and 0.1, $v_{\max}$ is the maximum velocity of all particles in the computational domain, $r_{ij} = r_i - r_j$

which is the distance between particles i and j, and $\bar{r}_i$ is the average distance between the neighboring particles of particle i:

$$\bar{r}_i = \frac{1}{N}\sum_{j=1}^{N} r_{ij}. \tag{23}$$

It is noted that the problem-dependent parameter α should be selected carefully. This value should be small enough not to affect the physics of the flow but also large enough to avoid the accumulation of particles to form groups. In the present work, the value of artificial particle displacement is less than 0.1% of the physical particle displacement for a given time step, which is consistent with the magnitude in [40].

After moving the particles, the pressure and velocity components should be corrected by Taylor expansion.

2.5. Boundary Conditions

Several layers of virtual particles are used to implement the boundary condition. Similar treatments can be observed in the SPH and MPS simulations. On a flat wall, virtual particles are obtained by extending the boundary particles to the outside of the computational region, and the distribution of virtual particles is regular. The number of layers can be chosen according to the scale of the support domain. Figure 2 is a sketch of the treatment of particles near the wall. i represents the particle number, and Δx is the particle spacing.

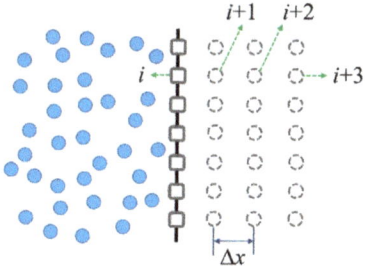

● fluid particles ☐ boundary particles ○ virtual particles

Figure 2. A sketch of a simulation of an acoustic boundary using virtual particles. Fluid particles are inside the computational domain and boundary particles are fixed on the boundary.

For a flat wall, both the no-slip and free-slip boundary conditions can be implemented using virtual particles. For no-slip walls, the particle-based boundary conditions are as follows

$$p_{i+3} = p_{i+2} = p_{i+1} = p_i, \ v_{i+3} = v_{i+2} = v_{i+1} = 0, \tag{24}$$

For free-slip walls, the tangential velocity component of virtual particles is maintained the same as the boundary particles.

For a round surface, virtual particles are established based on a radial distribution inside the object domain with particle spacing Δx. The particle distribution is shown in Figure 3.

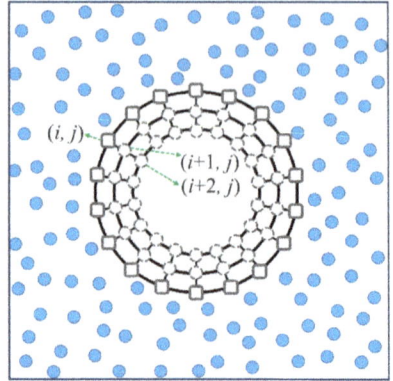

● fluid particles ☐ boundary particles ○ virtual particles

Figure 3. A sketch of a simulation of an acoustic boundary with a curved surface using virtual particles. Fluid particles are inside the computational domain and boundary particles are fixed on the boundary.

The boundary condition for a rigid wall satisfies the following equations.

$$p_{i+2} = p_{i+1} = p_i,\ v_{i+3} = v_{i+2} = v_{i+1} = 0. \qquad (25)$$

Since the FDPM simulation is still based on the local support domain, most boundary conditions in the existing Lagrangian particle-based methods, such as the SPH and MPS methods, can be used directly or implemented with minor changes. The particle representation of no-slip, free-slip and superhydrophobic surfaces [41–43] can be found in [44–46].

3. Applications of the Finite Difference Particle Method

3.1. Fundamental Definition

Several test cases are simulated with the FDPM based on second-order Taylor series expansion. The numerical accuracy is evaluated by the root mean square errors (ε_{RMS}) and the maximum errors (ε_{MAX}), which are defined as

$$\varepsilon_{RMS}(S) = \frac{\sqrt{\frac{1}{NT}\sum_{k=1}^{NT}|S_{num}(k)-S_{ana}(k)|^2}}{\sqrt{\frac{1}{NT}\sum_{k=1}^{NT}|S_{ana}(k)|^2}}, \qquad (26)$$

$$\varepsilon_{MAX}(S) = \max_{1 \ll k \ll NT}|S_{num}(k)-S_{ana}(k)|, \qquad (27)$$

where $S_{num}(k)$ and $S_{ana}(k)$ are the numerical and analytical results of variable k, respectively. k could be the velocity or pressure.

The convergence rate of the FDPM is evaluated based on the root mean square error convergence rate (R_{ERMS}) and the maximum error convergence rate (R_{EMAX}) as follows:

$$R_{ERMS} = \left|\frac{\ln(\varepsilon_{RMS}(NT_{max}))-\ln(\varepsilon_{RMS}(NT_{min}))}{\ln(NT_{max})-\ln(NT_{min})}\right|, \qquad (28)$$

$$R_{EMAX} = \left|\frac{\ln(\varepsilon_{MAX}(NT_{max}))-\ln(\varepsilon_{MAX}(NT_{min}))}{\ln(NT_{max})-\ln(NT_{min})}\right|. \qquad (29)$$

3.2. Unsteady Flow in a Pipeline

Unsteady flow field in the pipeline is simulated to verify the FDPM method. The theoretical solutions of unsteady flow in a 1D pipeline (chapter 3 in book [47]) are

$$u = \frac{2}{\gamma+1} \frac{x}{t} + C_1, \tag{30}$$

$$p = C_3 \left(\frac{\gamma-1}{C_3 \gamma} \left(-\frac{1}{2} \left(C_1 + \frac{2x}{(\gamma+1)t} \right)^2 + \frac{C_1^2(\gamma+1)}{2(\gamma-1)} + \frac{C_2(\gamma-3)}{\gamma+1} t^{\frac{2-2\gamma}{\gamma+1}} + \frac{x^2}{(\gamma+1)t^2} \right) \right)^{\gamma/(\gamma-1)}, \tag{31}$$

$$\rho = \left(\frac{\gamma-1}{C_3 \gamma} \left(-\frac{1}{2} \left(C_1 + \frac{2x}{(\gamma+1)t} \right)^2 + \frac{C_1^2(\gamma+1)}{2(\gamma-1)} + \frac{C_2(\gamma-3)}{\gamma+1} t^{\frac{2-2\gamma}{\gamma+1}} + \frac{x^2}{(\gamma+1)t^2} \right) \right)^{1/(\gamma-1)}, \tag{32}$$

$$c^2 = \left(\frac{\gamma-1}{\gamma+1} -C_1 \right)^2 - \frac{3-\gamma}{\gamma+1}(\gamma-1)C_2 t^{-2(\gamma-1)(\gamma+1)}, \tag{33}$$

where u_0 is the flow velocity distribution and x is the coordinate along the length of the pipeline. The coefficients (the unit can be obtained from dimensional analysis) $C_1 = 30.0$, $C_2 = -1.0 \times 10^6$, $C_3 = 82571.0$, and $\gamma = 1.4$; moreover, the initial time is 12.5 s, and the pipe length x is 700 m.

The FDPM algorithm with a second-order Taylor truncation is used, and the time step (Δt) is 0.0029 s. The effect of viscosity in the process of fluid motion is not considered. Dirichlet boundary conditions are used at both ends of the boundary. The velocity, pressure and density of four particles at both ends are set based on theoretical values.

The space of the initial particle (Δx) is 5.0 m, and r_s is 3.2 times Δx. The cubic spline kernel function is used in the calculations. Table 1 provides data for comparing the numerical velocity and theoretical solution of a particular particle at different times and positions. Notably, although the particle moves from $x = 2.48$ m to $x = 25.64$ m, the FDPM results agree well with the theoretical values, and this result verifies the algorithm.

Table 1. FDPM results and theoretical solutions of the position and velocity of a particle (particle number: 70) at different times.

Time (s)	Particle Method: x (m)	Theoretical Solution: u (m/s)	Particle Method: u (m/s)	Error (10^{-8})
0.25	2.48	30.16201462	30.16201544	2.74
0.50	10.08	30.64615858	30.64615942	2.75
0.75	17.80	31.11956025	31.11956110	2.75
1.00	25.64	31.58265402	31.58265470	2.13

The convergence verification of the FDPM method for unsteady flow in a pipeline is shown in Figure 4. The numerical error curves at three different moments are given using different Δx values, and r_s is 3.2 times Δx in the computation.

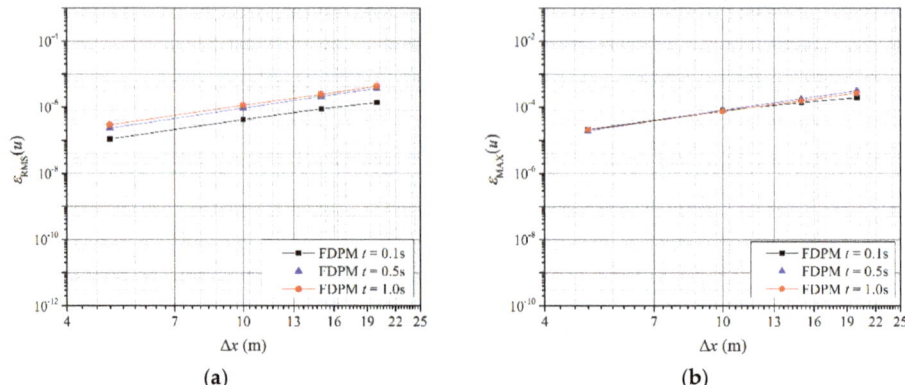

Figure 4. Convergence curves of the FDPM method for 1D unsteady flow simulation: (**a**) Root mean square error, see Equation (26), of FDPM simulation using different particle spacing Δx at different time; (**b**) maximum error, see Equation (27), of FDPM simulation using different particle spacing Δx at different time.

Figure 4 shows that the FDPM method displays good convergence at different times. When $t = 1.0$ s, the convergence curve yields a R_{ERMS} value of 1.7 and R_{EMAX} value of 1.8.

Given that the FDPM is a KGF method, the effect of the type of kernel function on this method is evaluated. Four types of kernel functions, including $\frac{1}{r^3}$, Gaussian, cubic spline and quintic spline functions, are compared through two types of errors with different r_s conditions, as shown in Figure 5. The figure shows that the maximum error of the Gaussian kernel function is larger than that of the other methods. The errors of other types of functions are similar.

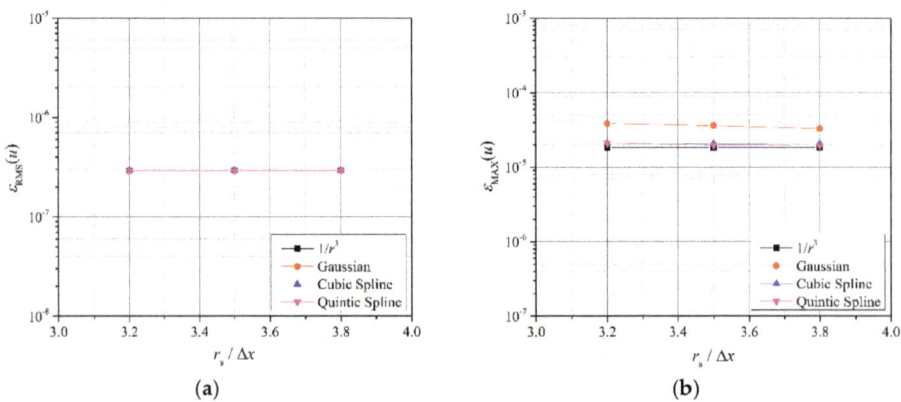

Figure 5. Velocity errors of the computations with different kernel functions in pipe flow modeling: (**a**) Root mean square error, see Equation (26), of FDPM simulation using different smoothing length and kernel functions; (**b**) maximum error, see Equation (27), of FDPM simulation using different smoothing length and kernel functions.

3.3. Poisseuille Flow

Steady, axisymmetric Poisseuille flow between two infinite plates is a classical test model in hydrodynamics. In this section, the model is used to verify the governing equations and the rigid wall boundaries. Assuming that the distance between two infinite plates is L, the volume force F is loaded on the fluid between plates in the x direction from time $t = 0$. The theoretical solution of the velocity distribution of the flow at a given time [48] is as follows.

$$u(y,t) = \frac{F}{v_k}y(y-L) + \sum_{n=0}^{\infty}\frac{4FL^2}{v_k\pi^3(2n+1)^3}\sin\left[\frac{\pi y}{L}(2n+1)\right]\exp\left[-\frac{(2n+1)^2\pi^2 v_k}{L^2}t\right]. \quad (34)$$

The numerical simulation for Poisseuille flow is obtained under weakly compressible ($Ma = 0.0125$) conditions. Based on reference [48], the parameters of the Poisseuille field are chosen as $v_k = 10^{-6}$ m^{2}s^{-1}, $L = 10^{-3}$ m, $\rho = 10^3$ kgm^{-3}, and $F = 10^{-4}$ ms^{-2}, so the maximum velocity is 1.25×10^{-5} ms^{-1} and the Reynold number is $Re = 1.25 \times 10^{-2}$. The plate boundaries at the upper and lower ends are established using rigid walls. One layer of boundary particles and three layers of virtual particles are used. The FDPM with second-order Taylor truncation is utilized to perform the computation. The speed of sound c is taken as 0.001 m/s, as suggested in [49], and the time step Δt is 3.0×10^{-4} s. The initial particle spacings Δx and Δy are both set as 5×10^{-5} m, r_s is 3.2 times Δx, and the kernel function is selected as a cubic spline function.

A comparison of the numerical and theoretical solutions of the velocity of the flow field in the x direction at different times is shown in Figure 6. The particle velocity is obtained by bilinear interpolation. As time increases, the positions of particles gradually change until the uniformly distributed particles at the initial stage are completely mixed in disorder. At this time, the numerical solution of the particle velocity is still consistent with the theoretical solution.

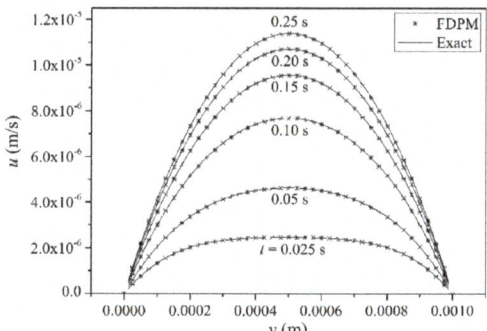

Figure 6. Velocity profiles of the FDPM results (stars) and theoretical solutions (lines) at different times along the y direction (from the bottom plate to the top plate).

During the computation, the particle velocity remains symmetrically distributed and gradually increases at different times before reaching a steady state. The velocity of particles in the middle of the two plates is the largest due to the viscous force, and the velocity is small near the plates. The FDPM solution is in good agreement with the theoretical solution.

Figure 7 shows the numerical error curves of the FDPM method at different times and is used to analyze the convergence of the FDPM method. During the computation, r_s remains 3.2 times Δx.

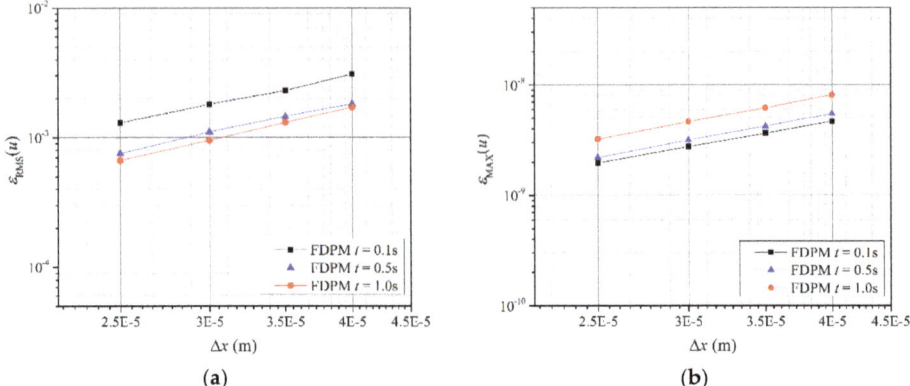

Figure 7. Convergence curves of the FDPM computation in Poisseuille flow modeling: (**a**) Root mean square error, see Equation (26), of FDPM simulation using different particle spacing Δx at different time; (**b**) maximum error, see Equation (27), of FDPM simulation using different particle spacing Δx at different time.

Figure 7 shows that ε_{RMS} and ε_{MAX} decrease as Δx decreases, which indicates that the numerical accuracy converges with the initial particle spacing at different times. ε_{RMS} is on the order of 10^{-3}, indicating that the computational results agree well with the theoretical solutions. For different error evaluation indexes, the R_{ERMS} and R_{EMAX} values of the FDPM method are approximately 1.7 and 1.8, respectively, with good convergence at $t = 1.0$ s. Since the second-order Taylor expansion-based FDPM is implemented in the test, the convergence rate is reasonable.

An error analysis of the four different types of kernel functions is conducted to analyze the sensitivity of the FDPM method to the kernel function, as shown in Figure 8.

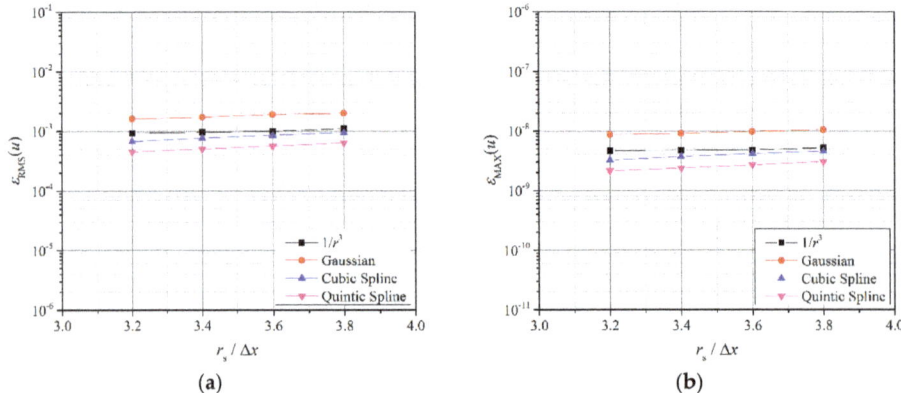

Figure 8. Velocity errors of the computations with different kernel functions in Poisseuille flow modeling: (**a**) Root mean square error, see Equation (26), of FDPM simulation using different smoothing length and kernel functions; (**b**) maximum error, see Equation (27), of FDPM simulation using different smoothing length and kernel functions.

Figure 8 shows that different types of kernel functions can be used in the FDPM method and that the differences in the calculation errors are insignificant. The calculation errors of the four types of kernel functions from large to small exhibit the following order: Gaussian, r^{-3}, cubic spline, and quantic spline.

3.4. Couette Flow

Couette flow considers the fluid flow between a stationary plate and a sliding plate. To accurately solve the flow distribution, the viscous term and boundary flow must be solved correctly. Initially, the two plates and the fluid between them remain stationary. At a constant speed, the upper plate begins to slide parallel to the lower plate. Assuming that the plate spacing is L and the sliding velocity is u_0, the theoretical solution of the flow velocity over time in the direction perpendicular to the plate [48] is as follows:

$$u(y,t) = \frac{u_0}{L}y + \sum_{n=1}^{\infty} \frac{2u_0}{n\pi}(-1)^n \sin(\frac{n\pi}{L}y)\exp(-\frac{n^2\pi^2 v_k}{L^2}t). \tag{35}$$

Couette flow is numerically simulated under weakly compressible ($Ma = 0.0125$) conditions. The parameters for Couette flow are $v_k = 10^6$ m^{2}s^{-1}, $L = 10^{-3}$ m, $\rho = 10^3$ kgm^{-3}, and $u_0 = 1.25 \times 10^{-5}$ m/s.

The plate boundaries at the upper and lower ends are obtained using rigid walls, and the upper plate is set with a constant velocity u_0. One layer of boundary particles and three layers of virtual particles are used. The FDPM with second-order Taylor truncation is utilized to perform the computation. The speed of sound c is taken as 0.001 m/s, and the time step Δt is 5.0×10^{-5} s. The initial particle spacings Δx and Δy are both set as 2.5×10^{-5} m, r_s is 3.2 times Δx, and the kernel function selected is the cubic spline function. Figure 9 shows a comparison between the FDPM method and the theoretical solution for the flow velocity at different times.

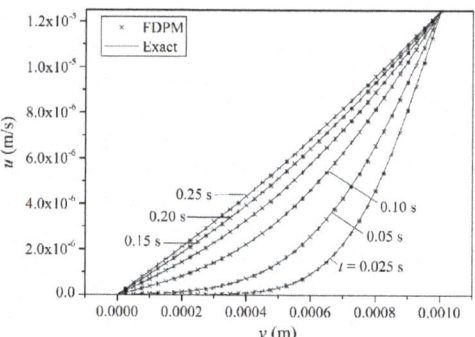

Figure 9. Comparison of the FDPM result (stars) and the theoretical solution (lines) for the flow velocity at different times along the y direction (from the stationary plate to the sliding plate).

Before reaching a steady state, the particle velocity near the upper plate rapidly increases due to the viscous force, and that near the lower plate increases in a relatively slow manner. The velocity distribution of the particles between the two plates is nonlinear.

The velocity error (ε_{RMS} and ε_{MAX}) at different times and at different Δx values is used to evaluate the convergence of the FDPM, as shown in Figure 10.

From the numerical results, both error indexes gradually decrease with decreasing Δx, which suggests that the numerical accuracy converges with the initial particle spacing at different times. ε_{RMS} is on the order of 10^{-2}, indicating that the computational results agree well with the theoretical solution. When $t = 1.0$ s, the two errors result in an R_{ERMS} value of 1.7 and R_{EMAX} value of 1.8. Since the second-order Taylor expansion-based FDPM is implemented in the test, the convergence rate is reasonable.

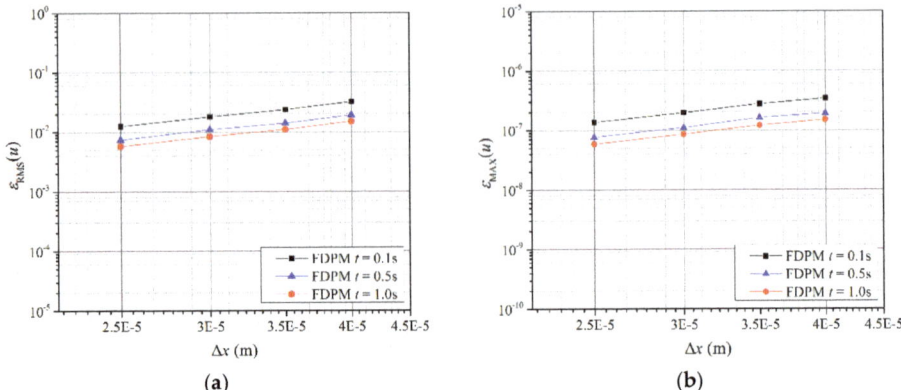

Figure 10. Convergence curves of the FDPM computations in Couette flow modeling: (**a**) Root mean square error, see Equation (26), of FDPM simulation using different particle spacing Δx at different time; (**b**) maximum error, see Equation (27), of FDPM simulation using different particle spacing Δx at different time.

To analyze the sensitivity of the FDPM method to the kernel function, four different types of kernel functions with different $r_s/\Delta x$ values are applied, as shown in Figure 11. Different types of kernel functions can be used in the FDPM method, and the calculation error of the Gaussian kernel function is larger than that of the other methods.

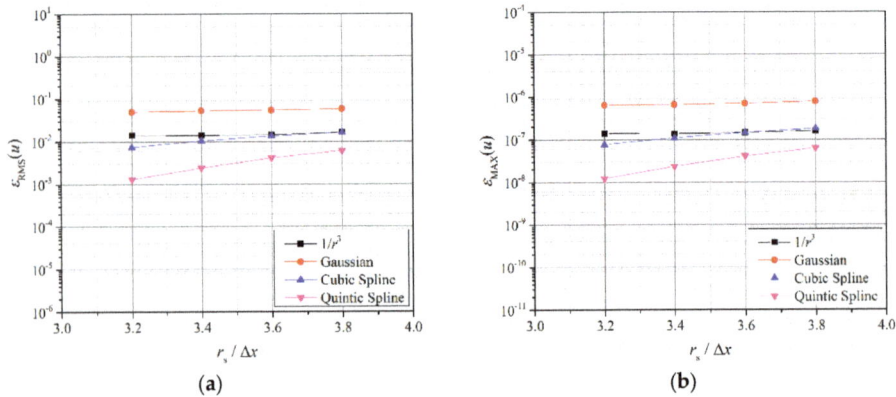

Figure 11. Velocity errors of the computations with different kernel functions in Couette flow modeling: (**a**) Root mean square error, see Equation (26), of FDPM simulation using different smoothing length and kernel functions; (**b**) maximum error, see Equation (27), of FDPM simulation using different smoothing length and kernel functions.

3.5. Flow in Porous Media

In this section, the FDPM algorithm is used to simulate the flow in a simplified model of porous media [50]. The simplified model can be seen as flow around a circular cylinder, as shown in Figure 12, and four sides of the domain are periodic boundaries.

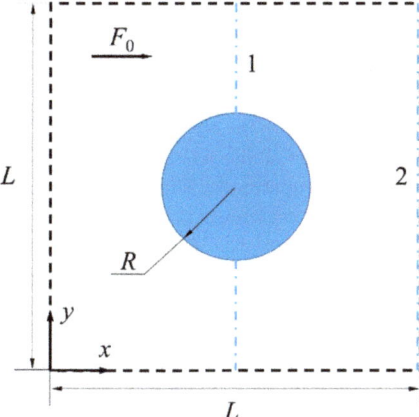

Figure 12. Simplified model of porous media. The solid circle is a circular cylinder and four sides of the domain are periodic boundaries. L is the size of the computational domain, R is the cylindrical radius, and F_0 is the volume force.

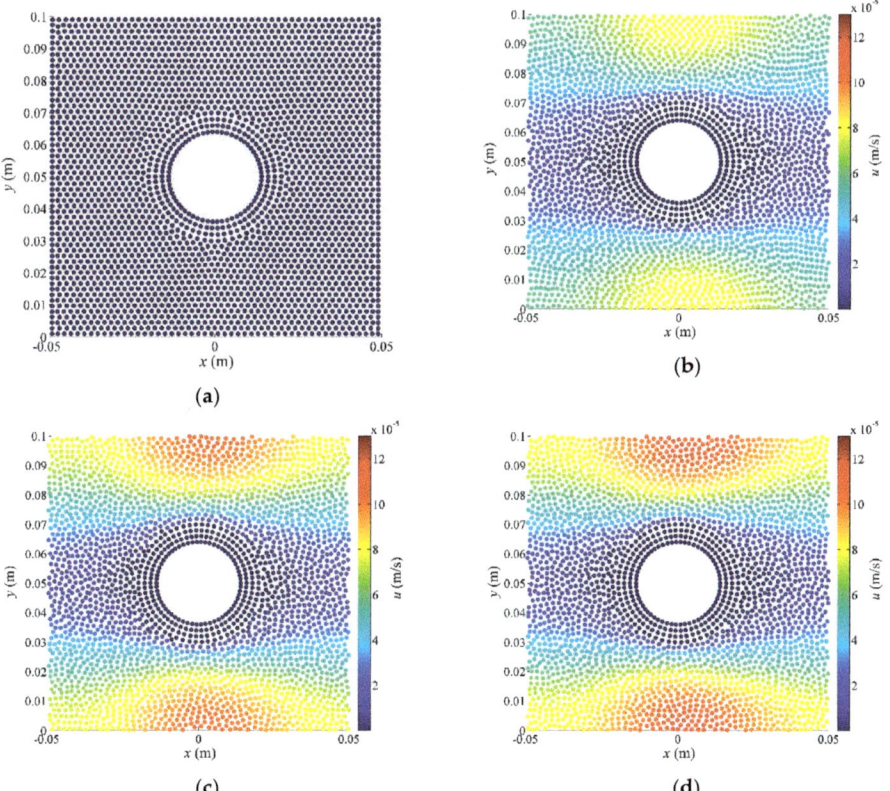

Figure 13. Particle distribution (solid circles) and velocity contours from the initial time to the steady state: (**a**) $t = 0$, (**b**) $t = 693$ s, (**c**) $t = 1386$ s and (**d**) $t = 2080$ s.

The size of the computational domain $L = 0.1$ m, the kinematic viscosity $v_k = 10^{-6}$ m^{2}s^{-1}, the cylindrical radius $R = 2 \times 10^{-2}$ m, the volume force $F_0 = 1.5 \times 10^{-7}$ ms^{-2}, and the speed of sound $c = 5.77 \times 10^{-4}$ ms^{-1}. Δx and Δy are 0.003 m, r_s is 3.2 times Δx, $\Delta t = 1.04$ s with 2000 steps, and the coefficient of artificial particle displacement is 0.05. A rigid wall boundary is used for the cylindrical boundary, and a periodic boundary is used on the four sides of the computational domain. One layer of boundary particles and three layers of virtual particles are used. The FDPM with second-order Taylor truncation is utilized to perform the computations. The particle distribution and velocity contours at the initial time and the final steady state are shown in Figure 13.

At the initial time, the particle distribution is regular. Then, the fluid begins to flow, and particles are gradually scattered and evenly distributed in the computational domain.

The velocity distributions along lines 1 and 2 (dotted-dashed lines in Figure 12) are shown in Figure 14. Both the FDPM results and finite element method (FEM) results are given to evaluate the accuracy of the numerical method. The FEM results come from the data of figure 6 in [48].

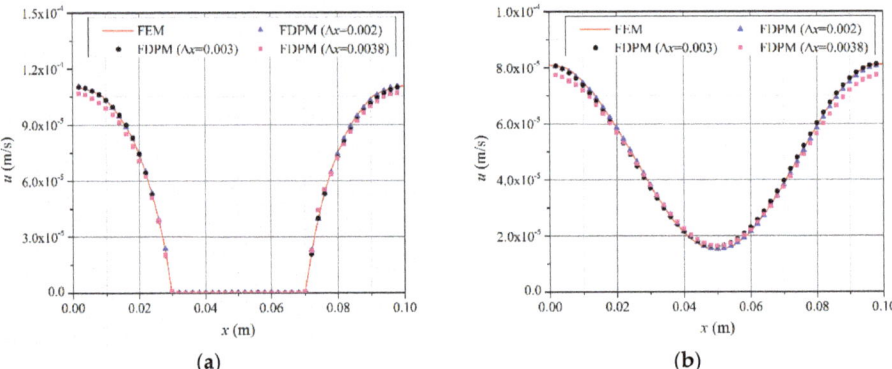

Figure 14. Velocity distributions along observation lines 1 and 2 (dotted-dashed lines in Figure 12): (a) Observation line 1 and (b) observation line 2. Lines are FEM results and solid points are FDPM results with different particle spacing Δx.

When $\Delta x > 0.004$, the computation is not sufficiently stable and can easily collapse, so this condition is not shown in Figure 14. When $\Delta x = 0.0038$ m, the FDPM calculation results and the FEM results at $x = 0$ and 0.1 m produce significant differences. When $\Delta x = 0.002$ m, the FDPM results are similar to the FEM results. After convergence is obtained, the results of the FDPM method are consistent with the FEM results, which verifies the correctness of the numerical method and the boundary conditions.

A comparison of the FDPM results ($\Delta x = 0.002$ m) with different kernel functions and the FEM reference results is shown in Figure 15. The figure shows that the FDPM results with different types of kernel functions are comparable to the reference results and exhibit only minor differences.

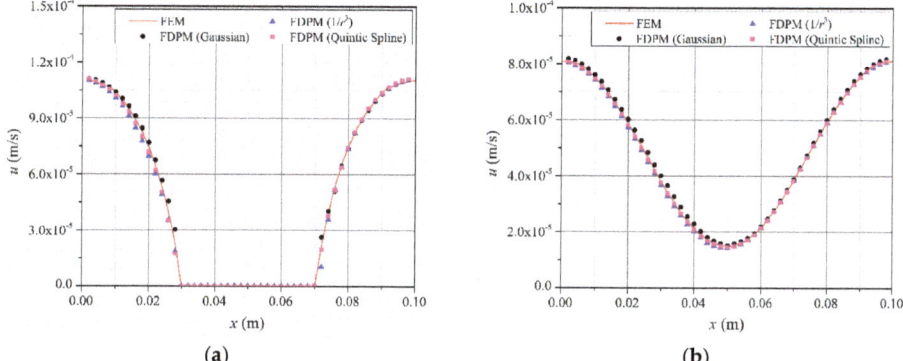

Figure 15. Comparison along observation lines 1 and 2 (dotted-dashed lines in Figure 12) between the FDPM results (solid points) with different kernel functions and the FEM results (lines): (**a**) Observation line 1 and (**b**) observation line 2.

3.6. Lid-Driven Cavity Flow

The lid-driven cavity flow is widely used as a benchmark test case and the model in [51] is used to verify the method. The case is in a square cavity with a sliding plate on the upper side and three fixed rigid walls around, as shown in Figure 16. Initially, the fluid in the cavity remain stationary. At a constant speed, the upper plate begins to slide horizontally and the simulation is at $Re = 100$.

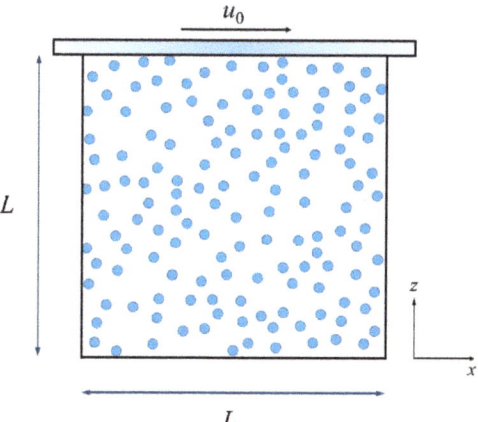

Figure 16. Schematic of the lid-driven cavity flow. The solid circle is a circular cylinder and four sides of the domain are periodic boundaries. L is the size of the computational domain, R is the cylindrical radius, and F_0 is the volume force.

Rigid wall boundary condition is used for four sides of the cavity, and the upper plate is set with a constant velocity u_0. The size of the cavity $L = 1.0$ m, the kinematic viscosity $v_k = 0.01$ m^{2}s^{-1}, the sliding velocity $u_0 = 1.0$ m/s, and the speed of sound $c = 10.0$ ms^{-1}. Δx and Δy are 0.025 m, r_s is 2.7 times Δx, $\Delta t = 0.001$ s with 3000 steps, the coefficient of artificial particle displacement is 0.05, and the kernel function selected is the cubic spline function. One layer of boundary particles and three layers of virtual particles are used. The FDPM with second-order Taylor truncation is utilized to perform the computations. The particle distribution and velocity contours at the initial time and the final steady state are shown in Figure 17.

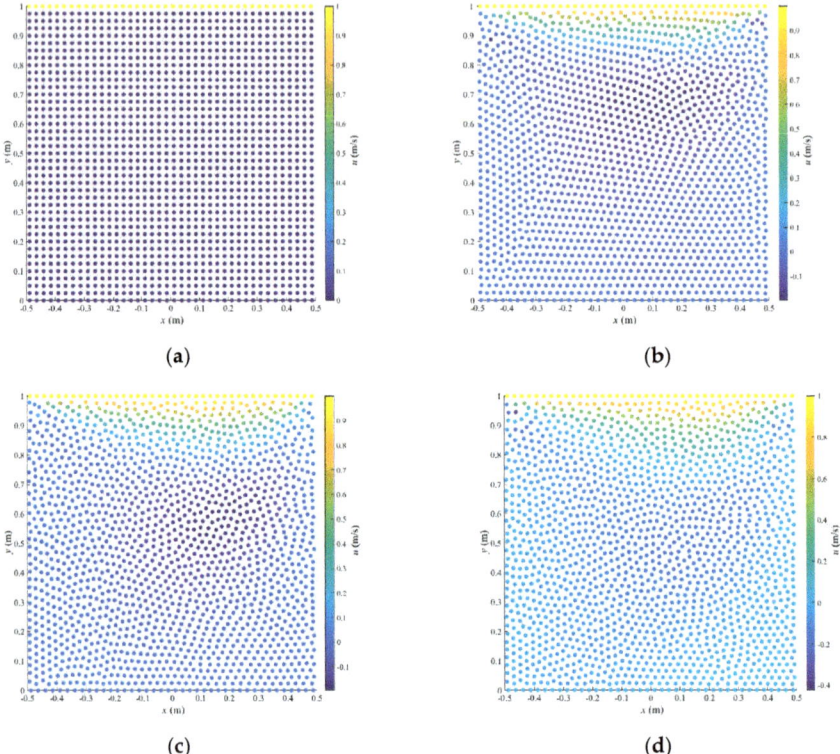

Figure 17. Particle distribution (solid circles) and velocity contours from the initial time to the steady state: (**a**) $t = 0$, (**b**) $t = 1.0$ s, (**c**) $t = 2.0$ s and (**d**) $t = 3.0$ s.

At the initial time, the particle distribution is regular. Then, the fluid begins to flow, and particles are gradually scattered and evenly distributed in the computational domain.

Horizontal velocity component profiles along horizontal and vertical geometric centerlines at $t = 3.0$ s, respectively, are shown in Figure 18. Although the present FDPM computation employed only 2/5 particles in the work [51], these profiles are in good agreement with the reference results. The particle distribution shows the method works well in geometries with corners.

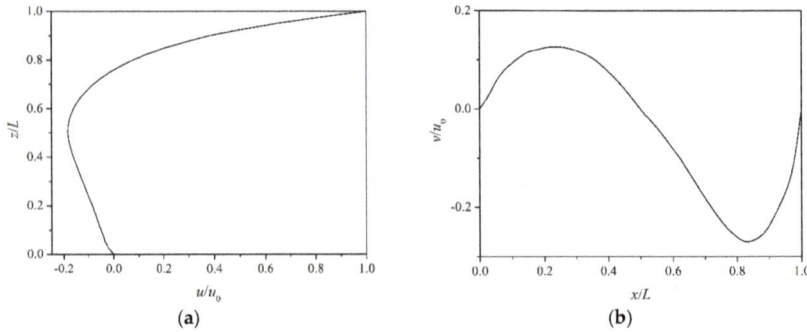

Figure 18. Horizontal Velocity distributions along horizontal and vertical geometric centerlines at $t = 3.0$ s: (**a**) Along vertical geometric centerlines and (**b**) along horizontal geometric centerlines.

4. Conclusions

In this paper, a particle method based on the GFD scheme is proposed to simulate weakly compressible viscous flow. This approach represents the partial differential terms in the Navier-Stokes equations as the solution of a symmetric system of linear equations. The convergence and accuracy of the symmetric particle-based method are tested by modeling flow in a pipeline, Poisseuille flow, Couette flow, flow in porous media, and lid-driven cavity flow. The numerical results exhibit close agreement with the theoretical solutions and finite element results. The particle method utilizes the kernel function itself instead of its gradient, which avoids using artificial parameters to solve for the viscous term and reduces the limitations on the choice of kernel function. Moreover, the order of the Taylor series expansion can easily be improved in the meshless algorithm. The convergence rate of the particle-based calculations with second-order Taylor truncation is approximately 1.7 in the tests, and four different kernel functions are tested and determined to be reliable.

Author Contributions: These authors contributed equally to this work.

Funding: This research was funded by National Natural Science Foundation of China (grant number 51809208 and 51741910) and Fundamental Research Funds for the Central Universities (grant number WUT: 2018IVA032).

Conflicts of Interest: The authors declare no conflict of interest.

References

1. Kim, K.S. A Mesh-Free Particle Method for Simulation of Mobile-Bed Behavior Induced by Dam Break. *Appl. Sci.* **2018**, *8*, 1070. [CrossRef]
2. De Padova, D.; Mossa, M.; Sibilla, S. SPH numerical investigation of characteristics of hydraulic jumps. *Environ. Fluid Mech.* **2018**, *18*, 849–870. [CrossRef]
3. Zuo, J.; Tian, W.; Chen, R.; Qiu, S.; Su, G. Two-dimensional numerical simulation of single bubble rising behavior in liquid metal using moving particle semi-implicit method. *Prog. Nucl. Energy* **2013**, *64*, 31–40. [CrossRef]
4. Zhang, A.; Sun, P.; Ming, F. An SPH modeling of bubble rising and coalescing in three dimensions. *Comput. Methods Appl. Mech. Eng.* **2015**, *294*, 189–209. [CrossRef]
5. Wang, X.; Ban, X.; He, R.; Wu, D.; Liu, X.; Xu, Y. Fluid-Solid Boundary Handling Using Pairwise Interaction Model for Non-Newtonian Fluid. *Symmetry* **2018**, *10*, 94. [CrossRef]
6. Liu, S.; Ban, X.; Wang, B.; Wang, X. A Symmetric Particle-Based Simulation Scheme towards Large Scale Diffuse Fluids. *Symmetry* **2018**, *10*, 86. [CrossRef]
7. Lucy, L.B. A numerical approach to the testing of the fission hypothesis. *Astron. J.* **1977**, *82*, 1013–1024. [CrossRef]
8. Gingold, R.A.; Monaghan, J.J. Smoothed particle hydrodynamics: Theory and application to non-spherical stars. *Mon. Not. R. Astron. Soc.* **1977**, *181*, 375–389. [CrossRef]
9. Monaghan, J. Smoothed Particle Hydrodynamics and Its Diverse Applications. *Annu. Rev. Fluid Mech.* **2012**, *44*, 323–346. [CrossRef]
10. Violeau, D.; Rogers, B.D. Smoothed particle hydrodynamics (SPH) for free-surface flows: Past, present and future. *J. Hydraul. Res.* **2016**, *54*, 1–26. [CrossRef]
11. Wang, Z.-B.; Chen, R.; Wang, H.; Liao, Q.; Zhu, X.; Li, S.-Z. An overview of smoothed particle hydrodynamics for simulating multiphase flow. *Appl. Math. Model.* **2016**, *40*, 9625–9655. [CrossRef]
12. Zhang, A.-M.; Sun, P.-N.; Ming, F.-R.; Colagrossi, A. Smoothed particle hydrodynamics and its applications in fluid-structure interactions. *J. Hydrodyn.* **2017**, *29*, 187–216. [CrossRef]
13. Liu, G.R.; Liu, M.B. *Smoothed Particle Hydrodynamics—A Meshfree Particle Method*; World Scientific Publishing: Singapore, 2003; pp. 103–176.
14. Liu, M.B.; Liu, G.R. Smoothed Particle Hydrodynamics (SPH): An Overview and Recent Developments. *Arch. Comput. Methods Eng.* **2010**, *17*, 25–76. [CrossRef]
15. Cleary, P.W.; Prakash, M.; Ha, J.; Stokes, N.; Scott, C. Smooth particle hydrodynamics: Status and future potential. *Prog. Comput. Fluid Dyn. Int. J.* **2007**, *7*, 70–90. [CrossRef]
16. Cummins, S.J.; Rudman, M. An SPH projection method. *J. Comput. Phys.* **1999**, *152*, 584–607. [CrossRef]

17. Chen, J.K.; Beraun, J.E.; Jih, C.J. An improvement for tensile instability in smoothed particle hydrodynamics. *Comput. Mech.* **1999**, *23*, 279–287. [CrossRef]
18. Chen, J.K.; Beraun, J.E.; Carney, T.C. A corrective smoothed particle method for boundary value problems in heat conduction. *Int. J. Numer. Methods Eng.* **1999**, *46*, 231–252. [CrossRef]
19. Zhang, G.M.; Batra, R.C. Modified smoothed particle hydrodynamics method and its application to transient problems. *Comput. Mech.* **2004**, *34*, 137–146. [CrossRef]
20. Liu, M.; Liu, G.; Lam, K. Constructing smoothing functions in smoothed particle hydrodynamics with applications. *J. Comput. Appl. Math.* **2003**, *155*, 263–284. [CrossRef]
21. Huang, C.; Lei, J.M.; Liu, M.B.; Peng, X.Y. A kernel gradient free (KGF) SPH method. *Int. J. Numer. Methods Fluids* **2015**, *78*, 691–707. [CrossRef]
22. Lei, J.M.; Peng, X.Y. Improved kernel gradient free-smoothed particle hydrodynamics and its applications to heat transfer problems. *Chin. Phys. B* **2015**, *25*, 020202. [CrossRef]
23. Huang, C.; Lei, J.M.; Liu, M.B.; Peng, X.Y. An improved KGF-SPH with a novel discrete scheme of Laplacian operator for viscous incompressible fluid flows. *Int. J. Numer. Methods Fluids* **2016**, *81*, 377–396. [CrossRef]
24. Koh, C.G.; Gao, M.; Luo, C. A new particle method for simulation of incompressible free surface flow problems. *Int. J. Numer. Meth. Eng.* **2012**, *89*, 1582–1604. [CrossRef]
25. Luo, M.; Koh, C.G.; Bai, W.; Gao, M. A particle method for two-phase flows with compressible air pocket. *Int. J. Numer. Methods Eng.* **2016**, *108*, 695–721. [CrossRef]
26. Jensen, P.S. Finite difference techniques for variable grids. *Comput. Struct.* **1972**, *2*, 17–29. [CrossRef]
27. Perrone, N.; Kao, R. A general finite difference method for arbitrary meshes. *Comput. Struct.* **1975**, *5*, 45–57. [CrossRef]
28. Liszka, T.; Orkisz, J. The finite difference method at arbitrary irregular grids and its application in applied mechanics. *Comput. Struct.* **1980**, *11*, 83–95. [CrossRef]
29. Ding, H.; Shu, C.; Yeo, K.; Xu, D. Simulation of incompressible viscous flows past a circular cylinder by hybrid FD scheme and meshless least square-based finite difference method. *Comput. Methods Appl. Mech. Eng.* **2004**, *193*, 727–744. [CrossRef]
30. Li, P.-W.; Chen, W.; Fu, Z.-J.; Fan, C.-M. Generalized finite difference method for solving the double-diffusive natural convection in fluid-saturated porous media. *Eng. Anal. Bound. Elem.* **2018**, *95*, 175–186. [CrossRef]
31. Gavete, L.; Ureña, F.; Benito, J.; García, A.; Ureña, M.; Salete, E.; Corvinos, L.A.G. Solving second order non-linear elliptic partial differential equations using generalized finite difference method. *J. Comput. Appl. Math.* **2017**, *318*, 378–387. [CrossRef]
32. Benito, J.; Ureña, F.; Gavete, L. Influence of several factors in the generalized finite difference method. *Appl. Math. Model.* **2001**, *25*, 1039–1053. [CrossRef]
33. Gavete, L.; Gavete, M.; Benito, J. Improvements of generalized finite difference method and comparison with other meshless method. *Appl. Math. Model.* **2003**, *27*, 831–847. [CrossRef]
34. Benito, J.; Ureña, F.; Gavete, L. Solving parabolic and hyperbolic equations by the generalized finite difference method. *J. Comput. Appl. Math.* **2007**, *209*, 208–233. [CrossRef]
35. Alamri, S.Z.; Ellahi, R.; Shehzad, N.; Zeeshan, A. Convective radiative plane Poiseuille flow of nanofluid through porous medium with slip: An application of Stefan blowing. *J. Mol. Liq.* **2019**, *273*, 292–304. [CrossRef]
36. Ellahi, R.; Shivanian, E.; Abbasbandy, S.; Hayat, T. Numerical study of magnetohydrodynamics generalized Couette flow of Eyring-Powell fluid with heat transfer and slip condition. *Int. J. Numer. Methods Heat Fluid Flow* **2016**, *26*, 1433–1445. [CrossRef]
37. Shehzad, N.; Zeeshan, A.; Ellahi, R.; Rashidi, S. Modelling Study on Internal Energy Loss Due to Entropy Generation for Non-Darcy Poiseuille Flow of Silver-Water Nanofluid: An Application of Purification. *Entropy* **2018**, *20*, 851. [CrossRef]
38. Bašić, J.; Degiuli, N.; Werner, A. Simulation of water entry and exit of a circular cylinder using the ISPH method. *Trans. Famena* **2014**, *38*, 45–62.
39. Nestor, R.M.; Basa, M.; Lastiwka, M.; Quinlan, N.J. Extension of the finite volume particle method to viscous flow. *J. Comput. Phys.* **2009**, *228*, 1733–1749. [CrossRef]
40. Ozbulut, M.; Yildiz, M.; Goren, O. A numerical investigation into the correction algorithms for SPH method in modeling violent free surface flows. *Int. J. Mech. Sci.* **2014**, *79*, 56–65. [CrossRef]
41. Rothstein, J.P. Slip on superhydrophobic surfaces. *Annu. Rev. Fluid Mech.* **2010**, *42*, 89–109. [CrossRef]

42. Gentili, D.; Bolognesi, G.; Giacomello, A.; Chinappi, M.; Casciola, C.M. Pressure effects on water slippage over silane-coated rough surfaces: Pillars and holes. *Microfluid. Nanofluidics* **2014**, *16*, 1009–1018. [CrossRef]
43. Bolognesi, G.; Cottin-Bizonne, C.; Pirat, C. Evidence of slippage breakdown for a superhydrophobic microchannel. *Phys. Fluids* **2014**, *26*, 082004. [CrossRef]
44. Marrone, S.; Antuono, M.; Colagrossi, A.; Colicchio, G.; Le Touzé, D.; Graziani, G. δ-SPH model for simulating violent impact flows. *Comput. Methods Appl. Mech. Eng.* **2011**, *200*, 1526–1542. [CrossRef]
45. Adami, S.; Hu, X.; Adams, N.; Adams, N. A generalized wall boundary condition for smoothed particle hydrodynamics. *J. Comput. Phys.* **2012**, *231*, 7057–7075. [CrossRef]
46. Sun, P.; Ming, F.; Zhang, A. Numerical simulation of interactions between free surface and rigid body using a robust SPH method. *Ocean Eng.* **2015**, *98*, 32–49. [CrossRef]
47. Von Mises, R.; Geiringer, H.; Ludford, G.S.S. *Mathematical Theory of Compressible Fluid Flow*; Chapter 3; Academic Press: New York, NY, USA, 2004.
48. Morris, J.P.; Fox, P.J.; Zhu, Y. Modeling Low Reynolds Number Incompressible Flows Using SPH. *J. Comput. Phys.* **1997**, *136*, 214–226. [CrossRef]
49. Liu, M.; Xie, W.; Liu, G. Modeling incompressible flows using a finite particle method. *Appl. Math. Model.* **2005**, *29*, 1252–1270. [CrossRef]
50. Fang, J.; Parriaux, A. A regularized Lagrangian finite point method for the simulation of incompressible viscous flows. *J. Comput. Phys.* **2008**, *227*, 8894–8908. [CrossRef]
51. Khorasanizade, S.; Sousa, J.M.M.; De Sousa, J.M. A detailed study of lid-driven cavity flow at moderate Reynolds numbers using Incompressible SPH. *Int. J. Numer. Methods Fluids* **2014**, *76*, 653–668. [CrossRef]

© 2019 by the authors. Licensee MDPI, Basel, Switzerland. This article is an open access article distributed under the terms and conditions of the Creative Commons Attribution (CC BY) license (http://creativecommons.org/licenses/by/4.0/).

MDPI
St. Alban-Anlage 66
4052 Basel
Switzerland
Tel. +41 61 683 77 34
Fax +41 61 302 89 18
www.mdpi.com

Symmetry Editorial Office
E-mail: symmetry@mdpi.com
www.mdpi.com/journal/symmetry

www.ingramcontent.com/pod-product-compliance
Lightning Source LLC
LaVergne TN
LVHW071935080526
838202LV00064B/6609